NUMBER THEORY

Canadian Mathematical Society
Société Mathématique du Canada

CONFERENCE PROCEEDINGS · Volume 7

NUMBER THEORY

Proceedings of the
1985 Montreal Conference
held June 17–29, 1985

H. Kisilevsky and
J. Labute, Editors

Published by the American Mathematical Society
for the Canadian Mathematical Society
PROVIDENCE, RHODE ISLAND

Volume 7
1985 Conference in Number Theory
held at the Loyola Campus of Concordia University, Montreal, Quebec
June 17–29, 1985

Proceedings of a conference sponsored by the Natural Sciences and Engineering Research Council of Canada, Concordia University and McGill University

Accepted for publication by the Canadian Mathematical Society

Published by the American Mathematical Society

1980 *Mathematics Subject Classification* (1985 *Revision*). Primary 11F55, 11R, 11F41, 11F30, 12A70, 11G05, 11G40, 11S31, 14L05.

Library of Congress Cataloging-in-Publication Data

Montreal Conference in Number Theory (1985)
Proceedings of the 1985 Montreal Conference in Number Theory, held June 17-29, 1985.
(Canadian Mathematical Society conference proceedings, ISSN 0731-1036; v. 7)
1. Numbers, Theory of—Congresses. I. Kisilevsky, H. II. Labute, J. III. Canadian Mathematical Society. IV. Title. V. Series: CMS conference proceedings; v. 7.
QA241.M63 1985 512′.7 87-1307
ISBN 0-8218-6012-7 (alk. paper)

Printed in the United States of America.
Copying and Reprinting information can be found at the back of this volume.
This volume was printed directly from author-prepared copy.
The paper used in this book is acid-free and falls within the guidelines established to ensure permanence and durability. ∞

CONTENTS

INTRODUCTION

The following Proceedings is composed of research articles presented to the Canadian Mathematical Society's 1985 Summer Seminar in Number Theory held at the Loyola Campus of Concordia University from June 17 to June 29. There were two main lecture series given by B. Gross and H. Stark and shorter lecture series and hour lectures given by other prominent number theorists. These lectures exposed some of the recent advances in the subject.

The Conference was supported by grants from Concordia University, McGill University and the Natural Sciences and Engineering Research Council of Canada. Special thanks go to the members of the Quebec-Vermont Number Theory Group and to the staff of the Mathematics Department of Concordia University for their help in making this Conference a success.

H. KISILEVSKY
J. LABUTE

PARTICIPANTS

Don Blasius
Columbia University

Peter Borwein
Dalhousie University

Nigel Boston
Harvard University

Mohamed Bouchoum
Université Laval

David Boyd
University of British Columbia

Priscilla Bremser
Middlebury College

Armand Brumer
Fordham University

Ted Chinburg
University of Pennsylvania

Richard Crew
Boston University

Ehud de Shalit
Harvard University

Julianna Davidoff
Mount Holyoke College

Aissa Derhem
Université Laval

David Dorman
Middlebury College

David Dummit
University of Vermont

Isaac Efrat
Massachusetts Institute of Technology

Jacek Fabrykowski
University of Alberta

Paul Feit
Princeton University

Richard Freije
Brown University

Robert Gold
Ohio State University

Brent Gordon
Purdue University

D. Goss
Ohio State University

David Grant
University of California, San Diego

B. Gross
Harvard University

R. Gupta
University of British Columbia

David Hayes
University of Massachusetts

Rhonda Hatcher
Harvard University

John Hsia
Ohio State University

John Jones
Harvard University

Kevin Keating
Harvard University

H. Kisilevsky
Concordia University

Kenneth Kramer
Queens College

John Labute
McGill University

Claude Levesque
Université Laval

John Minardi
University of Washington

R. A. Mollin
University of Calgary

M. R. Murty
McGill University

V. K. Murty
Concordia University

Margaret Napolitano
Brown University

J. Nemchenok
University of California, San Diego

Toby Orloff
Yale University

Patrice Philippon
Harvard University

Dinakar Ramakrishnan
Johns Hopkins University

David Rohrlich
Rutgers University

Michael Rosen
Brown University

K. Rubin
Ohio State University

Richard Sarkisian
University of Toronto

Chad Schoen
Duke University

Peter Schumer
Middlebury College

Andrew Schwartz
Harvard University

Robert Sczech
University of Maryland

Thomas Shemanske
Dartmouth College

Alice Silverberg
Ohio State University

W. H. Simons
Oregon State University

David Solomon
Brown University

H. Stark
Massachusetts Institute of Technology

Jacob Sturm
Johns Hopkins University

Leon Tatevossian
Brown University

Olga Taussky-Todd
California Institute of Technology

John Todd
California Institute of Technology

Doug Ulmer
Brown University

Lawrence Washington
University of Maryland

Noriko Yui
University of Toronto

Canadian Mathematical Society
Conference Proceedings
Volume 7 (1987)

Intersection Theory and Capacity Theory on Arithmetic Surfaces

T. Chinburg*

I. Introduction

In this paper we define via adelic capacity theory a pairing $[\ ,\ ]_{Cap}$ on pairs of coprime elements of a subgroup $Div_A^*(X)$ of the group $Div_A(X)$ of Arakelov divisors on an arithmetic surface X. Such an X is a regular model over $Spec(\underline{O}_K)$ of a smooth, projective, geometrically irreducible curve over a number field K, where $\underline{O}_K$ denotes the ring of integers of K. An Arakelov divisor on X is a formal sum of a Weil divisor on X with a real linear combination of the fibres of X over the infinite places of K. The group $Div_A^*(X)$ is the subgroup of $Div_A(X)$ generated by effective horizontal divisors which do not pass through multiple points of reducible fibres of the projection morphism $\pi: X \to Spec(\underline{O}_K)$. Two divisors in $Div_A^*(X)$ are coprime if they have no irreducible constituent in common.

In [1] Arakelov defined a real, symmetric bilinear pairing $[\ ,\ ]_A$ on $Div_A(X)$, and showed that this pairing respected linear equivalence and base change from K to a finite extension of K. The pairing $[\ ,\ ]_{Cap}$ in this paper is also real and bilinear, and respects base change from K to a finite extension

*Partially supported by NSF grant MCS-82-43648 and the Sloan Foundation.

field. If D_1 and D_2 are distinct sections of π, $\exp([D_1,D_2]_{Cap})$ is the capacity, in the sense of Cantor [3] and Rumely [9,10], of the set of adelic points of X which are at least as close to D_1 as to D_2 in every adelic fibre of X. Here distance in the adelic fibres is measured by means of certain canonical distance functions defined by Rumely.

We will show that when $X = P^1_{\underline{O}_K}$ is the projective line over $\mathrm{Spec}(\underline{O}_K)$, $[\ ,\ ]_{Cap}$ may be extended to a pairing on $\mathrm{Div}_A(P^1_{\underline{O}_K})$ which respects linear equivalence in the Arakelov sense, and which is alternating:

$$[D_1,D_2]_{Cap} = -[D_2,D_1]_{Cap}$$

for all D_1 and D_2 in $\mathrm{Div}_A(P^1_{\underline{O}_K})$. Thus $[\ ,\ ]_{Cap}$ descends to an alternating pairing on the Arakelov divisor class group $\mathrm{Pic}_A(P^1_{\underline{O}_K})$ of $P^1_{\underline{O}_K}$. If, for example, $K = \mathbf{Q}$, then $\mathrm{Pic}_A(P^1_{\mathbf{Z}})$ is isomorphic to $\mathbf{R} \oplus \mathbf{Z}$, and $[\ ,\ ]_{Cap}$ is the restriction of the usual determinant pairing on $\mathbf{R} \oplus \mathbf{R}$ to $\mathbf{R} \oplus \mathbf{Z}$. (A similar explicit description for $K \neq \mathbf{Q}$ is given after the statement of Theorem 4.1.)

For arbitrary X and admissible choices of canonical distance functions, the results of Cantor and Rumely relate $[D_1,D_2]_{Cap}$ to arithmetic properties of D_1 and D_2. Suppose, for example, that D_1 and D_2 are distinct sections of $\pi\colon X \to \mathrm{Spec}(\underline{O}_K)$, and that $[D_1,D_2]_{Cap} \neq 0$. Then there will be infinitely many complete sets of $\mathrm{Gal}(\overline{K}/K)$-conjugate points of $X(\overline{K})$ as close to D_1 as to D_2 in every adelic fibre of X if and only if $[D_1,D_2]_{Cap} > 0$. In particular, if $[D_1,D_2]_{Cap}$ and $[D_2,D_1]_{Cap}$ have opposite signs, then there will be infinitely many such sets iff there are <u>not</u> infinitely many such sets when D_1 and D_2 are interchanged. Because $[\ ,\ ]_{Cap}$ is alternating when $X = P^1_{\underline{O}_K}$, this last remark applies to all distinct sections D_1 and D_2 for which $[D_1,D_2]_{Cap} \neq 0$ if $X = P^1_{\underline{O}_K}$.

This paper is organized in the following way. In §II we recall the definition of Arakelov's pairing $[\ ,\ ]_A$, since $[\ ,\ ]_{Cap}$ for $X = P^1_{\underline{O}_K}$ will be computed in terms of $[\ ,\ ]_A$. The main elements of Cantor and Rumely's adelic capacity theory are recalled in §III. For our purposes, we will need only the capacity $\gamma(\mathbf{E},\chi)$ of an adele of sets $\mathbf{E}$ with respect to a set χ which consists of a single point of $X(K)$. However, it does not take significantly more work to state Cantor and Rumely's definition of $\gamma(\mathbf{E},\chi)$ when χ is a finite $Gal(\overline{K}/K)$-stable subset of $X(\overline{K})$, and the functorial properties of $\gamma(\mathbf{E},\chi)$ are clearer in ths generality. In §IV we define $[\ ,\ ]_{Cap}$ and then state our main result, Theorem 4.1. After discussing some questions concerning the behavior of $[\ ,\ ]_{Cap}$ for other X than $P^1_{\underline{O}_K}$, we prove Theorem 4.1 in §V.

The author would like to thank D. Harbater for suggesting the interest of Arakelov intersection theory and for valuable conversations. He would also like to thank D. Blasius and B. Gross for conversations leading to a better understanding of the case of P^1. Finally, special thanks to R. Rumely for sending the early outline [9] of his work on capacity theory, as well as for the final manuscript [10], without which this paper could not have been written.

Added in proof (4/86): After this paper was accepted for publication, the author and R. Rumely obtained the following result. Let X be an arithmetic surface whose general fibre has positive genus. Then there are admissible canonical distance functions on the adelic fibres of X and Hermitian metrics on the infinite fibres of X for which $[\ ,\ ]_{Cap}$ extends to a pairing on $\mathrm{Div}_A(X)$ which respects linear equivalence. This extension is uniquely determined by the above distance functions and metrics, and the resulting pairing is in general neither alternating nor symmetric. Details will appear in a forthcoming paper by the author and Rumely.

II. _Definition of the Arakelov intersection pairing._

As in the introduction, let K be a number field and let X be a regular two-dimensional scheme with a projective morphism π: $X \to \mathrm{Spec}(\underline{O}_K)$ whose generic fibre is a smooth, geometrically irreducible curve over K. Let S_{inf} denote the set of infinite places of K.

For $\infty \in S_{inf}$, let X_∞ be the Riemann surface $X(\overline{K}_\infty)$, where $\overline{K}_\infty = \mathbb{C}$. We will regard X_∞ as the fibre of X above ∞. An Arakelov divisor on X (c.f. [1]) is a formal linear combination

$$D = D_{fin} + D_{inf} = \sum k_i C_i + \sum_{\infty \in S_{inf}} r_\infty X_\infty$$

in which the k_i are integers, each C_i is an irreducible closed subscheme of codimension 1 in X, and the r_∞ are real numbers. If $\pi(C_i) = \mathrm{Spec}(\underline{O}_K)$ then C_i is called horizontal.

Otherwise $\pi(C_i) = \mathrm{Spec}(\underline{O}_K/\mathfrak{p})$ for some prime ideal $\mathfrak{p}$ of $\underline{O}_K$, and C_i is called vertical. The fibres X_∞ for $\infty \in S_{inf}$ will also be called vertical, irreducible divisors. The additive group of Arakelov divisors will be denoted $\mathrm{Div}_A(X)$.

Let $\infty \in S_{inf}$, let ds_∞ be a Hermitian metric on the Riemann surface X_∞. Normalize ds_∞ so that the associated volume element $d\mu_\infty$ is a probability measure on X_∞. Let $|\ |_\infty$ be the Euclidean absolute value on K_∞, whether or not ∞ is real or complex. The Arakelov divisor of a function $f \in K(X)$ is

$$(f) = (f)_{fin} + \sum_{\infty \in S_{inf}} v_\infty(f)\, X_\infty$$

where $(f)_{fin}$ is the usual Weil divisor of f, and where

$$v_\infty(f) = \int_{X_\infty} -\log|f|_\infty \, d\mu_\infty$$

is a real number. (In [1] Arakelov uses embeddings of K into $\mathbb{C}$ rather than S_{inf}. To translate between embeddings and places, one may view the fibre X_∞ as the sum of the fibres of X over the embeddings of K into $\mathbb{C}$ which induce the place ∞.)

We now recall the definition of the Arakelov intersection $[D_1,D_2]_A$ of two distinct irreducible Arakelov divisors D_1 and D_2.

For $\infty \in S_{inf}$ define $\varepsilon_\infty = 1$ (resp. $\varepsilon_\infty = 2$) if ∞ is real (resp. complex). Define $[D_1,X_\infty]_A = [X_\infty,D_1]_A$ to be 0 (resp. $\varepsilon_\infty m$) if D_1 is vertical (resp. if D_1 is horizontal of degree m on the generic fibre of X).

Suppose now that D_1 and D_2 are distinct, finite, irreducible divisors. Let

$$[D_1,D_2]_A = [D_1,D_2]_{fin} + [D_1,D_2]_{inf}$$

where $[D_1,D_2]_{fin}$ is the usual Weil intersection of divisors on X, and where $[D_1,D_2]_{inf}$ will be defined below. (Recall that

$$[D_1,D_2]_{fin} = \sum_P \log(\#(\underline{O}_{X,P}/(f_{1,P},f_{2,P}))$$

where the sum is over the zero-dimensional points P of X, and where $f_{i,P}$ is a local equation of D_i at P.)

Define $[D_1,D_2]_{inf} = 0$ if either D_1 or D_2 is a component of a vertical fibre.

We now suppose that D_1 and D_2 are horizontal. The function field L_i of D_i is a finite extension of K. Let P_i be the point of $X(L_i)$ which is determined by D_i. For $\infty \in S_{inf}$, let $\tau_\infty: K \to ¢$ be an embedding which induces ∞. For each $\infty \in S_{inf}$, the point P_i determines a set of points $\{P^\infty_{i,j}: j = 1,\ldots,[L_i:K]\}$ on X_∞ corresponding to the embeddings of L_i into $¢$ over τ_∞. Define

$$[D_1,D_2]_{inf} = \sum_{\infty \in S_{inf}} \varepsilon_\infty \sum_{j,k} (P^\infty{}_{i,j},P^\infty{}_{2,k})_\infty$$

where $(P,Q)_\infty = -\log \phi_P(Q)$ for distinct points P and Q on X_∞ is defined by requiring $\phi_P(z)$ to be the unique function of $z \in X_\infty$ having the following properties.

(2.1) $\phi_P(z)$ _is a smooth non-negative real valued function of_ z _with a unique zero at_ $z = P$, _which is of first order._

(2.2) $$\frac{i}{\pi} \frac{\partial}{\partial z} \frac{\partial}{\partial \bar{z}} \log \phi_P(z) \, dz \wedge d\bar{z} = - d\mu_\infty \quad \text{for} \quad z \neq P .$$

(2.3) $$\int_{X_\infty} \log \phi_P(z) \, d\mu_\infty = 0 .$$

(The existence and uniqueness of $\phi_P(z)$ is shown in [1]; $g(P,Q) = \log \phi_P(Q)$ is called in [1] the Green's function of X_∞ with respect to $d\mu_\infty$.)

The first main result of [1] is

Theorem 1.1 (Arakelov): _The pairing_ $[D_1,D_2]_A$ _extends in a unique way to a symmetric bilinear pairing on_ $\mathrm{Div}_A(X)$ _which respects linear equivalence._

A definition of $[D_1,D_2]_A$ in terms of Hermitian line bundles which also works for non-proper intersections is given in [1]. An exposition of some basic results of Arakelov theory and a list of further references is given in [4].

III. Adelic capacity theory.

We recall in this section the main elements of the adelic capacity theory developed by Cantor in [3] for P^1 and generalized to arbitrary curves of Rumely in [9,10]. The notations K and X are as in §II.

Suppose v is a place of K. Let $\mathbb{C}_v$ be the completion of $\overline{K}_v$. Rumely develops local capacity theory for both subsets of $X(\overline{K}_v)$ and of $X(\mathbb{C}_v)$. We will state Rumely's results only for $X(\overline{K}_v)$, since this is what is needed for global capacity theory.

A. Canonical and chordal distance functions.

For v a place of K, define $p(v)$ to be the rational prime determined by v is v is non-archimedean, and let $p(v) = e$ if v is archimedean. Define $\log_v(r) = \log(r)/\log(p(v))$ for real numbers $r > 0$. If v is non-archimedean, let $|\ |_v$ be the absolute value on $\overline{K}_v$ normalized so that $|p(v)|_v = p(v)^{-1}$. If v is archimedean, let $|\ |_v$ be the Euclidean absolute value on $\overline{K}_v \cong \mathbb{C}$, whether or not v is real or complex.

Let X be an arithmetic surface over $\mathrm{Spec}(\underline{O}_K)$. Fix a point $\zeta \in X(\overline{K}_v)$, and let $f(z) \in \overline{K}_v(X)$ be a uniformizing parameter at ζ. Let $<\ ,\ >_v$ denote the negative of Neron's pairing (c.f. [10, chapter 2.2], [8], [5]) on the cycles of degree zero of $X(\overline{K}_v)$ which have disjoint support. A canonical distance function with respect to ζ is a real valued, non-negative function $[x,y]_\zeta$ of $x,y \in X(\overline{K}_v) - \{\zeta\}$ of the following kind. Let $c(\zeta)$ be a real constant depending on ζ. Define $[x,x]_\zeta = 0$, and if $x \neq y$, define $[x,y]_\zeta$ by

$$-\log_v[x,y]_\zeta = c(\zeta) + \lim_{w\to\zeta} (<(x)-(w),(y)-(\zeta)>_v + \log_v|f(w)|_v)$$

If one chooses a different uniformizer $f(z)$, this equality holds when one changes $c(\zeta)$ appropriately. Thus different canonical distance functions associated to ζ differ from one another by a real scalar.

The properties of canonical distances are analyzed in detail by Rumely in [10]. He shows, for example, that $[x,y]_\zeta = [y,x]_\zeta$, and that for a fixed x, $[x,y]_\zeta$ has a unique zero at $y = x$ and a simple pole as y approaches ζ. Thus one can regard $[x,y]_\zeta$ as measuring the distance between x and y in some sense, even though $[\ ,\ ]_\zeta$ need not satisfy the triangle inequality in general, and thus may not define a metric on $X(\overline{K}_v)$.

We will later also need the definition the chordal distance $\|x,y\|_v$ between points $x = (x_0,\ldots,x_N)$ and $y = (y_0,\ldots,y_N)$ of $P^N(\overline{K}_v)$ when N is a positive integer. If v is archimedean, define $\langle x,y\rangle = \Sigma\, x_i\, y_i$, $|x| = \sqrt{\langle x,x\rangle}$ and let

$$\|x,y\|_v = \frac{\left(\sum_{i<j} |x_i y_j - x_j y_i|_v^2\right)^{1/2}}{|x|\quad |y|}$$

Then $\|x,y\|_v$ is the chordal distance associated to the Fubini-Study metric on $P^N(\overline{K}_v)$ normalized so that the maximal geodesic distance between two points is π (c.f. [10, section 1.1]). Suppose now that v is non-archimedean. Define $|x| = \max_i |x_i|_v$ and

$$\|x,y\|_v = \frac{\max_{i<j} |x_i y_j - x_j y_i|_v}{|x|\quad |y|}$$

Then $\| \ , \ \|_v$ is the usual v-adic metric on $P^N(\overline{K}_v)$ coming from the sup norm on the standard affine patches of $P^N(\overline{K}_v)$.

Returning to the arithmetic surface X and K , let $X/\overline{K}_v$ be the curve over $\overline{K}_v$ which is determined by the generic fibre of X . Suppose $\phi: X/\overline{K}_v \to P^N/\overline{K}_v$ is a non-singular embedding defined over $\overline{K}_v$. The chordal distance $\|x,y\|_{v,\phi}$ on $X(\overline{K}_v)$ with respect to ϕ is defined to be the pullback of the chordal distance on $P^N(\overline{K}_v)$.

In an important special case, one can describe Rumely's canonical distances using chordal distances. Suppose that v is non-archimedean. One says that $X/\overline{K}_v$ has good reduction at v with respect to the embedding ϕ (c.f. [11, section 10]) if the special fibre of $X/\overline{K}_v$ consists of a single component which is non-singular and which has multiplicity one. If this is the case, Rumely shows in [10, section 2.4] that

$$(3.1) \qquad [z_1 z_2]_\zeta = \frac{\|z_1,z_2\|_{v,\phi}}{\|z_1,\zeta\|_{v,\phi}\|z_2,\zeta\|_{v,\phi}}$$

is a canonical distance function. We will call (3.1) the spherical canonical distance function at ζ with respect to the embedding $\phi: X/\overline{K}_v \to P^N/\overline{K}_v$.

B. Local capacity theory.

Let v be a place of K , and let E_v be a compact subset of $X(\overline{K}_v)$. Suppose that $\zeta \in X(\overline{K}_v) - E_v$ and that $[z_1,z_2]_\zeta$ is a canonical distance function with respect to ζ . Rumely defines (c.f. [10, sections 3.1 and 4.1])

$$V_\zeta(E_v) = \inf \iint_{E_v \times E_v} -\log_v[z_1,z_2]_\zeta \, d\mu(z_1)\, d\mu(z_2)$$

where the minimum is over probability measures μ supported on E_v. A measure $\mu_{E_v,\zeta}$ which achieves $V_\zeta(E_v)$ will be called an equilibrium distribution. Rumely shows there is always an equilibrium distribution, and that this distribution is unique if $V_\zeta(E_v) < \infty$. The capacity of E_v with respect to ζ is defined to be $\gamma_\zeta(E_v) = \exp(-\log(p(v))V_\zeta(E_v))$. Notice that $\gamma_\zeta(E_v)$ depends on the choice of a canonical distance function $[\ ,\ ]_\zeta$.

To define the inner and outer capacity of non-compact sets, we will need the notions of RL-domains and PL_ζ-domains. An RL-domain is one of the form

$$U_v = \{z \in X(\overline{K}_v): |f(z)|_v \leq 1\}$$

for some rational function $f(z) \in \overline{K}_v(X)$. If in addition $f(z)$ has all its poles at ζ, then U_v is called a PL_ζ-domain.

Let F_v be an arbitrary subset of $X(\overline{K}_v)$. The inner capacity of F_v with respect to ζ is defined to be

$$\underline{\gamma}_\zeta(F_v) = \sup \{\gamma_\zeta(E_v): \ E_v \ \underline{\text{is a compact subset of}} \ F_v\} .$$

The outer capacity of F_v is

$$\overline{\gamma}_\zeta(F_v) = \inf \{\gamma_\zeta(U_v): \ U_v \ \underline{\text{is a}}\ PL_\zeta\underline{\text{-domain containing}} \ F_v\} .$$

One says F_v is admissible if it is closed, and if $\underline{\gamma}_\zeta(F_v) = \overline{\gamma}_\zeta(F_v)$ for each $\zeta \notin F_v$. In this case, one defines the capacity $\gamma_\zeta(F_v)$ of F_v with respect to ζ to be $\underline{\gamma}_\zeta(F_v)$.

We will later need the following basic result, which is proved in [10, Proposition 4.4.6].

Proposition 3.1 (Rumely): *Compact subsets of* $X(\overline{K}_v)$ *and RL-domains in* $X(\overline{K}_v)$ *are admissible.*

Let E_v be a compact subset of $X(\overline{K}_v)$ such that $\gamma_\zeta(E_v) > 0$ if $\zeta \notin E_v$. For $\zeta \notin E_v$, define the Green's function $G(z,\zeta;E_v)$ by

$$G(z,\zeta,E_v) = V_\zeta(E_v) + \int_{E_v} \log_v[z,x]_\zeta \, d\mu_{E_v,\zeta}(x)$$

where $\mu_{E_v,\zeta}$ is the equilibrium distribution on E_v with respect to ζ. For $\zeta \in E_v$, define $G(z,\zeta;E_v) = G(\zeta,z;E_v)$ if $z \notin E_v$ and let $G(z,\zeta;E_v) = 0$ if $z \in E_v$. Recall that $[z_1,z_2]_\zeta$ is determined up to multiplication by a positive real constant $c(\zeta)$. The equilibrium distribution $\mu_{E_v,\zeta}$ and the Green's function $G(z,\zeta;E_v)$ are independent of the choice of the constant $c(\zeta)$.

Suppose now that F_v is an admissible subset of $X(\overline{K}_v)$ such that $\gamma_\zeta(F_v) > 0$ if $\zeta \notin F_v$. The Green's function of F_v is defined to be

$$G(z,\zeta;F_v) = \begin{cases} \inf\{G(z,\zeta;E_v) : E_v \text{ is compact, } E_v \subseteq F_v \text{ and } \gamma_\zeta(E_v) > 0\} & \text{if both } z, \zeta \notin F_v, \\ 0 & \text{if either } z \in F_v \text{ or } \zeta \in F_v. \end{cases}$$

C. Definition of global capacity.

Let $\chi = \{x_1, \ldots, x_r\}$ be a finite set of global algebraic points of $X(\overline{K})$ which is stable under the action of $\mathrm{Gal}(\overline{K}/K)$. An adele of sets is a collection $\mathbf{E} = \{E_v\}_v$ of subsets $E_v \subseteq X(\overline{K}_v)$ indexed by the places v of K.

Let $X \to P^N$ be a fixed embedding over K. For each place v we then have a chordal distance function $\| \, , \|_v$ on $X(\overline{K}_v)$ induced by the chordal distance on $P^N(\overline{K}_v)$.

A subset $E_v \subseteq X(\overline{K}_v)$ is called χ-trivial if X has good reduction at v, if the points of χ reduce to distinct points modulo the maximal ideal of the ring of integers of $\overline{K}_v$, and if

$$E_v = X(\overline{K}_v) - \bigcup_{i=1}^{r} \{z \in X(\overline{K}_v): \|z, x_i\|_v < 1\} .$$

An adele of sets $\mathbf{E} = \{E_v\}_v$ is called χ-admissible if for all v, E_v is an admissible subset of $X(\overline{K}_v)$ which is disjoint from χ and stable under $\mathrm{Gal}(\overline{K}_v/K_v)$, and if E_v is χ-trivial for all but finitely many v.

Suppose that L is a finite extension of K and that X_L is a regular model of $X \times_{\mathrm{Spec}(\underline{O}_K)} \mathrm{Spec}(\underline{O}_L)$ over $\mathrm{Spec}(\underline{O}_L)$. Let w be a place of L, and let v be the place of K under w. We may identify L_w with K_v. Suppose $\mathbf{E} = \{E_v\}_v$ is an adele of sets of X. Let $E'_w \subseteq X_L(\overline{L}_w)$ be the set E_v when $X_L(\overline{L}_w)$ is identified with $X(\overline{K}_v)$. We call $\mathbf{E}' = \{E'_w\}_w$ the adele of sets for X_L which is induced by $\mathbf{E}$. If $\mathbf{E}$ is χ-admissible then $\mathbf{E}'$ will be χ-admissible.

We assume for the rest of this section that $\mathbf{E} = \{E_v\}_v$ is χ-admissible. We are going to associate to $\mathbf{E}$ and χ a real number $\gamma(\mathbf{E}, \chi)$ which will be invariant on replacing K by a finite extension of K and $\mathbf{E}$ by the induced adele of sets over the extension. To do this we may thus replace K by a finite extension of K large enough so that $x_1,\ldots,x_r$ become rational over K. For each x_i let $g_i(z) \in K(X)$ be a rational function over K which has a simple zero at x_i.

For each place v the local Green's matrix $\Gamma_v = (\Gamma_{v,i,j})$ is the $r \times r$ matrix defined as follows. If there is a point $\zeta \in X(\overline{K}_v) - E_v$ such that $\gamma_\zeta(E_v) = 0$, let every entry of Γ_v be ∞. Otherwise the Green's function $G(z,z';E_v)$ is defined, and we let

$$\Gamma_{v,i,j} = \begin{cases} G(x_i,x_j;E_v) & \underline{\text{if}} \quad i \neq j \text{ , } \underline{\text{and}} \\ \lim\limits_{z\to x_i} G(z,x_i;E_v) + \log_v |g_i(z)|_v & \underline{\text{if}} \quad i = j \text{ .} \end{cases}$$

The off diagonal entries of Γ_v are non-negative, and Γ_v is symmetric. All but finitely many of the Γ_v will be the zero matrix.

For each place v of K let $w(v)$ be the place of $\mathbf{Q}$ under v. The global Green's matrix is defined to be

$$\Gamma = \sum_v ([K_v : \mathbf{Q}_{w(v)}]/[K:\mathbf{Q}]) \log(p(v)) \Gamma_v$$

where the sum is over all places v of K. By the product formula, Γ is independent of the choice of the functions $g_i(z)$. Like the Γ_v, Γ is a symmetric $r \times r$ matrix with non-negative off diagonal entries.

Let $P = \{(s_1,\ldots,s_r) \in \mathbf{R}^r:\ s_i > 0 \ \underline{\text{for all}}\ \underline{i},\ \Sigma_i s_i = 1\}$ be the set of r-element real probability vectors. Define

$$V(\mathbf{E},\chi) = \max_{s\varepsilon P} \min_{u\varepsilon P} {}^t s \,\Gamma\, u$$

(This number is the value of Γ as a matrix game.) The global capacity of the adele of sets $\mathbf{E} = \{E_v\}_v$ with respect to χ is defined by Cantor [3] for $X = P^1$ and by Rumely [10] for all X, to be

$$\gamma(\mathbf{E},\chi) = \exp(- V(\mathbf{E},\chi)) .$$

D. Properties of global capacity and a special case.

Let L be a finite extension of K, and let X_L be a regular model of $X \times_{\mathrm{Spec}(\underline{O}_K)} \mathrm{Spec}(\underline{O}_L)$ over $\mathrm{Spec}(\underline{O}_L)$. Suppose $\mathbf{E} = \{E_v\}_v$ is an χ-admissible adele of sets of X. Let $\mathbf{E'} = \{E'_w\}_w$ be the χ-admissible adele of sets for X_L which is induced by $\mathbf{E}$.

The following two properties of $\gamma(\mathbf{E},\chi)$ are proved in [10], and will be used in later computations.

(3.2) Adelic capacity is invariant under base change from K to a finite extension of K, i.e. $\gamma(\mathbf{E},\chi) = \gamma(\mathbf{E'},\chi)$.

(3.3) Suppose $f\colon X_1 \to X$ is a rational function between arithmetic surfaces X_1 and X over K, and that f has degree m as a function on the generic fibre of X_1. Then $f^{-1}(\mathbf{E}) = \{f^{-1}(E_v)\}_v$ is an $f^{-1}(\chi)$-admissible adele of sets for X_1, and

$$\gamma(f^{-1}(\mathbf{E}),f^{-1}(\chi)) = \gamma(\mathbf{E},\chi)^{1/m} .$$

An adelic neighborhood $\mathbf{U} = \{U_v\}_v$ of $\mathbf{E} = \{E_v\}_v$ is a collection of open sets $U_v \subseteq X(\overline{K}_v)$ such that $E_v \subseteq U_v$. The following generalization of the Fekete-Szego theorem was proved by Cantor when $X = P^1_{\underline{O}_K}$ and generalized by Rumely to all X:

Theorem 3.1 ([3], [10]): *Let* $\mathbf{E}$ *be an* χ *-admissible adele of sets for* X. *Then*

(a) *If* $\gamma(\mathbf{E},\chi) < 1$, *there is an adelic neighborhood of* $\mathbf{E}$ *which contains only finitely many complete sets of* $\mathrm{Gal}(\overline{K}/K)$ *-conjugate points of* $X(\overline{K})$.

(b) *If* $\gamma(\mathbf{E},\chi) > 1$, *every adelic neighborhood of* $\mathbf{E}$ *contains infinitely many such sets.*

To illustrate the theory of the previous sections, we will discuss a particularly simple example relevant to §V.

Suppose that $X = P^1_{\underline{O}_K}$. Write $X(L)$ for all fields $L \supseteq K$ as $A^1(L) \cup \{\underline{\infty}\}$, where $\underline{\infty}$ is defined over K. Let v be a place of K and let x and y be points of $A^1(\overline{K}_v)$. The chordal distance $\|x,y\|_v$ is

$$(3.4)\quad \begin{cases} \dfrac{|x-y|_v}{((1+|x|_v^2)(1+|y|_v^2))^{1/2}} & \text{if } v \text{ is archimedean}, \\[2em] \dfrac{|x-y|_v}{\max(1,|x|_v)\max(1,|y|_v)} & \text{if } v \text{ is non-archimedean}. \end{cases}$$

(Recall that $|\ |_v$ is the Euclidean absolute value if v is archimedean.) These formulas also hold when x and y is $\underline{\infty}$ if one interprets the right side as a suitable limit.

The spherical distance function $[z_1,z_2]_\zeta$ associated to a point $\zeta \in X(\overline{K}_v)$ at the end of §III.A is

$$[z_1,z_2]_\zeta = \begin{cases} |z_1 - z_2|_v & \underline{\text{if}}\ \ \zeta = \underline{\infty}\ , \\ \dfrac{|z_1 - z_2|_v\,(1 + |\zeta|_v^{\,2})}{|z_1 - \zeta|_v\ |z_2 - \zeta|_v} & \underline{\text{if}}\ \ \zeta \neq \underline{\infty}\ . \end{cases}$$

Suppose now that for each place v , r_v is a positive real number and that almost all of the r_v equal 1 . Let $\chi = \{\underline{\infty}\}$ and let $E_v = \{z\colon |z|_v \leq r_v\} \subseteq A^1(\overline{K}_v)$ for all v . The Green's function of E_v is given by

$$G(z,\underline{\infty};E_v) = \begin{cases} 0 & \underline{\text{if}}\ \ z \in E_v\ , \\ \log_v(|z|_v) - \log_v(r_v) & \underline{\text{if}}\ \ z \notin E_v\ . \end{cases}$$

On taking the local parameter $g(z) = z^{-1}$ at $\underline{\infty}$, one has local 1 by 1 Green's matrices

$$\Gamma_v = [\,-\log_v(r_v)]$$

Let $r = \pi\, r_v^{[K_v:\,Q_{w(v)}]}$. The global Green's matrix is simply

$$\Gamma = \frac{1}{[K:\mathbf{Q}]}\,[\,-\log r]\ .$$

From this the global capacity of $\mathbf{E} = \{E_v\}_v$ with respect to $\chi = \{\underline{\infty}\}$ is

$$(3.5) \qquad \gamma(\mathbf{E},\chi) = r^{1/[K:\mathbf{Q}]}\ .$$

IV. *The capacity pairing.*

A. Definition and general properties of $[\ ,\ ]_{\mathrm{Cap}}$.

The pairing will be normalized via the following definition.

Definition 4.1: A family $\mathcal{F}$ *of canonical distances is a choice for each place* v *of* K *and for each* $\zeta \in X(\overline{K}_v)$ *of a canonical distance function* $[z_1,z_2]_\zeta$ *with respect to* ζ . *We call* $\mathcal{F}$ *a continuous family if for each* v *and every pair of points* $z_1, z_2 \in X(\overline{K}_v)$, *the function* $\zeta \to [z_1,z_2]_\zeta$ *is a continuous function on* $X(\overline{K}_v) - \{z_1,z_2\}$.

In [10, sections 2.3 and 2.4] Rumely shows that if the generic fibre of X has non-zero genus, there is a natural choice of a continuous family of distances $\mathcal{F}$. However, it may be also useful to consider other possible $\mathcal{F}$.

We now fix a family $\mathcal{F}$ of canonical distances for X .

Suppose that D_1 and D_2 are distinct, irreducible horizontal divisors on X which do not pass through any multiple points of reducible fibres of X . The subgroup $\mathrm{Div}_A^*(X)$ of $\mathrm{Div}_A(X)$ is the group generated by such D_1 and D_2 . Let L

be a finite Galois extension of K containing the function fields of D_1 and D_2. Let X_L be a regular model over $\mathrm{Spec}(\underline{O}_L)$ of $X_O = X \times_{\mathrm{Spec}(\underline{O}_K)} \mathrm{Spec}(\underline{O}_L)$. We construct X_L by Lipman's method in ([7],[2]), in which X_L is the final term in a finite sequence of two-dimensional schemes starting with X_O. The (j+1)-st term in this sequence is obtained from the j-th term by blowing up the singular locus of the j-th term and then normalizing. Since L/K is Galois, and the D_i avoid multiple points of reducible fibres of X, the pullback of each D_i to X_L will be a sum of sections of $X_L \times \mathrm{Spec}(\underline{O}_L)$, and these sections will lie in $\mathrm{Div}_A^*(X_L)$. Hence on replacing K by L, X by X_L, and $\mathcal{F}$ by the induced set of canonical distances for X_L, we may reduce to the case in which D_1 and D_2 are distinct sections of $X \to \mathrm{Spec}(\underline{O}_K)$. Let P_i be the point of $X(K)$ determined by D_i. Define

$$(4.1) \qquad F_v = \{z \in X(\overline{K}_v): [z,P_1]_{P_2} \leq [z,P_2]_{P_1}\}$$

where $[\ ,\ ]_\zeta$ is the canonical distance function in $\mathcal{F}$ for $\zeta \in X(\overline{K}_v)$. Let $\mathbf{F} = \mathbf{F}(D_1,D_2)$ be the adele of sets $\{F_v\}_v$. We will call $\mathcal{F}$ an admissible set of canonical distances if the above construction always leads to a $\{P_2\}$-admissible adele of sets $\mathbf{F}$. In ths case we define

$$(4.2) \qquad [D_1,D_2]_{\mathrm{Cap}} = \log \gamma(\mathbf{F},\{P_2\})$$

and we then extend $[\ ,\ ]_{\mathrm{Cap}}$ to all coprime pairs of elements of $\mathrm{Div}_A^*(X)$ by bilinearity and base change. The resulting pairing is well-defined and respects base change because adelic capacity respects base change (c.f. §III.D).

Recall from §III.A that for distinct, fixed z_2 and ζ in $X(\overline{K}_v)$, the canonical distance function $[z,z_2]_\zeta = [z_2,z]_\zeta$ is non-negative and continuous in z, has a unique pole (resp. a unique zero) at $z = \zeta$ (resp. at $z = z_2$). Thus the adele of sets $\mathbf{F} = \{F_v\}_v$ may be pictured as a strip of adelic points of X which are at least as close to D_1 as to D_2 in every adelic fibre of X.

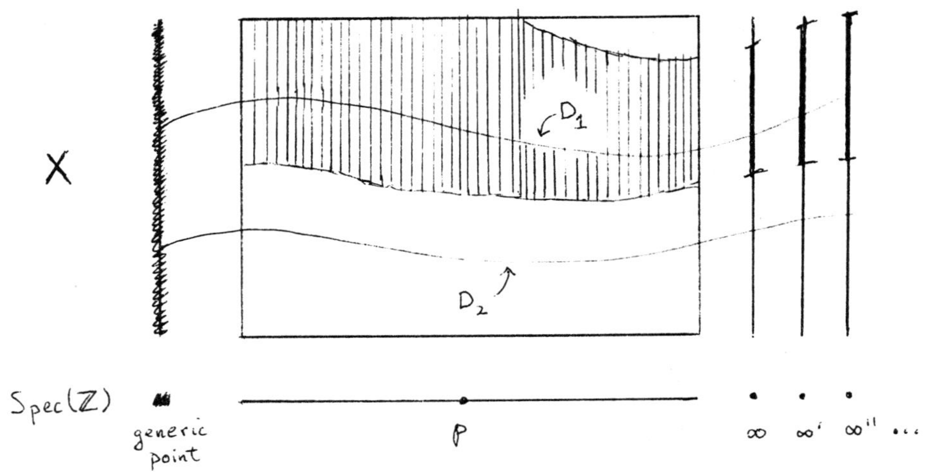

figure 4.1

Call $\mathbf{F}$ <u>narrow</u> if $\mathbf{F}$ has an open adelic neighborhood $\mathbf{U}$ which contains only finitely many complete sets of $\mathrm{Gal}(\overline{K}/K)$ - conjugate global algebraic points of $X(\overline{K})$. Call $\mathbf{F}$ <u>wide</u> if every open adelic neighborhood $\mathbf{U}$ of $\mathbf{F}$ contains infinitely many such sets. Theorem 3.1 implies

(4.3) $\mathbf{F}(D_1,D_2)$ is narrow if $[D_1,D_2]_{\mathrm{Cap}} < 0$ and wide if $[D_1,D_2]_{\mathrm{Cap}} > 0$.

Viewing $\mathbf{F}(D_1,D_2)$ as the strip pictured in figure 4.1, we will call $\mathbf{F}(D_1,D_2)$ the $\mathcal{F}$-complementary strip to $\mathbf{F}(D_1,D_2)$. From (4.3) one has the following corollary.

(4.4) If $[D_1,D_2]_{\mathrm{Cap}}$ and $[D_2,D_1]_{\mathrm{Cap}}$ have different, non-zero signs, then $\mathbf{F}(D_1,D_2)$ is narrow iff its $\mathcal{F}$-complement $\mathbf{F}(D_2,D_1)$ is wide.

<u>Question 4.1</u>: Is there an admissible family $\mathcal{F}$ of canonical distances and set of Hermitian metrics $\{ds_\infty\}_\infty$ at infinity such that $[\ ,\ ]_{\mathrm{Cap}}$ may be extended from $\mathrm{Div}_A^*(X)$ to a pairing on $\mathrm{Div}_A(X)$ which respects linear equivalence or which is alternating?

When $X = \mathbb{P}^1_{\underline{O}_K}$ this question is answered affirmatively in the next section.

To show that the definition of $[\ ,\ ]_{\mathrm{Cap}}$ is non-empty, we now exhibit some admissible families $\mathcal{F}$ of canonical distances.

<u>Proposition 4.1</u>: <u>Let</u> $\phi: X/K \to \mathbb{P}^N/K$ <u>be a projective nonsingular embedding defined over</u> K . <u>Let</u> $\mathcal{F}$ <u>be a family of canonical distances such that for almost places</u> v <u>of</u> K , <u>the canonical distance function</u> $[\ ,\ ]_\zeta$ <u>associated to every point</u> ζ <u>of</u> $X(\overline{K}_v)$ <u>is the spherical canonical distance</u>

$$[z_1,z_2]_\zeta = \frac{\|z_1,z_2\|_{v,\phi}}{\|z_1,\zeta\|_{v,\phi}\ \|z_2,\zeta\|_{v,\phi}} .$$

Then $\mathcal{F}$ is an admissible family of canonical distances.

Proof: Let P and P' be distinct points of $X(K)$. Define

$$F_v = \{z \in X(\overline{K}_v): [z,P]_{P'} \leq [z,P']_P\} .$$

where $[z,P']_P$ and $[z,P]_{P'}$ are the canonical distance functions in $\mathcal{F}$ for the place v . We are to show that $\mathbf{F} = \{F_v\}_v$ is a $\{P'\}$ -admissible adele of sets for X . This amounts to showing that each F_v is an admissible, $Gal(\overline{K}_v/K_v)$ -stable subset of $X(\overline{K}_v)$ which is disjoint from P' , and that almost all of the F_v are $\{P'\}$ -trivial.

Suppose $[\ ,\ ]_\zeta$ is a canonical distance function at $\zeta \in X(K_v)$, z and z' are points of $X(\overline{K}_v) - \{\zeta\}$, and that $\sigma \in Gal(\overline{K}_v/K_v)$. In [10, Theorem 2.1.1] it is shown that $[z,z']_\zeta = [\sigma z,\sigma z']_\zeta$. Since P and P' are rational over K , this shows that F_v is $Gal(\overline{K}_v/K_v)$ -stable. If v is archimedean, then F_v is a closed subset of the compact set $X(\overline{K}_v)$. Hence F_v is compact, and therefore admissible by Proposition 3.1. For the remainder of the proof we will suppose v is non-archimedean.

For simplicity, denote $\|\ ,\ \|_{v,\phi}$ by $\|\ ,\ \|_v$. For $z \in X(\overline{K}_v)$ and $\partial > 0$, let $B_v(z,\partial)$ (resp. $B_v(z,\partial)^-$) be the closed ball (resp. the open ball) of points $z' \in X(\overline{K}_v)$ such that $\|z,z'\|_v \leq \partial$ (resp. $\|z,z'\|_v < \partial$). From the continuity in z of $[z,P']_P$ and $[z,P]_{P'}$, we conclude that for $\partial_0 > 0$ sufficiently small, F_v contains $B_v(P,\partial_0)$ and $F_v \cap B_v(P',\partial_0)^-$ is empty.

By [10, section 2.1, eq. 12], we can decrease $\partial_0 > 0$

further if necessary so that the following is true. There are functions $f(z)$ and $g(z)$ in $\overline{K}_v(X)$, of degrees n and m respectively, such that $[z,P]_{P'} = |f(z)|_v^{1/n}$ for $z \notin B_v(P',\partial_0)$ and $[z',P']_P = |g(z')|_v^{1/m}$ for $z' \notin B_v(P,\partial_0)$. Let $h(z) = f(z)^m/g(z)^n$. The set $U_0 = \{z \in X(\overline{K}_v) : |h(z)|_v \leq 1\}$ is an RL-domain in the sense of [10] and §III.B. We have shown furthermore that

$$F_v = (U_0 \cup B_v(P,\partial_0)) \cap (X(\overline{K}_v) - B_v(P',\partial_0)^-)$$

Here $B_v(P,\partial_0)$ and $X(\overline{K}_v) - B_v(P',\partial_0)^-$ are island domains, in the sense of [10, section 4.3]. From [10, Corollary 4.39] it follows that island domains are RL-domains, while in [10, Corollary 4.3.10] it is shown that the finite union or intersection of RL-domains is an RL-domain. Hence F_v is an RL-domain, so F_v is admissible by Proposition 3.1.

It remains to be proved that almost all of the F_v are $\{P'\}$-trivial. There will be a finite set S of places of K, containing the archimedean places, such that the following statements are true if $v \notin S$: (i) X_v has good reduction at v with respect to $\phi: X/K \to P^N/K$, (ii) The canonical distance functions in $\mathfrak{F}$ for the place v are the spherical canonical distances stated in Proposition 4.1, (iii) $\|P,P'\|_v = 1$, and (iv) the points $\phi(P)$ and $\phi(P')$ have unit coordinates in every affine patch of $P^N(\overline{K}_v)$. Because of (ii) and (iii), one has $z \in F_v$ iff $\|z,P'\|_v > \|z,P\|_v$. Since $\|z,P\|_v \leq 1$ for all z and P, we will certainly have $z \in F_v$ if $\|z,P'\|_v = 1$. On the other hand, if $\|z,P'\|_v < 1$, we conclude from (iii), (iv) and the definition of $\|z,P\|_v$ that

$\|z,P\|_v = 1$. Hence $F_v = \{z \in X(\overline{K}_v): \|z,P'\|_v > 1\}$ if $v \notin S$. Thus f_v is $\{P'\}$-trivial for $v \notin S$, so $\mathbf{F} = \{F_v\}_v$ is $\{P'\}$-admissible.

B. The case $X = P^1_{\underline{O}_K}$.

Theorem 4.1: Let $X = P^1_{\underline{O}_K}$ and let $\mathcal{F}$ and the ds_∞ be as follows. For all v and $\zeta \in X(\overline{K}_v)$ let the corresponding canonical distance in $\mathcal{F}$ be the spherical distance

$$[z_1,z_1]_\zeta = \frac{\|z_1,z_2\|_v}{\|z_1,\zeta\|_v \ \|z_2,\zeta\|_v}$$

where $\| \ , \ \|_v$ is the chordal distance on $P^1(\overline{K}_v) = X(\overline{K}_v)$. Let

$$d\mu_\infty = \frac{1}{2\pi i} \ \frac{dz \wedge d\overline{z}}{(1 + |z|^2)^2}$$

be the volume form of the Fubini-Study metric ds_∞ on $X_\infty = P^1(\mathbb{C})$ for each infinite place ∞ of K . Define

$$[D_1,D_2]_{Cap} = - [D_2,D_1]_{Cap} = [D_1,D_2]_A \ / \ [K:\mathbf{Q}]$$

and

$$[D_2,D_2]_{Cap} = 0$$

for all distinct irreducible Arakelov divisors D_1 and D_2 for which D_1 is vertical, and extend $[\ , \ ]_{Cap}$ to all of $\mathrm{Div}_A(P^1_{\underline{O}_K})$ by bilinearly. Then $[\ , \ ]_{Cap}$ is alternating and respects linear equivalence.

We now describe explicity the pairing on the Arakelov divisor class group of $P^1_{\underline{O}_K}$ which is induced by that of Theorem 4.1.

For any arithmetic surface X over $\mathrm{Spec}(\underline{O}_K)$, let $\mathrm{Prin}_A(X)$ be the group of principal Arakelov divisors, and let $\mathrm{Pic}_A(X) = \mathrm{Div}_A(X)/\mathrm{Prin}_A(X)$ be the Arakelov divisor class group. Let $\mathrm{Pic}(X)$ be the divisor class group of X as a two-dimensional scheme, and let $\beta\colon \mathrm{Pic}_A(X) \to \mathrm{Pic}(X)$ be the homomorphism induced by forgeting the infinite components of an Arakelov divisor. Let $t = \#S_{\inf}$. Then β induces an exact sequence

$$(4.5) \qquad 0 \to R^t/\gamma(\underline{O}_K^*) \overset{\sigma}{\to} \mathrm{Pic}_A(X) \overset{\beta}{\to} \mathrm{Pic}(X) \to 0$$

in which $\lambda\colon \underline{O}_K^* \to R^t$ is the usual logarithm map defined by $\lambda(e) = (\varepsilon_\infty \log|e|_\infty)_{\infty \in S_{\inf}}$. (Recall ε_∞ is 1 or 2 depending on whether ∞ is real or complex.) Here σ is induced by the homorphism $\sigma(r) = \Sigma(r_\infty/\varepsilon_\infty)X_\infty$ for $r = (r_\infty)_\infty \in \mathbf{R}^t$.

When $X = P^1_{\underline{O}_K}$, write $P^1(K) = A^1(K) \cup \{\underline{\infty}\}$. The closure of the point $\underline{\infty}$ on the generic fibre of X is a divisor $D(\underline{\infty})$ of X. Let $\mathrm{Cl}(\underline{O}_K)$ be the ideal class group of K. There is a unique isomorphism $\tau\colon \mathrm{Pic}(P^1_{\underline{O}_K}) \overset{\sim}{\to} \mathbf{Z} \oplus \mathrm{Cl}(\underline{O}_K)$ for which $\tau(D(\underline{\infty})) = 1 \oplus 0$ and $\tau(P^1{}_v) = 0 \oplus [v]$ if v is a finite place of K, where $P^1{}_v$ is the fibre of $P^1_{\underline{O}_K}$ over v and $[v]$ is the ideal class of the prime ideal of K determined by v. By modifying the divisors $P^1{}_v$ for finite v by suitable linear combinations of infinite fibres, it is not difficult to see that the sequence (4.5) splits:

$$\text{(4.6)} \qquad \mathrm{Pic}_A(P^1_{\underline{O}_K}) = \mathbf{R}^t/\lambda(\underline{O}_K^*) \oplus \mathbf{Z} \oplus \mathrm{Cl}(\underline{O}_K)$$

Define the trace $\mathrm{Tr}(r)$ of an element $r \in \mathbf{R}^t/\lambda(\underline{O}_K^*)$ to be $\Sigma_\infty r_\infty$ for any vector $(r_\infty)_\infty \in \mathbf{R}^t$ representing r. Suppose $\alpha = (r,z,c)$ and $\alpha' = (r',z',c')$ are two elements of $\mathrm{Pic}_A(P^1_{\underline{O}_K})$ relative to the isomorphism (4.6), where z and z' are integers and c and c' are ideal classes. In the course of proving Theorem 4.1 we will show that the pairing induced on $\mathrm{Pic}_A(P^1_{\underline{O}_K})$ is

$$\text{(4.7)} \qquad [\alpha,\alpha']_{\mathrm{Cap}} = (\mathrm{Tr}(r)z' - \mathrm{Tr}(r')z)/[K:\mathbf{Q}] .$$

Thus if $K = \mathbf{Q}$, one has $\mathrm{Pic}_A(P^1_{\underline{O}_K}) = \mathbf{R} \oplus \mathbf{Z}$, and $[\ ,\ ]_{\mathrm{Cap}}$ is the determinant pairing.

V. Proof of Theorem 4.1.

Write $P^1(L)$ for all fields L K as $A^1(L)$ $\{\underline{\infty}\}$, where $\underline{\infty}$ is defined over K. For $a \in P^1(K)$, let $D(a)$ be the irreducible horizontal Arakelov divisor which is the closure in $X = P^1_{\underline{O}_K}$ of the point a on the generic fibre of X.

Lemma 5.1: Let $\mathcal{G}$ and $d\mu_\infty$ be as in the statement of Theorem 4.1, and suppose $a,b \in P^1(K)$.

(a) $[D(a),D(\underline{\infty})]_{\mathrm{Cap}} = -[D(\underline{\infty}),D(a)]_{\mathrm{Cap}} = [D(a),D(\underline{\infty})]_A/[K:\mathbf{Q}] + 1/2$.

(b) $[D(a),D(b)]_{\mathrm{Cap}} = [D(a),D(\underline{\infty})]_{\mathrm{Cap}} - [D(b),D(\underline{\infty})]_{\mathrm{Cap}}$.

(c) $[D,D(a)]_{Cap} = [D,D(\underline{\infty})]_{Cap}$ <u>if</u> D <u>is an irreducible vertical Arakelov divisor</u>.

<u>Proof</u>. Supppose first that $a \neq \underline{\infty}$. By definition, $[D(a),D(\underline{\infty})]_{Cap} = \log \gamma(\mathbf{F},\{\underline{\infty}\})$ when $\mathbf{F} = \{F_v\}_v$ is defined as follows. Let F_v be the set of $z \in P^1(\overline{K}_v)$ for which

$$\frac{\|z,a\|_v}{\|z,\underline{\infty}\|_v \ \|a,\underline{\infty}\|_v} = [z,a]_{\underline{\infty}} \leqslant [z,\underline{\infty}]_a = \frac{\|z,\underline{\infty}\|_v}{\|z,a\|_v \ \|a,\underline{\infty}\|_v}$$

Define $r_v(a) = (1 + |a|_v^2)^{1/2}$ if v is archimedean, and let $r_v(a) = \max(1,|a|_v)$ if v is non-archimedean. Formula (3.4) of §III.D now shows that $F_v = \{z \in P^1(\overline{K}_v): |z - a|_v \leqslant r_v(a)\}$.

Define $f: X \to X$ to be the rational function $f(z) = z + a$. Then f has degree one on the generic fibre of X , $f^{-1}(\underline{\infty}) = \underline{\infty}$ and $f^{-1}(F_v)=\{z \in X(\overline{K}_v):|z|_v \leqslant r_v(a)\}$. Define $w(v)$ to be the place of $\mathbf{Q}$ determined by v , and let $N_v = [K_v: Q_{w(v)}]$ be the local degree of K_v . By (3.3) and (3.5) of §III.D, we have

$$(5.1) \qquad [D(a),D(\underline{\infty})]_{Cap} = \log \gamma(\{F_v\}_v,\{\underline{\infty}\})$$

$$= \log \gamma(\{f^{-1}(F_v)\}_v, \{f^{-1}(\underline{\infty})\})$$

$$= \sum_v \frac{N_v}{N} \log r_v(a)$$

where the sum is over all places of K , and $N = [K:\mathbf{Q}]$.

By definition, $[D(\underline{\infty}),D(a)]_{Cap} = \log \gamma(\{F'_v\}_v,\{a\})$ where

$F'_v = \{z \in P^1(\overline{K}_v): |z - a|_v > r_v(a)\}$. Define $h: X \to X$ by $h(w) = a + 1/w$, so that $h^{-1}(z) = 1/(z-a)$. Then h has degree 1 on the generic fibre of X , $h^{-1}(a) = \underline{\infty}$ and $h^{-1}(F'_v) = \{z \in P^1(\overline{K}_v): r_v(a)^{-1} > |z|_v\}$. Again by (3.3) and (3.5) of §III.D we have

$$\begin{aligned}[D(a),D(\underline{\infty})]_{Cap} &= \log \gamma(\{F'_v\}_v,\{a\}) \\ &= \log \gamma(\{h^{-1}(F'_v)\}_v, \{h^{-1}(a)\}) \\ &= \sum_v \frac{N_v}{N} \log (r_v(a)^{-1}) \\ &= - [D(a),D(\underline{\infty})]_{Cap} .\end{aligned}$$

This proves the first equality in Lemma 5.1(a), since it is true by definition that $[D(a),D(\underline{\infty})]_{Cap} = - [D(\underline{\infty}),D(a)]_{Cap} = 0$ if $a = \underline{\infty}$.

We now compute

$$(5.2) \qquad [D(a),D(\underline{\infty})]_A = [D(a),D(\underline{\infty})]_{fin} + [D(a),D(\underline{\infty})]_{inf}$$

if $a \neq \underline{\infty}$. The zero-dimensional points P of X where $D(a)$ and $D(\underline{\infty})$ intersect are the intersections of $D(\underline{\infty})$ with the fibres of X over finite places v of K where $|a|_v > 1$. For such P one has local equations $f_a(z) = z-a$ and $f_{\underline{\infty}}(z) = z^{-1}$ for $D(a)$ and $D(\underline{\infty})$, respectively. The maximal ideal of $\underline{O}_{X,P}$ is generated by z^{-1} and any element of K which has order 1 at v . Hence

$$\#(\underline{O}_{X,P}/(f_a,f_{\underline{\infty}})) = \#(\underline{O}_K/a^{-1}\underline{O}_K) = |a|_v^{N_v} .$$

This leads to the formula

$$(5.3) \qquad [D(a),D(\underline{\infty})]_{fin} = \sum_{v \text{ finite}} N_v \log r_v(a)$$

Arakelov remarks in [1, p. 1176] that the Green's function of the Riemann surface $X_\infty = P^1(\mathbb{C})$ with respect to the $d\mu_\infty$ of Theorem 4.1 is $\log \phi_P(Q)$ when

$$\phi_P(Q) = \frac{\sqrt{e}\ |P - Q|_\infty}{(1 + |P|_\infty{}^2)^{1/2}\ (1 + |Q|_\infty{}^2)^{1/2}}$$

Hence if $a_\infty \in X_\infty$ is the point determined by a , one has

$$(5.4) \qquad [D(a),D(\underline{\infty})]_{inf} = -\sum_{\infty \in S_{inf}} \varepsilon_\infty \log \phi_{\underline{\infty}}(a_\infty)$$

$$= \sum_{\infty \in S_{inf}} \varepsilon_\infty\ (\log((1 + |a|_\infty{}^2)^{1/2}) - 1/2)$$

$$= (\sum_{v \in S_{inf}} N_v \log r_v(a)) - [K:\mathbf{Q}]/2$$

Substituting (5.4) and (5.3) into (5.2) and then comparing with (5.1) completes the proof of Lemma 5.1(a) when $a \neq \underline{\infty}$.

To evaluate $[D(\underline{\infty}),D(\underline{\infty})]_A$, we observe that the Arakelov divisor of the function $f(z) = z$ is

$$(z) = D(O) - D(\underline{\infty}) + \sum_{\infty \in S_{inf}} t_\infty X_\infty$$

where

$$t_\infty = \int_{X_\infty} - \log|z|_\infty d\mu_\infty \quad .$$

The measure $d\mu_\infty$ is invariant under the transformation $z \to z^{-1}$, while $\log|z^{-1}|_\infty = - \log|z|_\infty$. Hence $t_\infty = 0$ and z has divisor

$$(z) = D(0) - D(\underline{\infty}) \quad .$$

Because $[\ ,\ ]_A$ respects linear equivalence, we have

$$(5.5) \qquad 0 = [(z),D(\underline{\infty})]_A = [D(0),D(\underline{\infty})]_A - [D(\underline{\infty}),D(\underline{\infty})]_A$$

$$= ([D(0),D(\underline{\infty})]_{Cap} - 1/2)\ [K:\mathbf{Q}] - [D(\underline{\infty}),\ D(\underline{\infty})]_A$$

The formula (5.1) shows $[D(0),D(\underline{\infty})]_{Cap} = 0$, so (5.5) shows $[D(\underline{\infty}),D(\underline{\infty})]_A = - [K:\mathbf{Q}]/2$. Since $[D(\underline{\infty}),D(\underline{\infty})]_{Cap} = 0$ by definition, we have now completed the proof of Lemma 5.1(a) for all $a \in P^1(K)$.

Lemma 5.1(b) is trivial if either a or b is $\underline{\infty}$, or if $a = b$. Suppose now that $a \neq b$, and that a and b are both in $A^1(K)$. Computations similar to those used in determining $[D(a),D(\underline{\infty})]_{Cap}$ show that $[D(a),D(b)]_{Cap} = \log \gamma(\{F_v\}_v,\{b\})$ when $F_v = \{z \in P^1(\overline{K}_v):\ |(z-a)|_v < r_v(a)/r_v(b)\}$. Let $f\colon X \to X$ be the rational function $f(z) = (bz - a)/(z-1)$, so that $f^{-1}(z) = (z - a)/(z - b)$. Then f has degree one on the general fibre of X , $f^{-1}(b) = \underline{\infty}$ and $f^{-1}(F_v) = \{z \in P^1(\overline{K}_v): |z|_v < r_v(a)/r_v(b)\}$. Now by (3.3) and (3.5) of §III.D we have

$$[D(a),D(b)]_{Cap} = \log \gamma(\{F_v\}_v,\{b\})$$

$$= \log \gamma(\{f^{-1}(F_v)\}_v,\{f^{-1}(b)\})$$

$$= \sum_v \frac{N_v}{N} \log(r_v(a)/r_v(b)) .$$

From (5.1) we conclude $[D(a),D(b)]_{Cap}=[D(a),D(\underline{\infty})]_{Cap}-[D(b),D(\underline{\infty})]_{Cap}$, so Lemma 5.1(b) is proved.

To show Lemma 5.1(c), recall that in Theorem 4.1 we defined

$$[D,D(a)]_{Cap} = [D,D(a)]_A/[K:\mathbf{Q}]$$

if D is an irreducible, vertical Arakelov divisor. We thus must show

$$(5.6) \qquad [D,D(a)]_{Cap} - [D,D(\underline{\infty})]_{Cap} = [D,D(a) - D(\underline{\infty})]_A/[K:\mathbf{Q}]$$

is zero for all such D . This is clearly so if $a = \underline{\infty}$, so suppose $a \neq \underline{\infty}$. The function $f(z) = z - a$ has divisor $D(a) - D(\underline{\infty}) + D'$ where D' is a real linear combination of infinite vertical fibres. Since $[D,D']_A = [D,(z-a)]_A = 0$, the right hand side of (5.6) is trivial and the proof is complete.

Proof of Theorem 4.1.

Let D_1 and D_2 be irreducible divisors. If either D_1 is vertical then $[D_1,D_2]_{Cap} = -[D_2,D_1]_{Cap}$ by definition. Suppose now that the D_i are horizontal. After replacing K by a suitable finite extension of K, we may assume that $D_1 = D(a)$ and $D_2 = D(b)$ for some $a,b \in P^1(K)$. Lemma 5.1(b) now shows that $[\ ,\ ]_{Cap}$ is alternating.

We must now show $[(f),D]_{Cap} = 0$ for an arbitrary irreducible divisor D and principal Arakelov divisor (f). If D is vertical, this is so because $[(f),D]_{Cap} = -[(f),D]_A/[K:\mathbf{Q}] = 0$. After replacing K by a finite extension, we may thus reduce to the case in which $D = D(a)$ for some $a \in P^1(K)$.

We may write

$$(f) = \Sigma_i k_i C_i + D_{vert}$$

where the sum is finite, and k_i are integers, the C_i are irreducible and horizontal, and D_{vert} is a linear combination of finite and infinite vertical fibres. On replacing K by a suitable finite extension of K, we may assume that ech C_i has the form $D(b_i)$ for some $b_i \in P^1(K)$.

If $a = \underline{\infty}$ we have from Lemma 5.1(a) that

$$\begin{aligned}[(f),D(\underline{\infty})]_{Cap} &= (\Sigma_i k_i [D(b_i),D(\underline{\infty})]_{Cap}) + [D_{vert},D(\underline{\infty})]_{Cap} \\ &= \{\Sigma_i k_i([D(b_i),D(\infty)]_A/[K:Q] + 1/2)\} \\ &\quad + [D_{vert},D(\underline{\infty})]_A/[K:\mathbf{Q}] .\end{aligned}$$

The degree of the divisor on the generic fibre of X which is determined by (f) is $\Sigma_i k_i = 0$. Hence

$$[(f),D(\underline{\infty})]_{Cap} = [(f),D(\underline{\infty})]_A/[K:\mathbf{Q}] = 0$$

We now suppose that $a \neq \underline{\infty}$. From Lemma 5.1(b) one has

$$\begin{aligned}[(f),D(a)]_{Cap} &= (\Sigma_i k_i [D(b_i),D(a)]_{Cap}) - [D_{vert},D(a)]_{Cap} \\ &= \{\Sigma_i k_i ([D(b_i),D(\underline{\infty})]_{Cap} - [D(a),D(\underline{\infty})]_{Cap})\} \\ &\qquad - [D_{vert},D(\underline{\infty})]_{Cap} .\end{aligned}$$

Because $\Sigma_i k_i = 0$, we conclude from Lemma 5.1(a) that

$$\begin{aligned}[(f),D(a)]_{Cap} &= \Sigma_i k_i [D(b_i),D(\underline{\infty})]_{Cap} - [D_{vert},D(\underline{\infty})]_{Cap} \\ &= [(f),D(\underline{\infty})]_{Cap}\end{aligned}$$

which has already been shown equal to zero. This completes the proof of Theorem 4.1. The description of $[\ ,\ ]_{Cap}$ in equation (4.8) follows from $[D(\underline{\infty}),D(\underline{\infty})]_{Cap} = 0$ and $[X_\infty,D(\underline{\infty})]_{Cap} = -[D(\underline{\infty}),X_\infty]_{Cap} = \varepsilon_\infty/[K:\mathbf{Q}]$ if ∞ is an infinite place of K.

References

1. Arakelov, S.: An intersection theory for divisors on an arithmetic surface, I. z. v. Akad. Nauk. 38 (1974), 1179 - 1192.

2. Artin, M.: Lipman's proof of resolution of singularities for surfaces, to appear in Arithmetic Geometry, Proceedings of the 1984 Storrs Conference on Arithmetic Algebraic Geometry, G. Cornell and J. Silverman editors, Springer-Verlag, New York (1986).

3. Cantor, D.: On an extension of the definition of transfinite diameter and some applications, J. reine angew. Math. 316 (1981), 160-207.

4. Chinburg, T.: An introduction to Arakelov intersection theory, to appear in Arithmetic Geometry, Proceedings of the 1984 Storrs Conference on Arithmetic Algebraic Geometry, G. Cornell and J. Silverman editors, Springer-Verlag, New York (1986).

5. Gross, B.: Local heights on curves, to appear in Arithmetic Geometry, Proceedings of the 1984 Storrs Conference on Arithmetic Algebraic Geometry, G. Cornell and J. Silverman editors, Springer-Verlag, New York (1986).

6. Lichtenbaum, S.: Curves over discrete valuation rings, Amer. J. Math. 15, n. 2 (1968), 380-405.

7. Lipman, J.: Desingularization of two-dimensional schemes, Annals of Math. 107 (1978), 151-207.

8. Neron, A.: Quasi-fonctions et hauteurs sur les varieties abeliennes, Annals of Math. (2) 82 (1965), 299-331.

9. Rumely, R.: Capacity theory on algebraic curves and canonical heights, manuscript (9/8/82).

10. Rumely, R.: Capacity theory on algebraic curves, to appear in the Springer Lecture Notes in Mathematics.

11. Shimura, G. and Taniyama, Y.: Complex multiplication of abelian varieties and its applications to number theory, Math. Soc. of Japan, Tokyo (1961).

Department of Mathematics
University of Pennsylvania
Philadelphia, PA 19104-6395
U.S.A.

Canadian Mathematical Society
Conference Proceedings
Volume 7 (1987)

Geometric Iwasawa Theory and a Conjecture of Katz

Richard Crew*

Introduction

Let $X_0/\mathbf{F}_q$ be a separated scheme of finite type over a finite field of characteristic p. More than fifteen years ago N. Katz made a conjecture on the behavior of the L-functions attached to p-adic representations of the fundamental group of X. I would like to describe this conjecture and one of its arithmetic consequences.

In brief, the conjecture asserts that the L-function of a p-adic representation of π_1 is p-adically meromorphic, and that its zeroes and poles or the unit circle are determined by the action of Frobenius on p-adic étale cohomology of the étale sheaf associated to the given representation. In treating our two versions of the conjecture (2.1 and 4.7) we will find it simpler to deal systematically with étale sheaves rather than representations. In §1 we give the dictionary between the two points of view and review some technicalities; the reader familiar with ℓ-adic cohomology will probably want to skip most of it. Some of the cases in which it has been proven are described in §3; these seem to have been known to the experts for some time now, but lack a reference in the literature. As they are used in [4] without much in the way of explanation, I have seen fit to go into some of the details here. In §4 we describe another case of the conjecture which was proven in [4] and can be viewed as a geometric analogue of the Mazur Wiles theorem on $\mathbf{Z}_p$-extensions (i.e. the "main conjecture" of Iwasawa theory). It is interesting to note that in their proof of the main conjecture, Mazur and Wiles are led to study a particular "geometric" $\mathbf{Z}_p$-extension (the "tower of Igusa curves," c.f. [12]) and the "geometric main conjecture" allows one to obtain some information on the μ and λ-invariants of this tower. We do not describe this here, but the reader can consult [4]§4. The last section was an afterthought; in it we give a new proof of a theorem of Deligne on the L-function of a modular representation of π_1 (first proven by Katz for the trivial representation). I have included it because it suggests a possible approach to Conjecture 4.7 below, and because it is an amusing application of the results of §3 and the theory of modular characters, analogous in some ways to Brauer's proof of the meromorphy of Artin L-functions.

* NSF Postdoctoral Fellow

0731-1036/87 $1.00 + $.25 per page

§1 L-functions

Fix a prime p and a finite field $\mathbf{F}_q$ of cardinality $q = p^f$. If S is an $\mathbf{F}_q$-scheme of finite type, we denote by $|S|$ its set of closed points, and for $x \in |S|$ we denote the (absolute) degree of x by $\deg(x) = [\mathbf{F}_q(x) : \mathbf{F}_p]$.

Let us first recall how one attaches an L-function to a representation of $\pi_1(X_0)$ of a connected separated $\mathbf{F}_q$-scheme X_0 of finite type. For any $x \in |X_0|$, the inclusion $x \to X_0$ gives rise to a conjugacy class of maps $\mathrm{Gal}(\overline{\mathbf{F}_q(x)})/\mathbf{F}_q(x) \simeq \pi_1(x) \to \pi_1(X_0)$ (the ambiguity arises from the choice of base points). If $\phi \in \mathrm{Gal}\ (\overline{\mathbf{F}}_p/\mathbf{F}_p)$ denotes the absolute arithmetic Frobenius ($x \mapsto x^p$), then we denote by ϕ_x any element of the conjugacy class consisting of the images of $\phi^{-\deg(x)}$ under the maps $\pi_1(x) \to \pi_1(X_0)$. Suppose now that V is any finite free module over a topological ring R and that $\rho : \pi_1(X_0) \to GL(V)$ is a continuous homomorphism. Then for any $x \in |X_0|$, the local factor $\det(1 - T^{deg\ (x)}\rho(\phi_x))^{-1}$ is well-defined and the L-function $L(\rho, T)$ is defined to be

1.0.1 $$L(\rho, T) = \prod_{x \in |X_0|} \det(1 - T^{deg\ (x)}\rho(\phi_x))^{-1} \in R[[T]]$$

In using $\deg(x) = [\mathbf{F}_q(x) : \mathbf{F}_p]$ instead of the degree relative to $\mathbf{F}_q$ we are following Deligne ([SGA $4\frac{1}{2}$] Rapport); it makes $L(\rho, T)$ depend only on ρ and X_0, not on the structure map $X_0 \to \mathrm{Spec}\mathbf{F}_q$. This has the effect of putting a T^f in many formulas where one usually has a T (e.g. 1.0.7).

When R is a local field of residual characteristic $\ell \neq \rho$, then it is well known that $L(\rho, T)$ is a rational function of T. This is a consequence of Grothendieck's cohomological formula for $L(\rho, T)$ (theorem 1.1 below), which we will now attempt to describe.

We first recall that a sheaf $\mathfrak{F}$ on the étale site of X_0 is said to be *locally constant constructible* (abbreviated l.c.c.) if it is represented on the étale site by a finite étale cover of X_0, and *constructible* if X_0 is a union of locally closed subschemes U_α such that each $\mathfrak{F}|U_\alpha$ is l.c.c. When X_0 is connected, the theory of the fundamental group provides us with an equivalence of categories

1.0.2 $$\begin{pmatrix}\text{l.c.c. sheaves of} \\ \text{sets on } X_0\end{pmatrix} \xrightarrow{\sim} \begin{matrix}\text{finite sets with an} \\ \text{action of } \pi_1(X_0, \bar{x})\end{matrix}$$
$$\mathfrak{F} \mapsto \mathfrak{F}_{\bar{x}} + \pi_1(X_0, \bar{x})\text{-action}$$

for every geometric point $\bar{x}$ of X_0. Thus when R is a finite ring, the category of l.c.c. sheaves of R-modules on X_0 whose fibers are free R-modules is equivalent to the category of representations of $\pi_1(X_0)$ in free R-modules.

For $X_0/\mathbf{F}_q$ separated and of finite type and $\mathfrak{F}_0$ a set $X = X_0 \otimes_{\mathbf{F}_q} \overline{\mathbf{F}}_q$ and let $\mathfrak{F}$ be the restriction of $\mathfrak{F}_0$ to X. If $F_q : X_0 \to X_0$ is the qth-power Frobenius and $F : X \to X$ is the map obtained by extension of scalars, then there is a natural isomorphism (cf [SGA 4-$\frac{1}{2}$] pp 77-78) $F^*\mathfrak{F} \xrightarrow{\sim} \mathfrak{F}$ known as the *Frobenius correspondence*. If $x \in |X_0|$ is of degree n and $\bar{x} \in X_0(\overline{\mathbf{F}}_q)$ is a geometric point lying above n, then $\bar{x} \in X(\overline{\mathbf{F}}_q)^{F^n}$, and F^n induces an automorphism of the fiber $\mathfrak{F}_{\bar{x}}$ which we call F_x If $\mathfrak{F}_0$ is l.c.c. and if we choose $\bar{x}$ as a base point for X_0, then F_x can be identified with the action of $\phi_x \in \pi_1(X_0, \overline{x})$ via the equivalence 1.0.2. In particular, if $\mathcal{F}_0$ is a l.c.c. sheaf of free R-modules (i.e. the fibers of $\mathcal{F}_0$ are finite free R-modules) and $\rho : \pi_1(X_0) \to GL(V)$ is the corresponding representation of $\pi_1(X_0)$, then we have $\det(1 - T^{deg\,(x)}\rho(\phi_p)) = \det(1 - TF_x|\mathcal{F}_{\overline{x}})$, which suggests the definition

1.0.3 $$L(X_0, \mathcal{F}_0, T) = \prod_{x \in |X_0|} \det(1 - T^{deg\,(x)}F_x|\mathcal{F}_{\overline{x}})^{-1} \in R[[T]]$$

for any constructible sheaf of R-modules $\mathcal{F}_0$; we clearly have $L(X_0, \mathcal{F}_0, T) = L(\rho, T)$ if $\mathcal{F}_0$ is l.c.c. and corresponds to ρ.

So far R has been a finite ring. If R is a complete local noetherian ring with maximal ideal m and finite reside field, then a *constructible R-sheaf* on X_0 is a projective system $(\mathcal{F}_n)_n$ of sheaves on X_0 such that each $\mathcal{F}_n$ is a constructible sheaf of R/m^n-modules, and such that the transition maps $\mathcal{F}_{n+1} \to \mathcal{F}_n$ factor through isomorphisms $\mathcal{F}_{n+1} \otimes R/m^n \xrightarrow{\sim} \mathcal{F}_n$. $\mathcal{F}$ is a *lisse R-sheaf* if each $\mathcal{F}_n$ is l.c.c. We define Hom $(\mathcal{F}, \mathcal{G})$ for two constructible R-sheaves $\mathcal{F}, \mathcal{G}$, by $\mathrm{Hom}(\mathcal{F}, \mathcal{G}) = \varprojlim_n \mathrm{Hom}(\mathcal{F}_n, \mathcal{G}_n)$, and for any geometric point $\overline{x}$ of X_0, we define the fiber $\mathcal{F}_{\overline{x}}$ of $\mathcal{F}$ at $\overline{x}$ by $\mathcal{F}_{\overline{x}} = \varprojlim_n (\mathcal{F}_n)_{\overline{x}}$. Clearly the constructions of the previous paragraph carry over without change; in particular we can define $L(X_0, \mathcal{F}_0, T)$ for any constructible R-sheaf $\mathcal{F}_0$ by the formula 1.0.3. When X_0 is connected, the category of continuous representations of $\pi_1(X_0)$ in finite free R-modules is equivalent to the category of lisse R-sheaves with free fibers.

Finally, when K is a local field with integer ring R, then one defines the category of constructible K-sheaves (resp. lisse K-sheaves) to be the category whose objects are the constructible (resp. lisse) R-sheaves, and whose morphisms are given by
Hom $(\mathcal{F} \otimes K, \mathcal{G} \otimes K) = \mathrm{Hom}(\mathcal{F}, \mathcal{G}) \otimes K$, where $\mathcal{F} \otimes K$ is the constructible (resp. lisse) R-sheaf $\mathcal{F}$ viewed as a K-sheaf. The fiber of the K-sheaf $\mathcal{F} \otimes K$ at a geometric point $\overline{x}$ of X_0 is defined by $(\mathcal{F} \otimes K)_{\overline{x}} = \mathcal{F}_{\overline{x}} \otimes K$, and one defines $L(X_0, \mathcal{F}_0 \otimes K, T)$ again by 1.0.3. If the fibers of $\mathcal{F}_0$ are free over R then of course we have $L(X_0, \mathcal{F}_0, T) = L(X_0, \mathcal{F}_0 \otimes K, T)$; in particular, the L-function of a constructible K-sheaf is always an element of $R[[T]]$. Again, when X_0 is connected, the category of continuous representations of $\pi_1(X_0)$ is finite dimensional K-vector spaces is equivalent to the category of lisse K-sheaves, and if $\mathcal{F}_0, \rho$ correspond to each other by this equivalence we have $L(X_0, \mathcal{F}_0, T) = L(\rho, T)$.

We shall often refer to constructible R-sheaves (or K-sheaves) as "constructible sheaves of R-modules" (resp. K-vector spaces); strictly speaking, R-sheaves and K-sheaves are not sheaves on the étale site of X_0. A constructible sheaf of free (torsion, etc.) R-modules will be constructible R-sheaf whose fibers are free (torsion, etc.).

Suppose now that S is, say, any scheme of finite type over a field. If $\mathcal{F} = (\mathcal{F}_n)_n$ is a constructible R-sheaf on S (R complete, local, noetherian, with finite residue field) then recall that the cohomology groups $H^i_c(S, \mathcal{F})$ are defined by

1.0.4 $$H^i_c(S, \mathcal{F}) = \varprojlim_n H^i_c(S, \mathcal{F}_n)$$

where $H_c^i(S, \mathfrak{F}_n)$ is the étale cohomology with compact supports with coefficients in the constructible sheaf $\mathfrak{F}_n$. If K is a local field with integer ring R, $\mathfrak{F} = (\mathfrak{F}_n)$ is a constructible R-sheaf, and $\mathfrak{F} \otimes K$ is the corresponding K-sheaf, then one defines

1.0.5
$$\begin{aligned} H_c^i(S, \mathfrak{F} \otimes K) &= H_c^i(S, \mathfrak{F}) \otimes K \\ &= (\varprojlim_n H_c^i(S, \mathfrak{F}_n)) \otimes K \end{aligned}$$

When $X_0/\mathbf{F}_q$ is separated and of finite type and $\mathfrak{F}_0$ is a constructible sheaf on X_0 (or a constructible R-sheaf, or K-sheaf) the Frobenius correspondence $F^*\mathfrak{F} \xrightarrow{\sim} \mathfrak{F}$ (as before, $\mathfrak{F} = \mathfrak{F}_0|X$, $X = X_0 \otimes \overline{\mathbf{F}}_q$) gives rise to an isomorphism induced by functoriality

1.0.6
$$F : H_c^i(X, \mathfrak{F}) \xrightarrow{\sim} H_c^i(X, \mathfrak{F})$$

We can now define the "cohomological L-function" attached to a $\mathfrak{F}_0$ on X_0. The definition is the simplest when $\mathfrak{F}_0$ is a constructible K-sheaf, for then the $H_c^i(X, \mathfrak{F})$ are finite-dimensional K-vector spaces, zero for almost all i, and so we can set

1.0.7
$$H(X_0, \mathfrak{F}_0, T) = \prod_i \det(1 - T^f F | H_c^i(X, \mathfrak{F}))^{(-1)^{i+1}} \in K[[T]]$$

This is *not* standard notation. In order to define $H(X_0, \mathfrak{F}_0, T)$ when $\mathfrak{F}_0$ is a sheaf of R-modules we must use a standard trick: If $\mathfrak{F}_0$ is a sheaf of projective R-modules of finite type, then there is a bounded complex $M^\cdot$ of projective R-modules such that

1.0.8
$$H^i(M^\cdot) \simeq H_c^i(X, \mathfrak{F})$$

and the isomorphism $F : H_c^i(X, \mathfrak{F}) \xrightarrow{\sim} H_c^i(X, \mathfrak{F})$ can be represented by an endomorphism $F : M^\cdot \to M^\cdot$. When R is local noetherian the M^i are free R-modules and one sets

1.0.9
$$H(X_0, \mathfrak{F}_0, T) = \prod_i \det(1 - T^f F | M^i)^{(-1)^{i+1}}$$

which can be shown to be independent of the choice of $M^\cdot$ (so long as it is isomorphic, in the derived category, to $R\Gamma(X, \mathfrak{F})$ to $R\Gamma(X, \mathfrak{F})$, cf [SGA $4\frac{1}{2}$] p.115, Cor. 1.13) In SGA $4\frac{1}{2}$, objects of this sort are denoted $\det(1 - T^f | R\Gamma(X, \mathfrak{F}))$. Note that in virtue of 1.0.9 and 1.0.5 we actually have $H(X_0, \mathfrak{F}_0, T) \in R[[T]]$ if $\mathfrak{F}_0$ is a K-sheaf. When R is finite and not necessarily local, the M^i are at least projective, and $\det(1 - TF|M^i)$ can be defined by another trick: if M is a projective R-module and $\alpha \in Aut(M)$, we choose an R-module N such that $M \oplus N$ is free, and define $\det \alpha = \det(\alpha \oplus 1)$; this is clearly independent of N.

We can now, finally state Grothendieck's formula for $L(X_0, \mathfrak{F}_0, T)$:

1.1 **Theorem** ([SGA $4\frac{1}{2}$]Rapport 3.1, [SGA 5]XV 3.2) *Suppose that $X_0/\mathbf{F}_q$ is separated and of finite type, and $\mathcal{F}_0$ is one of*

(i) *a constructible sheaf of R-modules, where R is a finite ring in which $p =$ char $\mathbf{F}_q$ is invertible*

(ii) *a constructible R-sheaf, where R is a complete local noetherian ring with finite residue field of characteristic different from p*

(iii) *a constructible K-sheaf, where K is a local field of residual characteristic different from p*

Then

$$L(X_0, \mathcal{F}_0, T) = H(X_0, \mathcal{F}_0, T)$$

1.1.1 **Remark** When $\mathcal{F}_0$ is a constructible K-sheaf, the zeroes and poles of the rational function $H(X_0, \mathcal{F}_0, T)$ are all ℓ-adic units, where ℓ is the residue characteristic of K; in fact if $\mathcal{F}_0 = \mathcal{G}_0 \otimes R$ where $\mathcal{G}_0$ is a constructible R-sheaf, then the isomorphism comes from an isomorphism $F : H^i_c(X, \mathcal{G}) \xrightarrow{\sim} H^i_c(X, \mathcal{G})$ of finite generated R-modules. Therefore if $\ell = p$, we cannot expect to have $L(X_0, \mathcal{F}_0, T) = H(X_0, \mathcal{F}_0, T)$, since even when $L(X_0, \mathcal{F}_0, T)$ is a rational function, its zeroes and poles are not necessarily p-adic units. Such is the case when $\mathcal{F} = \mathbf{Q}_p$, i.e. the L-function is the zeta function $Z(X_0, T)$ of $X_0/\mathbf{F}_q$; then the functional equation of $Z(X_0, T)$ guarantees that there are always nonunit zeroes or poles of $Z(X_0, T)$ when $\dim X_0 > 0$. The extent to which $L(X_0, \mathcal{F}_0, T)$ and $H(X_0, \mathcal{F}_0, T)$ differ when the residual characteristic of K is p is the subject of the Katz conjecture.

1.1.2 **Remark** There is one case when one has $L(X_0, \mathcal{F}_0, T) = H(X_0, \mathcal{F}_0, T)$ without hypotheses on R or K: when the support of $\mathcal{F}_0$ is finite. Then one has

$$\begin{aligned} H^0_c(X, \mathcal{F}) &= \oplus_{\bar{x} \in X(\bar{\mathbf{F}}_q)} \mathcal{F}_{\bar{x}} \\ H^i_c(X, \mathcal{F}) &= 0 \quad i > 0 \end{aligned}$$

and the equality follows from the definitions.

1.2 Let us now record some of the formal properties of L, H that we shall need later. When not specified, "sheaf" means a constructible sheaf, an R-sheaf, or a K-sheaf.

First, if

1.2.1
$$0 \longrightarrow \mathcal{F}'_0 \longrightarrow \mathcal{F}_0 \longrightarrow \mathcal{F}''_0 \to 0$$

is an exact sequence of sheaves on X_0, then L and H are additive:

1.2.2
$$\begin{aligned} L(X_0, \mathcal{F}_0) &= L(X_0, \mathcal{F}'_0) L(X_0, \mathcal{F}''_0) \\ H(X_0, \mathcal{F}_0) &= H(X_0, \mathcal{F}'_0) H(X_0, \mathcal{F}''_0) \end{aligned}$$

For any open subscheme $j : U_0 \hookrightarrow X_0, D = X_0 - U_0 \xhookrightarrow{i} X_0$, they satisfy the excision property

1.2.3
$$\begin{aligned} L(X_0, \mathfrak{F}_0) &= L(U_0, j^*\mathfrak{F}_0)L(D_0, i^*\mathfrak{F}_0) \\ H(X_0, \mathfrak{F}_0) &= H(U_0, j^*\mathfrak{F}_0)H(D_0, i^*\mathfrak{F}_0) \end{aligned}$$

and the homotopy property

1.2.4
$$\begin{aligned} L(X_0, \mathfrak{F}_0) &= L(X_0^{red}, \mathfrak{F}_0|X_0^{red}) \\ H(X_0, \mathfrak{F}_0) &= H(X_0^{red}, \mathfrak{F}_0|X_0^{red}) \end{aligned}$$

where X_0^{red} is the reduced subscheme associated to X_0.

Suppose now that R is the integer ring of a local field and $\mathfrak{F}_0 = (\mathfrak{F}_n)_n$ is a constructible sheaf of free R-modules on X_0. If m is the maximal ideal of R and $R/m = k$, we set $\mathfrak{F}_0 \otimes k = \mathfrak{F}_1$, and then we have

1.2.5
$$\left.\begin{aligned} L(X_0, \mathfrak{F}_0 \otimes k) &\equiv L(X_0, \mathfrak{F}_0) \\ H(X_0, \mathfrak{F}_0 \otimes k) &\equiv H(X_0, \mathfrak{F}_0) \end{aligned}\right. \mod m$$

For L this is clear, and for H it follows from the universal coefficient theorem: if $M^{\cdot}$; $F : M^{\cdot} \to M^{\cdot}$ "calculate" $H_c^i(X, \mathfrak{F})$ with its Frobenius action, then

$$H(X_0, \mathfrak{F}_0 \otimes k) = \prod_i \det(1 - T^f F|M^i \otimes k)^{(-1)^{i+1}}$$

Comparing this with 1.0.9 yields 1.2.5.

Finally, we must say a word about crystalline cohomology. There is at present no p-adic cohomology theory strong enough to prove a p-adic analogue of theorem 1.1, but crystalline cohomology suffices for the zeta function of a smooth proper $\mathbf{F}_q$-scheme:

1.3 Theorem (Berthelot [2] Ch.VII Théorème 3.2.3) *If $X_0/\mathbf{F}_q$ is smooth and proper, then*

1.3.1
$$Z(X_0, T) = \prod_i \det(1 - T^f F|H^i_{cris}(X_0/W(\mathbf{F}_q)) \otimes \mathbf{Q})^{(-1)^{i+1}}$$

We should briefly explain 1.3.1. For any perfect field k and proper smooth X/k, the $H^i_{cris}(X/W(k))$ are finitely generated $W(k)$-modules, functorial in the pair X/k. Thus if $a : X \to X$ is an automorphism of X over k, then it induces a $W(k)$-linear automorphism of the $H^i_{cris}(X/W(k))$, while if $F_p : X \to X$ is the absolute (p^{th} power) Frobenius morphism of X, it induces a σ-linear endomorphism

1.3.2
$$\Phi : H^i_{cris}(X/W(k)) \to H^i_{cris}(X/W(k))$$

where $\sigma : W(k) \to W(k)$ is the usual lifting of Frobenius to the Witt vectors. Φ is not in general an isomorphism, but $\Phi \otimes \mathbf{Q}$ is. When $k = \mathbf{F}_q$, $\sigma^f = 1$ and so $\Phi^f = F$ is $W(\mathbf{F}_q)$-linear; this is the F in formula 1.3.1.

§2 The Katz Conjecture

One version of it runs as follows:

2.1 **Conjecture** Let $X_0/\mathbf{F}_q$ be separated of finite type, and let $\mathcal{G}_0$ be a constructible sheaf of K-vector spaces on X_0, where K is a local field of mixed characteristic p. Then

2.1.1 $L(X_0, \mathcal{G}_0, T)$ is p-adically meromorphic, i.e. a quotient of two entire power series

2.1.2 The quotient

$$L(X_0, \mathcal{G}_0, T)/H(X_0, \mathcal{G}_0, T)$$

has no zeroes or poles on $|T| = 1$.

2.2 The original conjecture ([8] 6.1) is more general as regards 2.1.1; it asserts that the L-function of any F-crystal on $X_0/W(\mathbf{F}_q)$ is p-adically meromorphic. We are, however, primarily interested in 2.1.2, where the original conjecture makes essentially the same assertion given here (in fact the L-function defined in [8]§6 is, for a unit-root F-crystal, the same as the L-function as defined here of the corresponding étale sheaf).

In general one does not expect that $L(X_0, \mathcal{G}_0)$ is rational, or that it satisfies a functional equation (although no examples of an irrational $L(X_0, \mathcal{G}_0)$ is known). There are of course $X_0, \mathcal{G}_0$ for which $L(X_0, \mathcal{G}_0)$ is rational; the cases of the conjecture that we shall treat in §3 are of this sort, and in all of them $\mathcal{G}_0$ is lisse corresponding to a representation with finite image. In these cases $L(X_0, \mathcal{G}_0)$ can be computed by means of crystalline cohomology, whose relation to p-adic étale cohomology is known (Theorem 3.2 below).

The difficult case is when $\mathcal{G}_0$ is lisse and corresponds to a representation with infinite image. There is no cohomology theory powerful enough to compute $L(X_0, \mathcal{G}_0)$ in general, although Dwork's theory can be used in some specific cases (e.g. [1],[5]). Dwork was able to show that if X_0 is an affine space $\mathbf{A}^n$ and $\mathcal{G}_0$ is lisse, then $L(X_0, \mathcal{G}_0)$ is meromorphic in the closed disk $|T| \leq 1$ (unpublished). However the relation between Dwork's theory and p-adic étale cohomology is not well understood.

Finally, one can use a kind of "p-adic interpolation" argument to show (in [4]) that if X_0 is a smooth affine curve and $\mathcal{G}_0$ is lisse, corresponding to an *abelian* representation of $\pi_1(X_0)$, then $L(X_0, \mathcal{G}_0)$ is meromorphic for $|T| \leq 1$ and that 2.1.2 (which then has a sense!) is true. We shall discuss this case, and its applications, in §4.

§3 Some elementary cases

We shall now see what can be done when $\mathcal{G}_0$ is lisse corresponding to a representation with finite image. We first discuss the trivial representation:

3.1 **Proposition** *If $X_0/\mathbf{F}_q$ is smooth and proper and $\mathcal{G}_0$ is trivial, then 2.1 is true for $X_0, \mathcal{G}_0$.*

When X_0 is a smooth projective curve or an abelian variety, this was first proven by Serre [13].

Proof. In fact, from 1.3.1 we see that $Z(X_0,T)$ is rational (which we already know) and that 2.1.2 is equivalent to the following assertion: the eigenvalues of F acting on $H^i(X,\mathbf{Q}_p)$ are the same as the p-adic unit eigenvalues of F acting on $H^i_{cris}(X_0/W(\mathbf{F}_q))\otimes\mathbf{Q}$. There seems to be no reference for this fact, but it can be deduced from the following well known result of Bloch [3] and Illusie ([7]II5.2.2):

3.2 **Theorem** *Let X/k be a scheme that is smooth and proper over an algebraically closed field. There are exact sequences*

3.2.1
$$0 \longrightarrow H^i(X,\mathbf{Z}_p) \to H^i_{cris}(X/W(k)) \xrightarrow{1-\Phi} H^i_{cris}(X/W(k)) \to 0$$

natural in X/k.

Here Φ is the σ-linear endomorphism of 1.3.2. In the situation of 3.1, both $H^i_{cris}(X_0/W(\mathbf{F}_q))$ and $H^i_{cris}(X/W(\overline{\mathbf{F}}_q))$ carry such a Φ, and the two can be identified by means of the canonical inclusion and isomorphism

3.2.2
$$H^i_{cris}(X_0/W(\mathbf{F}_q)) \hookrightarrow H^i_{cris}(X_0/W(\mathbf{F}_q))\otimes W(\overline{\mathbf{F}}_q) \xrightarrow{\sim} H^i_{cris}(X/W(\overline{\mathbf{F}}_q))$$

In other words the action of Φ on $H^i_{cris}(X/W(\overline{\mathbf{F}}_q))$ is the σ-linear extension of its action on $H^i_{cris}(X_0/W(\mathbf{F}_q))$. We have to compare Φ with the action on $H^i(X,\mathbf{Z}_p)$ and $H^i_{cris}(X/W(\overline{\mathbf{F}}_q))$ induced by the morphism $F: X \to X$ which, we recall from §1, is obtained from the qth power morphism $F_q: X \to X_0$ by the extension of scalars $X = X_0\otimes\overline{\mathbf{F}}_q \to X_0$ (i.e. F acts as the identity on $\overline{\mathbf{F}}_q$). On $H^i(X,\mathbf{Z}_p)$ we obtain the endomorphism F of 1.0.6, while 3.2.2 shows that the induced action on $H_{cris}(X/W(\overline{\mathbf{F}}_q))$ is the $W(\overline{\mathbf{F}}_q)$-linear extension of the action of Φ^f on $H^i_{cris}(X_0/W(\mathbf{F}_q))$ (again denoted F in 1.3.2). The assertion we need concerning the eigenvalues of F then follows from the naturality of 3.2.1 and the following lemma in σ-linear algebra:

3.3 **Lemma** *Let K be a local field with residue field $\mathbf{F}_q, q = p^f$, and let $\sigma: K \to K$ be a lifting of the pth power automorphism of $\mathbf{F}_q$ fixing a uniformizer π of K. Denote by L the completion of the maximal unramified extension of K. Furthermore let V be a finite-dimensional K-vector space with a σ-linear automorphism $\Phi: V \to V$, and denote by F (resp. Φ) the linear extension of Φ^f to $V\otimes_K L$ (resp. the σ-linear extension of Φ to $V\otimes_K L$). Then $W = \mathrm{Ker}(1-\Phi|V\otimes_K L)$ is F stable, and the eigenvalues of F on W are the p-adic unit eigenvalues of F on V*

Proof: The theory of σ-linear endomorphisms shows that there is a unique Φ-stable direct sum decomposition

$$V\otimes_K L = \bigoplus_{\lambda\in\mathbf{Q}} V_\lambda$$

such that $V_\lambda = 0$ for all but finitely many λ, and each V_λ has a basis $\{v_i\}$ for which $\Phi^b(v_i) = \pi^a v_i$ if $\lambda = a/b$ in lowest terms (for $\lambda = 0$, we take $a = 0, b = 1$). Note that $W \otimes_{\mathbf{Q}_p} L = V_0$. Furthermore the sum U of the generalized eigenspaces for F on V corresponding to p-adic unit eigenvalues is Φ-stable and contained in V_0, while the other eigenspaces are contained in the sum of the V_λ for $\lambda \neq 0$. We can therefore assume that $U = V$, whence $V \otimes_K L = V_0 = W \otimes_{\mathbf{Q}_p} L$. The lemma will then follow if we can show that W is F-stable. Now in W, we have $F\Phi - \Phi F = 0$, and since $F\Phi - \Phi F$ is σ-linear, it is zero on all of V. Thus F and Φ commute, and W is F-stable.

3.4 **Proposition** *If $X_0/\mathbf{F}_q$ is a smooth curve and $\mathcal{G}_0$ is lisse, coming from a representation $\rho : \pi_1(X_0) \to GL(V)$ with finite image, then 2.1 is true for $X_0, \mathcal{G}_0$.*

Proof: Let us first consider the case when X_0 is complete. If $G = \operatorname{Im} \rho$, then corresponding to the surjection $\pi_1(X_0) \twoheadrightarrow G$ is a principal étale G-cover $\pi : Y_0 \to X_0$ and one has $\mathcal{G}_0 \simeq (\pi_* \mathbf{Q}_p)^\rho$, where the $(\ )^\rho$ means that we tensor over $\mathbf{Q}_p[G]$ with the dual of V, viewed as a right G-module. As π is finite, we have

3.4.1 $$H^i(X, \mathcal{G}) \simeq H^i(X, \pi_* \mathbf{Q}_p)^\rho \simeq H^i(Y, \mathbf{Q}_p)^\rho$$

by the Leray spectral sequence. Since π is finite étale, $Y/\overline{\mathbf{F}}_q$ is smooth and proper, and 3.2 is applicable. If K is large so that $W(\mathbf{F}_q) \subset K$, then 3.2 and 3.1.1 yield.

3.4.2 $$0 \to H^i(X, \mathcal{G})^\rho \to H^i_{cris}(Y/W(\overline{\mathbf{F}}_q))^\rho \xrightarrow{1-\Phi} H^i_{cris}(Y/W(\overline{\mathbf{F}}_q))^\rho \to 0$$

If, in addition, K is large enough so that the assumptions of 3.3 hold (which we can assume), we obtain from 3.4.2 and 3.3 that

$$\prod_i \det(1 - T^f F | H^i_{cris}(Y/W(\mathbf{F}_q))^\rho)^{(-1)^{i+1}} / H(X_0, \mathcal{G}, T)$$

has no zeroes or poles on $|T| = 1$. We must therefore check that

$$L(X_0, \mathcal{G}_0, T) = \prod_i \det(1 - T^f F | H^i_{cris}(Y/W(\mathbf{F}_q))^\rho)^{(-1)^{i+1}}$$

For this we can use the ℓ-adic formula. In fact, since ρ has finite image, it is the extension of scalars by $K_0 \hookrightarrow K$ of some representation $\rho_0 : \pi_1(X_0) \to GL(V_0)$ defined over a number field K_0. A choice of an ℓ-adic completion $K_0 \hookrightarrow K_\lambda$ then gives an ℓ-adic model ρ_λ of ρ (or ρ_0) and a corresponding lisse ℓ-adic sheaf $\mathcal{G}_{\lambda_0}$ on X_0 for which we clearly have $L(X_0, \mathcal{G}_0) = L(X_0, \mathcal{G}_{\lambda_0})$.

Since

3.4.4 $$H^i(X, \mathcal{G}_\lambda) \simeq H^i(Y, \mathbf{Q}_\ell)^{\rho_\lambda}$$

(same argument as before), we see from 3.4.4 that

3.4.5 $$L(X_0, \mathcal{G}_0) = \prod_i \det(1 - T^f F | H^i(Y, \mathbf{Q}_\ell)^{\rho_\lambda})^{(-1)^{i+1}}$$

Comparing 3.4.3 with 3.4.5, we see that we must check that for each i, the spaces $H^i(Y_0/W(\mathbf{F}_q))$ and $H^i(Y, \mathbf{Q}_p)$ are, after semi-simplifying, isomorphic (say, over some common extension of K on K_λ) as G-modules. Formally, this is a consequence of the comparison theorem ([9] Theorem 2) of Katz-Messing: if $X/\mathbf{F}_q$ is smooth and projective (here we use $\dim X_0 = 1$ and $H^i(\ )$ is any "Weil cohomology" theory (cf. Kleiman [10]) then the characteristic polynomial of any correspondence acting on $H^i(X)$ is independent of the cohomology theory. In the present case, however, a direct proof of the assertion we need can be extracted from the paper of Katz-Messing, which we leave as an exercise.

In the general case, where $X_0/\mathbf{F}_q$ is not complete, one first finds a smooth compactification $X_0 \overset{j}{\hookrightarrow} \overline{X}_0$ over $\mathbf{F}_q$. With ρ, G as above, one obtains as before a principal étale G cover $\pi : Y_0 \to X_0$ which again gives $\mathcal{G}_0 \simeq (\pi_* \mathbf{Q}_p)^\rho$. Since 3.2 can only be applied to complete varities, however, we must choose a smooth completion $Y_0 \hookrightarrow \overline{Y}_0$ dominating $\overline{X}_0$ by a finite map $\overline{\pi} : \overline{Y}_0 \to \overline{X}_0$. We then have $j_* \mathcal{G}_0 \simeq (\overline{\pi}_* \mathbf{Q}_p)^\rho$ and 3.4.1 should be replaced by

3.4.6 $$H^i(\overline{X}, j_* \mathcal{G}) \simeq H^i(\overline{Y}, \mathbf{Q}_p)^\rho$$

and the same argument that led to 3.4.3 shows that

$$L(X_0, j_* \mathcal{G}_0, T) = \prod_i \det(1 - T^f F | H^i_{cris}(\overline{Y}_0/W(\mathbf{F}_q))^\rho)^{(-1)^{i+1}}$$

We can now apply 3.2 (as $\overline{Y}/\mathbf{F}_q$ is smooth and proper) which, together with 3.4.6 and 3.3 shows that the quotient $L(\overline{X}_0, j_*, \mathcal{G}_0)/H(\overline{X}_0, j_* \mathcal{G}_0)$ has no zeroes or poles on $|T| = 1$. To conclude, we need only show that

3.4.7 $$L(\overline{X}_0, j_* \mathcal{G}_0)/H(\overline{X}_0, j_* \mathcal{G}_0) = L(X_0, \mathcal{G}_0)/H(X_0, \mathcal{G}_0)$$

If we let $\overline{X}_0 - X_0 = D_0 \overset{i}{\hookrightarrow} \overline{X}_0$, then 1.1.2 and 1.2.3 applied to $j_* \mathcal{G}_0$ yield

$$L(\overline{X}_0, j_* \mathcal{G}_0)/L(X_0, \mathcal{G}_0) = H(X_0, j_* \mathcal{G}_c)/H(X_0, \mathcal{G}_0) = L(D_0, i^* j_* \mathcal{G}_0)$$

from which 3.4.7 follows.

We want to make one consequence of the proof explicit:

3.5 Corollary *If $X_0/\mathbf{F}_q, \mathcal{G}_0$ are as in 3.1 or 3.4, then*

$$L(X_0, \mathcal{G}_0, T)/H(X_0, \mathcal{G}_0, T) \in R[T]^\times_{(p)}$$

where $R[T]_{(p)}$ denotes the p-adic completion of $R[T]$.

Proof: In the proofs of 3.1 and 3.4 we have seen that the quotient is a quotient of polynomials which can be expressed as products $\prod_i (1 - \alpha_i T)$, where α_i has valuation > 0. Then any of the $1 - \alpha_i T$ is a unit in $R[T]_{(p)}$

§4 Iwasawa Theory

We now describe one consequence of Proposition 3.4 which can be interpreted as a geometric analogue of the "Main Conjecture" of Iwasawa theory. In this section $X_0/\mathbf{F}_q$ is a smooth affine *curve* and Λ is a complete local noetherian ring with maximal ideal m and finite residue field of characteristic p. As usual $X = X_0 \otimes \overline{\mathbf{F}}_q$, etc.

4.1 **Lemma** *If $\mathcal{G}$ is a lisse sheaf of free Λ-modules on X, then $H_c^i(X, \mathcal{G}) = 0$ for $i \neq 1$, and $H_c^1(X, \mathcal{G})$ is a free Λ-module of finite rank.*

Proof See [4] 2.1

It follows from 4.1 that

4.1.1 $$H(X_0, \mathcal{G}_0, T) = \det_\Lambda(1 - T^f F | H_c^1(X, \mathcal{G})).$$

The next assertion, in which $\Lambda[T]_m$ denotes the m-adic completion of the polynomial ring $\Lambda[T]$, is a partial strengthening of 3.4 and 3.5:

4.2 **Theorem** *If $X_0/\mathbf{F}_q$ is a smooth affine curve and $\mathcal{G}_0$ is a lisse sheaf of Λ-modules corresponding to a character $\rho : \pi_1(X_0) \to \Lambda^\times$, then*

$$L(X_0, \mathcal{G}_0, T)/H(X_0, \mathcal{G}_0, T) \in \Lambda[T]_m^\times$$

There is *no assumption* about the image of ρ. When $\Lambda = R =$ the integer ring of a local field K, we observe that $R[T]_{(\pi)}^\times$ is convergent and nonvanishing for $|T| \leq 1$; then 4.2 yields the rank one case of

4.3 **Corollary** *If $X_0/\mathbf{F}_q$ is a smooth affine curve and $\mathcal{G}_0$ is a lisse sheaf of K-vector spaces on X_0 corresponding to an abelian representation of $\pi_1(X_0)$, then*

4.3.1 $L(X_0, \mathcal{G}_0, T)$ is convergent for $|T| \leq 1$

4.3.2 $L(X_0, \mathcal{G}_0, T)/H(X_0, \mathcal{G}_0, T)$ has no poles or zeroes on $|T| = 1$.

In fact, any abelian representation of $\pi_1(X_0)$ can be upper-triangularized after making a finite extension of K, which changes neither $L(X_0, \mathcal{G}_0)$ nor $H(X_0, \mathcal{G}_0)$.

Theorem 4.2 can be deduced from Proposition 3.4 by a simple argument using the base-change properties of $H_c^i(X, \ )$. Since the details will soon appear in print, we will just refer the reader to [4].

4.4 Let us now take a look at the characters which will give rise to a "geometric" Iwasawa theory. Let R be the integer ring of a local field of residue characteristic p, Γ a profinite p-group, and $\Lambda = R[[\Gamma]]$, the completed R-group ring of Γ. Let $\rho : \pi_1(X_0) \to \Lambda^\times$ be the product $\rho_0 = \rho_0 \chi$ of a character of the form

$$\rho_0 : \pi(X_0) \twoheadrightarrow \Gamma \hookrightarrow \Lambda^\times$$

by a character of the form

$$\chi : \pi_1(X_0) \to R^\times \hookrightarrow \Lambda^\times$$

of finite prime-to-p order. Finally, let $\mathcal{G}_0$ be the corresponding lisse sheaf of Λ-modules on χ. Let $\Delta = \text{Im }\chi$ and $S = \mathbf{Z}_p[\Delta]$. Then for every finite quotient H of $\pi_1(X_0)$, the map $\pi_1(X_0) \to \Delta \times H$ gives rise to a (not necessarily connected) principal $\Delta \times H$-cover $X_H \to X_0$, and for every surjective map $H' \twoheadrightarrow H$ of quotients of Γ there is a corresponding étale surjective map $X_{H'} \to X_H$. If we choose, for each H, a nonsingular compactification $X_H \subset \overline{X}_H$, then each $X_{H'} \to X_H$ extends uniquely to a finite map $\overline{X}_{H'} \to \overline{X}_H$, and the action of $\Delta \times H$ on X_H extends to $\overline{X}_H$.

Now set

$$M_H = \text{Hom}(\text{Pic}^o\overline{X}_H(\mathbf{F}_q), \mathbf{Q}_p/\mathbf{Z}_p) \otimes_S R_\chi$$
$$\overline{M}_H = \text{Hom}(\text{Pic}^o\overline{X}_H(\overline{\mathbf{F}}_q), \mathbf{Q}_p/\mathbf{Z}_p) \otimes_S R_\chi$$

In view of what we have just said, the natural functoriality of Pic^o induces maps

$$M_{H'} \longrightarrow M_H \qquad \overline{M}_{H'} \longrightarrow \overline{M}_H$$

for every surjective $H' \to H$, and the inverse limits

$$M = \varprojlim_H M_H \qquad \overline{M} = \varprojlim_H \overline{M}_H$$

have natural Λ-module structures. Furthermore since $\text{Pic}^o\overline{X}_H(\mathbf{F}_q) = \text{Pic}^o\overline{X}_H(\overline{\mathbf{F}}_q)^F$ and R_X is a flat S-module we have (cf. [4]3.0.3 or [12] Proposition 3) an exact sequence

4.4.1 $$0 \to \overline{M} \xrightarrow{1-F} \overline{M} \to M \to 0$$

The Λ-module M is the analogue of the inverse limit of p-ideal-class-groups studied in "classical" Iwasawa theory. One can show that it is a torsion Λ-module. For $\overline{M}$, on the other hand, we have

4.5 **Proposition** *if ρ is ramified of every point of $\overline{X}_0 - X_0$, then there is a natural isomorphism*

$$\overline{M} \simeq H_c^1(X, \mathcal{G})$$

of Λ-modules with Frobenius action.

Proof: See [4] 3.1.

From 4.1.1, 4.4.1 and 4.5 we deduce

4.6 **Corollary** *If ρ is ramified at every point of $\overline{X}_0 - X_0$, then $\overline{M}$ is a free Λ-module of finite rank, and M is a torsion Λ-module whose characteristic element (in the sense of [12]§2) is $H(X_0, \mathcal{G}_0, 1)$.*

4.6.1 **Remark** Without further assumptions on ρ, we do *not* in general have $\overline{M}_H \simeq \overline{M} \otimes_\Lambda \mathbf{Z}_p[H]$, and $\overline{M}_H$ is in general *not* free over $\mathbf{Z}_p[H]$.

Let us now reconsider theorem 4.2. The element $L(X_0, \mathcal{G}_0, 1) \in \Lambda$ can be viewed as a geometric analogue of the "analytic p-adic L-function" of Iwasawa theory; in fact for each finite quotient H of Γ, the image of $L(X_0, \mathcal{G}_0, 1)$ in $R[H]$ is a "Stickelberger element" associated to the cover $X_H \to X_0$ (c.f. [15]V.1). Setting $T = 1$ in 4.2 yields

$$L(X_0, \mathcal{G}_0, 1)/H(X_0, \mathcal{G}_0, 1) \in \Lambda^\times$$

i.e., $L(X_0, \mathcal{G}_0), 1)$ is also a characteristic element of the Λ-module M. This is our analogue of the "Main Conjecture." A weaker version of this had been proven by Goss and Sinnott ([6] Theorem 1.2).

In view of corollary 3.5 and theorem 4.2, it is natural to extend the second assertion of conjecture 2.1 as follows:

4.7 **Conjecture** *Let $X_0/\mathbf{F}_q$ be separated of finite type, Λ a complete noetherian local ring with maximal ideal m and finite residue field of characteristic p, and $\mathcal{G}_0$ a constructible sheaf of Λ-modules on X_0 whose fibers are projective Λ-modules. Then*

$$L(X_0, \mathcal{G}_0, T)/H(X_0, \mathcal{G}_0, T) \in \Lambda[T]_m^\times$$

where $\Lambda[T]_m$ is the m-adic completion of $\Lambda[T]$.

§5 The Congruence Formula

When Λ is a field, conjecture 4.7 is a theorem:

5.1 **Theorem** (Katz[SGA 7II]XXII 3.4, Deligne[SGA $4\frac{1}{2}$] p.116 2.2(b)) *Let $X_0/\mathbf{F}_q$ be separated and of finite type, k a finite field of characteristic p, and $\mathcal{G}_0$ a constructible sheaf of k-vector spaces on X_0. Then*

5.1.1
$$L(X_0, \mathcal{G}_0) = H(X_0, \mathcal{G}_0)$$

In fact, in the notation of 4.7, we have $\Lambda[T]_m^\times = 1$ in this case. When $\mathcal{G}_0 = \mathbf{F}_p$, 5.1 was first proven by Katz; one reduces immediately to the situation in which $X_0/\mathbf{F}_q$ is proper, in which case 5.1.1 is equivalent to the so-called congruence formula for the zeta function

5.1.2
$$Z(X_0, T) = \prod_i \det(1 - TF|H^i(X_0, \mathcal{O}_{X_0}))^{(-1)^{i+1}}$$

as one sees using the same sort of argument that we used to deduce 3.1 from 3.1.1 (see [SGA 7II]XXII§2). Katz proved 5.1.2 by reducing to the case of a hypersurface and using Dwork theory. Deligne's proof, which works for any (constructible) $\mathcal{G}_0$, uses the trace formula in coherent cohomology and the symmetric form of the Kunneth formula. In Deligne's version one can take k to be any reduced noetherian ring of characteristic p, and $\mathcal{G}_0$ a constructible sheaf of flat k-modules (or even more generally, an element of the derived category $D_{ctf}(X_0, k)$ but this more general version can be reduced immediately to 5.1 as stated ([SGA $4\frac{1}{2}$]p.125 4.2), which is more or less how Deligne proceeds. Finally, it is possible to deduce 5.1 from corollary 3.5 using the theory of modular characters and in this section we shall explain how to do this.

It is more convenient to work with an (apparently) more general statement:

5.2 Theorem *Let $X_0/\mathbf{F}_q, Y/\mathbf{F}_q$, and $f : X_0 \to Y_0$ be separated and of finite type, and $\mathcal{G}_0$, k be as in 5.1. Then*

$$L(X_0, \mathcal{G}_0) = \prod_i L(Y_0, R^i f_! \mathcal{G}_0)^{(-1)^i}$$

When $Y_0 = \text{Spec } \mathbf{F}_q$, this is just 5.1. First, we make some reduction steps:

5.2.1 Lemma *If 5.2 is true for morphisms $f : X_0 \to Y_0$, $g : Y_0 \to Z_0$ and for all $\mathcal{G}_0$ on X_0, Y_0, then it is true for the composite $g \circ f$, and all $\mathcal{G}_0$ on X_0.*

Proof: We have

$$\begin{aligned} L(X_0, \mathcal{G}_0) &= \prod_i L(Y_0, R^i f_! \mathcal{G}_0)^{(-1)^i} \\ &= \prod_{i,j} L(Z_0, R^j g_! R^i f_! \mathcal{G}_0)^{(-1)^{i+j}} \\ &= \prod_i L(Z_0, R^i (g \circ f)_! \mathcal{G}_0)^{(-1)^i} \end{aligned}$$

where the last equality follows from the Leray spectral sequence.

5.2.2 Lemma *If 5.1 is true when* $\dim X_0 = 1$, *then 5.2 is true is general.*

Proof: By excision and induction on dimension, we can assume that X_0 and Y_0 are affine. Then (cf [SGA $4\frac{1}{2}$]p.36 4.1) f can be factored $f = f_1 \circ \cdots \circ f_r$, where the fibers of each f_i have dimension ≤ 1. By 5.2.1 it is then enough to prove 5.2 when f has fiber dimension ≤ 1. In this case we have

$$\begin{aligned} L(X_0, \mathcal{G}_0) &= \prod_{y \in |Y_0|} L(f^{-1}(y), \mathcal{G}_0 | f^{-1}(y), T^{deg(y)}) \\ &\quad \text{by comparing Euler factors} \\ &= \prod_{y \in |Y_0|} \prod_i \det(1 - T^{deg(y)} F | H^i_c(f^{-1}(\bar{y}), \mathcal{G} | f^{-1}(\bar{y})))^{(-1)^{i+1}} \\ &\quad \text{by assumption} \\ &= \prod_{y \in |Y_0|} \prod_i \det(1 - T^{deg(y)} F_y | (R^i f_! \mathcal{G})_y)^{(-1)^{i+1}} \\ &\quad \text{by the proper base change theorem} \\ &= \prod_i L(Y_0, R^i f_! \mathcal{G}_0)^{(-1)^i} \end{aligned}$$

5.3 For the proof of 5.1 in the case $\dim X_0 = 1$, we can, by 1.2.3, 1.2.4, and 1.1.2, assume that X_0 is a smooth curve and that $\mathcal{G}_0$ is lisse. Corresponding to $\mathcal{G}_0$ is a representation

$\rho_0 : \pi_1(X_0) \to GL(V)$ where V is a finite-dimensional k-vector space, and $G = \text{Im } \rho_0$ is finite. Let R be a mixed characteristic complete discrete valuation ring with residue field k, maximal ideal m, and fraction field K. We shall prove 5.1 in the present case by expressing $L(X_0, \mathcal{G}_0)$ in terms of L-functions of lisse sheaves of K-vector spaces and applying Corollary 3.5.

Let $\text{Rep}_K(g)$ (resp. $\text{Rep}_k(G)$) denote the category of representations of G in finite-dimensional K-vector spaces (resp. k-) and $R_K(G), R_K(G)$ be the corresponding groups of virtual representations. There are fully faithful exact functors

5.3.1
$$\text{Rep}_K(G) \to \begin{pmatrix} \text{lisse sheaves of} \\ \text{K-vector spaces of } X_0 \end{pmatrix}$$
$$\text{Rep}_k(G) \to \begin{pmatrix} \text{lisse sheaves of} \\ \text{k-vector spaces on } X_0 \end{pmatrix}$$
$$\psi \to \mathcal{F}(\psi)$$

obtained by composing ψ with our given $\rho_0 : \pi_1(X_0) \to G$ and passing to the associated étale sheaf. If $\mathbf{W}(R)$ denotes the "big Witt group" $(1 + TR[[T]])^\times$ of power series with constant term one, then we can define maps

5.3.2
$$L_k, H_k : R_k(G) \to \mathbf{W}(k)$$

by

$$L_k(\psi) = L(X_0, \mathcal{F}(\psi)) \; H_k(\psi) = H(X_0, \mathcal{F}(\psi))$$

That the maps 5.3.2 are group homomorphisms is evident from 1.2.2 and the exactness of 5.3.1. In a similar fashion one obtains group homomorphisms $R_K(G) \to \mathbf{W}(K)$, which factor through $\mathbf{W}(R)$. Reducing mod m gives us group homomorphisms

5.3.3
$$L_K, H_K : R_K(G) \to \mathbf{W}(k)$$

Now recall that for any finite group G, one defines the decomposition homomorphism

$$d : R_K(G) \to R_k(G)$$

as follows: given a representation $\rho : G \to GL(V)$ of G on a K-vector space, there is a G-stable lattice $V_0 \subset V$; the induced action of G on $V_0 \otimes k$ yields an element of $R_k(G)$ which is independent of the choice of $V_0 \subset V$ ([14]15.2). A well known theorem of Brauer asserts that d is *surjective* ([14] Théorème 33).

5.3.4 Lemma $L_K = L_k \circ d$, $H_K = H_k \circ d$

Proof: Let $\rho \in \text{Obj}(\text{Rep}_K(G))$ and let $\mathcal{F}$ be a lisse R-sheaf representing $\mathcal{F}(\rho)$. Then $\mathcal{F}(d(\rho)) \simeq \mathcal{F} \otimes k$, and the lemma follows from 1.2 5.

Conclusion of the proof: It is enough to show $L_k = H_k$, which will prove what we need not just for $\mathcal{G}_0 = \mathcal{F}(\rho_0)$ but for all the $\mathcal{F}(\rho)$. By corollary 3.5 we have $L_K = H_K$, so the result follows from 5.3.4 and the surjectivity of d.

References

[1] Adolphson, A., *A p-adic theory of Hecke polynomials*, Duke Math. J. **43** No.1 (1976), 115-145.

[2] Berthelot, P., *Cohomologie cristalline des Schemas de caracteristique $p > 0$.* Lecture Notes in Mathematics **407**, Springer 1974.

[3] Bloch, S., *Algebraic K-theory and Crystalline Cohomology*, Publ. Math. IHES **47** (1978), pp. 187-268.

[4] Crew, R., *L-functions of p-adic characters and geometric Iwasawa theory*, to appear in Inv. Math.

[5] Dwork, B., *On Hecke polynomials*, Inv. Math **12** (1971), 249-256.

[6] Goss, D., and Sinnot, W., *Class-groups of function fields*, Duke Math. J. **52** No.2 (1985) 507-516.

[7] Illusie, L., *Complexe de De Rham-Witt et cohomologie cristalline*, Ann. sc. Ec. Norm. Sup. 4^e ser. **12** (1979), 501-661.

[8] Katz, N., *Travaux de Dwork*, Sem. Bourb. 409, in LNM 317, (Springer-Verlag).

[9] Katz, N., and Messing, W., *Some consequences of the Riemann hypothesis for varieties over finite fields*, Inv. Math. **23** (1974), 73-77.

[10] Kleiman, S., *Algebraic cycles and the Weil conjectures*, in Dix Exposés sur la cohomologie des schemas, North Holland (1968) pp. 359-386.

[11] Yu. Manin, *The theory of commutative formal groups over fields of finite characteristic*, Russian Math Surveys **18** (1963) 1-83.

[12] Mazur, B., and Wiles, A., *Analogies between function fields and number fields*, Am. J. Math. **105** No.2 (1983), 507-521.

[13] Serre, J.-P. *Quelque propriétés des varietés abéliennes en caracteristique* p, Am. J. Math. **80** No 3 (1958), 715-739.

[14] Serre, J.-P., *Representations lineaires des groupes finis,* 2nd edn., Paris, Hermann (1971).

[15] Tate, J., *Les Conjectures de Stark sur les Fonctions L d'Artin en $s = 0$*, Birkhauser-Boston, 1984.

Treatises:

[SGA $4\frac{1}{2}$] Deligne, P., Cohomologie étale, Lecture Modes in Mathematics **569**, Springer (1977).

[SGA 5] Grothendieck, A., Cohomologie ℓ-adique ét fonctions L, Lecture Notes in Mathematics **589**, Spring (1977).

[SGA 7II] Deligne, P. and Katz, N., Groupes de monodromie en géometrie algebrique II, Lecture Notes in Mathematics **340**, Springer (1973).

Department of Mathematics
Harvard University
1 Oxford St.
Cambridge, MA 02138

Canadian Mathematical Society
Conference Proceedings
Volume 7 (1987)

MAKING CLASS FIELD THEORY EXPLICIT

Ehud de Shalit

1. The two main theorems of class field theory are the existence theorem and the reciprocity law. The first classifies abelian extensions of a global or a local field in terms of groups constructed from the base. The reciprocity law - in its various forms - describes the Galois groups of those extensions in a similar fashion. Both theorems are in general non explicit. What we mean by "explicit" is to give generators for these abelian extensions and an analytic formula describing the action of the Galois group on these generators.

In this note we shall concentrate on the case where the ground field k is a local field of characteristic 0 and residual characteristic p. Here Lubin-Tate theory gives an explicit description of all abelian extensions of k. Let π be a uniformizer of k, $\mathcal{O}$ and p the ring of integers and its maximal ideal. Let $\mathcal{O}/p = \mathbb{F}_q$. With any power series

$$\ell(T) \in \mathcal{O}[[T]] \qquad \ell(T) \equiv \pi T \bmod \deg 2, \qquad \ell(T) \equiv T^q \bmod \pi$$

Lubin and Tate associate a certain one dimensional formal group F_ℓ with ring of endomorphisms $\mathcal{O}$. See [L-T] or [Se]. If W_n denotes the kernel of $[\pi^n]$ in the valuation ideal of $\bar{k}$ (in general $[a]$ denotes the endomorphism whose differential is a), then W_n is a cyclic $\mathcal{O}$-module of cardinality q^n. Let $k_\pi^n = k(W_n)$. Then k_π^n/k is abelian, totally ramified, of degree $(q-1)q^{n-1}$, and k_π^n is class field to $\langle\pi\rangle U_n$, where $U_n = 1 + p^n$. More generally, using "relative" Lubin-Tate groups one can construct (i.e. give generators of) class fields to $\langle\xi\rangle U_n$ where $\xi \in k^\times$ is of order $d \geq 1$ [dS 1].

2. To address the second question, of making the reciprocity law explicit, we must first have an "explicit" picture of our Galois group in mind. One such example is provided again by Lubin-Tate theory. If $u \in U_0$ is a unit in k, and $\omega \in W_n$, the Artin symbol of u on k_π^n is given by the formula

$$(u, k_\pi^n/k)(\omega) = [u^{-1}](\omega).$$

The other case where we can describe the Galois group explicitly is <u>Kummer theory on Lubin Tate groups</u>.

Let $\mathcal{O}_n$ and p_n be the valuation ring and ideal of k_π^n. Let $\alpha \in p_n$ and choose a in the valuation ideal of $\bar{k}$ satisfying $[\pi^n](a) = \alpha$. Then $k_\pi^n(a)/k_\pi^n$ is abelian, and we have a "Kummer pairing"

$$(\alpha,\beta)_n = \sigma_\beta(a)[-]a \qquad\qquad p_n \times (k_\pi^n)^\times \longrightarrow W_n$$

σ_β being the Artin symbol of $\beta \in (k_\pi^n)^\times$. For fixed α the pairing induces an isomorphism of the Galois group with a subgroup of W_n. It is bilinear, and even $\mathcal{O}$-linear in α.

3. Our aim is to give an explicit analytic formula for $(\alpha,\beta)_n$. Such formulas ("explicit reciprocity laws") were first discovered by Artin and Hasse [Ar-Ha], and then by Iwasawa [Iw] and Shafarevitch among others. They worked with the multiplicative group. Wiles [Wi] gave the first treatment of Lubin-Tate groups in general. All the results mentioned so far were partial, and imposed restrictions on β (and sometimes on α too). In [C2] Coleman conjectured a general formula, and in [C3] he proved it for the multiplicative group. We have managed to prove his conjecture, which we explain below. The proof is quite involved, and will appear elsewhere [dS 2].

One may ask for a similar result when the Kummer pairing is computed on any <u>extension</u> of k_π^n. This is still an open problem. For the multiplicative group it was answered by Sen (see bibliography).

4. We recall two results of Coleman [C1].

(a) There exists a unique multiplicative operator

$$N: \quad \mathcal{O}((T))^\times \longrightarrow \mathcal{O}((T))^\times$$

satisfying

$$Nh \circ [\pi](T) = \prod_{\omega \in W_1} h(T[+]\omega).$$

(b) Fix a choice of $\omega_n \in W_n - W_{n-1}$ for each n with $[\pi](\omega_n) = \omega_{n-1}$. Let $\beta = \beta_n \in (k_\pi^n)^\times$ and $\beta_i = N_i^n \beta_n$, N_i^n being the norm from k_π^n to k_π^i. Then there exists $g \in \mathcal{O}((T))^\times$, $\mathrm{ord}_T g = \mathrm{ord}_{p_n} \beta_n$, and $g(\omega_i) = \beta_i$ for all $1 \le i \le n$.

Let $\lambda(T)$ be the logarithm of F, $\frac{d\lambda}{dT} = 1 + \ldots \in \mathcal{O}[[T]]$. Define $\delta g = \frac{1}{d\lambda/dT} \frac{dg/dT}{g} \in T^{-1} \cdot \mathcal{O}[[T]]$, the (Kummer) logarithmic derivative of g.

5. THEOREM. Let $\alpha \in p_n$, $\beta \in (k_\pi^n)^\times$. Let $f \in T\mathcal{O}[[T]]$, $f(\omega_n) = \alpha$, and choose g as above, $g(\omega_i) = \beta_i$, $1 \leq i \leq n$. Let

$$(f,g)_n = \pi^{-n}\left\{\sum_{\omega \in W_n} (\lambda\circ f - \frac{\lambda\circ f\circ[\pi]}{\pi})\cdot\delta g\ (\omega) + \frac{df}{dT}(0)(1 - \frac{Ng}{g}(0))\right\}.$$

Then (i) $(f,g)_n \in \mathcal{O}$ and $(f,g)_n \bmod \pi^n$ depends only on α and β, and not on the choice of f or g. Put $[\alpha,\beta]_n = [(f,g)_n](\omega_n)$.

(ii) (Explicit reciprocity law). $(\alpha,\beta)_n = [\alpha,\beta]_n$.

REMARK. $(f,g)_n$ is a sum of two terms. The first looks like the symbol $\int \log f\ .d(\log g)$ that appears in the study of Riemann surfaces. The integral is replaced here by an average over the torsion points. The second term is a correction term. It vanishes if either $\alpha \in p_n^2$ or $\beta_n \in N_n^{2n}(k_\pi^{2n})^\times$ (the case proved by Iwasawa and Wiles).

6. In view of the last remark it is amusing to see an application where the correction term assumes the major role, and the main term disappears. Following Hasse we shall show how to derive Wiefrich's criterion for the first case of Fermat's last theorem.

We shall apply our result in the following set-up. The field k will be $\mathbb{Q}_p$, $\pi = p$, and the formal group just $\mathbb{G}_m$. Let $n = 1$, $k_\pi^1 = \mathbb{Q}_p(\zeta)$, ζ a primitive p^{th} root of unity, and $\omega_1 = \zeta - 1$. Since the formal group is the multiplicative group, $(\alpha,\beta)_1$ is nothing but the local power residue symbol. See [CF], p. 348-353, for Hilbert's power reciprocity law. Let ℓ be a rational prime different from p, and take $\beta = \ell$. Then we may take $g(T) = \ell$, so $\delta g = 0$ and $Ng = \ell^p$.

Now suppose Fermat's last theorem is false and $x^p+y^p+z^p = 0$, x,y,z in $\mathbb{Z}$, and $p \nmid yz$. Let

$$u = \frac{y}{x+y} \in \mathbb{Z}_p^\times \quad \text{and} \quad \alpha = \frac{x+\zeta y}{x+y} - 1 = u\omega_1 .$$

As is well known, $(1+\alpha)$ must be a pth power of an ideal in $\mathbb{Q}(\zeta)$. If p_1 is the unique prime above p in $\mathbb{Q}(\zeta)$, it follows from Hilbert's power reciprocity law, and from the last remark, that

$$\left(\frac{1+\alpha}{(\ell)}\right)_p^{-1} = \left(\frac{1+\alpha}{(\ell)}\right)_p^{-1}\cdot\left(\frac{\ell}{(1+\alpha)}\right)_p = \left(\frac{1+\alpha,\ell}{p_1}\right) = (\alpha,\ell)_1 = \zeta^{u(\frac{1-\ell^{p-1}}{p})} .$$

If furthermore $\alpha \equiv 0 \bmod \ell$, the left hand side is 1, so we must have $\ell^{p-1} \equiv 1 \bmod p^2$. In the first case $(p \nmid xyz)$ we can always assume that y is even, so with $\ell = 2$, $\alpha \equiv 0 \bmod \ell$. We have arrived at

<u>Wiefrich's</u> <u>criterion</u>: Suppose $2^{p-1} \not\equiv 1 \bmod p^2$. Then there are no first case solutions to Fermat's equation.

BIBLIOGRAPHY

[Ar-Ha] Artin, E., Hasse, H., "Die beiden Erganzungssatze zum Reziprozitatsgesetz der ℓ^n-ten Potenzreste im Korper der ℓ^n-ten Einheitswurzeln," Abh. Math. Sem. Hamburg, 6 (1928).

[C1] Coleman, R., "Division values in local fields," Inv. Math., 53 (1979), 91-116.

[C2] Coleman, R., "The arithmetic of Lubin Tate division towers," Duke Math. J., 48 (1981), 449-466.

[C3] Coleman, R., "The dilogarithm and the norm residue symbol," Bull. Soc. Math. France, 109 (1981), 373-402.

[CF] Cassels, J.W.S., Fröhlich, A., Algebraic Number Theory, Academic Press, 1967.

[dS1] de Shalit, E., "Relative Lubin Tate groups," Proc. A.M.S., 95 (1985), 1-4.

[dS2] de Shalit, E., "The explicit reciprocity law in local class field theory," to appear in Duke Math. J.

[Iw] Iwasawa, K., "Explicit formulas for the norm residue symbol," J. Math. Soc. Japan, 20 (1968), 151-164.

[L-T] Lubin, J. and Tate, J., "Formal complex multiplication in local fields," Ann. of Math., 81 (1965), 380-387.

[Se] Serre, J-P., "Local class field theory". In: Algebraic Number Theory, Cassels and Frohlich, eds., Academic Press, 1967.

[Sen] Sen, S., "On explicit reciprocity laws, I, II," J. reine ang. Math., 313 (1980), 1-26; 323 (1981), 68-87.

[Wi] Wiles, A., "Higher explicit reciprocity laws," Ann. Math., 107 (1978), 235-254.

Department of mathematics,
Harvard University,
Cambridge, MA, 02138, USA.

Canadian Mathematical Society
Conference Proceedings
Volume 7 (1987)

THE POLES AND RESIDUES OF EISENSTEIN SERIES ON SYMMETRIC DOMAINS

by Paul Feit

ABSTRACT. We discuss, without proofs, recent results on certain non-holomorphic Eisenstein series for symplectic groups over a totally real field and for special unitary groups of signature (m,m) over a CM-field. Much of the classical theory for the group $SL_2(\mathbf{Q})$ has been generalized. If Γ is an arbitrary congruence subgroup and k is a sufficiently large weight, specializing an argument in a non-holomorphic series produces an analytic automorphic form with a cyclotomic Fourier expansion; moreover, we can describe the field generated by the Fourier coefficients. There are also new results concerning small weights (including weight 0) which generalize the classical theorems on the "exceptional" weights k = 0,1 and 2.

We will be discussing new results concerning the poles and residues of Eisenstein series. We hope to give the reader a feeling for the types of theorems which have now been proved. These theorems come from a series of papers of which my thesis is the latest, but not the last. The objective is to generalize the theory of non-holomorphic Eisentein series for the classical arithmetic group $SL_2(\mathbf{Q})$ (which is also the group $Sp(1,\mathbf{Q})$) to include the Eisenstein series of other arithmetic groups which act on symmetric domains. In addition to the theorems concerning the algebraic and analytic properties of the series of "large" weights, we have extended the theorems on "exceptional" weights. (The Eisenstein series of weights k = 0, 1, and 2 are the classical exceptional series.)

1980 Mathematics Subject Classification. (1985 Revision). 11F55

The blueprint for setting up and proving theorems is given by Shimura in [17], although many of the ideas can be traced back to Hecke. In this paper, Shimura shows how to combine the technical results from other papers to characterize the Fourier expansions of the Eisenstein series. He analyzes the algebraic and analytic properties of the series whose weight exceeds a specific bound. For certain smaller weights, he conjectures but does not prove that the Fourier coefficients have poles only at certain prescribed points; he lists several "theorems" which depend on the conjecture. Kitaoka proved the conjecture for Sp(m,F), where F is a totally real field, in [7]. The rest of the conjecture, which characterizes poles for Eisenstein series on SU(m,m;K) (defined below) and for Eisenstein series of half-integral weight on Sp(m,F) is proved in [2]. In [2], the author deduces theorems for the series of arbitrary integral weights, including weight 0. In particular, we construct an explicit polynomial over **Z** whose product with the Eisenstein series has no poles outside of a critical strip; we believe that the constructed function is the smallest polynomial which will cancel the poles, but we can not prove it. In addition, we generalize an argument by Sturm in [21] concerning the rationality of Fourier series expansions.

Let us begin with a very naive definition of the Eisenstein series. We start with the following data:

(1) G is a semi-simple linear algebraic group over **Q**.

P is a parabolic subgroup,

H is an analytic manifold with an action by $G(\mathbf{R})$ so that

$H \approx G(\mathbf{R})/K$, for K a maximal compact subgroup of $G(\mathbf{R})$.

$j: G(\mathbf{R}) \times H \to \mathbf{C}^*$ is a factor of automorphy (that is,

$j(\alpha\cdot\beta,z) = j(\alpha,\beta\cdot z)j(\beta,z)$).

We assume that $G(\mathbf{R})$ and the function j act analytically with respect to the variable z on H. Let $\mathbf{A}_f$ denote the ring of finite adeles over **Q**. A subgroup Γ

of $G(\mathbf{Q})$ is called a congruence subgroup if there is a compact-open subgroup U of $G(\mathbf{A}_f)$ such that $\Gamma = G(\mathbf{Q}) \cap U$. For k an integer and Γ a congruence subgroup, define the Eisenstein series of weight k over Γ to be

$$E(z,s;k,\Gamma) = Y(z)^s \sum_{\alpha \varepsilon (P \cap \Gamma)\backslash\Gamma} j(\alpha,z)^{-k} |j(\alpha,z)|^{-2s} \tag{2}$$

where z varies over H and s is a complex variable. The function $Y(z)$ is an explicit non-vanishing real-valued function on H with the property that

$$Y(\alpha \cdot z) = Y(z)\ |j(\alpha \cdot z)|^{-2}. \tag{3}$$

We must assume that the summand in (2) depends only on the $P \cap \Gamma$-coset of α. Throughout this discussion, we tacitly assume that all the appropriate conditions for definition are met, for this summation and for others.

The paper [2] looks at two classes of groups:

$$G = \begin{cases} SP(m,F) & \text{where } F \text{ is a totally real field,} \\ SU(m,m;K) & \text{where } K \text{ is a CM-field.} \end{cases} \tag{4}$$

In the second case, $SU(m,m;K)$ is the special unitary group of a m-dimensional hyperbolic hermitian space with respect to the involution over the maximal totally real subfield. It is known that the series E for a congruence subgroup converges uniformly as the pair (z,s) varies over any compact set in $H \times \mathbf{C}$ on which

$$k + 2\mathrm{Re}(s) > B \tag{5}$$

for some explicit bound B. Furthermore, Langlands and Arthur have shown that the function E has a unique continuation to all of $H \times \mathbf{C}$ which is meromorphic in the variable s.

For any one of these groups, we can take H to be a product of spaces of complex matrices; moreover, an analytic automorphic form has a "Fourier" expansion. For example, if $G = Sp(m,\mathbf{Q})$, we take H to be Siegel upper half-space of $m \times m$ matrices and $Y(z) = \det({}^t\overline{z} - z)$. Then an analytic automorphic form f has an expansion

$$(6) \qquad f(z) = \sum c(h)e^{2\pi i\, tr(hz)},$$

where tr is the matrix trace function and h ranges over a lattice of symmetric matrices in $M_m(\mathbf{Q})$. The coefficient c(h) is non-zero only if h is a positive semi-definite matrix. For other groups, we can define Fourier expansions indexed by symmetric or hermitian matrices such that only positive semi-definite terms appear in the expansion of an analytic form. An automorphic form is called cyclotomic or arithmetic if all of its Fourier coefficients are cyclotomic numbers. There exist systems of canonical models for the group actions which we are studying, and so there is a well-developed theory of arithmetic automorphic forms.

Property (3) implies that E is an automorphic form of weight k in the sense that

$$(7) \qquad E(\alpha\cdot z,s;k,\Gamma) = E(z,s;k,\Gamma)j(\alpha,z)^k \quad \text{for } \alpha \in \Gamma,$$

for any value s. If E(z,s) has a pole with respect to s, then the residue will satisfy the same equation. We ask three questions:

(8.A) Where are the poles of E(z,s) (in the variable s), and what are their orders?

(8.B) For which values of s is E(z,s) or its residue an analytic automorphic form? More generally, which values determine forms with a positive semi-definite Fourier expansion?

(8.C) If E(z,s) is analytic in z, what is the field generated by its Fourier coefficients?

We answer (8.A), (8.B), and (8.C) by introducing a family of "twisted" Eisenstein series, and asking the same questions about them.

For a group of the form (4), Shimura has introduced a family of "twisted" Eisenstein series $E(z,s;k,\psi,\mathbf{b})$. Here, k is an integer, $\mathbf{b}$ is an ideal of the number field F, and ψ is a Hecke character of the idele ring of F. (In the SU Case, F denotes the maximal totally real subfield of K.) There is also a technical condition which k and ψ must satisfy [See 17; (1.13.a,b)]. When $G = Sp(m,\mathbf{Q})$, the series is

(9)

$$E(z,s;k,\psi,\mathbf{b}) = Y(z)^s \sum_{\begin{pmatrix} a & b \\ c & d \end{pmatrix} \in P\cap\Gamma_0(\mathbf{b})\backslash\Gamma_0(\mathbf{b})} \psi(\det(d))\det(cz+d)^{-k}|\det(cz+d)|^{-2s}$$

where a,b,c, and d are $m \times m$ submatrices, and

$$(10) \qquad \Gamma_0(\mathbf{b}) = \left\{ \begin{pmatrix} a & b \\ c & d \end{pmatrix} \varepsilon\ Sp(m,\mathbf{Z}) \ :\ c \equiv 0 \bmod (\mathbf{b}) \right\}.$$

The definition of the twisted series gets very complicated when the class number is bigger than 1. The twisted and un-twisted Eisenstein series are related by the equation

$$(11) \qquad r\cdot E(z,s;k,\Gamma) = \sum_{\tau\ \varepsilon\ \Gamma(\mathbf{b})\backslash\Gamma} \ \sum_{\psi\varepsilon X} E(z,s;k,\psi,\mathbf{b})|_k\tau ,$$

where $\Gamma(\mathbf{b})$ is the congruence subgroup of level $\mathbf{b}$, Γ is a congruence group which contains $\Gamma(\mathbf{b})$, X is a set of characters determined by k and $\mathbf{b}$, and r is a product of group indices. The twisted series for $G = SL_2(\mathbf{Q})$ were defined and studied by Hecke.

There is an $\eta\ \varepsilon\ G(\mathbf{Q})$ such that Fourier coefficients for $E(z,s;k,\psi,\mathbf{b})|\eta$ can be calculated. In fact, the coefficients can be expressed as products of integrals on local spaces. Let us try to describe why the Fourier expansion of the usual series is so difficult to calculate and why the series for $E(z,s;k,\psi,\mathbf{b})|\eta$

has such a convenient form. Consider the case $G = Sp(m,F)$ where F is a totally real field. In this situation, let $\eta = \begin{pmatrix} 0 & -1 \\ 1 & 0 \end{pmatrix} \in G(\mathbf{Q})$.

There is a standard way to define automorphic functions on $G(\mathbf{A})$ so that there is a bijection between automorphic functions on H and automorphic functions on $G(\mathbf{A})$. Each $E(z,s;k,\psi,\mathbf{b})$ is constructed as an adelic automorphic function. For a character ψ, we produce a function ν such that for $\alpha \in G(\mathbf{A})$ and $s \in \mathbf{C}$,

$$(12) \qquad \nu(q\alpha,s) = \nu(\alpha,s) \quad \text{for } q \in P(\mathbf{Q}),$$

$$\nu(\alpha,s) = \prod_{\mathfrak{P} \in M} \nu_{\mathfrak{P}}(\alpha_{\mathfrak{P}},s),$$

where M is the set of primes of F, $\alpha_{\mathfrak{P}}$ is the projection of α into $Sp(m,F_{\mathfrak{P}})$, and $\nu_{\mathfrak{P}}$ is a function on $Sp(m,F_{\mathfrak{P}}) \times \mathbf{C}$. Furthermore

$$(13) \qquad E(x,s;k,\psi,\mathbf{b}) = \sum_{\alpha \in P(\mathbf{Q})\backslash G(\mathbf{Q})} \nu(\alpha x,s),$$

$$E(x,s;k,\psi,\mathbf{b})|\eta = \sum_{\alpha \in P(\mathbf{Q})\backslash G(\mathbf{Q})} \nu(\alpha x\eta_f^{-1},s),$$

where x is an adelic variable and $E(x,s;k,\psi,\mathbf{b})$ and $E(x,s;k,\psi,\mathbf{b})|\eta$ are regarded as functions on $G(\mathbf{A})$. For a congruence subgroup Γ, the adelic form of $E(z,s;k,\Gamma)$ is given by a sum indexed by $P(\mathbf{Q}) \cap \Gamma\backslash\Gamma \simeq P(\mathbf{Q})\backslash P(\mathbf{Q})\Gamma$. By summing terms like those in (13) over certain sets of characters, we can arrange for all terms to vanish except those corresponding to cosets of $P(\mathbf{Q})\backslash P(\mathbf{Q})\Gamma$.

Let S be the additive group of F-symmetric matrices regarded as an algebraic group over $\mathbf{Q}$. The unipotent radical U of the parabolic group is isomorphic to S under that map

$$(14) \qquad \tau: S \to U, \qquad \tau(x) = \begin{pmatrix} 1 & x \\ 0 & 1 \end{pmatrix}.$$

The set $\eta U(\mathbf{Q})$ is a family of irredundant representatives of $P(\mathbf{Q})$-cosets in $G[\mathbf{Q})$, although it does not represent all of $P(\mathbf{Q})\backslash G(\mathbf{Q})$. However, for our choice of function ν, when $\mathbf{b}$ is a proper ideal and $x \in P(\mathbf{A})C$ for C a particular compact subgroup of $G(\mathbf{A})$, then $\nu(\alpha x \eta_f^{-1}, s) = 0$ unless $\alpha \in P(\mathbf{Q})\eta U(\mathbf{Q})$. Fix $\omega \in P_\infty$ and consider $E(x,s;k,\psi,\mathbf{b})|\eta$ restricted to $x \in U(\mathbf{A})\omega$. The function is invariant under $U(\mathbf{Q})$. Thus, it has a Fourier expansion indexed by $U(\mathbf{Q}) \approx S(\mathbf{Q})$ whose coefficients are functions in s and ω. Each coefficient is the integral of $E(x,s;k,\psi,\mathbf{b})|\eta$ times a character over $U(\mathbf{A})/U(\mathbf{Q})$. Since $E(x,s;k,\psi,\mathbf{b})|\eta$ is a sum over $U(\mathbf{Q})$, the coefficient is an integral of $\nu(\alpha x \eta_f^{-1}, s)$ over $U(\mathbf{A})$. Since $\nu(\alpha x \eta_f^{-1}, s)$ is a product of local functions, and any character is a product of local characters, the coefficient is simply a product of local integrals.

Each integral at an infinite prime yields a hypergeometric function. The product of integrals over the finite primes is just a ratio of L-functions. The local integrals depend on the quantity $k + 2s$. For this reason, the set of values s at which the series has interesting properties depends on the choice of weight k.

Let us give some of the consequences of the technical work. We only give the statements for $G = Sp(m,\mathbf{Q})$; there are analogues for all of the groups. First, there is an explicit product $\Lambda(s;k,\psi,\mathbf{b})$ of L-functions and Γ-factors such that

$$(15) \quad \Lambda(s;k,\psi,\mathbf{b})E(z,s;k,\psi,\mathbf{b}) \text{ is } \begin{cases} \text{finite} & \text{if } \psi^2 \neq 1, \\ \text{has finitely many poles (in s) which can occur only at explicitly given points.} & \text{otherwise} \end{cases}$$

For example, when m is odd and $k \geq (m-1)/2$, then

$$(16) \quad \Lambda(s;k,\psi,\mathbf{b}) = \Gamma(k+1)L_{\mathbf{b}}(k+2s,\psi) \prod_{j=1}^{(m-1)/2} \{\Gamma(2k+2s-2j)L_{\mathbf{b}}(2k+4s-2j,\psi^2)\}$$

The function $L_{\mathbf{b}}$ is the L-function determined by summing $\psi(n)/n^s$ over all positive integers n prime to $\mathbf{b}$.

Theorem 1: The function $E(z,s;k,\psi,\mathbf{b})$ is finite at $s = 0$, holomorphic in the variable z, and has a cyclotomic Fourier expansion when

(1) $k \geq [\frac{m+5}{2}]$, or

(2) $k = [\frac{m+3}{2}]$ and $\psi^2 \neq 1$.

Theorem 2: Suppose m is odd and $\psi^2 = 1$. Then the function $E(z,s;(m+3)/2,\psi,\mathbf{b})$ is finite at $s = 0$ and has the form

$$\Delta(p_1(z) + p_2(z)\log Y(z))$$

where Δ is a differential operator, p_1 has a cyclotomic expansion which generates the positive definite part of the expansion of E, and p_2 has a cyclotomic Fourier expansion which determines the positive semi-definite terms of rank m-1.

Suppose m is even and $\psi^2 = 1$. Then the function $E(z,s;(m/2)+1,\psi,\mathbf{b})$ is finite at $s = -(1/2)$ and is a power of π times a cyclotomic form.

In the classical case $G = SL_2(\mathbf{Q})$, the operator Δ is $\partial/\partial z$. Then $E(z,0;2) = a/y + \Sigma_r c(r)e^{2\pi irz}$, where a and each c(r) is a constant.

Theorem 3: The function $E(z,s,k,\psi,\mathbf{b})$ has a simple pole in s whose residue is an explicit constant times a cyclotomic automorphic form for

(1) $0 < k \leq [\frac{m}{2}]$, $s = \frac{m+1}{2} - k$, and $\psi^2 = 1$, or

(2) $k = 0$, $s = \frac{m+1}{2}$, and $\psi = 1$.

Moreover, in case (2) the residue is a non-zero constant function of z.

Theorem 4: If $k \geq (m+2)/2$ and $\psi^2 \neq 1$, then the function $\Lambda(s;k,\psi,\mathbf{b})E(z,s;k,\psi,\mathbf{b})$ is finite at $s = (m+1)/2-k$, and the product determines a cyclotomic form times a power of π.

There are other substitutions of s for which $E(z,s;k,\psi,\mathbf{b})$ has a positive Fourier expansion. However, these are all derived from the above substitutions. The differential operators $\{Y(z)^r \Delta^d Y(z)^{-r}: r,d \in \mathbf{Z}\}$ transform Eisenstein series of one weight to series of another. (In the classical case, these are the familiar operators $y^r(\partial/\partial z)^d y^{-r}$.) We can classify several families of substitutions as non-holomorphic derivatives of cyclotomic forms.

Using the summation formula (11), we can prove similar theorems about the Eisenstein series for a general congruence subgroup. For example, one can compute an explicit polynomial $Q(k,s)$ such that $Q(k,s)^n E(z,s;k,\Gamma)$ has only a finite set of explicit simple poles outside of a critical strip. From the theory of canonical models, we can get the following result.

Theorem 5: Let $k \geq [\frac{m+5}{2}]$, and let Γ be a congruence subgroup so that $j(\sigma,z)^k = 1$ for all $\alpha \in P \cap \Gamma$. Let $\mathbf{Q}_{ab}$ be the abelian closure of $\mathbf{Q}$, and suppose $\sigma \in Gal(\mathbf{Q}_{ab}/\mathbf{Q})$. Then $E(z,0;k,\Gamma)$ is a cyclotomic form, and $E(z,0;k,\Gamma)^\sigma = E(z,0;k,\Gamma^\sigma)$.

If Γ is a level group, then $\Gamma^\sigma = \Gamma$ for all $\sigma \in Gal(\mathbf{Q}_{ab}/\mathbf{Q})$. Thus, the Eisenstein series at $s = 0$ for a level group has rational Fourier coefficients.

The congruence subgroup Γ^σ arises naturally in the theory of canonical models. The construction is easy, but we do not give it here.

REFERENCES

1 Hel Braun, Hermitian Modular Functions, I, II, Ann. of Math. (1) **50, 51** (1949, 1950) 827-855, 92-104.

2. P. Feit, Poles and Residues of Eisenstein Series for Symplectic and Unitary Groups, [to appear in Memoirs of the A.M.S.].

3. P. Garrett, Arithmetic Properties of Fourier-Jacobi Expansions of Automorphic Forms in Several Variables, Amer. J. Math. **103** (1981) 1103-1134.

4. F. P. Greenleaf and L. Corwin, Character Formulas and Spectra of Compact Nilmanifolds, J. Funct. Anal. **21**, (1976) no. 2 123-154

5. M. Harris, Maass Operators and Eisenstein Series, Math. Ann. **258**, (1981/82) no. 2 135-144.

6. V. I. Kalinin, Eisenstein Series on the Symplectic Group, Math. U.S.S.R. Sbornik **32** (1977) 449-476

7. Y. Kitaoka, Dirichlet Series in the Theory of Siegel Modular Forms, Nagoya Math. J. **95** (1984), 73-84.

8. H. Klingen, On Eisenstein Series and some applications, in Automorphic Forms of Several Variables, Katata Conference, 1983, Birkhauser, 1984.

9. R. P. Langlands, On the Functional Equation satisfied by Eisenstein Series, Lecture Notes in Mathematics 544, Springer-Verlag, 1976.

10. H. Maass, Springer Lecture Notes, Vol. 216, 300-318.

11. I. Reiner, Maximal Orders, Academic Press, New York, 1975.

12. G. Shimura, On Certain Reciprocity Laws for Theta Functions and Modular Forms, Acta Math. **141** (1978) 35-71.

13. G. Shimura, The Arithmetic of Automorphic Forms with respect to a Unitary Group, Ann. of Math. **107** (1978) 560-605.

14. G. Shimura, The Special Values of the Zeta Functions associated with Hilbert Modular Forms, Duke Math, J. **45** (1978) 637-679.

15. G. Shimura, Arithmetic of Differential Operators on Symmetric Domains. Duke Math. J. **48** (1981) 813-843.

16. G. Shimura, Confluent Hypergeometric Functions on Tube Domains, Math. Ann. **260** (1982) 269-302.

17. G. Shimura, On Eisenstein Series, Duke Math. J. **50** (1983) 417-476.

18. G. Shimura, Differential Operators and the Singular Values of Eisenstein Series, Duke Math. **51** (1984) 261-329.

19. G. Shimura, On Eisenstein Series of Half-Integral Weight, Duke Math. J. **52** (1985) 281-314

20. C. L. Siegel, Gesammelte Abh., Vol. 11, 125-137.

21. J. Sturm, The Critical Values of the Zeta Functions Associated to the Symplectic Group, Duke Math. J. **48** (1981) 327-350.

22. R. Weissauer, Eisensteinreihen von Gewicht n+1 sur Siegelschen Modulgruppe n-ten Grades. Math. Ann. **268** (1984) 357-377.

Paul Feit
University of Chicago
5734 University Ave.
Chicago, Ill. 60637

Canadian Mathematical Society
Conference Proceedings
Volume 7 (1987)

Kida's Theorem and p-adic L-functions

Robert Gold

§0. The CMS presentation consisted of some material which appears in [M1], [M2], and [M3], in addition to what is described below. The content of §1 is joint work with H. Kisilevsky while that of §2 is based on discussions with Childress and Sinnott.

§1. Let F be a finite number field and for each rational prime ℓ let $F^{(\ell)}$ denote the cyclotomic Z_ℓ- extension of F. Let $\hat{F}$ be the composite of $F^{(\ell)}$ over all ℓ. Then $\hat{F}$ is an abelian extension of F with galois group isomorphic to $\hat{Z}$. For each natural number N let F_N denote the unique extension of F of degree N in $\hat{F}$; F_N/F is cyclic. If F is totally real then so is F_N. If F is of CM-type, then so is F_N.

Assume now that F is of CM-type and for a fixed odd prime p let $\lambda_p^-(F_N)$ be the Iwasawa λ^--invariant associated to the cyclotomic Z_p-extension of F_N [W,p. 127]. We want to ask if the collection of $\lambda_p^-(F_N)$ over all N prime to p is bounded.

1980 Mathematics Subject Classification. 11R

For any CM-field k, let k^+ be the maximal real subfield of k and let s(k) denote the number of places of $k^{+(p)}$ which lie over p and split in $k^{(p)}/k^{+(p)}$. Then it is known (Greenberg [Gr]) that $\lambda_p^-(k) \geq s(k)$. It is clear that $s(F_N)$ is on the order of S(F) times the number of primes of Q_N lying over p. This second factor is unbounded iff there exist infinitely many prime ℓ such that

$$p^{\ell-1} \equiv 1 \pmod{\ell^2}$$

(Cf.[F]).

We will assume that this is indeed the case and conclude that if s(F) > 0, then $\lambda_p^-(F_N)$ are unbounded over all N prime to p.

Consider instead the invariant $\lambda'_p(k) = \lambda_p^-(k) - s(k)$. Is $\lambda'_p(F_N)$ bounded over N prime to p? Let K be a CM-field such that K/F is cyclic of degree p. Kida's theorem [K] gives $\lambda_p^-(k_N) = p\cdot\lambda_p^-(F_N) + (\tau_N - \delta_N)(p-1)$ where $\delta_N = 1$ or 0 as F_N does or does not contain a pth root of unity and τ_N is the number of non-p-places of $F_N^{+(p)}$ which split in $F_N^{(p)}/F_N^{+(p)}$ and ramify in $K_N^{(p)}/F_N^{(p)}$. Note that any non-p-place of $F_N^{(p)}$ is either ramified or split in $K_N^{(p)}$. An easy computation then shows that in this situation $\lambda'_p(K_N) = p\cdot\lambda'_p(F_N) + (\tau'_N - \delta_N)(p - 1)$ where τ'_N is the number of places of $F_N^{+(p)}$ which split in $F_N^{(p)}/F_N^{+(p)}$ and do not split in $K_N^{(p)}/F_N^{(p)}$.

Let Q be a prime of F which divides the rational prime q, $q \neq p$, is split in F/F^+, and ramifies in K/F. Then τ_N is unbounded if there exist infinitely many ℓ such that

$$q^{\ell-1} \equiv 1 \pmod{\ell^2}.$$

Assuming this is the case, we see from Kida's formula for λ'_p that $\lambda'_p(K_N)$ will be unbounded. Thus eliminating the trivial zeros of the p-adic L-functions or, equivalently, the genus contribution of p-places accounted for by $s(F_N)$ we still have too large a λ-invariant. Is it possible to eliminate the contribution to λ_p' due to nonsplit places (i.e. the term $(\tau_N' - \delta_N)(p - 1))$ in a natural way to obtain a λ_p'' bounded over F_N?

§2. Fix an odd rational prime p. Let χ be a nontrivial even Dirichlet character with conductor not divisible by p^2 and k/Q the associated real cyclic extension. Set $L_p(s,\chi)$ be the Kubota-Leopoldt L-function and $f_\chi(T) \in Z_p'[\chi][[T]]$ the corresponding Iwasawa power series characterized by

$$L_p(s,\chi) = f_\chi((1+p)^s-1).$$

Using the Weierstrass Preparation Theorem and the Ferrero-Washington Theorem, we may write

$$(*)\qquad f_\chi(T) = \prod_{i=1}^{\lambda} (T-\alpha_i)\, U(T)$$

where $\alpha_i \in C_p$, $|\alpha_i| < 1$ and $U(T) \in Z_p[\chi][[T]]^\times$.

Let Q_m be the m-th level of the cyclotomic Z_p-extension Q_∞ of Q and $k_m = k\cdot Q_m$. Then χ can be viewed as a character of the real cyclic extension k_m/Q_m and by Deligne-Ribet [DR] there is a p-adic L-function defined over Q_m for χ and a corresponding Iwasawa power series:

$$L_p(s,\chi, Q_m) = f_\chi^{(m)}((1 + p)^{sp^m} -1).$$

The factor p^m appears in the exponent since a generator of $Gal(Q_\infty/Q_m)$ is a p^m-th power of a generator of $Gal(Q_\infty/Q)$.

Since $L_p(s,\chi, Q_m) = \prod_\psi L_p(s,\chi\psi)$ where ψ runs over all characters of $Gal(Q_m/Q)$ and

$$L_p(s,\chi\psi) = f_\chi(\zeta(1 + p)^s-1) \text{ for } \zeta = \psi(1 + p)^{-1}$$

we see that $f_\chi^{(m)}((1+T)^{p^m}-1) = \prod_{\zeta^{p^m}=1} f_\chi(\zeta(1+T)-1)$.

The right hand side is, by definition, $(N^m f_\chi)(T)$ where N is Coleman's norm operator ([C],[W, §13.7]). It follows that $f_\chi^{(m)} = N^m f_\chi$. Using the factorization (*) we have $f_\chi^{(m)}(T) = \prod ((1+T)-(1+\alpha_i)^{p^m}).(N^m U)(T)$.

The limit $f_\chi^\infty(T) = \lim_{m\to\infty} f_\chi^{(m)}(T) = T^\lambda .(N^\infty U)(T) \in Z_p[\chi][[T]]$ exists and is invariant under the norm operator. Observe also that

$$\lim \frac{1}{p^{\lambda m}} f_\chi^{(m)}(0) = \lim \left(\prod_{i=1}^{\lambda} \left(\frac{1-(1+\alpha_i)^{p^m}}{p^m}\right) (N^m U)(0)\right)$$

$$= (-1)^\lambda \left(\prod_i \log(1+\alpha_i)\right).(N^\infty U)(0).$$

The factor $(-1)^\lambda (N^\infty U)(0)$ can be interpreted as the p-adic limit of the non-p-part of class numbers up the Z_p-tower. The existence of such a a limit was observed earlier by Sinnott.

References

[C] Coleman, R. , Division values in local fields, Inv. Math. 53 (1979).

[DR] Deligne, P. and K. Ribet, Values of abelian L-functions at negative integers over totally real fields, Inv. Math. 59 (1980).

[F] Friedman, E., Iwasawa theory for several primes and a connection to Wieferich's criterion, in Number Theory Related to Fermat's Last Theorem, ed. N. Koblitz, Birkhauser, 1982.

[M1] Gold, R. and M. Madan, Iwasawa invariants, Communications in Algebra 13 (1985).

[M2] ________________, Galois representations of Iwasawa modules, to appear in Acta Arithmetica.

[M3] ________________, Generalization of Kida's formula, preprint.

[Gr] Greenberg, R., On a certain ℓ-adic representation, Inv. Math. 21 (1973)

[K] Kida, Y., ℓ-extensions of CM-fields and cyclotomic invariants, J. Number Theory 12 (1980).

[W] Washington, L., Introduction to Cyclotomic Fields, Springer Verlag, N. Y., 1982.

Department of Mathematics
Ohio State University
Columbus, OH 43210
U.S.A.

Canadian Mathematical Society
Conference Proceedings
Volume 7 (1987)

ALGEBRAIC CYCLES OF HIGHER WEIGHT AND MODULAR FORMS

B. Brent Gordon

1. MOTIVATION. The question that got me interested in algebraic cycles of higher weight and modular forms was the following: In 1976 Hirzebruch and Zagier [6] defined a family of algebraic cycles $(T_n^c)_{n\in\mathbb{N}}$ in the smoothly compactified Hilbert modular surface $\tilde{X}$ associated to the real quadratic field $\mathbb{Q}(\sqrt{p})$, for p a prime congruent to $1 \bmod 4$. They then computed the pairwise intersection multiplicities $(T_m^c \cdot T_n^c)$ of these cycles, and discovered the remarkable fact that the generating function for these intersection numbers $\sum_{n\geq 0}(T_m^c \cdot T_n^c)e[n\tau]$ was an elliptic modular form of weight 2 on $\Gamma_0(p)$ with a character. A short time later Zagier [17] noticed that if certain weighting factors were attached to these intersection multiplicities, then the new, weighted intersection numbers were again the Fourier coefficients of an elliptic modular form with the same level and character, but now of higher weight. Thus he was led to ask if these weighted intersection numbers could somehow be realized as the ordinary geometric intersection multiplicities of some algebraic cycles in some appropriate homology theory over the Hilbert modular surface.

In an attempt to answer Zagier's question, Tong [19] defined certain classes in the cohomology of $\tilde{X}$ with coefficients in a vector bundle whose intersection multiplicities were Zagier's weighted intersection numbers. However, he did not consider the question of whether the classes he defined were algebraic. On the other hand, when $\tilde{X}$ is replaced by a quaternionic modular surface, on which there are no cusps, then we can answer Zagier's question completely and positively. Thus, the purpose of this talk is to describe in fairly general terms what an algebraic cycle of higher weight should be, and then describe as an example the answer to Zagier's question in

---

1980 Mathematics Subject Classification (1985 Revision). 11F41, 11F30

the quaternionic case. I will restrict myself here to stating the results; additional details and proofs will appear elsewhere.

To put Zagier's question into a more general perspective, first of all observe that in the case studied by Hirzebruch and Zagier a Hirzebruch-Zagier cycle $T_m^c \subset \tilde{X}$ represents, via Poincaré duality, a class $[T_m^c] \in H^2(\tilde{X})$. (all cohomology groups have coefficients in $\mathbb{Q}$ unless otherwise indicated.) On the other hand, there is also a natural injection j from Hilbert modular forms of weight 2 into $H^2(\tilde{X},\mathbb{C})$, by associating to such a cusp form a differential 2-form on $\tilde{X}$ [4, 12], and it turns out that the component $[T_m^{co}]$ of $[T_m^c]$ lying in the image of j differs from $[T_m^c]$ by a multiple of the first Chern class. So now the intersection map $T_m^{co} \longrightarrow \sum_{n\geq 0} (T_m^{co}\cdot T_n^c)e[n\tau]$ can be thought of as lifting of a Hilbert modular cusp form of weight 2 to an elliptic modular (cusp) form of weight 2. (In fact the Chern class maps to an Eisenstein series.) Nowadays such liftings of modular forms are usually understood in terms of integral kernels, especially the theta-functions that arise from the restriction of the Weil representation to the groups of a dual reductive pair; see, for example, [18, 7, 13, 14]. In the present example, the dual pair would be $SO(2,2) \times SL_2(\mathbb{R})$, since $SO(2,2)$ has $SL_2(\mathbb{R}) \times SL_2(\mathbb{R})$ for a 2-fold cover; and the formal Fourier series $\sum_{n\geq 0} T_n^c e[n\tau]$ is somewhat analogous to the theta-kernal (compare, for example [9]). In any case, in the context of lifting modular forms by theta-kernels it is a simple matter to start with a Hilbert modular form of higher weight, say, and end up with an elliptic modular form of higher weight. The question here is when and how these modular forms and this lifting can be realized and interpreted in terms of algebraic-geometric objects. I hope it is clear that this question makes sense much more generally than just for Hilbert and elliptic modular forms.

2. ALGEBRAIC CYCLES OF HIGHER WEIGHT. So let G be a semisimple algebraic group over $\mathbb{Q}$, of $\mathbb{Q}$-rank zero, such that the symmetric space $\mathcal{D} \simeq G_{\mathbb{R}}/K$, for a maximal compact $K \subset G_{\mathbb{R}}$, can be given the structure of a hermitian domain. Assume also for simplicity that $G_{\mathbb{R}}$ has no compact factors. Let $\Gamma \subset G$ be a torsion-free arithmetic subgroup. Then suppose there exists a symplectic representation $\sigma: G \longrightarrow Sp(W,\beta)$, where W is a vector space and β is a nondegenerate alternating form on W, defined over

$\mathbb{Q}$. Let $\Lambda \subset W$ be a lattice such that $\sigma(\Gamma)\Lambda = \Lambda$ and $\beta(\Lambda,\Lambda) \subseteq \mathbb{Z}$. Finally, suppose that the map of symmetric domains $\tau\colon \mathfrak{D} \longrightarrow \mathfrak{h}(W,\beta)$ induced by σ , i.e. so that the diagram

$$\begin{array}{ccc} G_{\mathbb{R}} & \xrightarrow{\ \sigma\ } & Sp(W,\beta)_{\mathbb{R}} \\ \downarrow & & \downarrow \\ \mathfrak{D} & \xrightarrow{\ \tau\ } & \mathfrak{h}(W,\beta) \end{array}$$

commutes, is holomorphic. Then Kuga [10, 12] has shown that the torus bundle $\mathcal{A} := \Gamma\backslash(\mathfrak{D} \times (W_{\mathbb{R}}/\Lambda))$ over the arithmetic variety $X := \Gamma\backslash\mathfrak{D}$ can be embedded as a complex projective variety with the structure of a family of abelian varieties.

Now let (π,E) denote a finite dimensional representation of G defined over $\mathbb{Q}$, and let $\mathcal{E} \longrightarrow X$ be the associated locally constant sheaf. Then, since the Leray spectral sequence for the fibration $\varphi : A \longrightarrow X$ degenerates at the E_2-term, it is easy to see that if $E \subset \bigwedge^b W$, for some b , then there is a natural embedding of $H^a(X,\mathcal{E})$ into $H^{a+b}(A)$. Call the image $H^{a+b}(M)$; this is just a certain subspace of $H^{a+b}(A)$ which is isomorphic to $H^a(X;\mathcal{E})$. However, what is interesting is that when (π,E) satisfies a certain technical condition we can prove that $H^{a+b}(M)$ can be obtained algebraically.

THEOREM 1. <u>If the $\mathbb{Q}$-algebra generated by the image of</u> $\mathrm{End}_G(W)$ <u>in</u> $\mathrm{End}_G(\bigwedge^b W)$ <u>contains a projection from</u> $\bigwedge^b W$ <u>to</u> E , <u>then there exists an algebraic cycle</u> $\mathcal{P} \subset A \times A$, <u>of dimension equal to that of</u> A , <u>which induces a projection</u> $P\colon\ H^{a+b}(A) \longrightarrow H^{a+b}(M)$.

In particular, an algebraic class in $H^{a+b}(A)$ projects to an algebraic class in $H^{a+b}(M)$, and a class in $H^{a+b}(M)$ is algebraic if and only if it is algebraic as a class in $H^{a+b}(A)$.

DEFINITION. When (π,E) is nontrivial and satisfies the condition of Theorem 1, an algebraic cycle in A which represents a class in $H^{a+b}(M)$ will be called an <u>algebraic cycle of higher weight</u>.

The philosophy behind this definition is that $H^{a+b}(M)$ is the Betti realization of a motive ([1],§0), so that it should be possible to reformulate any question about algebraic cycles and the cohomology of a complex projective variety, such as the Hodge conjecture, for $H^{a+b}(M)$. Moreover, some extra structure is known for $H^{a+b}(M)$, since

$$H^{a+b}(M) \simeq H^a(X\mathbb{E}) \simeq H^a(\Gamma,E) ,$$

while for some (absolutely irreducible) (π,E) there is also a natural identification between $H^{a+b}(M)$ and a space of cusp forms for Γ on $\mathfrak{D}$. Thus in these cases we also have an algebraic-geometric setting for cusp forms of higher weight, as sought earlier.

To say just a word about the proof of Theorem 1, let x_0 be a generic point of X, so that the abelian variety $\varphi^{-1}(x_0)$ is a generic fiber of A. Then note (as in [11, 12]) that

$$\textstyle\bigwedge^b W \simeq H^b(\varphi^{-1}(x_0))$$

and

$$\mathrm{End}_G(W) \simeq \mathrm{End}(\varphi^{-1}(x_0)) \otimes \mathbb{Q} .$$

So E may be identified with a subspace of $H^b(\varphi^{-1}(x_0))$, and then the algebraic projector $\mathfrak{P}$ can be taken to be a $\mathbb{Q}$-linear combination of correspondences of A which come from endomorphisms of $\varphi^{-1}(x_0)$. For details, see [3].

3. AN EXAMPLE. To return now to the example of a quaternionic modular surface, the set-up is this: Let V be a 4-dimensional vector space over $\mathbb{Q}$ equipped with a quadratic form q which is anisotropic over $\mathbb{Q}$ but has signature (2,2) over $\mathbb{R}$. Let L be a lattice in V on which q is integer-valued, and assume that the discriminant D of L is not a square. Then the even Clifford algebra $C^+(V)$ of V is the totally indefinite division quaternion algebra over the real quadratic field $\mathbb{Q}(\sqrt{D})$ which is ramified at exactly those rational primes at which q is anisotropic. The even Clifford algebra $C^+(L)$ of L over $\mathbb{Z}$ is an order in $C^+(V)$. Furthermore, as V is 4-dimensional, the group G of norm 1 units of $C^+(V)$ is the spin group of V; over $\mathbb{R}$ it is isomorphic to $SL_2(\mathbb{R}) \times SL_2(\mathbb{R})$. Hence the symmetric space $\mathfrak{D}$ is isomorphic to a product of two complex upper half-

planes. Next let $W := C^+(V) \otimes M$ and $\Lambda := C^+(L) \otimes M$, where M is any $\mathbb{Z}$-lattice of rank $k \geq 0$. Then G acts on W on the left, and any torsion-free $\Gamma \subset C^+(L) \cap G$ preserves Λ. Since Satake [15] has shown that W can be given a G-invariant symplectic structure with which the "Eichler map" τ is holomorphic, we get as before a family A of abelian 4k-folds whose endomorphism ring tensored with $\mathbb{Q}$ always contains and is generically equal to $M_k(C^+(V))$, lying over a quaternionic modular surface $X := \Gamma\backslash\mathfrak{D}$.

Now, for $v \in L$ with $q(v) > 0$, let V_v denote the orthogonal complement of v in V, and let $L_v := L \cap V_v$. Then in exactly the same way we can construct a family A_v of abelian 2k-folds lying over a Shimura curve: Let $C^+(V_v)$ be the even Clifford algebra of V_v; let $G_v := G \cap C^+(V_v)$; the symmetric space D_v is isomorphic to an upper half-plane; $W_v := C^+(V_v) \otimes M$, while $\Lambda_v := \Lambda \cap W_v$; and $\Gamma_v := \Gamma \cap G_v$. Moreover, the natural inclusion of V_v into V induces an immersion $h_v : A_v \longrightarrow A$ which is compatible with the fiber structure on both sides.

Now let $E \subset \bigwedge^{4k} W$ be the absolutely irreducible representation of highest weight. Then in fact E is defined over $\mathbb{Q}$ and it occurs in $\bigwedge^{4k} W$ with multiplicity one. Moreover E is isomorphic to the space of harmonic polynomials of degree $2k$ on V, on which G acts via its action on V, as well as to the tensor product of two symmetric tensor representations of degree $2k$, if G is viewed as embedded in $SL_2(\mathbb{R}) \times SL_2(\mathbb{R})$. As before, let $H^{4k+2}(M) \simeq H^2(X,\mathbf{E})$ denote the corresponding subspace of $H^{4k+2}(A)$. It is not hard to check that E satisfies the conditions of Theorem 1, so $H^{4k+2}(M)$ is indeed algebraically defined.

Then using Theorem 1 we may define an algebraic cycle of higher weight $\mathfrak{C}_v$ so that

$$[\mathfrak{C}_v] = n_L(v)^{2k} P([h_v(A_v)]) \in H^{4k+2}(M) ,$$

where $P: H^{4k+2}(A) \longrightarrow H^{4k+2}(M)$ is the projection and $n_L(v)^2 := 2q(v)D/D_v$ when D_v is the discriminant of L_v. Now the cycles $\mathfrak{C}_n$ which generalize the Hirzebruch-Zagier cycles are defined by

$$\mathfrak{C}_n := \sum_{v \in \Gamma\backslash L(n)} \mathfrak{C}_v ,$$

where $L(n) := \{w \in L | q(w) = n\}$. It follows from reduction theory that this is a finite sum. Define $\mathfrak{C}_0$ formally by

$$[\mathfrak{C}_0] = \begin{cases} \frac{1}{2}c_1 & \text{if } k = 0 \\ 0 & \text{if } k \neq 0 , \end{cases}$$

where c_1 is the first Chern class on X .

Then the intersection multiplicities of these Hirzebruch-Zagier cycles of higher weight have precisely the same form as Zagier's weighted intersection numbers, as indicated in the following theorem.

THEOREM 2.

$$(\mathfrak{C}_m \cdot \mathfrak{C}_n) = \sum_{\substack{(v,w)\in\Gamma\backslash(L(m)\times L(n)) \\ D(v,w)<0}} P_{2k}(v,w) + \sum_{\substack{(v,w)\in\Gamma\backslash(L(m)\times L(n)) \\ D(v,w)=0}} E(X_v)P_{2k}(v,w)$$

where

$$P_{2k}(v,w) := \sum_{j=0}^{k} (-1)^j \binom{2k-j}{j} q(v)^j q(w)^j (v,w)^{2k-2j}$$

with

$$(v,w) := q(v+w) - q(v) - q(w) ,$$

while $E(X_v)$ *is the Euler volume of* X_v , *and*

$$D(v,w) := (v,w)^2 - 4q(v)q(w) .$$

In order to prove Theorem 2, we first write down the harmonic differential form ξ_m on A which represents $[\mathfrak{C}_m]$. When coordinates are chosen for A coming from an identification of $\mathfrak{D}$ with the product of two complex upper half-planes, then ξ_m is very similar in form to the function $\omega_{m,D}$ in [17, 18]. Then the intersection number $(\mathfrak{C}_m \cdot \mathfrak{C}_n)$ is obtained by integrating ξ_m over $\mathfrak{C}_n$; the actual computations are similar to those in [17, Thm. 6; 18, Thm. 2]. For further details, see [4].

Now, to finish off the example, it only remains to check that these intersection multiplicities are the Fourier coefficients of an elliptic modular form of higher weight.

THEOREM 3. The Fourier series

$$F_m(\tau) := \sum_{n\geq 0} (\mathcal{T}_m \cdot \mathcal{T}_n) e[n\tau]$$

is an elliptic modular form of weight $2k + 2$ on $\Gamma_0(N) \subset SL_2(\mathbb{Z})$ with character χ, where N and χ are the level and character, respectively, of L. If $k > 0$ then $F_m(\tau)$ is a cusp form.

Here level and character are to be understood in the sense of Hecke [5].

To deduce Theorem 3 from Theorem 2, we first show, by extending the ideas of [8], that for $v \in L$ with $q(v) = m$, the function

$$\varphi_v(t,z) := \sum_{\substack{w\in\Gamma_v\backslash L \\ D(v,w)<0}} e[q(w)t + (v,w)z]$$

$$+ E(X_v) \sum_{\substack{w\in\Gamma_v\backslash L \\ D(v,w)=0}} e[q(w)\tau + (v,w)z]$$

is a Jacobi form of weight 2 and index m, in the language of [2], on $\Gamma_0(N)$ with character χ. Then Theorem 3 follows immediately from [2, Thm. 3.1]. For more detail, again see [4].

BIBLIOGRAPHY

[1] P. Deligne, Valeurs de fonctions L et périods d'integrales, in Automorphic Forms, Representations and L-functions, AMS Proc. Symp. Pure Math 33 (1979), 313-346.

[2] M. Eichler and D. Zagier, The Theory of Jacobi Forms, Progress in Math 55, Birkhäuser, Boston, 1985.

[3] B.B. Gordon, Algebraically defined subspaces in the cohomology of a Kuga fiber variety, preprint.

[4] ___________, Intersections of higher weight cycles and modular forms, thesis, Univ. of Maryland, 1984.

[5] E. Hecke, Analytische Arithmetic der positiven quadratischen Formen, Mathematische Werke, Venderhoeck and Ruprecht, Gottingen, 1950, 789-918.

[6] F. Hirzebruch and D. Zagier, Intersection numbers of curves on Hilbert modular surfaces and modular forms of Nebentypus, Invent. Math 36(1976), 57-113.

[7] S.S. Kudla, Theta functions and Hilbert modular forms, Nagoya Math J. 69(1978), 97-106.

[8] __________, Intersection numbers for quotients of the complex 2-ball and Hilbert modular forms, Invent. Math 47(1978), 189-208.

[9] __________ and J. Millson, Geodesic cycles and the Weil representation I; quotients of hyperbolic space and Siegel modular forms, Comp. Math. 45(1982), 207-271.

[10] M. Kuga, Fibre varieties over a symmetric space whose fibres are abelian varieties, vols. I and II, Univ. of Chicago, 1964.

[11] ______, Fibre varieties over symmetric space whose fibres are abelian varieties, in Proc. of the U.S. - Japan Seminar in Differential Geometry, Kyoto, Japan, 1965, pp. 72-81, Nippon Hyoronsha, 1966.

[12] ______ and G. Shimura, On the zeta function of a fibre variety whose fibres are abelian varieties, Annals of Math. 82(1965), 478-539.

[13] T. Oda, On modular forms associated with indefinite quadratic forms of signature (2,n-2), Math. Ann. 231(1977), 97-144.

[14] S. Rallis and G. Schiffman, Automorphic forms constructed from the Weil representation: holomorphic case, Amer. J. of Math. 100(1978), 1049-1122.

[15] I. Satake, Clifford algebras and families of abelian varieties, Nagoya Math. J. 27(1966), 435-446. Corrections ibid 31(1968), 295-296.

[16] Y.-L.L. Tong, Weighted intersection numbers on Hilbert modular surfaces, Comp. Math. 38(1979), 299-310.

[17] D. Zagier, Modular forms whose Fourier coefficients involve zeta-functions of quadratic fields, in Modular Functions of One Variable VI, Bonn, Lecture Notes in Math. 627, Springer-Verlag, New York, 1976.

[18] ______, Modular forms associated to real quadratic fiels, Invent. Math. 30(1975), 1-46.

[19] ______, The Eichler-Selberg trace formula, appendix to S. Lang, Introduction to Modular Forms, Springer-Verlag, New York, 1976.

DEPARTMENT OF MATHEMATICS
PURDUE UNIVERSITY
WEST LAFAYETTE, IN 47906
U.S.A.

Current Address:
Department of Mathematics
University of Oklahoma
Norman, OK 73019
U.S.A.

Canadian Mathematical Society
Conference Proceedings
Volume 7 (1987)

Analogies Between Global Fields

(Expanded Version of Lectures Given at The Canadian Mathematical Society Summer Seminar in Number Theory)

David Goss*

Introduction

The past few years have seen progress in the arithmetic theory of function fields over finite fields in many areas. These include: The theory of modular curves and units (E.-U. Gekeler); abelian modules associated to "totally-real" (totally-split at a prime ∞) extensions, i.e., the analogue of Hilbert-Blumenthal abelian varieties (G. Anderson, U. Stuhler); transcendency results (J. Yu); etc. It is our purpose in this paper to review the progress made in the study of the abelian extensions of function fields that arise from division points of Drinfeld modules.

The theory of such extensions is now fairly well developed and most of the basic objects of the study seem to be in place, e.g., we now have a satisfactory construction of "cyclotomic" extensions in complete generality - due to D. Hayes - etc.. Moreover, we are now beginning to fully understand the role played by the special-values of the L-series of ([G.1]). As our purpose here is to provide a guide to these ideas, we will work in the simplest case of $A = \mathbb{F}_r[T]$ and $k = \mathbb{F}_r(T)$, $r = p^m$. Thus this paper should be viewed as a sequel to ([G.1]) and the reader can refer there for most of the background not covered here. In the appendix to Chapter 2 we will explain what is needed

*partially supported by N.S.F. Grant #MCS-8306739

for arbitrary k and A and we give a dictionary between number fields and function fields.

We have not provided detailed proofs of those results which appear elsewhere. However enough ideas are provided so that we may hope that the results seem reasonable to the reader.

In our first chapter we review the theory of zeta-functions of totally-real L/k, ([G.3]). These functions are defined in a completely classical manner by taking norms of ideals in the ring, $\mathcal{O}_L$, of A-integers of L and then exponentiating a monic generator. The resulting function, $\zeta_{\mathcal{O}_L}(s)$, is then shown to be "entire" on a certain character space S_∞. We focus on the special-values of $\zeta_{\mathcal{O}_L}(s)$ at the negative and positive integers.

Let $i \in \mathbb{N}$. We find that $\zeta_{\mathcal{O}_L}(-i) \in A$. But, in fact, much more is true. We find that $\zeta_{\mathcal{O}_L}(s)$ gives rise to a sequence, $\bar{\zeta}_{\mathcal{O}_L}(x,-i)$, of A-polynomials which have $\zeta_{\mathcal{O}_L}(-i)$ as their value at $x = 1$. The main point being that the reduction of $\bar{\zeta}_{\mathcal{O}_L}(x,-i)$ at a prime of A can be seen to be the reduction mod p of the complex L-series of the i-th power of a Teichmüller character. One is then able to deduce that $\zeta_{\mathcal{O}_L}(s)$ has certain trivial zeroes, etc. Indeed, this relationship is so strong that it allows us to deduce the Artin conjecture for non-abelian L-series on S_∞, and this is also beiefly reviewed in the chapter.

We turn next to the theory of $\zeta_{\mathcal{O}_L}(s)$ at positive i. Here we are unfortunately forced to assume that our totally-real L is also abelian over k (but see 1.11.C). Let ξ be the "period" of the Carlitz-module, C, and let $t = [L:k]$. We show that $(\zeta_{\mathcal{O}_L}(i)/\xi^{ti})^2 \in k$, when $i \equiv 0\ (r-1)$. This result, whose analogue in number fields is very well-known, becomes more intriguing when one realizes that for general A, one can only show that the square lies in some Hilbert Class-field.

Very little is known about the above values when L is arbitrary. Thus we finish the chapter by recalling what is known when $L = k$.

Our second chapter is concerned with cyclicity of the Galois components, $Cl_L(\wp^n)(\omega_\wp^i)$, of the p-primary class groups of the fields $L(\wp^n)$ obtained by adjoining the $\wp^n$-division points of C to L. Here $\wp \in Spec(A)$ and $\omega_\wp$ is its Teichmüller character. By Nakayama's lemma, we need only restrict ourselves to the first level $n = 1$. We then discuss two kinds of criterion: an <u>arithmetic</u> one involving $\{\bar{\zeta}_{\mathcal{O}_L}(x,-i)\}$ for $Cl_L(\wp)(\omega_\wp^{-i})$ and a <u>geometric</u> one involving $\{\zeta_{\mathcal{O}_L}(i), i \equiv 0(r-1)\}$ for $Cl_L(\wp)(\omega_\wp^{-i})$, $i \equiv 0(r-1)$.

The arithmetic criterion amounts to having the Frobenius possess a unique <u>1-unit</u> root on a certain Tate space; cyclicity then follows easily from the theory of modules over a local p.i.d. .

The geometric criterion (established only for $L = k$ at this time), involves the <u>cyclotomic units</u> associated to C. By using a version of the Kummer homomorphism, due to S. Okada, one can show that if $i \equiv 0(r-1)$ and $\wp \nmid \zeta_A(i)/\xi^i$, then $Cl_k(\wp)(\omega_\wp^i)$ is thrown off to infinity and is thus visibly cyclic. Remarkably, one can show this non-divisibility for all i of the form $p^j \cdot (r^h - 1)$. Thus we deduce cyclicity for <u>some</u> components and <u>all</u> primes $\wp$. Finally we show that the geometric criterion implies the arithmetic for $i \equiv 0(r-1)$.

All of the above results for function fields seem to have analogues of one form or another in algebraic number fields. Thus, as in ([G.1]), we make these comparisons whenever possible. We also offer some speculation on possible analogues of $\bar{\zeta}_{\mathcal{O}_L}(x,-i)$ in the theory of number fields. We hope these comparisons stimulate the reader.

Finally we wish to express our appreciation to the hosts of this seminar on number theory for their kind invitation to participate. We also thank W. Sinnott for much help in the writing of this paper.

§ 1. In this chapter we review the known results on the L-series involved in the cyclotomic theory of curves when $A = \mathbb{F}_r[T]$, $r=p^m$. We put $k = \mathbb{F}_r(T)$, ∞ = the prime at infinity on $\mathbb{P}^1$ and $K = k_\infty = \mathbb{F}_r((\frac{1}{T}))$. We let $\bar{k} \subseteq \bar{K}$ be fixed algebraic closures. An algebraic extension L/k is said to be <u>totally-real</u> if it splits totally at ∞. Such an extension is easily seen to be <u>seperable</u> over k. We set $\mathcal{O}_L$ = the ring of A-integers in L.

Let $\mathfrak{A} \subseteq \mathcal{O}_L$ be an ideal. We abuse notation and let $N(\mathfrak{A})$ be the <u>monic</u> generator of the ideal norm of $\mathfrak{A}$ and we put $t = t_L = [L:k]$. We also set $\delta = \delta_L$ = the monic generator of the discriminant ideal of L. Unless otherwise stated, L will be totally-real in all that follows.

If $\alpha \in A$, then we set $D(\alpha)$ to be its degree. We use the convention that lower case Roman letters refer to <u>monic</u> elements of A.

<u>1.1 Definition:</u> a) We set $S_\infty = \overrightarrow{K}^* \times \mathbb{Z}_p$ as an additive topological group.

b) We fix π to be uniformizer in $K \equiv \frac{1}{T} \pmod{\frac{1}{T^2}}$.

c) If $a \in A$ then we put $\langle a\rangle = \pi^{D(a)} \cdot a \in U_1$ = (group of 1-units at ∞).

d) If $a \in A$ and $s = (s_0, s_1) \in S_\infty$, then we put $a^s = s_0^{D(a)} \langle a\rangle^{s_1}$.

As U_1 is a topological p-group, the element $\langle a\rangle^{s_1}$ has an obvious meaning. <u>Note that for $i \in \mathbb{Z}$, $a^i = a^s$ for $s = (\pi^{-i}, i)$ in S_∞</u>. We let "i" also denote this element. One checks easily that a^s has all the usual properties.

<u>1.2 Definition:</u> We put $\zeta_{\mathcal{O}_L}(s) = \sum_{\mathfrak{A} \subseteq \mathcal{O}_L} N(\mathfrak{A})^{-s}$

$$= \prod_{\mathfrak{P}\ \text{prime}} (1 - N(\mathfrak{P})^{-s})^{-1}.$$

It is trivial to see that $\zeta_{\mathcal{O}_L}(s)$ will converge for all $s \in S_\infty$ with s_0 not in the integers of $\bar{K}$, (in particular it converges for $s = i \in \mathbb{N}^+$).

Let $J \subseteq \bar{K}$ be a finite extension of K. Let $h(x) = \sum_{n=0}^{\infty} a_n x^n \in J[[x]]$ be a power series. We set $\|h(x)\| = \sup\{|a_n|_\infty\}$; where $|\ |_\infty$ is the canonical extension to $\bar{K}$ of the normalized absolute value at ∞ on $\mathbb{P}^1$.

1.3 Definition: a) Let $f : S_\infty \to \bar{K}$ be a continuous function. We say f is entire if and only if

1) for fixed $s_1 \in \mathbb{Z}_p$, the function $g_{s_1}(s_0) = f(s_0, s_1)$ is given by an everywhere convergent power series in s_0^{-1};

2) the coefficients of all such $g_{s_1}(s_0)$ belong to a fixed finite extension of K;

3) let $\{\alpha_i\}$ be a collection of elements in $\mathbb{Z}_p$ converging to s_1. We then require that $\|g_{s_1}(s_0^{-1}) - g_{\alpha_i}(s_0^{-1})\|$ tend to 0.

b) We say an entire function is essentially algebraic if, for $i \in \mathbb{N}$ and $x \in \bar{K}^*$, the function

$$h_f(x,-i) = f(x\pi^i, -i)$$

is given by a polynomial in x^{-i} with coefficients in $\bar{k}$ that are integral over A. The coefficients of all such polynomials are further required to generate a finite extension of k.

As a consequence of the Riemann-Roch theorem and some estimates in multilinear algebra, one has ([G.3]):

1.4 Theorem: $\zeta_{\mathcal{O}_L}(s)$ can be extended to an essentially algebraic entire function.

In concrete terms, 1.4 implies that $\zeta_{\mathcal{O}_L}(s)$ gives rise to a 1-parameter family of everywhere convergent power-series. Moreover, it implies that for $i \in \mathbb{N}^+$

$$\bar{\zeta}_{\mathcal{O}_L}(x,-i) = h_{\zeta_{\mathcal{O}_L}}(x,-i) = \sum_{j=o}^{\infty} x^{-j}\Big(\sum_{\substack{\mathfrak{A} \subseteq \mathcal{O}_L \\ D(N(\mathfrak{A}))=j}} N(\mathfrak{A})^i\Big)$$

is a polynomial, i.e. the coefficient of x^{-j} is 0 for $j>>0$.

Such polynomials - which are really slight twists of $\zeta_{\mathcal{O}_L}(s_o,-i)$ - will be of central importance in the next chapter.

Now let F be the r-th power mapping in $\bar{k}$; $F(z) = z^r$. We put $F_{(o)} = F$ and for $i \in \mathbb{N}^+$, we put $F_{(i)} = r^i$-th power mapping. We let $k\{F\}$ be the ring of "polynomials" in the $F_{(i)}$. We set $C : A \to k\{F\}$ to be the Carlitz-module ([G.1],[G.2]) with $C(T) = T F_{(o)} + F_{(1)}$. For $I \subseteq A$ an ideal, we set $C[I] \subseteq \bar{k}$ to be the A-module of I-division points. We put $k(I) = k\{C[I]\}$; so $\mathrm{Gal}(k(I)/k) \simeq A/_I^*$ by the canonical Galois action on $C[I]$. We put $k(I)_+ =$ fixed field of $\mathbb{F}_r^* \subseteq A/_I^*$. One knows that $k(I)_+$ = maximal totally-real subfield of $k(I)$. In general, the ramification properties of these fields are what one would expect from the analogy to cyclotomic extensions of $\mathbb{Q}$.

We put $L(I) = L \cdot k(I)$, etc., and we set $G_L(I) = \mathrm{Gal}\,(L(I)/L)$ and $G_L(I)_+ = \mathrm{Gal}(L(I)_+/L)$.

From now on we assume $\mathfrak{P} \in \mathrm{Spec}(A)$ is prime to δ_L.

<u>1.5 Definition</u>: a) Let $W = W_{\wp}$ be the Witt ring of $A/\wp$; (W is thus a finite unramified extension of $\mathbb{Z}_p$) .

b) Let $\hat{\omega}_{\wp}$ be the canonical isomorphism of $G_L(\wp)$ with $A/\wp^*$ (as $(\wp, \delta_L) = 1$) .

c) We let $\omega_{\wp} : G_L(\wp) \to W^*$ be the composition of $\hat{\omega}_{\wp}$ and the canonical injection of $A/\wp^*$ into W^* .

Let $0 < i < r^{D(\wp)}-1$; so $\omega_{\wp}^i$ is a non-trivial character. Let $L(v,\omega_{\wp}^i)$ be the classical L-series of $\omega_{\wp}^i$ considered as an element of $W[v]$, $v = r^{-s}$ (in the usual sense). We let $\overline{L}(v,\omega_{\wp}^i)$ be the reduction of $L(v,\omega_{\wp}^i)$ modulo $pW(v)$. We consider $\overline{L}(v,\omega_{\wp}^i)$ as an element of $A/\wp(v)$.

For $i \in \mathbb{N}^+$ we also put

$$f_{\mathcal{O}_L}(x,-i) = \begin{cases} \overline{\zeta}_{\mathcal{O}_L}(x,-i) & \text{if } i \not\equiv 0(r-1) \\ \overline{\zeta}_{\mathcal{O}_L}(x,-i)/(1-x^{-1})^t & \text{if } i \equiv 0(r-1) . \end{cases}$$

We let $\overline{f}_{\mathcal{O}_L}(x,-i)$ be the element in $A/\wp(x^{-1})$ obtained by reducing mod in the obvious sense. The following basic result is then established directly from the definition of $\omega_{\wp}$:

<u>1.6 Theorem:</u> As elements of $A/\wp(x^{-1})$, we have

$$\overline{L}(x^{-1},\omega_{\wp}^i) = \overline{f}_{\mathcal{O}_L}(x,-i)$$

<u>1.7 Corollary:</u> $f_{\mathcal{O}_L}(x,-i)$ is a polynomial for all $i \in \mathbb{N}^+$.

Proof: Indeed, by 1.6, $f_{\mathcal{O}_L}(x,-i)$ is a polynomial modulo an infinite number of primes. The result is now clear since if $f_{\mathcal{O}_L}(x,-i)$ was not a polynomial, it would have poles (mod $\mathfrak{p}$) for almost all $\mathfrak{p}$.

Let $i \equiv 0 \ (r-1)$. Then 1.7 should be viewed as giving a "trivial zero" of order $\geq t$ at $-i \in S_\infty$. In fact, $\bar{\zeta}_{\mathcal{O}_L}(x,-i) = (1-x^{-1})^t \cdot f_{\mathcal{O}_L}(x,-i)$. Thus $\bar{\zeta}_{\mathcal{O}_L}(x,-i)$ has a zero of order at least t at 1. This immediately gives a zero of order at least t for $\zeta_{\mathcal{O}_L}(s)$ at $-i$. The analogous statement for number fields is, of course, an equality. In the appendix to this chapter we will establish equality for some abelian L/k.

1.8 Definition. Let $i \in \mathbb{N}^+$. We put $\beta_{\mathcal{O}_L}(i) = f_{\mathcal{O}_L}(1,-i) \in A$.

We refer the reader to ([G.1]),[G.3]) for the congruences satisfied by $\beta_{\mathcal{O}_L}(i)$.

1.9 Remarks: Let $R_p \subseteq \bar{\mathbb{Q}}_p$ be the ring of integers in some algebraic closure of $\mathbb{Q}_p$. Let $M_p \subseteq R_p$ be the maximal ideal and choose an injection of R_p/M_p into $\bar{k}$. Let G be the Galois group of the separable closure of L/L and let $\rho : G \to GL_n(\bar{\mathbb{Q}}_p)$ be a representation which factors thru a finite extension of L. Let $\mathfrak{B} \in \operatorname{Spec}(\mathcal{O}_L)$. One knows how to attach to $\mathfrak{B}$ a local Artin factor in the standard way. By reducing mod M_p and substituting $N(\mathfrak{B})^{-s}$, we can define an L-series $L(\rho,s)$ on S_∞ ([G.4]). One can show that such L-series can be continued to an essentially algebraic entire function on S_∞ (i.e., the Artin conjecture). The idea is to use Brauer induction, and the fact that 1.4 may be easily generalized to abelian L-series, to get a meromorphic continuation. One then uses Weil's proof of the Artin conjecture for complex L-series to show that $h_L(x,-i)$ is a polynomial modulo an infinite number of primes; thus is itself a polynomial. Strong continuity (i.e., 1.3.a.3) finishes the proof.

We turn now to the theory of $\zeta_{\mathcal{O}_L}(s)$ at positive integers. Here the assumption of total-reality seems to be essential in general. Further, at this point we will restrict ourselves to those L/k which are abelian; so that $\zeta_{\mathcal{O}_L}(s)$ is a product of abelian L-series.

Let ξ be the period of the module C (see appendix). Let $i \in \mathbb{N}^+$ with $i \equiv 0(r-1)$. We then have the following analogue of the classical result of Siegel on the special values at positive integers of Dedekind zeta-functions of totally-real number fields (see also 1.11.C).

1.10 Theorem: $\xi^{-ti}\zeta_{\mathcal{O}_L}(i)$ is a totally-real algebraic element whose square lies in k.

Proof: (Sketch) The Dedekind Determinant Formula immediately implies that $\xi^{-ti}\zeta_{\mathcal{O}_L}(i)$ is a k-multiple of the determinant of a matrix, M, of the following form: Let (f) be the conductor of L and let $(f^{-1}/A)^*$ be those classes in k/A of the form $\alpha/_f+A$, $(\alpha,f)=1$. Let $x \in (f^{-1}/A)^*$ and set $g_x(i) = (\sum_{\beta \in x} \beta^{-i})\xi^{-i}$. Then the elements of M are certain sums of the $g_x(i)$ which depend on $Gal(L/k)$. By using the logarithmic derivative of the exponential of the Carlitz-module, one sees that $g_x(i) \in k((f))$. Further $Gal(k((f))/k)$ permutes the rows of M; so the square of the determinant is fixed. As ξ^{-i} and $\zeta_{\mathcal{O}_L}(i)$ are in K, the result follows.

1.11 Remarks: a) It would be interesting to understand more precisely the nature of $\xi^{-ti}\zeta_{\mathcal{O}_L}(i)$; for instance, under what curcumstances does it actually lie in k?

b) For each $i \in \mathbb{N}^+$, the elements $g_x(i)/\xi^i$ form a distribution on $\lim\limits_{\to} (f^{-1}/A)^*$. For example, $g_x(1)/\xi$ can be seen to be the reciprocal of a certain (f)-division point for C. Moreover the cyclotomic units ([G.1]) are just the quotients of these special-values. We thus obtain a version of the classical $-\log|1-\zeta|$ distribution.

c) As an example, let $r = 2$ and let L/k be a quadratic extension ramified at one prime f. As $2 = [L:k] = p$, we can see easily that $\zeta_{\mathcal{O}_L}(s)=(1-f^{-s})\zeta_{\mathcal{O}_L}(s)^2$. Thus $\xi^{-2i}\zeta_{\mathcal{O}_L}(i)=(1-f^{-i})(\zeta_A(i)/\xi^i)^2=\frac{f^i-1}{f^i}(\zeta_A(i)/\xi^i)^2$. One can thus see that the discriminant of L/k enters the description of these values.

The phenomenon of $\zeta_{\mathcal{O}_L}(s)$ factoring essentially into a p-power of $\zeta_{\mathcal{O}_L}(s)$ has nothing to do with L being totally-real. In fact, by using non-abelian L-series (for instance), one can do the same thing for any L/k which is Galois and of p-power degree. For such L we thus obtain a version of 1.10!

Finally we put $L=k$ where we can say much more.

1.12 Definition: a) We set $[0] = 1$, and for $i \in \mathbb{N}^+$, $[i] = T^{r^i} - T$.

b) We set $D_0 = 1$, and for $i \in \mathbb{N}^+$, we put $D_i = [i][i-1]^r \cdots [1]^{r^{i-1}}$.

c) Let $i \in \mathbb{N}^+$ be written r-adically as $\sum_{e=o}^{v} \alpha_e r^e$, $0 \leq \alpha_e < r$. We put $\Gamma_i = \prod D_e^{\alpha_e}$.

d) We put $L_0 = 1$ and, for $i \in \mathbb{N}^+$, we put $L_i = [i][i-1]\cdots[1]$.

1.13 Definition: Let $i \in \mathbb{N}^+$. We put $B_i = 0$ if $i \not\equiv 0(r-1)$ and $B_i = \xi^{-i}\zeta_A(i)\Gamma_i$ if $i \equiv 0(r-1)$.

Thus, if $e(z)$ is the exponential of C, we find

$$z/e(z) = \sum_{j=0}^{\infty} (B_j/\Gamma_j)z^j .$$

Therefore, $\{B_i\} \subseteq k$.

Carlitz ([G.2]) has established that such B_i satisfy a von-Staudt-Clausen type result. We will only need the following immediate corollary:

<u>1.14 Proposition</u>: Let $\wp \in \mathrm{Spec}(A)$ and let $0 < i < r^{D(\wp)} - 1$, $i \equiv 0(r-1)$. Then B_i is $\wp$-integral.

<u>1.15 Proposition</u>: Let $i \in (r-1)\mathbb{N}^+$ and assume $pi < r^{D(\wp)} - 1$. Then $\wp \mid B_i$ if and only if $\wp \mid B_{pi}$.

<u>Proof</u>: One has $\zeta_A(pi) = \zeta_A(i)^p$. Further, Γ_i , Γ_{pi} are $\wp$-units. The result is now trivial.

The next result, also due to Carlitz, will be important in the next chapter. It follows upon comparing $e(z)$ with its logarithm.

<u>1.16 Proposition</u>: Let $i = r^h - 1$, $h > 0$. Then $\zeta_A(i)/\xi^i = (-1)^h/L_{h-1}$.

<u>1.17 Corollary</u>: Let $\wp \in \mathrm{Spec}(A)$ and let $i = p^j(r^h-1)$. Assume $i < r^{D(\wp)} - 1$. Then $\wp \nmid B_i$.

<u>1.18 Remarks</u>: In ([T.1]), E. Thomas presented a formula for $D(\beta_A(i))$ when $r = p$. It turns out that the obvious r-adic version of this result does <u>not</u> hold in general. Indeed, if $r = 8$ and $i = 37$, one can check that $\beta_A(i) = 1$ whereas Thomas' formula would imply that $D(\beta_A(i)) = 16$. However, if i is of the form $i = i(c,h) = cr^h + (r^h - 1)$, $0 < c < r$, then Thomas' result <u>does</u> hold. Thus, for such i one still has $D(\beta_A(i)) = D(\Gamma_i)$. Moreover, if we let $\overline{\beta_A(i)} = \beta_A(i)/\zeta$, where ζ is the coefficient of highest degree in $\beta_A(i)$, then one checks experimentally that $D(\beta_A(i)) - D(\Gamma_i - \overline{\beta_A(i)}) = cr^h$; a fact which certainly deserves explanation. Note that if $i(c,h) \equiv 0\ (r-1)$, then $i(c,h) = r^{h+1} - 1$ and so $\wp \nmid B_i$ by 1.17! All this suggests that much more may eventually be said.

Appendix to Chapter One

We establish here that the trivial zeros of $\zeta_{\mathcal{O}_L}(s)$ are of order $= [L:k]$ in certain limited instances. Let $\mathfrak{P} \in \operatorname{Spec}(A)$ and let $L = k(\mathfrak{P})_+$. We assume that is regular, i.e., $p \nmid h(\mathfrak{P})$, where $h(\mathfrak{P})$ is the class number of $k(\mathfrak{P})$. One knows $p \mid h(\mathfrak{P})$ iff $\mathfrak{P} \mid \prod_{j=1}^{r^{D(\mathfrak{P})}-2} \beta_A(j)$. Also, one can see that $\beta_A(j) \neq 0$ for $j > 0$.

<u>1.19Proposition</u>: Let be a regular prime with $p \nmid D(\mathfrak{P})$. If $0 \neq i \equiv 0(r-1)$, then $\zeta_{\mathcal{O}_L}(s)$ has a zero of order $[L:k]$ at $s = -i$. If $i \equiv 0(r-1)$, then $\zeta_{\mathcal{O}_L}(s)$ is non-zero at $s = -i$.

Proof: Let $t = [L:k]$. As is standard we have $\zeta_{\mathcal{O}_L}(s) = \prod_{j=0}^{t-1} L(\omega_{\mathfrak{P}}^{j(r-1)}, s)$; so $\overline{\zeta}_{\mathcal{O}_L}(x,-i) = \prod_{j=0}^{t-1} \overline{L}(\omega_{\mathfrak{P}}^{j(r-1)},(x,-i))$ in the obvious fashion. We need only examine the L-series for $\omega_{\mathfrak{P}}^{j(r-1)}$, $j \neq 0$.

We begin by assuming $i \not\equiv 0(r-1)$. Then $(\bmod \mathfrak{P})$ we find $\overline{L}(\omega_{\mathfrak{P}}^{j(r-1)}, (1,-i)) = \beta_A(i + j(r-1))$. Let v be the residue class of $i+j(r-1) \pmod{r^{D(\mathfrak{P})}-1}$; so $v < r^{D(\mathfrak{P})}-1$. One has $v \not\equiv 0(r-1)$ and $\beta_A(v) \equiv \beta_A(i + j(r-1)) \not\equiv 0(\bmod \mathfrak{P})$. The result for such i follows.

Now let $i \equiv 0(r-1)$. As in 1.7, we find that $\overline{L}(\omega_{\mathfrak{P}}^{j(r-1)}, (x,-1))$ has a zero at $x=1$. Consequently, we set

$$\overline{L}'(\omega_{\wp}^{j(r-1)}, (x,-i)) = \frac{d}{dx}\overline{L}(\omega_{\wp}^{j(r-1)}, (x,-i))$$

$$= \sum_{j=0}^{\infty} - jx^{-j-1} \left(\sum_{D(n)=j} \omega_{\wp}^{j(r-1)}(n)n^{i}\right) .$$

Thus, we have the congruence modulo $\wp$,

$$\overline{L}'(\omega_{\wp}^{j(r-1)}, (1,-i)) \equiv \sum_{j=0}^{\infty} - j \left(\sum_{\substack{D(n)=j \\ (n,\wp)=1}} n^{i+j(r-1)}\right) .$$

Let v again be the residue of $i + j(r-1) \pmod{r^{D(\wp)}-1}$. If $v \neq 0$, we find as before that $\overline{L}'(\omega^{j(r-1}, (1,-i)) \not\equiv 0 \pmod{\wp}$. If $v = 0$ we find

$$\overline{L}'(\omega_{\wp}^{j(r-1)}, (1,-i)) \equiv \sum - j \left(\sum_{\substack{D(n)=j \\ (n,\wp)=1}} n^{0}\right) \pmod{\wp}$$

$$= \frac{d}{dx}(1 - x^{-D(\wp)})|_{1} \pmod{\wp}$$

$$\neq 0 \qquad\qquad \pmod{\wp},$$

as $p \nmid D(\wp)$. The result is now proved.

<u>1.20 Remarks</u>: a) Under the same hypothesis one can extend the result to $k(\wp^{n})_{+}$. One can also generalize as in 1.11.c.

b) An example of such a $\wp$ is $\wp = (T^{4} + 2T + 2)$, $r = 3$.

§ 2. It is our purpose to discuss in this chapter what is known about the class groups of the fields $L(\mathcal{P})$, $L(\mathcal{P})_+$, and to compare with analogous (and potentially analogous) results on number fields.

We remind the reader that there are two types of "cyclotomic" extensions for a totally-real L; the <u>arithmetic</u> ones coming from extensions of constant fields, and the <u>geometric</u> ones L(I); (they contain <u>no</u> constant field extensions). It is the <u>interplay</u> between them that will be important here. For number fields, both types collapse into the usual cyclotomic extensions.

<u>2.1 Definition:</u> a) We set $\mathcal{O}_L(\mathcal{P}^n)$ (resp. $\mathcal{O}_L(\mathcal{P}^n)_+$) to be the ring of A-integers in $L(\mathcal{P}^n)$ (resp. $L(\mathcal{P}^n)_+$).

b) We set $Cl_L(\mathcal{P}^n)$ (resp. $Cl_L(\mathcal{P}^n)_+$) to be the <u>p-primary</u> class group of $L(\mathcal{P}^n)$ (resp. $L(\mathcal{P}^n)_+$).

c) We set $\widetilde{Cl}_L(\mathcal{P}^n)$ (resp. $\widetilde{Cl}_L(\mathcal{P}^n)_+$) to be the p-primary class group of $\mathcal{O}_L(\mathcal{P}^n)$ (resp. $\mathcal{O}_L(\mathcal{P}^n)_+$).

d) We set $r_L(\mathcal{P}^n)$ (resp. $r_L(\mathcal{P}^n)_+$) to be the p-primary component of the finite group consisting of divisors at infinity on $L(\mathcal{P}^n)$ (resp. $L(\mathcal{P}^n)$) modulo the divisors of units in $\mathcal{O}_L(\mathcal{P}^n)$ (resp. $\mathcal{O}_L(\mathcal{P}^n)_+$).

From now on we assume that $(\delta_L, \mathcal{P}) = 1$; so $L(\mathcal{P}^n)/L$ is totally ramified at the primes above $\mathcal{P}$. We let P_L denote the projection of $Cl_L(\mathcal{P}^n)$ to $\widetilde{Cl}_L(\mathcal{P}^n)$, etc. As degree $(\infty) = 1$ on $\mathbb{P}^1$, and L is totally-real, we see that these maps are <u>surjective</u>. The kernel is $r_L(\mathcal{P}^n)$ (resp. $r_L(\mathcal{P}^n)_+$).

The following result may be established as in ([G-R.1], [G.1])

<u>2.2 Proposition</u>: a) $\mathcal{O}_L(\mathcal{P}^n)_+^* = \mathcal{O}_L(\mathcal{P}^n)^*$

b) $r_L(\mathcal{P}^n)_+ \simeq r_L(\mathcal{P}^n)$.

Let M be a $\mathbb{Z}_p[G_L(\wp)]$ - module. Upon tensoring with $W_\wp$ we can form the isotypic components $M(\omega_\wp^i)$. The next result follows immediately from 2.2 and our ramification assumptions.

2.3 Proposition: a) Let $0 < i < r^{D(\wp)}-1$, $i \equiv 0(r-1)$. Then, via the above maps, $Cl(\wp^n)(\omega_\wp^i) \simeq \widetilde{Cl(\wp^n)}(\omega_\wp^i)$.

b) The norm map $Cl_L(\wp^n) \to Cl_L(\wp^n)_+$ (resp. $\widetilde{Cl_L(\wp^n)} \to \widetilde{Cl_L(\wp^n)_+}$) is surjective and gives rise to an isomorphism of components associated to $i \not\equiv 0(r-1)$.

We call $Cl_L(\wp^n)(\omega_\wp^i)$, etc., $i \equiv 0(r-1)$, an "even component". Otherwise, we say it is an "odd component". We use the same language for "i".

We shall be interested in the cyclicity of the components $Cl_L(\wp)(\omega_\wp^i)$ as $W_\wp$-module. Nakayama's lemma then allows us to lift cyclicity to level $\wp^n$ as $W[G_L(\wp^n)]$-module.

Our first result is the following refinement of 1.6. Let $\wp \in \mathrm{Spec}(A)$ and $0 < i < r^{D(\wp)}-1$, as before, and let $L(v,\omega_\wp^i)$ be written as $\prod_j(1-\alpha_j v)$ in $\overline{W}[v]$, where $\overline{W}$ is the integral closure of W in some algebraically closed field. We set $L_u(v,\omega_\wp^i) = \prod_{\alpha_j \text{ a unit}}(1-\alpha_j v)$ = the "unit-root" piece of $L(v,\omega_\wp^i)$. One now has trivially:

2.4 Proposition: As elements of $A/\wp^{-1}[x]$ we have

$$\overline{L}_u(x^{-1},\omega_\wp^i) = \overline{L}(x^{-1},\omega_\wp^i) = \overline{f}_{\mathcal{O}_L}(x,-i) .$$

Let $T_L(\wp)$ be the p-adic Tate module of $L(\wp)$, etc. One knows (say from crystalline theory) that $L_u(v,\omega_\wp^i)$ equals $\det(1 - vF|T_L(\wp)(\omega_\wp^{-i}))$; where F is the r-th power mapping. Furthermore, one may see without difficulty that

$$T_L(\wp)(\omega_\wp^{-i})/((F-1)T_L(\wp)(\omega_\wp^{-i})) = Cl_L(\wp)(\omega_\wp^{i}) ,$$

Thus, from 2.4 we have

<u>2.5 Theorem</u>: 1) The rank of $T_L(\wp)(\omega_\wp^{-i})$ <u>stabilizes</u> as $\wp$-varies.

2) $\mathrm{ord}_p(L_u(1,\omega_\wp^{i})) = \mathrm{ord}_p(L(1,\omega_\wp^{i}))$ = length of $Cl_L(\wp)(\omega_\wp^{-i})$ as W-module.

3) $Cl_L(\wp)(\omega_\wp^{-i}) \neq \{0\}$ if and only if $\wp \mid \beta_{\mathcal{O}_L}(i)$, $0 < i < r^{D(\wp)}-1$.

2.5.3 is the analogue of the result of Herbrand-Ribet for $\mathbb{Q}(\zeta_p)$, except that we can handle <u>all</u> components (not just odd ones). An analogue of 2.5.1 for number fields is that the component of the class-group associated to a <u>given</u> Bernoulli number eventually dies. Finally 2.5.2 has an analogue in the theory of number fields which arises as a corollary to Iwasawa's Main Conjecture - established for abelian fields over $\mathbb{Q}$ by Mazur-Wiles.

We refer the reader to ([K.1]) which uses 2.5.3 to show that, for $r > 2$, there are infinitely many $\wp$ with $Cl_k(\wp) \neq \{0\}$. One can also use the computations given at the end of Chapter 1 to show this.

We can now present, as in ([G.-S.1]), a <u>sufficient</u> condition for cyclicity of $Cl(\wp)(\omega_\wp^{-i})$, etc., as W-module. As before, let $L(v,\omega_\wp^{i}) = \prod(1-\alpha_j v)$ and let $M \subseteq \overline{W}$ be the maximal ideal.

<u>2.7 Proposition</u>: Assume that at most one α_j is $\equiv 1 \pmod M$. Then $Cl_L(\wp)(\omega_\wp^{-i})$ is cyclic as a W-module.

<u>Proof</u>: The hypothesis of the proposition means that (F-1) has at most one <u>non-unit</u> eigenvalue on $T_L(\wp)(\omega_\wp^{-i})$. As

$$T_L(\wp)(\omega_\wp^{-i})/((F-1)T_L(\wp)(\omega_\wp^{-i})) \simeq Cl_L(\wp)(\omega_\wp^{-i}) ,$$

a simple argument in the theory of finitely generated modules over a local p.i.d. gives the result.

Let $A_{\wp}$ be the completion of A at $\wp$ and let $k_{\wp}$ be its quotient field. Let $\overline{k}_{\wp}$ be an algebraic closure of $k_{\wp}$ and let $\overline{A}_{\wp} \subseteq \overline{k}_{\wp}$ be the integers. Let $M_{\wp} \subseteq \overline{A}_{\wp}$ be the maximal ideal. Factor $f_{\mathcal{O}_L}(x,-i)$ in $\overline{A}_{\wp}$ as $\prod(1-\beta_1 x^{-1})$. Then 2.4, gives the following reformulation of 2.7.

<u>2.8 Proposition</u>: Assume that there is at most one β_1 with $\beta_1 \equiv 1 \pmod{M_{\wp}}$. Then $Cl(\wp)(\omega_{\wp}^{-i})$ is cyclic as W-module.

Note that 2.8 holds for i if and only if it holds for pi. We call 2.8 the <u>arithmetic criterion</u> for cyclicity. We refer the reader to ([G.-S.1]) for a generalization when $L \neq k$.

<u>2.9 Remark:</u> a) Let $\overline{\mathbb{F}}_r \subseteq \overline{k}$ be the algebraic closure of $\mathbb{F}_r$ and let $\tilde{\mathbb{F}}_r \subseteq \overline{\mathbb{F}}_r$ be the unique $\mathbb{Z}_p$-extension of $\mathbb{F}_r$. Let $\tilde{k}(\wp) = k(\wp).\tilde{\mathbb{F}}_r$; so $\tilde{k}(\wp)/k(\wp)$ is a $\mathbb{Z}_p$-extension. Let $\tilde{J}_L(\wp)$ be the class-group of $\tilde{k}(\wp)$ and let $\tilde{T}_L(\wp)$ be the p-adic Tate module of $\tilde{k}(\wp) = \mathrm{Hom}(\mathbb{Q}_p/\mathbb{Z}_p, \tilde{J}_L(\wp))$. Under the assumption of 2.8 one finds readily that $\tilde{T}_L(\wp)(\omega_{\wp}^{-i})$ is a <u>rank-one</u> W-module.

b) An analogue of a) seems to exist in the theory of number fields: Let $L_p(\omega^i,s)$ be the classical p-adic L-series associated to the i-th power of the Teichmüller character ω, $i \equiv 0(2)$. Let $f(\omega^i,T) \in \mathbb{Z}_p[[T]]$ be the power series so that

$$L_p(\omega^i,s) = f(\omega^i, (1+p)^s-1) ,$$

and let $P(\omega^i,T)$ be its "Iwasawa polynomial" that arises from Weierstrass factorization.

Let $C = \varprojlim_{\text{norm}} CL(p^i)$; where $Cl(p^i)$ is the p-primary class-group of (ζ_p^i). Let $\alpha = \sigma_{1+p}-1$, where σ_{1+p} is the Galois element associated to $1+p$; so $\zeta^\alpha = \zeta^p$ for ζ a <u>p-power</u> root of one. By Mazur-Wiles, one knows that $P(\omega^i,T)$ is the characteristic polynomial of α on $C(\omega^{1-i})\otimes\mathbb{Q}_p$. All roots of $P(\omega^i,T)$ are non-units. Thus the analogue of 2.7 and a) would force $P(\omega^i,T)$ to have at most degree 1. All $P(\omega^i,T)$ that have been computed have this property.

<u>2.10 Example</u>: Let $r=3$ and $L=k$. We compute that $\bar{\zeta}_A(x,-5) = 1-(T^3-T)x^{-1}$; so $\beta_A(5) = 1 + T - T^3$. Let $\wp = (T^3 - T + 1)$; one sees trivially that $\wp$ is irreducible. The reduction of $\zeta_A(x,-5) \pmod{\wp}$ is linear. So by 2.4, 2.5 we find that $Cl_k(\wp)(\omega_\wp^{-5}) \neq \{0\}$ and that $T_k(\wp)(\omega_\wp^{-5})$ has rank one. Thus $Cl(\wp)(\omega_\wp^{-5})$ is a non-trivial cyclic W-module.

<u>2.11 Definition</u>: A) We let $J_L(\wp)[p]$ be the group of all p-division points on the Jacobian of $L(\wp)$. We let $J_L(\wp)[p](\omega_\wp^{-i})$ be defined by tensoring with W.

b) We put $J_L^i = \prod_{(\wp,\delta_L)=1} J_L(\wp)[p](\omega_\wp^{-i})$.

J_L^i is an A-module in the obvious fashion and it is <u>non-torsion</u>. Further as the eigenvalues of F on $T_L(\wp)$ are units, F acts invertibly on J_L^i. Thus, J_L^i is an $A[x^{-1}]$-module where $x \mapsto F$. By our congruences we find that $f_{\Theta_L}(x,-i)$ annihilates J_L^i.

<u>2.12 Question</u>: Is $(f_{\Theta_L}(x,-i))$ the annihilator of J_L^i ?

For instance, if the roots of $f_{\Theta_L}(x,-i)$ are <u>simple</u> the answer would be yes- this seems reasonable to expect when $L = k$. In fact, one can deduce this for i odd, if one can show the validity of the hypothesis of 2.7 for all odd i and all $\wp$, (see [G.-S.1]). The eigenvalues of F^{-1} on $J_L^i \otimes k$ are the roots of $f_{\Theta_L}(x,-i)$. We remind the reader that the zeroes of the Riemann zeta-function are expected to be simple; all known zeroes are simple.

2.13 Remark: The global elements $f_{\mathcal{O}_L}(x,-i)$ have their arithmetic interpretation based on the fact that, mod $\mathfrak{P}$, they are an L-series. If one tries to find an analogous phenomenon for number fields, one is lead to the following question (where we leave the appropriate modifications at $p = 2$ to the reader): Let $i \equiv 0(2)$. Does there exist a power series $P_i \in \mathbb{Q}[[T]]$ so that for all primes p with $i < p-1$, $P_i(T) \equiv f(\omega^i,T)(\text{mod } p\mathbb{Z}_p[[T]])$? Such a series would be unique and would, in fact, have $\zeta(1-i)$ as its constant term.

More generally, one can relax the hypothesis and vary the 1-unit, α_p, used at p to obtain $f(\omega^i,T)$ - one can then ask the same question in this more general context. The mapping $p \mapsto \alpha_p$ would give an isomorphism of $\hat{\mathbb{Z}}$ with $\prod_p U_p^1$; where U_p^1 is the group of 1-units at p. Thus we would obtain a sort of global "Frobenius" for number fields. The P_i would be analogues of $\overline{\zeta}_A(x,-i)$, $i \not\equiv 0(r-1)$.

Weil's proof of the Artin Conjecture for complex L-series of curves has its origin in his expression of L-series in terms of homology. The analogue of this in number fields is Iwasawa's Main Conjecture which implies the p-adic Artin Conjecture (see [Gr.1]). In 1.9 we used Weil locally at each prime $\mathfrak{P}$. Presumably the $\{P_i\}$ - if they exist - would allow us to attempt to copy 1.9 for global Artin L-series of number fields.

As a bit of evidence that some such thing may happen for number fields, we have the following: Let

$$g_i(x) = \sum_{a=1}^{p-1} \omega^i(a)x^a/(1-x^p) .$$

The function $g_i(x)$ is the essential part of the additive measure associated to $L_p(\omega^i,s)$. One has

$$g_i(x) \equiv \left(x\frac{d}{dx}\right)^i \left(\frac{1}{1-x}\right) \pmod p.$$

Thus mod p, $g_i(x)$ is global! (Added in proof: W. Sinnott and the author have found an object like P_i but where one uses additive measures.)

In order to go further, we must now set $L = k$. However, it seems reasonable that these results should generalize at least to abelian L/k .

We begin by recalling the theory of cyclotmic units for $k(\mathfrak{P})$: Let α_1, α_2 non-zero elements in $C[\mathfrak{P}]$. It is easy to see that $\alpha_1/\alpha_2 \in \mathcal{O}_k(\mathfrak{P})_+^*$. We let $\xi(\mathfrak{P})$ be the group of units generated by α_1/α_2 . Rosen and Galovich have established that $\mathcal{O}_k(\mathfrak{P})_+^*/\xi(\mathfrak{P})$ is a finite group of order equal to the class number of the ring $\mathcal{O}_k(\mathfrak{P})_+$. Let $H(\mathfrak{P})$ be the p-primary subgroup of $\mathcal{O}_k(\mathfrak{P})_+^*/\xi(\mathfrak{P})$.

Let $O(\mathfrak{P})$ be the p-primary subgroup of the group of divisors of degree 0 supported on the infinite primes of $k(\mathfrak{P})$ modulo the group of divisors of cyclotomic units. One knows that $O(\mathfrak{P})$ is finite and has the same cardinality as $Cl_k(\mathfrak{P})_+$. The same result is true on Galois components if we use length as W-module.

The next lemma is easy.

2.14 Lemma: The following sequence is exact,

$$0 \to H(\mathfrak{P}) \to Q(\mathfrak{P}) \to r_k(\mathfrak{P})_+ \to 0 .$$

So from the discussion before 2.2, we have the following result. It is and analogue of the Gras-Conjecture for cyclotomic fields - which has been established as a corollary of Iwasawa's Main Conjecture.

2.15 Proposition: Let $0 < i < r^{(D(\mathfrak{P})}-1$, $i \equiv 0(r-1)$. Then $\widetilde{Cl_k(\mathfrak{P})_+}(\omega_{\mathfrak{P}}^i)$ and $H(\mathfrak{P})(\omega_{\mathfrak{P}}^i)$ have the same W-length.

One checks that $H(\mathfrak{P})(\omega_{\mathfrak{P}}^i)$ has W rank at most 1 .

The extension $k(\wp)/k$ is totally ramified above $\wp$. Let $k_\wp$ be the completion of k at $\wp$ and let $k^1_\wp = k_\wp \cdot k(\wp)$. Let $0 \neq \lambda \in C[\wp]$. One knows that (λ) is the maximal ideal of the integers, $R^1_\wp$, of $k^1_\wp$.

<u>2.16 Definition</u>: Let $\alpha \in R^1_\wp$ be a unit and let $\alpha = f(\lambda)$, where $f(u) \in A[[u]]$ (this is possible because of the above remark). By ([G.2]) we can write

$$\frac{f'(e(z))}{f(e(z))} = \sum_{j=0}^{\infty} a_j z^j/\Gamma_j \ ;$$

where a_j is $\wp$-integral. We set $\phi_j(\alpha) = a_j + \wp$ for $j = 1,\ldots,r^{D(\wp)}-2$.

One checks that ϕ_j is a well-defined homomorphism from $(R^1_\wp)^*$ to $A/\wp$.

Let $0 < i < r^{D(\wp)}-1$, i even. As $H(\wp)(\omega^i_\wp)$ has rank at most one, we see that $H(\wp)(\omega^i_\wp) \neq \{0\}$ implies that λ^{β_i} is a <u>p-th power</u>; where

$$\beta_i = \sum_{\sigma \in G_k(\wp)_+} \omega^{-i}(\sigma)\sigma = \sum_{\sigma \in G_k(\wp)_+} \omega^{-i}(\sigma)(\sigma-1) \ .$$

Let $\sigma \in G_k(\wp)$ be associated to $\alpha + (\wp)$, $(\alpha,\wp) = 1$, under the isomorphism $G_k(\wp) \tilde{\to} A/\wp^*$. We have

$$\sigma(\lambda) = \lambda^\sigma = C(\alpha)(\lambda).$$

So for $\sigma \in G_k(\wp)_+ \simeq (A/\wp^*)/\mathbb{F}_r^*$, we have

$$\lambda^{\sigma-1} = \frac{C(a)(\lambda)}{\lambda}, \text{ for some } a \in A, \ D(a) < D(\wp) \ .$$

One computes that

$$\phi_i\,(\lambda^{\sigma-1}) = (a^i-1)\,\frac{B_i\Gamma_{i-1}}{\Gamma_i} + (\wp)\ ,$$

or

$$\phi_i(\lambda^{\beta_i}) = \sum_a (\alpha^i-1)\omega_{\wp}^{-i}(a)\,\frac{B_i\Gamma_{i-1}}{\Gamma_i} + (\wp)\ ,$$

$$= \frac{B_i\Gamma_{i-1}}{\Gamma_i} + (\wp)\ .$$

Note that Γ_i , $i < r^{D(\wp)}-1$, is easily seen to be a $\wp$-unit. Thus we have the following result of Okada ([O.1]):

2.17 Theorem: Let $0 < i < r^{D(\wp)}-1$, $i \equiv 0(r-1)$. Then $H(\wp)(\omega_{\wp}^i) \neq \{0\}$ implies $\wp \mid B_i$.

By 2.15 we then have

2.18 Theorem: $\widetilde{Cl_k(\wp)_+}(\omega_{\wp}^i) \neq \{0\}$ implies $\wp \mid B_i$.

We also recall that $\widetilde{Cl_k(\wp)}(\omega_{\wp}^{-i}) \neq \{0\}$ implies $\wp \mid \beta_A(r^{D(\wp)}-1-i)$. The reader should note the fact that "$s \to -s$" here whereas for number fields "$s \mapsto 1-s$" .

We now have the geometric criterion for cyclicity (for even components) as W-module:

2.19 Theorem: Let $0 < i < r^{D(\wp)}-1$, $i \equiv 0(r-1)$. Then $\wp \nmid B_i$ implies that $Cl_k(\wp)_+(\omega_{\wp}^i)$ is a cyclic W-module.

Proof: Indeed, by 2.18 we see that $Cl_k(\wp)_+(\omega_{\wp}^i)$ must then be $\simeq r_k(\wp)_+(\omega_{\wp}^i)$. Thus the result is easy to see.

By 1.17 we find,

2.20 Corollary: Let $0 \neq i = p^j(r^h-1)$ for some $j,h \in \mathbb{N}$. Assume $i < r^{D(\wp)}-1$. Then $Cl_k(\wp)_+(\omega_\wp^i)$ is cyclic.

2.21 Remarks: a) One has examples where $\wp | B_i$ but $Cl(\wp)_+(\omega_\wp^i) = \{0\}$ $(\wp \nmid \beta(r^{D(\wp)}-1-i))$.

b) As an example of 2.20, let $r = 3$ and $\wp = (T^3 - T + 1)$. We find $\wp | \beta(8)$. Thus $Cl_k(\wp)_+(\omega_\wp^{18}) \neq \{0\}$. As $18 = 3^2 \cdot (3-1)$, we find $\wp \nmid B_{18}$. So $Cl_k(\wp)_+(\omega_\wp^{18})$ is a non-trivial cyclic W-module.

c) In E. Artin's Thesis one finds a result that implies that $Cl_k(\wp)_+(\omega_\wp^{40}) \neq \{0\}$ for $\wp = (T^4 + T^2 - 1)$, $r = 3$. This component corresponds to a quadratic extension of k and is $\simeq W/(3)$ - in fact $Cl_k(\wp) \simeq \mathbb{Z}/(3)$. One has $\wp | \beta(40)$ and $\wp | B_{40}$. One is thus lead to an important question: If $\wp | B_i$ and $\wp | \beta(r^{D(\wp)}-1-i)$, is $Cl_k(\wp)_+(\omega_\wp^i) \neq \{0\}$? (See [I.-S.1] for some computations).

Ken Ireland has recently informed me that $r_k(\wp)_+(\omega^i)$ and $Cl_k(\wp)_+(\omega_\wp^i)$ are <u>both</u> non-zero for $\wp = (T^8 + 2T^5 + T^4 + T^3 + T^2 + 2)$, $r = 3$ and $i = (3^8 - 1)/2$. As above, these groups come from a quadratic extension of k and both are isomorphic to $W/(3)$. Their structure up to the $\wp^n$-tower is thus an interesting (and difficult) problem as the cyclicity lifts up the tower as Galois-module.

d) One analogue of 2.20 for number fields is the Kummer-Vandiver conjecture which does <u>not</u>, of course, involve divisibility of Bernoulli numbers. The following observation of Iwasawa, however, does furnish an analogue which does. I thank W. Sinnott for pointing it out to me. Let $4 \leq i \leq p-3$, $i \equiv 0(2)$ and $p \geq 5$. Let, $L_p(\omega^i,s)$, C, etc., be as before. We have

$$L_p(\omega^i,1-i) = -(1-p^{i-1})\frac{B_i}{i} = f(\omega^i,(1+p)^{1-i}-1) .$$

Let $f(\omega^i,T) = a_0 + a_1T + \cdots$. We find

$$\frac{-B_i}{i} \equiv a_0 + a_1(1-i)p \pmod{p^2} .$$

1) Suppose $p^2 \nmid B_i$. Then the Iwasawa polynomial $P(\omega^i,T)$ of $f(\omega^i,T)$ is irreducible. Indeed, if $p \nmid a_1$, then $P(\omega^i,T)$ is linear. If $p|a_1$, by our congruence we then have $p^2 \nmid a_0$. So $P(\omega^i,T)$ is then Eisenstein and, consequently, irreducible.

2) Suppose $p^2 \nmid B_i$, $4 \le i \le p-3$. Then $P(\omega^i,T)$ is the minimal (and characteristic) polynomial of the action of $\sigma_{1+p}-1$ on $C(\omega^{1-i}) \otimes \mathbb{Q}_p$. Indeed, this follows from 1) and the fact that the Iwasawa invariants of $C(\omega^{1-i})$ and $f(\omega^i,T)$ must match up. One can now deduce that $C(\omega^{1-i})$ is a free $\mathbb{Z}_p$-module and is cyclic as a Λ-module.

We finish these remarks by noting that there are no known examples where $p^2|B_i$ in the above range.

Our last result in this chapter will explain how the geometric criterion implies the arithmetic one. Let $i \in \mathbb{N}^+$ with i even and $i < r^{D(\wp)}-1$. Set $i_1 = r^{D(\wp)}-1-i$ and let $A_\wp, \bar{A}_\wp$, and $M_\wp$ be as before. Also, as before, we put $f_A(x,-i_1) = \prod(1-\beta_1 x^{-1})$, where $\{\beta_1\} \subseteq \bar{A}_\wp$.

2.22 Theorem: Suppose $\wp \nmid B_i$. Then there is at most one $\beta_1 \equiv 1 \pmod{M_\wp}$.

Proof: We will establish that on $T_k(\wp)(\omega_\wp^i)$, $(F-1)$ has a unique non-unit eigenvalue.

Let D be the p-primary class group of $\overline{\mathbb{F}_r} \cdot k(\wp)_+$ and let $\tilde{D}$ be the p-primary class group of $\overline{\mathbb{F}}_r \cdot \mathcal{O}_k(\wp)_+$. We still have an action of

$(A/\wp^*)/\mathbb{F}_r^*$ on these groups. For $i \equiv 0(r-1)$, as the primes at ∞ on $k(\wp)_+$ are $\mathbb{F}_r$-rational, we have the exact sequence

$$0 \to r_k(\wp)_+(\omega_\wp^i) \to D(\omega_\wp^i) \to \tilde{D}(\omega_\wp^i) \to 0$$

Note that the group on the left is a cydic W-module.

Next we take p-adic Tate modules, $T_p(\ )$, $(=\mathrm{Hom}(\mathbb{Q}_p/\mathbb{Z}_p, \_\_\_))$ of the above sequence. We obtain $0 \to \mathrm{Hom}\ (\mathbb{Q}_p/\mathbb{Z}_p, r_k(\wp)_+(\omega_\wp^i)) \to T_p(k(\wp)_+)(\omega_\wp^i) \to \mathrm{Hom}\ (\mathbb{Q}_p/\mathbb{Z}_p, \tilde{D}(\omega_\wp^i)) \to \mathrm{Ext}^1(\mathbb{Q}_p/\mathbb{Z}_p, r_k(\wp)_+(\omega_\wp^i)) \to \mathrm{Ext}^1(\mathbb{Q}_p/\mathbb{Z}_p, D(\omega_\wp^i))$.

The first term is zero as $r_k(\wp)_+(\omega_\wp^i)$ is finite and the last term is zero as $D(\omega^i)$ is divisible. We therefore have
$0 \to T_p(k(\wp)_+)(\omega_\wp^i) \to \mathrm{Hom}\ (\mathbb{Q}_p/\mathbb{Z}_p, \tilde{D}(\omega_\wp^i)) \to \mathrm{Ext}^1(\mathbb{Q}_p/\mathbb{Z}_p, r_k(\wp)_+(\omega_\wp^i)) \to 0$.

From the sequence $0 \to \mathbb{Z}_p \to \mathbb{Q}_p \to \mathbb{Q}_p/\mathbb{Z}_p \to 0$, we deduce that $\mathrm{Ext}^1(\mathbb{Q}_p/\mathbb{Z}_p, r_k(\wp)_+(\omega_\wp^i)) \simeq r_k(\wp)_+(\omega_\wp^i)$. If we put $\tilde{T}_k(\wp) = \mathrm{Hom}(\mathbb{Q}_p/\mathbb{Z}_p, \tilde{D})$, we have

$$(*) \qquad 0 \to T_k(\wp)(\omega_\wp^i) \to \tilde{T}_k(\wp)(\omega_\wp^i) \to r_k(\wp)_+(\omega_\wp^i) \to 0$$

Note that $\tilde{T}_k(\wp)(\omega_\wp^i)$ is torsion free; so $T_k(\wp)(\omega_\wp^i) \otimes \mathbb{Q}_p \simeq \tilde{T}_k(\wp)(\omega_\wp^i) \otimes \mathbb{Q}_p$ and $T_k(\wp)(\omega_\wp^i)$ is a <u>sublattice</u> of $\tilde{T}_k(\wp)(\omega_\wp^i)$ of maximal rank.

Clearly, F-1 acts on $(*)$ and has trivial action on $r_k(\wp)_+(\omega_\wp^i)$. Thus we have

$$(F-1)\ \tilde{T}_k(\wp)(\omega_\wp^i) \subseteq T_k(\wp)(\omega_\wp^i) .$$

We establish next that this is an equality.

By the assumption $\wp \nmid B_i$ (here is where it is used), we find

$$T_k(\wp)(\omega_\wp^i)/((F-1)T_k(\wp)(\omega_\wp^i)) \simeq r_k(\wp)_+(\omega_\wp^i).$$

Thus $\mathrm{ord}_p(\det(F-1))$ is the length of $r_k(\wp)_+(\omega_\wp^i)$. But $\tilde{T}_k(\wp)(\omega_\wp^i)/((F-1)\tilde{T}_k(\wp)(\omega_\wp^i))$ must then have the same length. This is also the length of $\tilde{T}_k(\wp)(\omega_\wp^i)/T_k(\wp)(\omega_\wp^i)$. Thus, $(F-1)\,\tilde{T}_k(\wp)(\omega_\wp^i) = T_k(\wp)(\omega_\wp^i)$.

Next, choose a basis for $\tilde{T}_k(\wp)(\omega_\wp^i)$, say $\{e_1,\ldots,e_h\}$ so that $\{qe_1,e_2,\ldots,e_h\}$ is a basis for $T_k(\wp)(\omega_\wp^i)$, for some q. Let $A = (a_{ij})$ be the matrix of $(F-1)$ with respect to this basis. Let B be the diagonal matrix with $\frac{1}{q}$ in the first row and all the other eigenvalues equal to 1. We find that AB is an automorphism of $T_k(\wp)(\omega_\wp^i)$ and BA is an automorphism of $\tilde{T}_k(\wp)(\omega_\wp^i)$. One thus sees that both the first row and the first column of A must have coefficients divisible by q.

Let A_1 be the minor obtained by deleting the first row and column of A. If we expand along the first row, we find

$$\det(A) \equiv a_{11}\det(A_1) \pmod{q^2}.$$

But $\mathrm{ord}_p(q) = \mathrm{ord}_p(\det(A))$; thus $\det(A_1)$ is <u>a unit</u>.

If we reduce $\det(tI-A)$ modulo (p), we obtain $tP(t)$, where $P(t)$ is the characteristic polynomial of the reduction of A_1. The result is now immediate.

<u>2.23 Corollary</u>: Let $0 \neq i = p^j(r^h-1)$. Assume $i < r^{D(\wp)}-1$. Then there is at most one $\beta_1 \equiv 1 \pmod{M_\wp}$.

Appendix to Chapter Two

In this appendix, we rapidly explain how one can generalize the results of the previous chapters, for more we refer this reader to (CH.1), [G.3]).

First of all, let k_1 be a global function field over $\mathbb{F}_r$ and let ∞ be a fixed prime of degree d_1. Let A be the Dedekind domain of functions regular outside ∞, and let $K = k_\infty$ be the completion of k at ∞. We begin by choosing a notion of "positivity" that generalizes the notion of "monic polynomial". To do so we choose a sign function, sgn, on K: this is a morphism of K^* to $\mathbb{F}^*$ ($\mathbb{F} \simeq \mathbb{F}_{r^d}$, is the constants of K) which is the identity on $\mathbb{F}^*$; so a positive element is just one with sign = 1.

One then constructs an abelian extension H_2/k which has the property that the norm of an ideal in the A-integers of H_2 is principal and generated by a <u>positive</u> element. Thus H_2 is an extension of degree $(r^{d_1}-1)/(r-1)$ of the Hilbert Class-field, H_1, which <u>splits totally</u> at ∞. Over H_2, David Hayes has shown how to construct an elliptic-module ρ associated to (an $\mathbb{F}_r$-twist of) sgn. This module plays a role just like C and via it one has the right "cyclotomic" extensions and the right exponential function.

It has been shown by Jing Yu ([Y.1]), at least when $d_1 = 1$, that the "period", ξ, of ρ is transcendental over k.

One still defines L to be "totally-real" by having ∞ split completely. One also assumes that $H_1 \subseteq L$. Then there are <u>two</u> (if $d_1 \neq 1$) types of zeta-functions that arise: The first type is associated to the A-integers in $H_2 \cdot L$ and the second to the integers in L. Both types can be shown to be essentially algebraic entire functions on S_∞.

The theory at the negative integers of the first type now works pretty much as described above. However, the first type does not seem to have an interesting theory at the positive integers (unless $d_1 = 1$). For this we look at the second type and we still need L/k abelian. The general result is as follows: let $i \equiv 0(r^{d_1}-1)$ and $t = [L:k]$. Then $\xi^{-ti}\zeta_{\mathcal{O}_L}(i)$ is an abelian totally-real element whose square is in H_1.

Of course for $A = \mathbb{F}_r[T]$, $H_1 = k$, but it does not seem possible to replace H_1 with k in general. Thus we see how the functional equation of classical L-series seems dependent on the fact that $\mathbb{Q}$ is its own Hilbert field.

Finally the generalizations of the Gras-conjecture and the Kummer-Okada map, etc., still need to be worked out. One would also like to extend these results to arbitrary totally-real L.

Dictionary

Function Fields	Number Fields
k = function field in one variable over $\mathbb{F}_r$, $r = p^m$	$\mathbb{Q}$
∞ = a fixed prime of degree d_1 in k	The archimedean valuation on $\mathbb{Q}$
A = Dedekind domain of functions regular at all primes $\neq \infty$	$\mathbb{Z}$
v = valuation corresponding to $\mathfrak{P} \subseteq \operatorname{Spec}(A)$	$p \in \operatorname{Spec}(\mathbb{Z})$
$K = k_\infty$ = completion of k w.r.t. ∞ (K/A is compact); sgn: $K^* \to \mathbb{F}^*_{r^{d_1}}$. If $\operatorname{sgn}(a) = 1$, we say a is "positive".	$\mathbb{R}$ ($\mathbb{R}/\mathbb{Z} \simeq S^1$) sign: $\mathbb{R}^* \to \pm 1$
$H_1 \subseteq \bar{k}$ = Hilbert class-field which is "totally-real" (= totally split at ∞) . $H_2 \subseteq \bar{k}$ is a certain extension of H_1 of degree $(r^{d_1}-1)/(r-1)$ which is totally ramified at ∞, unramified elsewhere and abelian over k.	$\mathbb{Q}$, as $\mathbb{Q}$ is its own Hilbert class-field.
ρ = a rank one Drinfeld module defined over the A-integers of H_2 and which is associated to a twist of sgn	$\mathbb{G}_m/\mathbb{Q}$
$\rho[\mathfrak{P}]$ = division values of ρ associated to $\mathfrak{P} \in \operatorname{Spec}(A)$	$\mathbb{G}_m[p] = \mu_p$ = p-th roots of unity

Function Fields	Number Fields
$k(\mathfrak{P})$ = extension of k obtained by adjoining $\rho[\mathfrak{P}]$	$\mathbb{Q}(\mu_p)$
$z^p = \rho_{\mathfrak{P}}(x/y)y^t$, $t = \#(A/\mathfrak{P})$: the "Fermat equation" associated to ρ at $\mathfrak{P}$.	$x^p - y^p = z^p \quad (z^p = y^p((\frac{x}{y})^p - 1))$
$\Gamma = \mathfrak{B}\cdot\xi$ = the lattice corresponding to ρ (Here $\mathfrak{B} \subseteq A$ is an ideal and $\xi \in \overline{K}^*$). If $A = \mathbb{F}_r[T]$ then $\xi = (T-T^r)^\alpha \prod_{i=1}^{\infty} (1 - (T^{r^i} - T)/(T^{r^{i+1}} - T))$; $\alpha = 1/(r-1)$	$\mathbb{Z}\cdot 2\pi i$; $(2i)^2 = -4$; $\pi = 2 \prod_{i=1}^{\infty} (1 - \frac{1}{(2i)^2})^{-1}$
$e_\Gamma(z) = z \prod_{0 \neq \alpha \in \Gamma} (1 - z/\alpha)$: The "exponential" function of Γ. ($e_\Gamma(z)$ is entire w.r.t. the infinite prime)	$e^{(z)}$
$S_\infty = \overline{K}^* \times \mathbb{Z}_p$ = space of "exponents" at ∞	$\mathbb{C}$; where $\mathbb{C}$ is viewed as those s for which $e^{(s)}$ makes sense i.e., is the space of "exponents".
Let π be a fixed uniformizer $\in K$ with $\mathrm{sgn}(\pi) = 1$. If $a \in A^*$, $\mathrm{sgn}(a) = 1$, then we set $\langle a\rangle = \pi^{-v_\infty(a)}\cdot a$; $\langle a\rangle$ is a 1-unit at ∞.	$e^{i\log(n)} \in S^1$ for $n \in \mathbb{N}^+$.
Let a be as above and let $s = (s_0, s_1) \in S_\infty$. We set $a^s = s_0^{\deg(a)}\langle a\rangle^{s_1}$; where $\deg(a) = -d_1 v_\infty(a)$. (One can now define L-series and obtain <u>special-values</u>)	$n^{x+iy} = e^{x\log n}\cdot e^{yi\log(n)}$

References

[G.-R.1] S. Galovich and M. Rosen, The Class Number of Cyclotomic Function Fields; Journal of Number Theory 13, No. 3 (1981), pp. 363 - 375

[G.1] D. Goss, The Arithmetic of Function Fields 2: The 'Cyclotomic' Theory; Journal of Algebra, Vol. 81, No. 1, (1983) pp. 107 - 149

[G.2] ____________, von-Staudt for $\mathbb{F}_q[t]$; Duke Mathematical Journal 45, (1978) pp. 885 - 910

[G.3] ____________, The Theory of Totally-real Function Fields; Proc. A.M.S. Conference on K-theory (1983)

[G.4] ____________, On the Holomorphy of Certain Non-abelian L-series; Math. Ann. 272, (1985) pp. 1-9

[G.-S.1] D. Goss and W. Sinnott, Class-groups of Function Fields; Duke Mathematical Journal, (Vol. 52), (1985), pp. 507 -516

[Gr.1] R. Greenberg, On p-adic Artin L-functions; Nagoya Math. Journal, vol 89, (1983), pp. 77 - 87

[H.1] D. Hayes, Stickelberger Elements in Function Fields, Compositio Mathematica, 55, (1985), pp, 209-239

[I.-S.1] K. Ireland and R. Small, Class Numbers of Cyclotomic Function Fields, (To appear)

[K.1] F. Kegin, A Note on Irregular Prime Polynomials in Cyclotomic Function Field Theory, (To appear in Journal of Number Theory)

[O.1] S. Okada, Kummer's Theory for Function Fields, (To appear)

[T.1] E. Thomas, On the Zeta Function for Function Fields over $\mathbb{F}_p$, Pacific Journal of Mathematics, vol. 107, (1983), pp. 251 - 256

[Y.1] J. Yu, Transcendence and Drinfeld Modules, (To appear)

Department of Mathematics
The Ohio State University
231 West 18th Avenue
Columbus, Ohio 43210

Canadian Mathematical Society
Conference Proceedings
Volume 7 (1987)

Heights and the Special Values of L-series

Benedict H. Gross

In this paper I will present a geometric approach to Eichler's arithmetic theory of definite quaternion algebras and to Waldspurger's results on the central critical values of L-series. The method uses the heights of special points on algebraic curves, and the arguments are similar to those that Zagier and I used to study central critical derivatives. Fortunately, the calculations are much less intimidating in this case; I have also restricted to forms of weight 2 and prime level N to simplify the exposition.

Table of contents:

0731-1036/87 $1.00 + $.25 per page

1. Brandt matrices and Eichler's trace formula.

In this section, we will review the arithmetic theory of maximal orders and ideals in definite quaternion algebras of prime discriminant. Almost all of these results are due to Eichler [2].

Let N be a rational prime, and let B be "the" quaternion algebra over $\mathbb{Q}$ which is ramified at the two places N and ∞. Let R be a fixed maximal order in B. A left ideal I of R is a lattice in B which is stable under left multiplication by R. The right order $\{b \in B : Ib \subset I\}$ of I is another maximal order in B. Furthermore, the set $I^{-1} = \{b \in B : IbI \subset I\}$ is a right ideal for R whose left order is the right order of I.

Two left ideals I and J are in the same class if $J = Ib$ with $b \in B^*$. The set of left ideal classes is finite, and its order n is independent of the choice of R. Let $\{I_1, I_2, \ldots, I_n\}$ be a set of left ideals representing the distinct ideal classes, with $I_1 = R$.

For $1 \le i \le n$ we let R_i be the right order of the ideal I_i. Then each conjugacy class of maximal orders in B is represented (once or twice) in the set $\{R_1, R_2, \ldots, R_n\}$. We let $t \le n$ be the number of distinct conjugacy classes of maximal orders in B. In the classical literature, n is the class number and t is the type number of B.

The groups $\Gamma_i = R_i^*/\mathbb{Z}^* = R_i^*/\langle\pm 1\rangle$ are all finite, as they embed as discrete subgroups of the compact Lie group $(B \otimes \mathbb{R})^*/\mathbb{R}^* \simeq SO_3(\mathbb{R})$. Let $w_i = [\Gamma_i]$, where for any finite set S we let $[S]$ denote its cardinality.

The integer

$$W = \prod_{i=1}^{n} w_i \tag{1.1}$$

is independent of the choice of R, and is equal to the exact denominator of the rational number $(\frac{N-1}{12})$. Eichler's mass formula states that

$$\sum_{i=1}^{n} \frac{1}{w_i} = \frac{N-1}{12}. \tag{1.2}$$

We therefore have the following values of n.

Table 1.3

N	W	$w_i > 1$	n
2	12	12	1
3	6	6	1
$\equiv 5\ (12)$	3	3	$\frac{N+7}{12}$
$\equiv 7\ (12)$	2	2	$\frac{N+5}{12}$
$\equiv 11(12)$	6	3,2	$\frac{N+13}{12}$
$\equiv 13(12)$	1		$\frac{N-1}{12}$

Now fix $1 \le i,j \le n$. The product $M_{ij} = I_j^{-1} I_i = \{\Sigma a_k b_k : a_k \in I_j^{-1}, b_k \in I_i\}$ is a left ideal of R_j with right order R_i. For any element $b \in M_{ij}$ we let $\mathbb{N}b$ be its reduced norm, and define $\mathbb{N}M_{ij}$ as the unique positive rational number such that the quotients $\mathbb{N}b/\mathbb{N}M_{ij}$ are all integers with no common factor. Define the theta series f_{ij} by

$$f_{ij} = \frac{1}{2w_j} \sum_{b \in M_{ij}} e^{2\pi i(\mathbb{N}b/\mathbb{N}M_{ij})\tau} \tag{1.4}$$

$$= \sum_{m \ge 0} B_{ij}(m) q^m \qquad q = e^{2\pi i \tau} .$$

These functions on the upper half plane are all modular forms of weight 2 for the group $\Gamma_0(N)$. Their Fourier coefficients $B_{ij}(m)$ give the entries of the Brandt matrix of degree m :

$$B(m) = ((B_{ij}(m)))_{1 \le i,j \le n} . \tag{1.5}$$

The matrix $B(0)$ has the form

$$B(0) = \frac{1}{2}\left(\left(\begin{matrix} \frac{1}{w_1} & \frac{1}{w_2} & & \frac{1}{w_n} \\ \vdots & \vdots & \cdots\cdots & \vdots \\ \frac{1}{w_1} & \frac{1}{w_2} & & \frac{1}{w_n} \end{matrix}\right)\right) \tag{1.6}$$

and $B(1)$ is the identity matrix. For $m \ge 1$ the matrix $B(m)$ has non-negative integral entries. An efficient algorithm for computing these matrices is given in Pizer [9].

We will now give a formula for the trace of $B(m)$ in terms of Hurwitz's class numbers $H(D)$. If d is a negative discriminant we let $h(d)$ be class number of binary quadratic forms of discriminant d, and let $u(d) = 1$ unless $d = -3,-4$ when $u(d) = 3,2$ respectively. If $\mathcal{O}$ is the order of discriminant d and rank 2 over $\mathbb{Z}$, then $h(d)$ is the order of the finite group $\mathrm{Pic}(\mathcal{O})$ and $u(d)$ is the order of the finite group $\mathcal{O}^*/\mathbb{Z}^* = \mathcal{O}^*/\langle\pm 1\rangle$. For $D > 0$ we define

(1.7)
$$H(D) = \sum_{df^2=-D} h(d)/u(d) ;$$

a short table is given below.

D	H(D)
3	1/3
4	1/2
7	1
8	1
11	1
12	4/3
15	2
16	3/2
19	1
20	2
23	3
24	2
27	4/3
28	2

D	H(D)
31	3
32	3
35	2
36	5/2
39	4
40	2
43	1
44	4
47	5
48	10/3
51	2
52	2
55	4
56	4

We use the prime N to define the modified invariant $H_N(D)$ as follows.

$$(1.8)\qquad H_N(D) = \begin{cases} 0 & \text{if } N \text{ splits in } \mathcal{O} = \mathcal{O}_{-D}, \\ H(D) & \text{if } N \text{ is inert in } \mathcal{O}, \\ \frac{1}{2}H(D) & \text{if } N \text{ is ramified in } \mathcal{O}, \text{ but does not divide the conductor of } \mathcal{O}, \\ H_N(D/N^2) & \text{if } N \text{ divides the conductor of } \mathcal{O}. \end{cases}$$

We also define $H_N(0) = \frac{N-1}{24}$. Then $H_N(D) = 0$ unless $D \geq 0$, $-D \equiv 0,1 \pmod 4$ (mod 4), and $(\frac{-D}{N}) \neq 1$. Using Table 1.3 one can show that $W \cdot H_N(D)$ is integral for $D > 0$, and $2WH_N(0)$ is an integer.

Eichler's trace formula is the following identity

<u>Proposition 1.9</u>. <u>For all</u> $m \geq 0$, $\text{Trace } B(m) = \sum_{\substack{s \in \mathbb{Z} \\ s^2 \leq 4m}} H_N(4m-s^2)$.

Before sketching the proof, we will give two examples. Taking $m = 1$ in Proposition 1.9 we find

$$n = \text{Trace } B(1) = H_N(4) + 2H_N(3) + 2H_N(0)$$

$$= \frac{1}{4}\left(1 - (\tfrac{-4}{N})\right) + \frac{1}{3}\left(1 - (\tfrac{-3}{N})\right) + \frac{N-1}{12}$$

which agrees with the entries in Table 1.3. Taking $m = N$ in Proposition 1.9 we find

$$(1.10)\qquad \text{Trace } B(N) = \begin{cases} 1 & \text{if } N = 2,3 \\ H_N(4N) & \text{if } N > 3, \end{cases}$$

as in the latter case $H_N(4n-s^2) = 0$ for all $s \neq 0$. Using (1.8) we find

(1.11)
$$H_N(4N) = \begin{cases} \frac{1}{2}h(-4N) & N \equiv 1 \ (4) \\ h(-N) & N \equiv 7 \ (8) \\ 2h(-N) & N \equiv 3 \ (8), \ N \geq 11 . \end{cases}$$

We now turn to the proof of Proposition 1.9. The result for $m = 0$ is equivalent to the mass formula, which is best proved using zeta functions, so we shall assume $m \geq 1$. The diagonal entry $B_{ii}(m)$ is equal to the number of elements $b \in R_i = M_{ii}$ of reduced norm m, modulo left multiplication by the $2w_i$ units in R_i^*. For every integer s, we define

$$A_i(s,m) = \{b \in R_i : \mathrm{Tr}(b) = s, \mathbb{N}(b) = m\} .$$

Then $A_i(s,m)$ is a finite set, which is empty once $s^2 > 4m$ (as every element $b \in R_i$ has discriminant $s^2-4m \leq 0$). We therefore have

$$\mathrm{Trace}\ B(m) = \sum_{i=1}^{n} \sum_{s^2 \leq 4m} \frac{[A_i(s,m)]}{[R_i^*]}$$

$$= \sum_{s^2 \leq 4m} \left(\sum_{i=1}^{n} \frac{[A_i(s,m)]}{[R_i^*]} \right) .$$

We will show that the inner sum is equal to $H_N(4m-s^2)$. This follows from the mass formula when $4m-s^2 = 0$, so $b = s/2$ is an integer. Hence we will assume that $D = 4m-s^2$ is positive.

In this case, every $b \in A_i(s,m)$ gives rise to an embedding of the order $\mathcal{O} = \mathcal{O}_{-D}$ into R_i. The group $\Gamma_i = R_i^*/\pm 1$ acts on $A_i(s,m)$ and the set of these embeddings by conjugation, and the Γ_i orbits of A_i correspond to embeddings of $\mathcal{O}$ up to conjugation by R_i^*. For each negative discriminant d, we let $h_i(d)$ be the number of optimal embeddings of the order of discriminant d into R_i, modulo conjugation by R_i^*; an embedding is optimal if it does not extend to any larger order in the quotient field. Then we have shown that

$$[A_i(s,m)/\Gamma_i] = \sum_{df^2=-D} h_i(d).$$

The order of the stabilizer of an element $b \in A_i(s,m)$ under the action of Γ_i is equal to 1 unless the corresponding embedding of $\mathcal{O}$ extends to $\mathbb{Z}[\mu_6]$ or $\mathbb{Z}[\mu_4]$, when it is equal to 3 or 2 respectively. Hence

$$[A_i(s,m)] = w_i \sum_{df^2=-D} h_i(d)/u(d).$$

On the other hand, Eichler calculated the sum $\sum_{i=1}^{n} h_i(d)$ of all optimal embeddings of the order of discriminant d into the n maximal orders R_i. His result is given by the formula:

$$(1.12) \qquad \sum_{i=1}^{n} h_i(d) = \begin{cases} (1 - (\frac{d}{N}))h(d) & \text{if } d \neq N^2 d' \\ 0 & \text{if } d = N^2 d'. \end{cases}$$

Combining this with our previous formula shows that

$$\sum_{i=1}^{n} \frac{[A_i(s,m)]}{2w_i} = H_N(D) = H_N(4m-s^2)$$

as desired.

2. Supersingular elliptic curves

It is known that the Brandt matrices for prime level N are related to isogenies between supersingular elliptic curves in characteristic N. We will review this connection, then use the theory of elliptic curves to establish some of the basic properties of these matrices. In this section, we shall assume that $m \geq 1$, so $B(m)$ lies in $M(n,\mathbb{Z})$.

Let $\mathbb{F}$ be an algebraically closed field of characteristic N. There are n distinct isomorphism classes of supersingular elliptic curves over $\mathbb{F}$, which may be ordered $E_1, E_2, \ldots, E_n$ so that $\mathrm{End}(E_i) \simeq R_i$. One then has an isomorphism

$$(2.1) \qquad M_{ij} \simeq \mathrm{Hom}(E_i, E_j)$$

as a left R_j and right R_i module. The degree of an isogeny $\phi_b : E_i \to E_j$ corresponding to a non-zero element $b \in M_{ij}$ is given by the formula

$$(2.2) \qquad \deg \phi_b = \mathbb{N} b / \mathbb{N} M_{ij} .$$

<u>Proposition 2.3</u>. <u>The entry</u> $B_{ij}(m)$ <u>is equal to the number of subgroup schemes</u> C <u>of order</u> m <u>in</u> E_i <u>such that</u> $E_i/C \simeq E_j$.

<u>Proof</u>. By (2.2) and the definition of $B_{ij}(m)$ in (1.4), $B_{ij}(m)$ is the number of equivalence classes of isogenies $\phi : E_i \to E_j$ of order m,

we identify ϕ' with ϕ if $\phi' = \alpha\phi$ and $\alpha \in \mathrm{Aut}(E_j) = R_j^*$. This has the effect of identifying two isogenies with the same kernel C, which is a subgroup scheme of order m in E_i.

It is also known that the orders R_i and R_j are conjugate in B if and only if the elliptic curves E_i and E_j are conjugate by an automorphism of the field $\mathbb{F}$. Since the modular invariants of all of the curves E_i lie in the field of N^2 elements, the curves are conjugate by an automorphism of $\mathbb{F}$ if and only if $i = j$ or $E_i^N \simeq E_j$. Since the kernel of the Frobenius morphism $E_i \to E_i^N$ is the only subgroup scheme of order N in E_i, we find by Proposition 2.3 that

Proposition 2.4. *The curves E_i and E_j are conjugate by an automorphism of $\mathbb{F}$ if and only if $i = j$ or $B_{ij}(N) = 1$. The number of supersingular moduli which lie in the prime field is equal to the trace of $B(N)$.*

As a corollary of Proposition 2.4, we find that the type number of our quaternion algebra is given by the formula

$$(2.5)\qquad \begin{aligned} t &= \mathrm{Trace}\, B(N) + \frac{n - \mathrm{Trace}\, B(N)}{2} \\ &= \mathrm{Trace}\left(\frac{B(1) + B(N)}{2}\right). \end{aligned}$$

This is equal to the number of distinct irreducible factors of the supersingular polynomial over the prime field $\mathbb{Z}/N$. By our formula for Trace $B(1)$ and Trace $B(N)$, given in (1.9-1.11), we see that

$t = \frac{N}{24} + O(N^{1/2+\varepsilon})$. Here is a table of the relevant invariants for small N .

Table 2.6

N	n	t	supersingular polynomial [1, pg. 143]
2	1	1	j
3	1	1	$j-1728$
5	1	1	j
7	1	1	$j-1728$
11	2	2	$j(j-1728)$
13	1	1	$j-5$
17	2	2	$j(j-8)$
19	2	2	$(j-1728)(j-7)$
23	3	3	$j(j-1728)(j+4)$
29	3	3	$j(j-2)(j+4)$
31	3	3	$(j-1728)(j-2)(j-4)$
37	3	2	$(j-8)(j^2-6j-6)$
41	4	4	$j(j-3)(j+9)(j+13)$
43	4	3	$(j-1728)(j+2)(j^2+19j+16)$
47	5	5	$j(j-1728)(j-9)(j-10)(j+3)$
53	5	4	⋮
59	6	6	
61	5	4	
67	6	4	
71	7	7	
73	6	4	
79	7	6	
83	8	7	
89	8	7	
97	8	5	

Proposition 2.7. 1) *The row sums* $\sum_j B_{ij}(m)$ *are independent of* i *and equal to* $\sigma(m)_N \underset{\text{defn.}}{=} \sum_{\substack{d|m \\ (d,N)=1}} d$.

2) *If* $(m,m') = 1$ *then* $B(m)B(m') = B(mm')$.

3) $B(N)$ *is a permutation matrix of order dividing* 2 *and for* $k \geq 1$ $B(N^k) = B(N)^k$.

4) *If* $p \neq N$ *is prime and* $k \geq 2$, $B(p^k) = B(p^{k-1})B(p) - pB(p^{k-2})$.

5) *The matrices* $B(m)$ *for* $m \geq 1$ *generate a commutative subring* $\mathbb{B}$ *of* $M(n,\mathbb{Z})$.

6) *We have the symmetry relation* $w_j B_{ij}(m) = w_i B_{ji}(m)$.

7) *The commutative algebra* $\mathbb{B} \otimes \mathbb{Q}$ *is semi-simple, and isomorphic to the product of totally real number fields.*

Proof. 1) The sum $\sum_j B_{ij}(m)$ is, by Proposition 2.3, the number of subgroup schemes C of order m in E_i . This is multiplicative in m , equal to 1 if $m = N^k$, and equal to $1+p+p^2+\ldots+p^k$ if $m = p^k$ with $p \neq N$. Hence $\sum_j B_{ij}(m) = \sigma(m)_N$.

2) By Proposition 2.3, $B_{ij}(mm')$ is the number of subgroup schemes. $C_{mm'}$ of order m in E_i with $E_i/C_{mm'} \simeq E_j$. Let C_m be the unique subgroup scheme of order m in $C_{mm'}$ and let $E_k = E_i/C_m$. Let $C_{m'}$ be the image of $C_{mm'}$ on E_k ; this has order m' and $E_k/C_{m'} \simeq E_j$. Since any isogeny of degree mm' may be uniquely factored in this fashion: $B_{ij}(mm') = \sum_k B_{ik}(m)B_{kj}(m')$. This proves the matrix identity.

3) Each E_i has a unique subgroup scheme C_N of order N , which is the kernel of the Frobenius mapping $E_i \to E_i^N$. Hence $B_{ij}(N) = 1$ if

$E_i^N \simeq E_j$ and $B_{ij}(N) = 0$ otherwise. This shows $B(N)$ is a permutation matrix of order dividing 2. Since the unique subgroup scheme of order N^k is the kernel of $Fr^k : E_i \to E_i^{N^k}$, this shows $B(N^k) = B(N)^k$.

4) Any isogeny $\phi : E_i \to E_j$ of order p^k can be factored as an isogeny $E_i \to E_k$ of degree p followed by an isogeny $E_k \to E_j$ of degree p^{k-1}. This factorization is unique if the kernel C of ϕ is cyclic. If the kernel is not cyclic, so $\phi = p\cdot\phi'$ with ϕ' of degree p^{k-2}, there are $p+1$ possible factorizations. Hence $B(p^k)_{ij} = \sum_{\ell} B_{i\ell}(p)B_{\ell j}(p^{k-1}) - pB(p^{k-2})_{ij}$, which proves the matrix identity.

5) By 3) and 4) the algebra $\mathbb{B}$ is generated over $\mathbb{Z}$ by the matrices $B(N)$ and $B(p)$ for $p \neq N$. These commute with each other by 2).

6) The duality $\phi \leftrightarrow \phi^{\vee}$ identifies $\mathrm{Hom}(E_i,E_j)$ with $\mathrm{Hom}(E_j,E_i)$ and preserves the degree. Since $w_j B_{ij}(m)$ is ½ the number of elements in $\mathrm{Hom}(E_i,E_j)$ of degree m, the symmetry follows.

7) Define an inner product on the group $\mathbb{Z}^n = \bigoplus_{i=1}^{n} \mathbb{Z}e_i$ by the formula $\langle e_i,e_j \rangle = w_i\delta_{ij}$. This is positive definite on $\mathbb{R}^n$, and by 6) the matrices $B(m)$ give self-adjoint endomorphisms of $\mathbb{Z}^n$. The result now follows from the spectral theorem.

We may interpret optimal embeddings of $\mathcal{O}$ into R_i as singular liftings of the supersingular elliptic curve E_i. Assume that $\mathcal{O} \otimes \mathbb{Z}_N$ is a discrete valuation ring, so local optimal embeddings exist at all finite primes, and let W be a complete discrete valuation ring containing $\mathcal{O} \otimes \mathbb{Z}_N$

with residue field $\mathbb{F}$. Then the $h_i(d)$ optimal embedding $\mathcal{O} \to R_i$, modulo conjugation by R_i^*, correspond to the elliptic curves E with complex multiplication by $\mathcal{O}$ over W which are isomorphic to E_i over $\mathbb{F}$, together with a fixed isomorphism $g : \mathcal{O} \xrightarrow{\sim} \mathrm{End}_W(E)$. We identify (E,g) with (E',g') if there is an isomorphism $i : E \xrightarrow{\sim} E'$ over $\mathbb{F}$ such that $g'(\alpha)\circ i = i\circ g(\alpha)$ for all $\alpha \in \mathcal{O}$.

3. Curves of genus zero and their special points.

We begin with an adèlic reinterpretation of Eichler's results. Let $\hat{\mathbb{Z}} = \varprojlim \mathbb{Z}/n\mathbb{Z} = \prod_p \mathbb{Z}_p$ be the profinite completion of $\mathbb{Z}$ and $\hat{\mathbb{Q}} = \hat{\mathbb{Z}} \otimes \mathbb{Q}$ the ring of finite adèles of $\mathbb{Q}$. For any prime p we let $R_p = R \otimes \mathbb{Z}_p$ be the local component of R in $B_p = B \otimes \mathbb{Q}_p$, and put $\hat{R} = R \otimes \hat{\mathbb{Z}} = \prod_p R_p$ and $\hat{B} = B \otimes \hat{\mathbb{Q}}$. Since every left ideal I for R is locally principal, we have $I_p = R_p g_p$ with $g_p \in R_p^* \backslash B_p^*$. The element $g_I = (\ldots g_p \ldots)$ then lies in $\hat{R}^* \backslash \hat{B}^*$; conversely any such coset determines a left ideal $I = \hat{R}g \cap B$ of R.

The left ideal classes correspond to the orbits of B^* acting on the right of $\hat{R}^* \backslash \hat{B}^*$, so the class number is the number of double cosets:

$$n = [\hat{R}^* \backslash \hat{B}^* / B^*] . \tag{3.1}$$

The choice of representative ideals $I_1, I_2, \ldots, I_n$ corresponds to a choice of cosets $g_1, g_2, \ldots, g_n$ in $\hat{R}^* \backslash \hat{B}^*$ such that

$$\hat{B}^* = \bigcup_{i=1}^{n} \hat{R}^* g_i B^* . \tag{3.2}$$

The right order R_i of the ideal I_i is given by the formula $R_i = B \cap g_i^{-1} \hat{R} g_i$. Since the maximal orders in B are all locally conjugate in $\hat{B}$, and the subgroup fixing $\hat{R}$ is the normalizer $N\hat{R}^*$ of $\hat{R}$, we have

(3.3) $$t = [N\hat{R}^* \backslash \hat{B}^* / B^*] .$$

We remark that $\hat{R}^*\hat{\mathbb{Q}}^*$ is a normal subgroup of index 2 in $N\hat{R}^*$, and the quotient is generated by the non-trivial class in $R_N^*\mathbb{Q}_N^*\backslash B_N^*$.

The optimal embeddings $f : \mathcal{O} \to R_i$ also admit an adèlic description. To give f is equivalent to giving a field homomorphism $f : K \to B$ such that $f(K) \cap g_i^{-1}\hat{R}g_i = f(\mathcal{O})$ in $\hat{B}$. The group B^* acts on the right of the set $\hat{R}^*\backslash\hat{B}^* \times \mathrm{Hom}(K,B)$ by the formula. $(g,f) \longmapsto (gb, b^{-1}fb)$. Since $\hat{B}^* = \bigcup \hat{R}^* g_i B^*$, the set of all optimal embeddings of $\mathcal{O}$ into the n orders R_i, modulo conjugation by R_i^*, is then identified with the classes (g,f) in the quotient space $(\hat{R}^*\backslash\hat{B}^* \times \mathrm{Hom}(K,B))/B^*$ which satisfy $f(K) \cap g^{-1}\hat{R}g = f(\mathcal{O})$. This quotient space admits a geometric interpretation, as the K-valued points of a curve X over $\mathbb{Q}$.

Indeed, let Y be the curve of genus zero over $\mathbb{Q}$ which is associated to the quaternion algebra B. The points of Y in any $\mathbb{Q}$-algebra E are given by $\{\alpha \in B \otimes E : \alpha \neq 0, \mathrm{Tr}\alpha = 0, \mathbb{N}\alpha = 0\}/E^*$. The group B^* acts on Y on the right by conjugation: $\alpha \to b^{-1}\alpha b$ and $\mathbb{Q}^*$ acts trivially; in fact, the automorphism group of Y is the $\mathbb{Q}$-form of PGL_2 associated to the quaternion algebra B. In particular, $\mathrm{Aut}_{\mathbb{Q}}(Y) \simeq B^*/\mathbb{Q}^*$. If K is a quadratic field we have a canonical identification $Y(K) = \mathrm{Hom}(K,B)$: to each embedding $f : K \to B$ we let $y = y_f$ be the image of the unique K-line on the quadric $\{\alpha \in B \otimes K : \mathrm{Tr}\alpha = \mathbb{N}\alpha = 0\}$ on which conjugation by $f(K^*)$ acts by multiplication by the character $k \longmapsto k/\bar{k}$. Then y_f is one of 2 fixed points of $f(K^*)$ acting on $Y(K)$; the other is the image of the line where conjugation acts by the character $k \mapsto \bar{k}/k$.

The curve X is defined as the double coset space

$$(3.4) \qquad X = \hat{R}^* \backslash \hat{B}^* \times Y/B^* ,$$

which is the disjoint union of n curves of genus zero over $\mathbb{Q}$. Indeed the decomposition $\hat{B}^* = \bigcup \hat{R}^* g_i B^*$ gives an isomorphism

$$(3.5) \qquad X \simeq \coprod_{i=1}^{n} Y/\Gamma_i$$

which takes the double coset $\hat{R}^* g_i \times y \pmod{B^*}$ to the coset $y \pmod{\Gamma_i}$ on the ith component $X_i = Y/\Gamma_i$ of X .

The special points on X over K are the image of $\hat{R}^* \backslash \hat{B}^* \times Y(K)$ in $X(K)$. We say the point $x = g \times y$ has discriminant d iff $f(K) \cap g^{-1} \hat{R} g = f(\mathcal{O})$ is the image of the order of discriminant d , where $f : K \to B$ is the embedding corresponding to y . If the component g of x is congruent to g_i in $\hat{R}^* \backslash \hat{B}^* / B^*$, then the special point x lies on the component $X_i(K)$. There are exactly $h_i(d)$ special points of discriminant d on this component, as they correspond to the number of optimal embeddings of $\mathcal{O}$ into R_i , modulo conjugation by R_i^* .

We can now prove Eichler's formula (1.12) for the total number $\sum_{i=1}^{n} h_i(d)$ of special points of discriminant d on X . We will show this number is divisible by $h(d) = [\mathrm{Pic}(\mathcal{O})]$ by exhibiting a free action of the group $\hat{\mathcal{O}}^*\backslash\hat{K}^*/K^* \simeq \mathrm{Pic}(\mathcal{O})$ on the set of special points of this discriminant; we will then count the number of orbits using the theory of local embeddings. Let $a \in \hat{K}^*$ be a finite idèle of K and $x = g\times y$ a special point of discriminant d. Let $f : K \to B$ be the embedding corresponding to y; this induces a homomorphism $\hat{f} : \hat{K}^* \to \hat{B}^*$ and we define

$$x_a = g\hat{f}(a) \times y \quad . \tag{3.6}$$

If $x \equiv g'\times y'$ then $g' = gb$ and $f' = b^{-1}fb$; hence $g'\hat{f}'(a)\times y' = gb(b^{-1}\hat{f}(a)b)\times y' = g\hat{f}(a)b\times y' \equiv x_a$ so the action is well-defined independent of our choice of representative for x.

Let us first verify that x_a has discriminant d. Since $\hat{f}(\hat{K}^*)$ is commutative, $f(K) \cap g\hat{f}(a)^{-1}\hat{R}g\hat{f}(a) = \hat{f}(a)^{-1}(f(K) \cap g^{-1}\hat{R}g)\hat{f}(a) = \hat{f}(a)^{-1} f(\mathcal{O})\hat{f}(a) = f(\mathcal{O})$. The subgroups K^* and $\hat{\mathcal{O}}^*$ of $\hat{K}^*$ act trivially; conversely if $x = x_a$ then a lies in the subgroup $K^*\hat{\mathcal{O}}^*$. Hence $\hat{\mathcal{O}}^*\backslash\hat{K}^*/K^* \simeq \mathrm{Pic}(\mathcal{O})$ acts freely. The orbit space is identified with the double cosets

$$\hat{R}^*\backslash\hat{N}^*/f(\hat{K}^*) \quad , \tag{3.7}$$

where $f : K \to B$ is a fixed embedding (if any exist) and $\hat{N}^* = \{g \in \hat{B}^* : f(K) \cap g^{-1}\hat{R}g = f(\mathcal{O})\}$. But the space in (3.7) is a product of local terms $R_p^*\backslash N_p^*/f(K_p^*)$, which classify the number of optimal embeddings of $\mathcal{O}_p$ into the maximal order R_p modulo conjugation by R_p^*. This number is 1 for all $p \neq N$, for $p = N$ it is 0,1, or 2 depending on the behavior of N in $\mathcal{O}$. This gives a geometric proof of (1.12)

Another description of the points of discriminant d on the component $X_i = Y/\Gamma_i$ as corresponding to optimal embeddings $f : \mathcal{O} \to R_i$ is useful in many computational contexts. Here we describe the action of $\mathrm{Pic}(\mathcal{O})$ on these points in terms of ideals. Let $\mathfrak{a}$ be the ideal (= projective module of rank 1 in K) which is determined by the idèle $a \pmod{\hat{\mathcal{O}}^*}$; specifically $\mathfrak{a} = K \cap a\hat{\mathcal{O}}$. Let R' be the right order of the left R_i module $R_i\mathfrak{a}$; since $\mathcal{O}$ also acts on the right of $\mathfrak{a}$, the map f induces an optimal embedding $\mathcal{O} \to R'$ which corresponds to the point x_a.

We may refine our modified Hurwitz class numbers by defining the rational divisor c_D on X for $D > 0$ by

$$c_D = \frac{1}{2} \sum_{-D=df^2} \frac{1}{u(d)} \sum_{\substack{x \text{ of} \\ \text{discriminant } d \\ \text{on } X}} (x) \qquad (3.8)$$

Then $\deg(c_D) = H_N(D)$ by formula (1.12). The element c_D lies in $\frac{1}{2W}\mathrm{Div}(X)$; we will analyze its class in $\mathrm{Pic}(X)$ in the next section.

The action of $\mathrm{Gal}(K/\mathbb{Q})$ on the special points in $Y(K)$ preserves the points of discriminant d and may be described as follows. If x corresponds to the embedding $\mathcal{O} \to R_i$ then $\bar{x}$ corresponds to the embedding $\alpha \to f(\bar{\alpha})$; in particular, $\bar{x}$ lies on the same component as x. When N is inert in $\mathcal{O}$ the group $\mathrm{Gal}(K/\mathbb{Q}) \times \mathrm{Pic}(\mathcal{O})$ acts simply transitively on the points of discriminant d. When N is ramified in $\mathcal{O}$, the group $\mathrm{Pic}(\mathcal{O})$ acts simply transitively; we have $x = \bar{x}$ if and only if x corresponds to an embedding $f : \mathcal{O} \to R_i$ where $f(\bar{\alpha}) = j\alpha j^{-1}$ for a 4th root of unity j in R_i^*. In this case, $w_i \equiv 0 \pmod 2$. Hence $w_i h_i(d)$ is always an <u>even</u> integer.

4. Correspondences and the height pairing.

The curve X has a large ring of correspondences over $\mathbb{Q}$, which come from the geometry of the coset space $\hat{R}^*\backslash\hat{B}^*$. Since the class number of $\mathbb{Q}$ is 1, we have $\hat{\mathbb{Q}}^* = \mathbb{Q}^*\hat{\mathbb{Z}}^*$ and

$$(4.1)\qquad X \simeq (\hat{R}^*\backslash\hat{B}^*/\hat{\mathbb{Q}}^*) \times Y / (B^*/\mathbb{Q}^*) .$$

The elements $g = (\dots g_p \dots)$ lie in the product of local spaces $R_p^*\backslash B_p^*/\mathbb{Q}_p^*$.

When $p \neq N$ the space $R_p^*\backslash B_p^*/\mathbb{Q}_p^* \simeq PGL_2(\mathbb{Z}_p)\backslash PGL_2(\mathbb{Q}_p)$ has the structure of the set of vertices in a homogeneous tree of degree $p+1$ [10, pg. 70]. When $p = N$ the space $R_p^*\backslash B_p^*/\mathbb{Q}_p^*$ has 2 elements, so may be viewed as the vertices in a homogeneous tree of degree 1. Let δ_p denote the distance function on the tree at the place p ; for $m \geq 1$ we define a correspondence t_m on the product of these trees by the formula

$$(4.2)\qquad t_m(g) = \sum_{\substack{\delta_p(g_p,h_p) \leq \mathrm{ord}_p(m) \\ \delta_p(g_p,h_p) \equiv \mathrm{ord}_p(m) \pmod 2}} (h)$$

This is self-dual of degree $\sigma(m)_N$. It preserves $\hat{R}^*\backslash\hat{B}^*/\hat{\mathbb{Q}}^*$, as m is divisible by only a finite number of primes. Since the right action of $B^*/\mathbb{Q}^*$ in (4.1) preserves distance, t_m induces a correspondence on X

$$(4.3)\qquad t_m(g\times y) = \sum_{h\in t_m(g)} (h\times y) .$$

The correspondences t_m on X commute, and satisfy the same relations as the Brandt matrices in Proposition 2.7 [10, pg. 73].

The group Pic(X) of line bundles on X is isomorphic to $\mathbb{Z}^n$, and is generated by the classes e_i of degree 1 on each component X_i. The correspondences t_m on X induce endomorphisms of the group Pic(X), and we have the following

Proposition 4.4. For all $m \geq 1$ and $i = 1,2,\ldots,n$

$$t_m e_i = \sum_{j=1}^{n} B_{ij}(m)e_j .$$

In other words, on the basis $\langle e_i \rangle_{i=1}^{n}$ of Pic(X), the action of t_m is given by the transpose $B(m)^{tr}$ of the mth Brandt matrix. This gives a geometric interpretation of Brandt matrices.

Proof. The components of X are indexed by the supersingular elliptic curves E_i. The number of points in the divisor class $t_m e_i$ which lie on the component X_j is equal to the number of isogenies $E_i \to E_j$ of degree m, which two isogenies identified if they have the same kernel. This follows from the definition of t_m in (4.2)-(4.3) and Tate's theorem that an isogeny is determined by its action on the Tate modules $T_p E_i$ and Dieudonné module $T_N E_i$. But the number of such isogenies is equal to $B_{ij}(m)$ by Proposition 2.3.

We define a height pairing $\langle , \rangle$ on Pic(X) with values in $\mathbb{Z}$ by setting

$$(4.5)\qquad \begin{aligned} &\langle e_i, e_j\rangle = 0 \quad \text{if} \quad i \neq j \\ &\langle e_i, e_i\rangle = w_i \end{aligned}$$

and extending bi-additively. If $e = \Sigma\, a_i e_i$ and $e' = \Sigma\, a_i' e_i$ are two divisor classes, then

$$\langle e, e'\rangle = \sum_{i=1}^{n} w_i a_i a_i' .$$

This pairing is clearly positive definite. It gives an isomorphism of $\mathrm{Pic}(X)^{\vee} = \mathrm{Hom}(\mathrm{Pic}(X), \mathbb{Z})$ with the subgroup of $\mathrm{Pic}(X) \otimes \mathbb{Q}$ with basis $\langle e_i^{\vee} = e_i/w_i\rangle_{i=1}^{n}$.

<u>Proposition 4.6. For all classes</u> e <u>and</u> e' <u>in</u> $\mathrm{Pic}(X)$

$$\langle t_m e, e'\rangle = \langle e, t_m e'\rangle .$$

<u>Proof</u>. It suffices to verify the equality when $e = e_i$ and $e' = e_j$. The left hand side is then $w_j B_{ij}(m)$ by Proposition 4.4, and the right hand side is $w_i B_{ji}(m)$. These are equal by Proposition 2.7.

We let e_D denote the class of the divisor c_D defined in (3.8). This lies in $\mathrm{Pic}(X) \otimes \mathbb{Q}$, but in fact we have

<u>Proposition 4.7. The class</u> e_D <u>lies in the subgroup</u> $\mathrm{Pic}(X)^{\vee}$ <u>of</u> $\mathrm{Pic}(X) \otimes \mathbb{Q}$.

Proof. By its definition, we have

$$e_D = \sum_{i=1}^{n}\left(\sum_{-D=df^2} \frac{h_i(d)}{2u(d)}\right) e_i = \sum_{i=1}\left(\sum_{-D=df^2} \frac{w_i h_i(d)}{2u(d)}\right) e_i^{\vee} \quad \text{in } \operatorname{Pic}(X) .$$

We must show that

$$\frac{w_i h_i(d)}{2u(d)} \in \mathbb{Z} .$$

This is clear if w_i is odd, as then $u(d)$ divides w_i and 2 divides $h_i(d)$. If $w_i \equiv 2 \pmod 4$, the only possible problem is when $u(d) = 2$; but in this case $d = -4$, N is odd, and $h_i(d)$ is even. If $N = 2$ and $w_i = 12$, then $w_i/2u(d)$ is always integral.

We define e_0 in $\operatorname{Pic}(X)^{\vee}$ as the class

(4.8) $$e_0 = \sum_{i=1} \frac{1}{w_i} e_i = \sum_{i=1}^{n} e_i^{\vee} .$$

The $\deg(e_0) = \frac{N-1}{12}$ by the mass formula. By proposition 4.4 we have

(4.9) $$t_m e_0 = \sigma(m)_N \cdot e_0 \qquad \text{for } m \geq 1 .$$

For any class e in $\operatorname{Pic}(X)$ we have the height formula:

(4.10) $$\langle e, e_0 \rangle = \deg e .$$

5. Modular forms of weight 2

We recall the slash notation for functions on the upper half plane $\mathfrak{h}$. Let $f : \mathfrak{h} \to \mathbb{C}$ be a function, $A = \begin{pmatrix} a & b \\ c & d \end{pmatrix}$ a matrix in $GL_2(\mathbb{R})$ with positive determinant, and $k \geq 0$ an integer. Then

$$(5.1) \qquad f|_k A = f\left(\frac{az+b}{cz+b}\right)(cz+d)^{-k} \ (\det A)^{k/2}$$

defines a right action of the matrices: $f|_k AB = (f|_k A)|_k B$. We say f is a modular form of weight k for $\Gamma_0(M)$ with character ε if f is holomorphic on $\mathfrak{h}$, regular at the cusps, and satisfies

$$(5.2) \qquad f|_k A = \varepsilon(d) f$$

for all $A = \begin{pmatrix} a & b \\ c & d \end{pmatrix}$ in $SL_2(\mathbb{Z})$ with $c \equiv 0 \pmod{M}$. In this section, we shall only consider modular forms of weight 2 and trivial character for $\Gamma_0(N)$.

The set of these modular forms forms a complex vector space $M_{\mathbb{C}}$, and it is well-known that $\dim M_{\mathbb{C}} = n$. Every function f in $M_{\mathbb{C}}$ has a Fourier expansion

$$f(\tau) = \sum_{m \geq 0}^{\infty} a_m q^m \quad \text{with} \quad q = e^{2\pi i \tau}.$$

We define M as the subgroup of those modular forms which satisfy

(5.3) $$\begin{cases} a_m \in \mathbb{Z} & \text{for all } m \geq , \\ Wa_0 \in \frac{1}{2}\mathbb{Z} . \end{cases}$$

Then M is a free $\mathbb{Z}$-module of rank n and $M \otimes \mathbb{C} = M_{\mathbb{C}}$. Some examples of elements in M are the theta-series f_{ij} defined in (1.4). We remark that the condition on a_0 is forced by the condition on the higher coefficients.

The Hecke algebra $\mathbb{T} = \mathbb{Z}[\ldots T_m \ldots]$ acts on the lattice M by the well-known formulae. This algebra is generated over $\mathbb{Z}$ by the operators of prime index, and the operators satisfy the same relations as the Brandt matrices in Proposition 2.7. If $p \neq N$ is prime we have

$$\sum a_m q^m \mid T_p = \sum \{a_{mp} + pa_{m/p}\} q^m .$$

If $p = N$ we have

$$\sum a_m q^m \mid T_N = \sum a_{mN} \, q^m .$$

We remark that as endomorphisms of M, $T_N + W_N = 0$ where W_N is the canonical involution $f \longmapsto f \mid \begin{pmatrix} 0 & -1 \\ N & 0 \end{pmatrix}$. Also, the subgroup M^+ on which $T_N = +1$ (or $W_N = -1$) has rank t. It is also know that $M \otimes \mathbb{Q}$ is a free $\mathbb{T} \otimes \mathbb{Q}$ module of rank 1, this is a restatement of the multiplicity one theorem, as every eigenform in $M \otimes \mathbb{R}$ is a new form of level N.

By computing the trace of T_m on the homology of the Riemann surface $\mathfrak{H}^*/\Gamma_0(N)$ using Lefschetz's fixed point formula, and comparing with Proposition 1.9, Eichler established the identity

$$\text{(5.4)} \qquad \text{Trace } T_m = \text{Trace } B(m) \quad \text{for all} \quad m \geq 1 .$$

The left hand side is the trace of T_m as an endomorphism of M, or equivalently, of $M_{\mathbb{C}}$. The semi-simplicity of the algebras $\mathbb{T} \otimes \mathbb{Q}$ and $\mathbb{B} \otimes \mathbb{Q}$ then shows that they are conjugate inside $M(n,\mathbb{Q})$. Hence the map $T_m \to B(m)$ induces a ring isomorphism $\mathbb{T} \simeq \mathbb{B}$. Since we have seen that $\mathbb{B}$ is isomorphic to the ring of correspondences $\mathbb{Z}[\cdots t_m \cdots]$ acting on $\text{Pic}(X)$ (Proposition 4.4), we may also identify this ring with the Hecke algebra $\mathbb{T}$.

We will now compute the action of the Hecke operators on our theta series f_{ij}.

<u>Proposition 5.5</u>. <u>For all</u> $m \geq 1$ <u>we have</u>

$$f_{ij} \mid T_m = \sum_k B_{ik}(m) f_{kj} = \sum_k B_{kj}(m) f_{ik} .$$

<u>Proof</u>. The second identity will follow from the first, as $w_j f_{ij} = w_i f_{ji}$ and $w_k B_{ik}(m) = w_i B_{ki}(m)$. It also suffices to calculate $f_{ij} \mid T_m$ for prime indices, as these operators generate $\mathbb{T}$ and satisfy the same relations as the matrices $B(m)$ in $\mathbb{B}$. Since $f_{ij} = \frac{1}{2w_j} + \sum B_{ij}(m) q^n$ we must show

$$B_{ij}(mp) + p\, B_{ij}(m/p) = \sum_k B_{ik}(p) B_{kj}(m)$$

$$B_{ij}(mN) = \sum_k B_{ik}(N) B_{kj}(m) .$$

These follow from Proposition 2.7.

In simple terms, Proposition 5.5 states that the subgroups

$$\begin{cases} N_j = \langle f_{1j}, f_{2j}, \ldots, f_{nj} \rangle \\ N'_j = \langle f_{j1}, f_{j2}, \ldots, f_{jn} \rangle \end{cases}$$

are stable under the action of $\mathbb{T}$, and that T_m acts on the spanning sets by $B(m)^{tr}$ and $B(m)$ respecticely. Since $w_j f_{ij} = w_i f_{ji}$, these two $\mathbb{Z}$-modules have the same rank n_j, which depends on the order R_j. The determination of n_j is Hecke's basis problem, which is still open and appears rather difficult. From (5.4) it follows that the n^2 theta-series $\{f_{ij}\}$ span $M \otimes \mathbb{Q}$.

We will now use the curve X to construct elements of M.

<u>Proposition 5.6</u>. <u>The map</u> $\phi(e,e^\vee) = \dfrac{\deg e \cdot \deg e^\vee}{2} + \sum_{m\geq 1} \langle t_m, e, e^\vee \rangle q^m$ <u>defines a</u> $\mathbb{T}$-<u>module homomorphism</u> $\phi : \mathrm{Pic}(X) \otimes_{\mathbb{T}} \mathrm{Pic}(X)^\vee \to M$ <u>which is an isomorphism over</u> $\mathbb{T} \otimes \mathbb{Q}$.

<u>Proof</u>. We first verify that $\phi(e,e^\vee)$ is an element of M. The definition is bi-additive in each variable, so it suffices to check this when $e = e_i$ and $e^\vee = e_j^\vee$. Then $\langle t_m, e, e^\vee \rangle = B_{ij}(m)$, so $\phi(e,e^\vee) = f_{ij}$.

The $\mathbb{T}$-linearity follows from Proposition 5.5 and Proposition 4.4. The fact that ϕ gives an isomorphism over $\mathbb{Q}$ results from the fact that $\mathrm{Pic}(X)\otimes\mathbb{Q} \simeq \mathrm{Pic}(X)^{\vee}\otimes\mathbb{Q}$ and $M\otimes\mathbb{Q}$ are free $\mathbb{T}\otimes\mathbb{Q}$ modules of rank 1, and the map ϕ is surjective (the n^2 theta series f_{ij} generate $M\otimes\mathbb{Q}$)

We now discuss the normalized Eisenstein series F of weight 2 for $\Gamma_0(N)$ and the cuspidal eigenforms. The Eisenstein series is given by

(5.7)
$$\begin{aligned} F &= \phi(e_i, e_0) \quad \text{for any } i = 1,2,\ldots,n \\ &= \sum_{j=1}^{n} f_{ij} \quad \text{for any } i = 1,2,\ldots,n \\ &= \frac{N-1}{24} + \sum_{m\geq 1} \sigma(m)_N q^m . \end{aligned}$$

It satisfies $F|T_m = \sigma(m)_N F$ for any $m \geq 1$. The other characters in the spectral decomposition of $\mathbb{T}\otimes\mathbb{R}$ correspond to normalized cusp forms f in M with $a_0 = 0$, $a_1 = 1$, and $a_m = O(m^{1/2+\varepsilon})$. The rank of the subgroup of cusp forms is equal to $n-1$, which is the genus of the modular curve $X_0(N)$.

6. Some examples

Consider first the case when $N = 2$, so $n = t = 1$. The algebra B was found by Hamilton: $B = \mathbb{Q} + \mathbb{Q}i + \mathbb{Q}j + \mathbb{Q}k$, $i^2 = j^2 = k^2 = -1$. $ij = -ji = k$. A maximal order, which is unique up to conjugacy, was found by Hurwitz:

(6.1) $$R = \mathbb{Z}i + \mathbb{Z}j + \mathbb{Z}k + \mathbb{Z}\left(\frac{1+i+j+k}{2}\right).$$

The supersingular elliptic curve E in characteristic 2 with $\mathrm{End}(E) = R$ has the equation $y^2+y = x^3$ and invariant $j = 0$. The group $\mathrm{Aut}(E) = R^*$ has order 24, and is given by

(6.2) $$R^* = \left\langle \pm 1, \pm i, \pm j, \pm k, \frac{\pm 1 \pm i \pm j \pm k}{2} \right\rangle .$$

By considering the action on the 3-division points of E, one obtains an isomorphism $R^* \simeq SL_2(\mathbb{Z}/3)$. Hence $\Gamma = R^*/\pm 1$ is isomorphic to $PSL_2(\mathbb{Z}/3)$, or to the alternating group on 4 letters. We have $W = w_1 = 12$. The theta series $f = f_{11}$ is given by

(6.3) $$\begin{aligned} f &= \frac{1}{24} \sum_{b \in R} q^{\mathbb{N}b} \\ &= \frac{1}{24} \sum_{x \equiv y \equiv z \equiv w \pmod 2} q^{(x^2+y^2+z^2+w^2)/4} \\ &= \frac{1}{24} + q + q^2 + 4q^3 + q^4 + 6q^5 + \cdots \end{aligned}$$

By (5.7), the mth Fourier coefficient of f is equal to $\sigma(m)_2$; in particular, this shows that every integer $\equiv 0\ (4)$ is the sum of 4 squares.

The first case where $n > 1$ and there are cusp forms is when $N = 11$. There $n = t = 2$, $w_1 = 2$ and $w_2 = 3$. The first few Brandt matrices are:

$$B(0) = \begin{pmatrix} 1/4 & 1/6 \\ 1/4 & 1/6 \end{pmatrix} \qquad B(3) = \begin{pmatrix} 2 & 2 \\ 3 & 1 \end{pmatrix}$$

$$B(1) = \begin{pmatrix} 1 & 0 \\ 0 & 1 \end{pmatrix} \qquad B(4) = \begin{pmatrix} 5 & 2 \\ 3 & 4 \end{pmatrix}$$

$$B(2) = \begin{pmatrix} 1 & 2 \\ 3 & 0 \end{pmatrix} \qquad B(5) = \begin{pmatrix} 4 & 2 \\ 3 & 3 \end{pmatrix}.$$

The unique normalized cusp form is

(6.4)
$$\begin{aligned} f &= f_{11} - f_{21} = f_{22} - f_{12} = 3f_{22} - 2f_{11} \\ &= q - 2q^2 - q^3 + 2q^4 + q^5 + 2q^6 - 2q^7 + \dots \\ &= q \prod_{m\geq 1} (1-q^m)^2 (1-q^{11m})^2 . \end{aligned}$$

Its Mellin transform is the zeta function of the elliptic curve $X_0(11)$ with equation $y^2+y = x^3-x^2-10x-20$ over $\mathbb{Q}$. We have the congruence

(6.5)
$$f \equiv F \pmod{5M}$$

where F is the Eisenstein series:

(6.6) $$F = f_{11} + f_{12} = f_{21} + f_{22} = 3f_{11} - 2f_{22}$$

$$= \frac{5}{12} + q + 3q^2 + 4q^3 + 7q^4 + 6q^5 + \dots$$

Indeed, $F - f = 5(f_{11} - f_{22})$, and $M = \mathbb{Z} f_{11} + \mathbb{Z} f_{22}$

7. The main identity

For the next five sections $f = \sum_{m\geq 1} a_m q^m$ will be a cusp form in $M_{\mathbb{C}}$, K will denote an imaginary quadratic field of discriminant $-D$ where the prime N is inert, and $\mathcal{O}$ will denote the ring of integers in K. We let A denote a fixed ideal class of $\mathcal{O}$ and ε the quadratic character of $(\mathbb{Z}/D\mathbb{Z})^*$ determined by $\varepsilon(p) = \left(\frac{-D}{p}\right)$ for primes p not dividing D. Also, $u = u(-D)$ and $h = h(-D)$.

Let $\omega_f = 2\pi i f(\tau)d\tau = \sum_{m\geq 1} a_m q^m \frac{dq}{q}$ be the holomorphic differential associated to f on the Riemann surface $X_0(N)$. If g is any element in $M_{\mathbb{C}}$, we define the Petersson product of f and g by the formula

$$(7.1) \qquad (f,g) = \iint_{X_0(N)} \omega_f \wedge \overline{i\omega_g}$$

$$= 8\pi^2 \iint_{\Gamma_0(N)\backslash\mathfrak{H}} f(z)\,\overline{g(z)}\,dxdy \qquad z = x+iy .$$

Let E_A be the theta series of weight 1 which is determined by the ideal class A:

$$(7.2) \qquad E_A(z) = \frac{1}{2u}\sum_{\lambda\in\mathfrak{a}} q^{\mathbb{N}\lambda/\mathbb{N}\mathfrak{a}}$$

$$= \frac{1}{2u} + \sum_{m\geq 1} r_A(m) q^m .$$

In this formula, $\mathfrak{a}$ is any ideal in the class A. Hecke showed that E_A is a modular form of weight 1 for $\Gamma_0(D)$ with character ε. The sum over all classes

$$(7.3) \qquad E = \sum_A E_A = \frac{h}{2u} + \sum_{m\geq 1} R(m)q^m$$

is the weight 1 Eisenstein series, whose Fourier coefficients $R(m)$ are the number of ideals of $\mathcal{O}$ of norm m. We put $R(0) = \frac{h}{2u} = \frac{h}{w}$ for consistency.

Define the L-function $L(f,A,s)$ as the product of the two Dirichlet series:

$$(7.4) \qquad L(f,A,s) = \sum_{\substack{m=1 \\ (m,N)=1}}^{\infty} \frac{\varepsilon(m)}{m^{2s-1}} \sum_{m=1}^{\infty} \frac{a_m r_A(m)}{m^s} .$$

The first is a modification of the Dirichlet L-function $L(\varepsilon,2s-1)$ and the second converges for $\mathrm{Re}(s) > 3/2$. In the next section, we will show that $L(f,A,s)$ admits an analytic continuation to the entire plane. It also satisfies the functional equation

$$(7.5) \quad L^*(f,A,s) \underset{\text{defn.}}{=} \left[(2\pi)^{-s}\Gamma(s)\right]^2 (ND)^s L(f,A,s) = L^*(f,A,2-s) .$$

Our main result gives the value of $L(f,A,s)$ at $s = 1$ in terms of the heights of special points of discriminant $-D$ on the curve X. Let x be a fixed point of this discriminant, and define the element g_A of M by taking the sum over all classes B in $\mathrm{Pic}(\mathcal{O})$

$$g_A = \sum_B \phi(x_B, x_{BA}) \tag{7.6}$$

and using Proposition 5.6. The main identity to be proved is then

<u>Proposition 7.7</u>. $L(f,A,1) = \dfrac{(f,g_A)}{u^2\sqrt{D}}$.

We will prove (7.7) by a method similar to Gross-Zagier [3]. First we will use Rankin's method to obtain the analytic identity

$$L(f,A,1) = \frac{(f,G_A)}{\sqrt{D}} \quad \text{with} \quad G_A(z) = \mathrm{Trace}\begin{Bmatrix}\Gamma_0(ND) \\ \Gamma_0(N)\end{Bmatrix} E_A(z)E(Nz) .$$

We will then explicitly compute the trace and obtain the Fourier coefficients a_m of G_A . (These computations are performed in greater detail in Chapter IV of [3]). Finally, we well explicitly compute the height pairing $\langle x_B, t_m x_{AB}\rangle$ and compare with our previous results to obtain the identity

$$u^2 a_m = \sum_B \langle x_B, t_m x_{AB}\rangle$$

for all $m \geq 1$. This shows that $u^2 G_A = g_A$ and completes the proof.

8. Rankin's method

We now discuss Rankin's integral representation for $L^*(f,A,s)$; more details can be found in [3, Ch. IV]. Let $\Gamma_\infty = \left\{\pm\begin{pmatrix}1 & n\\ 0 & 1\end{pmatrix} : n \in \mathbb{Z}\right\}$; a fundamental domain for Γ_∞ on $\mathfrak{H}$ is the region $0 \le x < 1$, $0 < y < \infty$. For $\mathrm{Re}(s)$ large, we therefore have

$$(4\pi)^{-s}\Gamma(s) \sum_{m=1}^{\infty} \frac{a_m r_A(m)}{m^s} = \int_0^\infty \left(\sum_{m=1}^{\infty} a_m r_A(m) e^{-4\pi m y}\right) y^s \frac{dy}{y}$$

$$= \int_0^\infty \left(\int_0^1 f(x+iy)\overline{E_A(x+iy)}\,dx\right) y^s \frac{dy}{y}$$

$$= \iint_{\Gamma_\infty\backslash\mathfrak{H}} f(z)\overline{E_A(z)}\, y^{s+1} \frac{dxdy}{y^2}$$

$$= \sum_{\gamma\in\Gamma_\infty\backslash\Gamma_0(ND)} \iint_{\gamma F_{ND}} f(z)\overline{E_A(z)}\, y^{s+1} \frac{dxdy}{y^2} .$$

where F_{ND} is any fundamental domain for the action of $\Gamma_0(ND)$ on $\mathfrak{H}$. Using the modular behavior of f and E_A under this subgroup:

$$\begin{cases} f(\gamma z) = f(z)(cz+d)^2 \\ E_A(\gamma z) = \overline{E_A(z)}\ (c\bar{z}+d)\ \varepsilon(d) \\ \mathrm{Im}(\gamma z) = \dfrac{y}{|c\bar{z}+d|^2} , \end{cases}$$

and the invariance of the measure $\frac{dxdy}{y^2}$ under $SL_2(\mathbb{R})$, we find that the last expression is equal to

$$\sum_{\gamma\in\Gamma_\infty\backslash\Gamma_0(ND)} \iint_{F_{ND}} f(\gamma z)\overline{E_A(\gamma z)}\ \mathrm{Im}(\gamma z)^{s+1}\ \frac{dxdy}{y^2} =$$

(8.1)

$$\sum_{\gamma=\pm\left(\begin{smallmatrix}\cdot & \cdot\\ c & d\end{smallmatrix}\right)\Gamma_\infty\backslash\Gamma_0(ND)} \iint_{F_{ND}} f(z)\overline{E_A(z)}\ \frac{\varepsilon(d)}{(c\bar{z}+d)}\ \frac{y^{s-1}}{|cz+d|^{2s-2}}\ dxdy\ .$$

We now introduce the Eisenstein series $E_{ND}(s,z)$ of weight 1, level ND, and character ε, defined by

(8.2)
$$E_{ND}(s,z) = \sum_{\substack{m=1\\(m,N)=1}} \frac{\varepsilon(m)}{m^{2s+1}} \sum_{\pm\left(\begin{smallmatrix}\cdot & \cdot\\ c & d\end{smallmatrix}\right)\in\Gamma_\infty\backslash\Gamma_0(ND)} \frac{\varepsilon(d)}{cz+d}\ \frac{y^s}{|cz+d|^{2s}}$$

$$= \frac{1}{2} \sum_{\substack{c,d\in\mathbb{Z}\\ c\equiv 0\,(ND)\\ (d,ND)=1}} \frac{\varepsilon(d)}{(cz+d)}\ \frac{y^s}{|cz+d|^{2s}}\ .$$

Then switching the order of integration and summation in (8.1) and multiplying by the Dirichlet L-function $\sum \frac{\varepsilon(m)}{m^{2s+1}}$ gives the identity

(8.3)
$$(4\pi)^{-s}\Gamma(s)L(f,A,s) = \iint_{F_{ND}} f(z)\ \overline{E_A(z)E_{ND}(\bar{s}-1,z)}\ dxdy\ .$$

Since the function $\pi^{-s}\Gamma(s)\,E_{ND}(s,z)$ can be continued to an entire function on the s-plane [3], this gives the analytic continuation of $L^*(f,A,s)$; the functional equation (7.5) also follows from a functional equation for the Eisenstein series. Setting $s = 1$ in (8.3) gives the formula

$$(8.4) \qquad L(f,A,1) = 4\pi \iint_{F_{ND}} f(z)\, \overline{E_A(z)E_{ND}(0,z)}\, dxdy \, .$$

We wish to express this as a Petersson product on $X_0(N)$; for any form g of wieght 2 on $\Gamma_0(ND)$ we define the trace to $\Gamma_0(N)$ by

$$(8.5) \qquad \mathrm{Tr}_N^{ND}\, g = \sum_{\gamma\in\Gamma_0(ND)\backslash\Gamma_0(N)} g|_2\, \gamma \, .$$

Then with our normalization of the Petersson product (7.1), formula (8.4) becomes

$$(8.6) \qquad L(f,A,1) = \frac{1}{2\pi}\left(f, \mathrm{Tr}_N^{ND}\left\{E_A(z)E_{ND}(0,z)\right\}\right) .$$

Finally, we remark that the imprimitive Eisenstein series $E_{ND}(s,z)$ can be expressed in terms of the Eisenstein series $E(s,z)$ of weight 1 and level D , defined by

$$(8.7) \qquad E(s,z) = \frac{1}{2} \sum_{\substack{c,d\in\mathbb{Z} \\ c\equiv 0\,(D)}} \frac{\varepsilon(d)}{(cz+d)} \frac{y^s}{|cz+d|^{2s}} \, .$$

At $s = 0$ this gives the identity:

$$E_{ND}(0,z) = E(0,Nz) + N^{-1}E(0,z) \, .$$

The second term contributes 0 to the trace, as $N^{-1}E_A(z)E(0,z)$ has weight 2 and level D, so $\mathrm{Tr}_N^{ND}(N^{-1}E_A(z)E(0,z) = \mathrm{Tr}_1^D(N^{-1}E_A(z)E(0,z)) = 0$ (there are no holomorphic forms of weight 2 and level 1). Hence

$$L(f,A,1) = \frac{1}{2\pi}\left(f, \mathrm{Tr}_D^{ND}\left\{E_A(z)E(0,Nz)\right\}\right).$$

On the other hand, Hecke proved that $E(0,z)$ was related to the holomorphic Eisenstein series $E = \sum_A E_A$ by the formula

$$E(0,z) = \frac{2\pi}{\sqrt{D}} E(z) . \tag{8.8}$$

Combining this with the previous formula, we see we have proved:

<u>Proposition 8.9</u>. <u>Let</u> $G_A(z) = \mathrm{Tr}_D^{ND}\left\{E_A(z)E(Nz)\right\}$. <u>Then</u>

$$L(f,A,1) = \frac{(f,G_A)}{\sqrt{D}}$$

<u>for any cusp form of</u> f <u>weight 2 on</u> $\Gamma_0(N)$.

9. The trace computation.

We put $g = E_A(z)E(Nz)$ and $G_A = Tr_N^{ND} g$. Our aim in this section is to compute the Fourier coefficients of the modular form G_A of weight 2 on $\Gamma_0(N)$. To simplify the exposition, we shall treat the case when D is prime in detail, and refer the reader to [3] for the proof in the general case.

Proposition 9.1. Assume that D is prime. Then

1) $G_A(z) = g(z) + \frac{1}{D} \sum_{j \pmod D} g\left(\frac{z+j}{D}\right)$.

2) *The Fourier coefficients of* $G_A = \sum_{m\geq 0} a_m q^m$ *are given by the formula:*

$$a_m = \sum_{n=0}^{Dm/N} r_A(Dm-nN)\ \delta(n)R(n) = \frac{r_A(m)h}{u} + \sum_{n=1}^{Dm/N} r_A(Dm-nN)\ \delta(n)R(n)$$

where $\delta(n) = \begin{cases} 1 & \text{if } (n,D) = 1 \\ 2 & \text{if } n \equiv 0 \pmod D \end{cases}$.

Proof. We will show how 2) follows from 1). Then we will derive 1) after a transformation lemma. If $g = \sum_{m\geq 0} b_m q^m$ then 1) gives the formula $a_m = b_m + b_{nD}$. By the definition of g, we have.

$$b_m = \sum_{\ell\geq 0} r_A(m-\ell N)\ R(\ell)$$

$$= \sum_{\substack{n\geq 0 \\ n\equiv 0\,(D)}} r_A(mD-nN)\ R(n)$$

where $n = D\ell$, as $r_A(k) = r_A(Dk)$ for all $k \geq 1$. Using the second formula for b_m and the first for b_{mD}, we get 2).

Lemma 9.2. 1) If $\gamma = \begin{pmatrix} a & b \\ c & d \end{pmatrix}$ is in $SL_2(\mathbb{Z})$ and $c \not\equiv 0 \pmod D$ then

$E_A\big|_1\gamma = \frac{\varepsilon(c)}{i\sqrt{D}}\, E_A\left(\frac{z+c^*d}{D}\right)$ where c^* is an inverse for $c \pmod D$

2) If γ is in $\Gamma_0(N)$ and $c \not\equiv 0 \pmod D$ then

$E(Nz)\big|_1\gamma = \frac{-\varepsilon(c)}{i\sqrt{D}}\, E\left(N\left(\frac{z+c^*d}{D}\right)\right)$.

Proof. 1) We will use the well-known formula $E_A\big|_1\begin{pmatrix} 0 & -1 \\ D & 0 \end{pmatrix} = \frac{1}{i} E_A$, which follows from Poisson summation. First consider the special case when $\gamma = \begin{pmatrix} 0 & -1 \\ 1 & j \end{pmatrix} = \frac{1}{D}\begin{pmatrix} 0 & -1 \\ D & 0 \end{pmatrix}\begin{pmatrix} 1 & j \\ 0 & D \end{pmatrix}$. Then

$$E_A\big|_1\gamma = \frac{1}{i} E_A\big|_1\begin{pmatrix} 1 & j \\ 0 & D \end{pmatrix} = \frac{1}{i\sqrt{D}} E_A\left(\frac{z+j}{D}\right).$$

This gives 1) in this case, as $c = 1$ and $c^*d = j$. The general case follows from this, as the matrices $\begin{pmatrix} 0 & -1 \\ 1 & j \end{pmatrix}$ represent the non-trivial cosets of $\Gamma_0(D)\backslash\Gamma_0(1)$. Hence any γ has the form $\gamma = \begin{pmatrix} \alpha & \beta \\ \gamma & \delta \end{pmatrix}\begin{pmatrix} 0 & -1 \\ 1 & j \end{pmatrix}$ with $\begin{pmatrix} \alpha & \beta \\ \gamma & \delta \end{pmatrix}$ in $\Gamma_0(D)$. Then

$$\begin{aligned} E_A\big|_1\gamma &= \varepsilon(\delta) E_A\big|_1\begin{pmatrix} 0 & -1 \\ 1 & j \end{pmatrix} \\ &= \frac{\varepsilon(c)}{i\sqrt{D}} E_A\left(\frac{z+c^*d}{D}\right) \end{aligned}$$

as $c = \delta$ and $c^*d \equiv j \pmod D$.

2) Clearly $E|\gamma = \frac{\varepsilon(c)}{i\sqrt{D}} E\left(\frac{z+c^*d}{D}\right)$ as $E = \sum_A E_A$. Now use the matrix identity

$$\begin{pmatrix} N & 0 \\ 0 & 1 \end{pmatrix}\begin{pmatrix} a & b \\ c & d \end{pmatrix} = \begin{pmatrix} a & bN \\ c/N & d \end{pmatrix}\begin{pmatrix} N & 0 \\ 0 & 1 \end{pmatrix}$$

to obtain:

$$\begin{aligned} E(Nz)\big|_1\gamma &= E\Big|_1\begin{pmatrix} a & bN \\ c/N & d \end{pmatrix} (Nz) \\ &= \frac{\varepsilon(c/N)}{i\sqrt{D}} E\left(\frac{Nz+Nc^*d}{D}\right) \\ &= \frac{-\varepsilon(c)}{i\sqrt{D}} E\left(N\left(\frac{z+c^*d}{D}\right)\right). \end{aligned}$$

The last identity follows from the fact that $\varepsilon(N) = -1$.

We are now ready to prove part 1) of Proposition 9.1. The $D+1$ cosets of $\Gamma_0(ND)\backslash\Gamma_0(N)$ are represented by $\begin{pmatrix} 1 & 0 \\ 0 & 1 \end{pmatrix}$ and matrices $\gamma = \begin{pmatrix} a & b \\ c & d \end{pmatrix}$ in $\Gamma_0(N)$ with $c \not\equiv 0 \pmod{D}$ and $j = c^*d$ running through the D residue classes $(\text{mod } D)$. We have

$$\begin{aligned} g|_2\gamma &= E_A\Big|_1\gamma\cdot E(Nz)\big|_1\gamma \\ &= \frac{\varepsilon(c)}{i\sqrt{D}} E_A\left(\frac{z+j}{D}\right)\cdot\frac{-\varepsilon(c)}{i\sqrt{D}} E\left(N\left(\frac{z+j}{D}\right)\right) \\ &= \frac{1}{D}\, g\left(\frac{z+j}{D}\right) \end{aligned}$$

by Lemma 9.2. Summing over the $D+1$ cosets gives 1).

The analog of part 2) of Proposition 9.1 in the general case when D is composite requires some genus theory. Let q be a prime with $q \equiv -N \pmod{D}$. Then $q = \mathfrak{q} \cdot \bar{\mathfrak{q}}$ is split in $\mathcal{O}$. Let $\mathfrak{a}$ be an ideal in the class of A; we say an ideal $\mathfrak{b}$ is in the genus $\{-NA\}$ if $\mathfrak{b}\mathfrak{a}\mathfrak{q}$ has square class in $\mathrm{Pic}(\mathcal{O})$. We let $R_{\{-NA\}}(n)$ be the number of integral ideals in the genus $\{-NA\}$ of norm n; this is equal to $R(n)$ or zero, as two ideals with the same norm lie in the same genus. Let $\delta(n)$ be equal to 2^k, where k is the number of primes which divide both n and D.

<u>Proposition 9.3.</u> <u>The Fourier coefficients of</u> $G_A = \sum_{m \geq 0} a_m q^m$ <u>are given by the formula</u>

$$a_m = \frac{r_A(m)h}{u} + \sum_{n=1}^{Dm/N} r_A(mD-nN)\ \delta(n) R_{\{-NA\}}(n) .$$

For example, we have $a_0 = \frac{1}{2}\frac{h}{u^2}$, so G_A is not a cusp form.

10. The heights of special points.

Let x be a special point of discriminant $-D$ on X. For A and B in $\mathrm{Pic}(\mathcal{O})$ we wish to compute $\langle x_B, t_m x_{AB} \rangle$. To do this, we must determine how many points in the divisor $t_m x_{AB}$ lie on the same component as x_B. The components of X are indexed by the supersingular elliptic curves in characteristic N, and we have

$$\langle x_B, t_m x_{AB} \rangle = \tfrac{1}{2}[\phi \in \mathrm{Hom}(E,E') : \deg \phi = m] \tag{10.1}$$

where E is the component of x_B and E' the component of x_{AB}. We will denote the group $\mathrm{Hom}(E,E')$ by $\mathrm{Hom}(x_B, x_{AB})$ in this section.

First we will consider the case when D is prime. Let $\mathcal{D} = (\sqrt{-D})$ be the different ideal of $\mathcal{O}$. The quaternion algebra has the form $B = K+Kj$, where $j^2 = -N$, $j\alpha = \bar{\alpha}j$ for all $\alpha \in K$. Let ε be a solution of the congruence $\varepsilon^2 \equiv -N \pmod{D}$. The point $x = x_1$ may be chosen so that

$$\mathrm{End}(x_1) = \{\alpha+\beta j : \alpha \in \mathcal{D}^{-1}, \beta \in \mathcal{D}^{-1}, \quad \alpha \equiv \varepsilon\beta \pmod{\mathcal{O}_{\mathcal{D}}}\}. \tag{10.2}$$

Let $\mathfrak{a}$ and $\mathfrak{b}$ be ideals in the classes of A and B which are relatively prime to $\mathcal{D}$.

<u>Proposition 10.3.</u> <u>We have a bijection</u>

$$\mathrm{Hom}(x_{AB}, x_B) \xrightarrow{\sim} \{\alpha+\beta j : \alpha \in \mathcal{D}^{-1}\mathfrak{a}, \beta \in \mathcal{D}^{-1}\mathfrak{b}^{-1}\bar{\mathfrak{b}}\bar{\mathfrak{a}}, \quad \alpha \equiv \varepsilon\beta \pmod{\mathcal{O}_{\mathcal{D}}}\}.$$

If ϕ *corresponds to* $\alpha+\beta j$ *then* $\deg\phi = (\mathbb{N}\alpha + N\,\mathbb{N}\beta)/\mathbb{N}\mathfrak{a}$.

Proof. The ring $\mathrm{End}(x_B)$ is the right order of the left $\mathrm{End}(x)$ module $\mathrm{End}(x)\cdot\mathfrak{b}$, and $\mathrm{Hom}(x_{AB},x_B)$ can be identified with the left $\mathrm{End}(x_B)$ module $\mathrm{End}(x_B)\cdot\mathfrak{a}$. The proposition now follows from a calculation in $K+Kj = B$.

By (10.1) and Proposition 10.3, we wish to count half the number of solutions to the identity

$$(10.4)\qquad \begin{gathered} \mathbb{N}\alpha + N\,\mathbb{N}\beta = m\,\mathbb{N}\mathfrak{a}, \\ \text{with } \alpha\in\mathcal{D}^{-1}\mathfrak{a},\ \beta\in\mathcal{D}^{-1}\mathfrak{b}^{-1}\bar{\mathfrak{b}}\bar{\mathfrak{a}},\quad \alpha\equiv\varepsilon\beta \pmod{\mathcal{O}_{\mathcal{D}}}. \end{gathered}$$

This is most easily done by introducing the integral ideals

$$(10.5)\qquad \begin{aligned} \mathfrak{c} &= (\alpha)\mathcal{D}\mathfrak{a}^{-1} \\ \mathfrak{c}' &= (\beta)\mathcal{D}\mathfrak{b}\bar{\mathfrak{b}}^{-1}\bar{\mathfrak{a}}^{-1}, \end{aligned}$$

which satisfy the identity

$$(10.6)\qquad \mathbb{N}\mathfrak{c} + N\,\mathbb{N}\mathfrak{c}' = mD\,.$$

The ideals $\mathfrak{c}$ and $\mathfrak{c}'$ lie in the classes of A^{-1} and AB^2 respectively, and the number of solutions to (10.6) with ideals in these classes is equal to

$$r_{A^{-1}}(mD) + \sum_{n>0} r_{A^{-1}}(mD-nN) r_{AB^2}(n) ,$$

where $n = \mathbb{N}\mathcal{L}'$ and $mD-nN = \mathbb{N}\mathcal{L}$.

But a solution to (10.6) in ideals gives a solution to (10.4) in elements (α,β) by inverting formula (10.5). There are <u>a priori</u> w^2 possible choices for (α,β) , except when $n = 0$ when there are w choices. These all satisfy the condition $\alpha \equiv \varepsilon\beta \pmod{\mathcal{O}_{\mathcal{D}}}$ in (10.4) when $n \equiv 0 \pmod{D}$, but when $n \not\equiv 0 \pmod{D}$ only half of them satisfy this final condition. Hence we find

$$\langle x_B, t_m x_{AB} \rangle = u^2 \sum_{n=0}^{mD/N} r_{A^{-1}}(mD-nN)\ \delta(n)\ r_{AB^2}(n) .$$

We sum this result over the classes B and use the fact that $\Sigma r_{AB^2}(n) = R(n)$, as D is prime and there are no elements of order 2 in $\mathrm{Pic}(\mathcal{O})$. Since $r_{A^{-1}}(k) = r_A(k)$, we have the final result

<u>Proposition 10.7</u>. <u>If</u> D <u>is prime, then for all</u> $m \geq 1$

$$\sum_B \langle x_B, t_m x_{AB} \rangle = u^2 \sum_{n=0}^{mD/N} r_A(mD-nN)\ \delta(n) R(n) .$$

In the case when D is composite, we let q be a rational prime with $q \equiv -N \pmod{D}$. Then $q = \mathfrak{q}\cdot\bar{\mathfrak{q}}$ is split in K and $B = K+Kj$ with $j^2 = -Nq$. We then find

$$\mathrm{Hom}(x_{AB}, x_B) \xrightarrow{\sim} \{\alpha+\beta j : \alpha \in \mathcal{D}^{-1}\mathfrak{a} , \beta \in \mathcal{D}^{-1}\mathfrak{q}^{-1}\mathfrak{b}^{-1}\bar{\mathfrak{b}}\bar{\mathfrak{a}}, \ \alpha \equiv q\beta \pmod{\mathcal{O}_{\mathcal{D}}}.$$

and

$$\langle x_B, t_m x_{AB} \rangle = u^2 \sum_{n=0}^{mD/N} r_{A^{-1}}(mD-nN)\ \delta(n) r_{A\mathfrak{q}B^2}(n) .$$

The sum over all classes B now only gives the ideals in the genus $\{A\mathfrak{q}\} = \{-NA\}$, so we find

Proposition 10.8. For all $m \geq 1$, we have

$$\sum_B \langle x_B, t_m x_{AB} \rangle = u^2 \sum_{n\ 0}^{mD/N} r_A(mD-nN)\ \delta(n)\ R_{\{-NA\}}(n) .$$

$$= u r_A(m) h + \sum_{n=1}^{mD/N} r_A(mD-nN)\ \delta(n)\ R_{\{-NA\}}(n) .$$

If we compare Proposition 10.8 with proposition 9.3, we see that we have established the identity

$$\sum_B \langle x_B, t_m x_{AB} \rangle = u^2 a_m \quad \text{for} \quad m \geq 1 .$$

Since the constant coefficient of $\sum_B \phi(x_B, x_{AB})$ is equal to $h/2 = u^2 a_0$, we have established the important:

Corollary 10.9. $u^2 G_A = \sum_B \phi(x_B, x_{AB})$ in M.

On the other hand, we have

$$L(f,A,1) = \frac{(f, G_A)}{\sqrt{D}}$$

by Proposition 8.9. Hence we have completed the proof of the main identity in Proposition 7.7. The rest of this paper will be devoted to a discussion of its corollaries.

11. Special values of L-series.

We now consider L-series with Euler products. Let χ be a complex character of the group $\mathrm{Pic}(\mathcal{O})$, and assume that f is a normalized eigenform for the Hecke algebra $\mathbb{T}$. Define

$$L(f,\chi,s) = \sum_{A} \chi(A)\, L(f,A,s) \quad . \tag{11.1}$$

This has an Euler product, where the general local term is of degree 4. If we write

$$L(f,s) = \prod_p (1-\alpha_p p^{-s})(1-\alpha'_p p^{-s})^{-1}$$

$$L(\chi,s) = \prod_p (1-\beta_p p^{-s})(1-\beta'_p p^{-s})^{-1} ,$$

then we have

$$L(f,\chi,s) = \prod_p (1-\alpha_p\beta_p p^{-s})(1-\alpha_p\beta'_p p^{-s})(1-\alpha'_p\beta_p p^{-s})(1-\alpha'_p\beta'_p p^{-s})^{-1} .$$

The product $\alpha_p\alpha'_p\beta_p\beta'_p$ is equal to $p\varepsilon(p)$ if p does not divide ND; otherwise this product is equal to zero.

To describe the value of $L(f,\chi,s)$ at $s = 1$, we let $c_{f,\chi}$ be the projection of the divisor $c_\chi = \sum_A \chi^{-1}(A) x_A$ to the f-isotypical component of $\mathrm{Pic}(X) \otimes \mathbb{C}$. Note that $\deg c_{f,\chi} = 0$, as f is a cusp form.

Proposition 11.2. $L(f,\chi,1) = \dfrac{(f,f)}{u^2\sqrt{D}} \langle c_{f,\chi}, c_{f,\chi} \rangle$.

Proof. We will adopt the convention that we extend $\mathbb{R}$-bilinear pairings, like $<,>$ and $\phi(\ ,\ )$ to complex pairings which are linear in the first argument and anti-linear in the second. Thus

$$<c_\chi, c_\chi> = <\sum_A \chi^{-1}(A)x_A, \sum_B \chi^{-1}(B)x_B>$$

$$= \sum_{A,B} \chi(A^{-1}B)<x_A, x_B> .$$

Combining the definition (11.1) with the main identity in Proposition 7.7, we find

$$L(f,\chi,1) = \frac{(f, \sum_A \chi(A) g_A)}{u^2\sqrt{D}} ,$$

so we are reduced to showing that the coefficient of f in the eigenvector expansion of $\Sigma\chi(A)g_A$ is equal to the (real) number $<c_{f,\chi}, c_{f,\chi}>$. But by definition

$$\Sigma\chi(A)g_A = \sum_A \chi(A) \sum_B \phi(x_B, x_{AB})$$

$$= \sum_{A,B} \chi(A)\ \phi(x_B, x_{AB})$$

$$= \sum_{A',B'} \chi(A'^{-1}B')\ \phi(x_{A'}, x_{B'}) \qquad \begin{matrix} A' = B \\ B' = AB \end{matrix}$$

$$= \phi(c_\chi, c_\chi) \quad .$$

Since ϕ is $\mathbb{T}$-bilinear, the f-eigencomponent of the modular form $\phi(c_\chi, c_\chi)$ is equal to

$$\begin{aligned}\phi(c_{\chi,f}, c_\chi) = \phi(c_{\chi,f}, c_{\chi,f}) &= \sum_{m\geq 1} \langle c_{\chi,f}, t_m c_{\chi,f}\rangle q^m \\ &= \sum_{m\geq 1} \langle c_{\chi,f}, c_{\chi,f}\rangle \cdot a_m(f) q^m \\ &= \langle c_{\chi,f}, c_{\chi,f}\rangle \cdot f .\end{aligned}$$

This completes the proof.

Corollary 11.3. 1) $L(f,\chi,1) \geq 0$, *with equality if and only if* $c_{f,\chi} = 0$.

2) *For any automorphism* α *of* $\mathbb{C}$

$$(L(f,\chi,1)\sqrt{D}/(f,f))^\alpha = L(f^\alpha,\chi^\alpha,1)\sqrt{D}/(f^\alpha,f^\alpha) .$$

In particular, the ratio lies in the numberfield generated by the values of χ *and the Fourier coefficients of the eigenform* f.

3) $L(f,\chi,1) = 0$ *if and only if* $L(f^\alpha,\chi^\alpha,1) = 0$.

Proof. 1) follows from the fact that $\langle , \rangle$ induces a positive definite Hermitian pairing on $\mathrm{Pic}(X) \otimes \mathbb{C}$ and Proposition 11.2. Since this pairing is rational on $\mathrm{Pic}(X) \otimes \mathbb{Q}$, we have

$$\langle c_{f,\chi}, c_{f,\chi}\rangle^{\alpha} = \langle c_{f^{\alpha},\chi^{\alpha}}, c_{f^{\alpha},\chi^{\alpha}}\rangle$$

which gives 2). Part 3) is an immediate corollary of 2).

In the case where $\chi = 1$ we have a decomposition:

(11.4) $$L(f,\chi,1) = L(f,1)L(f\otimes\varepsilon,1) ,$$

where $f \otimes \varepsilon = \sum_{m\geq 1} a_m\varepsilon(m)q^m$ is the twist of f, which has level ND^2. Also, we have

(11.5) $$c_{\chi} = \sum_{A} x_A \equiv u \cdot e_D \qquad \text{in } \operatorname{Pic}(X) ,$$

where e_D is the class of the rational divisor $c_D = \frac{1}{2u} \sum_{\operatorname{disc}(x)=-D} (x)$ de- in (3.8). This follows from the fact that $\operatorname{Pic}(\mathcal{O}) \times \operatorname{Gal}(K/\mathbb{Q})$ acts simply transitively on the special points, and the points x_A and $\overline{x}_A$ lie on the same component of X. Hence Proposition 11.2 becomes

<u>Corollary 11.6</u>. $L(f,1)L(f\otimes\varepsilon_D,1) = \frac{(f,f)}{\sqrt{D}}\langle e_{f,D}, e_{f,D}\rangle$.

We shall reinterpret this identity, using forms of weight 3/2, in section §13. Here we simply note that $e_{f,D}$ lies in the 1-dimensional space $(\operatorname{Pic}(X)\otimes\mathbb{R})^f$ for each value D, so the different values $\langle e_{f,D}, e_{f,D}\rangle$ have square ratios in the field generated by the fourier coefficients of f. This is a result due to Waldspurger.

Finally, we note that Proposition 11.2 can be extended to include the case where $f = F$ is the normalized Eisenstein series of weight 2. We have

$$L(F,s) = \sum_{m\geq 1} \sigma(m)_N m^{-s} = \zeta(s)\zeta(s-1) \cdot (1-N^{1-s}) .$$

and we define

$$(F,F) = \frac{-\pi\log N}{12}(N-1) .$$

This definition of the inner product is obtained by taking the residue of a Rankin L-function at $s = 2$, and was motivated by the considerations in Zagier [13]. Then the formula in Proposition 11.2 continues to hold, although for $\chi \neq 1$ both sides are equal to zero. When $\chi = 1$ we have

$$(11.7) \qquad \langle c_{F,\chi}, c_{F,\chi}\rangle = \frac{12h^2}{N-1} = \langle c_F, c_F\rangle$$

and both sides are equal to $\dfrac{-\pi\log N}{u^2\sqrt{D}} h^2$.

Formula (11.7) has some implications for cusp forms. Since $\langle c_\chi, c_\chi\rangle = \langle c_{F,\chi}, c_{F,\chi}\rangle + \sum_f \langle c_{f,\chi}, c_{f,\chi}\rangle$ is an integer, if p is a prime which divides the denominator of $\dfrac{12h^2}{N-1}$ it must also divide the denominator of some

$\langle c_{f,\chi}, c_{f,\chi}\rangle$ for $\chi = 1$. One then shows easily that $f \equiv F \pmod{pM}$, so we have obtained a result of Mazur.

Corollary 11.8. Assume p *divides* $\frac{N-1}{12}$ *but does not divide* h . *Then there is a cusp form* f *which is congruent to the Eisenstein series* $F \pmod{p}$ *and satisfies* $L(f,1)L(f\otimes\varepsilon,1) \neq 0$.

12. Modular forms of weight 3/2.

We begin by defining a subspace $M_{\mathbb{C}}^{*}$ of the space of modular forms of weight 3/2 on $\Gamma_0(4N)$ with trivial character, which is due to Kohnen [4]. (In his paper, Kohnen denotes the cusp forms in this subspace by $S_{3/2}(N)^{-}$.) Recall that a modular form of weight 3/2 and level $4N$ is a function $g(\tau)$ on the upper half-plane which is regular at the cusps and satisfies

(12.1) $$g(\tau)/\theta(\tau)^3 \quad \text{is invariant under } \Gamma_0(4N) .$$

where $\theta(\tau) = \sum q^{n^2}$ is the standard theta-series of weight 1/2 . Then g has a Fourier explansion

(12.2) $$g(\tau) = \sum_{D\geq 0} a_D q^D ,$$

and Kohnen's subspace $M_{\mathbb{C}}^{*}$ consists of those forms with

(12.3) $$a_D = 0 \quad \text{unless } -D \equiv 0,1 \pmod 4 \quad \text{and} \quad \left(\tfrac{-D}{N}\right) \neq 1 .$$

The space $M_{\mathbb{C}}^{*}$ has dimension t and is stable under Shimura's Hecke operators $T_{m^2}^{*}$ of degree m^2 for all m prime to $4N$. Kohnen defined operators $T_{m^2}^{*}$ on $M_{\mathbb{C}}^{*}$ for all m , and used the trace formula to prove that [4, pg. 47].

(12.4) $$\text{Trace } T^*_{m^2} = \text{Trace } T_m | M^+_{\mathbb{C}}$$

where T_m is the Hecke operator on forms of weight 2 for $\Gamma_0(N)$ and $M^+_{\mathbb{C}}$ is the subspace of $M_{\mathbb{C}}$ where $T_N = 1$ (or $W_N = -1$). The action of the operators with prime-squared index on Fourier coefficients is given by

(12.5) $$\begin{cases} \Sigma a_D q^D | T^*_{p^2} = \Sigma\{a_{Dp^2} + (\frac{-D}{N}) a_D + p\, a_{D/p^2}\} q^D, & p \neq N \\ \Sigma a_D q^D | T^*_{N^2} = \Sigma a_{DN^2} q^D . \end{cases}$$

On the subspace $M^*_{\mathbb{C}}$ the operator $T^*_{N^2}$ acts as the identity, so $a_D = a_{DN^2}$. There is a lattice M^* of rank t in $M^*_{\mathbb{C}}$ which is stable under the Hecke algebra $\mathbb{T}^*$; this consists of the forms g whose Fourier coefficients satisfy (12.3) as well as the integrality conditions

(12.6) $$a_D \text{ is integral for all } D > 0$$
$$a_0 \in \frac{1}{2}\mathbb{Z} .$$

We now use our maximal orders to construct elements in the lattice M^*. For each i with $1 \leq i \leq n$ we let S_i be the suborder of index 8 in R_i which is defined by

(12.7) $$S_i = \mathbb{Z} + 2R_i \quad .$$

Let S_i^o be the subgroup of elements of trace zero in S_i ; this has rank 3 over $\mathbb{Z}$. Let g_i be the theta-series of the lattice S_i^o with its norm form (which is positive definite):

(12.8) $$g_i(\tau) = \frac{1}{2} \sum_{b \in S_i^o} q^{\mathbb{N}b} = \frac{1}{2} + \sum_{D>0} a_i(D) q^D \quad .$$

Then $a_i(D)$ is one half the number of elements $b \in R_i$ with

$$\begin{cases} b \equiv 0,1 \mod 2R_i \\ \mathrm{Tr} b = 0 \\ \mathbb{N}b = D = -b^2 \end{cases} \quad .$$

This is an integer (as $b \neq -b$), which is zero unless $-D \equiv 0,1 \pmod 4$ and $(\frac{-D}{N}) \neq 1$. Since $g_i(\tau)$ is well-known to have weight 3/2 and level $4N$, the forms g_i all lie in Kohnen's subgroup M^* . Since the orders associated to the curves E_i and E_i^N are conjugate in B and give the same theta-series, only t of the modular forms $g_1, g_2, \ldots, g_n$ are distinct.

Proposition 12.9. *For* $1 \leq i \leq n$ *and* $D > 0$ *we have*

$$a_i(D) = \frac{w_i}{2} \sum_{-D=df^2} \frac{h_i(d)}{u(d)}$$

where $h_i(d)$ *is the number of optimal embeddings of the order of discriminant* d *into* R_i , *modulo conjugation by* R_i^* .

Proof. Let $\mathcal{O}$ be the order of discriminant $-D$. Any embedding $f : \mathcal{O} \to R_i$ gives rise to an element $b = f(\sqrt{-D})$ in R_i with $\mathrm{Tr}b = 0$ and $\mathrm{N}b = D$. Since $\mathcal{O} = \mathbb{Z} + \mathbb{Z}(\frac{D + \sqrt{-D}}{2})$ we have the congruence $b \equiv -D \pmod{2R_i}$, so b lies in S_i^0 and contributes to $c_i(D)$.

Conversely, if $b \in S_i^0$ then $b^2 = -D$, so $b \equiv -D \pmod{2R_i}$. Hence $\frac{b+D}{2}$ lies in R_i and we obtain an embedding $f : \mathcal{O} \to R_i$ by taking $\frac{\sqrt{-D}+D}{2}$ to $\frac{b+D}{2}$. This bijection completes the proof when $w_i = 1$. When $w_i > 1$ we take Γ_i orbits and analyse the stabilizers, as in the proof of Proposition 1.9.

Proposition 12.10. For all $m \geq 1$ *we have*

$$g_i | T_{m^2}^* = \sum_k B_{ik}(m) g_k = w_i \sum_k B_{ki}(m)(g_k/w_k) \quad .$$

In particular, the subgroups spanned by $\langle g_1, \ldots, g_n \rangle$ *and* $\langle g_1/w_1, \ldots, g_n/w_n \rangle$ *are stable under the Hecke algebra* $\mathbb{T}^*$, *and* $T_{m^2}^*$ *acts on the spanning sets by* $B(m)^{tr}$ *and* $B(m)$ *respectively*.

Proof. It suffices to check this when m is prime, as these operators generate $\mathbb{T}^*$ over $\mathbb{Z}$ and satisfy the same relations as the Brandt matrices in Proposition 2.7. Also, the second identity follows from the first, as $w_k B_{ik}(m) = w_i B_{ki}(m)$.

When $m = N$ we have $g_i | T^*_{N^2} = \frac{1}{2} + \sum_i a_i(DN^2) q^D$ by (12.5). But $a_i(DN^2) = a_i(D)$ using (12.9), as $h_i(DN^2) = 0$. Therefore $g_i | T^*_{N^2} = g_i$. On the otherhand, $\sum_k B_{ik}(m) g_k = g_j$ where $E_j \simeq E_i^N$ is the conjugate elliptic curve. Then $R_j \simeq R_i$ and $g_j = g_i$. This proves the identity in this case, as both sides are equal to g_i .

Now assume that $m = p$ is a prime not equal to N . According to (12.5), we must verify the identity:

$$c_i(Dp^2) + (\tfrac{-D}{p}) c_i(D) + p\ c_i(D/p^2) = \sum_k B_{ik}(p) c_k(D) .$$

First assume that $D \not\equiv 0 \pmod p$, so the third term on the left hand side is zero. Let a be an element of trace 0 and norm D in the order S_k and let $\phi : E_i \to E_k$ be an isogeny of degree p . Then the element $b = \phi^\vee \circ a \circ \phi$ lies in S_i , has trace zero and norm Dp^2 , and depends only on the kernel of ϕ .

In fact, all elements b of this trace and norm which are not divisible by p in S_i are obtained uiquely as $\phi^\vee \circ a \circ \phi$. Indeed, let $\phi : E_i \to E_k$

be the unique isogeny of degree p with $\ker\phi \subset \ker b$. Then the isogeny b factors through E_k : $E_i \underset{\phi}{\to} E_k \underset{\psi}{\to} E_i$. Since $b + b^\vee = 0$, $\ker\phi$ is the unique subgroup of order p in $\ker b^\vee$. Hence $\ker\phi \subset \ker\psi^\vee$ and the dual diagram factors

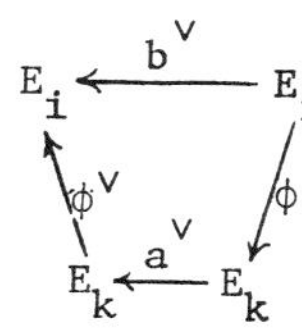

to define $a^\vee$ (and hence $a \in S_k$).

The element b is divisible by p in S_i if a stabilizes the subgroup $\phi(E_i[p]) = \ker\phi^\vee$. There are $(1 + (\frac{-D}{p}))$ subgroups of this type, and each gives a way to write $b = \phi^\vee \circ a \circ \phi$. Hence

$$\sum_k B_{ik}(p) c_k(D) = \{c_i(Dp^2) - c_i(D)\} + (1 + (\tfrac{-D}{p})) c_i(D)$$

$$= c_i(Dp^2) + (\tfrac{-D}{p}) c_i(D) .$$

We leave the case when p divides D to the reader.

As a corollary of Proposition 12.9 and Eichler's formula (1.12) the element G has Fourier expansion

$$(12.11)\qquad \begin{aligned} G &= \sum_{i=1}^{n} \frac{1}{w_i} g_i \\ &= \frac{N-1}{24} + \sum_{D>0} H_N(D) q^D \quad . \end{aligned}$$

As a corollary of Proposition 12.10, G is an eigenvector for the algebra $\mathbb{T}^*$ acting on $M^* \otimes \mathbb{Q}$ with eigenvalues $\sigma(m)_N$:

$$(12.12)\qquad G|T^*_{m^2} = \sigma(m)_N \cdot G \qquad \text{for all } m \geq 1 .$$

This is the normalized Eisenstein series of weight 3/2 and level $4N$; the multiple WG lies in the lattice M^*. Also, Eichler's trace formula in Proposition 1.9 is simply the identity

$$(12.13)\qquad G\cdot\theta|T_4 = \sum_{i=1}^{n} f_{ii}$$

among forms of weight 2 on $\Gamma_0(4N)$. We note that T_4 takes the product $G\cdot\theta$ to a form of level N , which generates $M \otimes \mathbb{Q}$ over $\mathbb{T} \otimes \mathbb{Q}$.

We conclude with a famous example, when $N = 2$. From (6.1) we find

$$S_1 = \mathbb{Z} + 2\mathbb{Z}i + 2\mathbb{Z}j + 2\mathbb{Z}k + \mathbb{Z}(1+i+j+k) .$$

Hence

$$S_1^o = \{b = xi+yj+zk \; : \; x,y,z \in \mathbb{Z} \; , \; x \equiv y \equiv z \pmod 2\}$$

$$g_1 = \frac{1}{2} \sum_{x \equiv y \equiv z (2)} q^{x^2+y^2+z^2}$$

$$= \frac{1}{2} + 4q^3 + 3q^4 + 6q^8 + 12q^{11} + \ldots \quad .$$

Since $g_1 = 12G = \frac{1}{2} + \sum_{D>0} 12H_2(D)q^D$, this gives the classical results on the number of representations of integers $D \equiv 3,4 \pmod 4$ as the sum of 3 squares.

13. Waldspurger's formula

Recall the classes e_D defined in $\mathrm{Pic}(X)^\vee$ by (3.8) and (4.7), and the class $e_0 = \sum_{i=1}^{n} e_i^\vee$. Define the formal series

$$g = \frac{1}{2} e_0 + \sum_{D>1} e_D q^D . \tag{13.1}$$

Then g may be viewed as a modular form of weight 3/2 with coefficients in $\mathrm{Pic}(X)^\vee$, or more precisely as an element of $\mathrm{Pic}(X)^\vee \otimes M^*$. Indeed, Proposition 12.9 gives the identity

$$g = \sum_{i=1}^{n} e_i^\vee \otimes g_i , \tag{13.2}$$

where g_i are the theta-series defined in (12.8) and $e_i^\vee = e_i/w_i$ is the basis of $\mathrm{Pic}(X)^\vee$.

Actually, a little more is true. We have the following.

Proposition 13.3. *g is an element of* $\mathrm{Pic}(X)^\vee \otimes_{\mathbb{T}} M^* = \mathrm{Hom}_{\mathbb{T}}(\mathrm{Pic}(X), M^*)$. *More precisely, for any class* $e \in \mathrm{Pic}(X)$ *the series*

$$g(e) = \frac{\deg e}{2} + \sum_{D>1} \langle e, e_D \rangle q^D$$

is an element of M^*, and $g(t_m e) = g(e)|T^*_{m^2}$ for all $m \geq 1$.

Proof. It suffices to check this when $e = e_i$. But $g(e_i) = g_i$ and $g(t_m e_i) = \sum_{j=1}^{n} B_{ij}(m)\, g_i$ by Proposition 4.4. This agrees with $g_i|T^*_m$ by Proposition 12.10.

Now let e_f be a non-zero element in the f-isotypical component of $\mathrm{Pic}(X) \otimes \mathbb{R}$, where f is an eigenform for $\mathbb{T}$. This component has dimension 1, so e_f is determined up to a real scalar multiple. The modular form

(13.4) $$g(e_f) = \sum m_D q^D$$

then lies in the f-isotypical component of $M^* \otimes \mathbb{R}$. It is clearly zero unless $f|T_N = f$, so the sign in the functional equation for $L(f,s)$ is $+1$.

Proposition 13.5. Let $-D$ be a fundamental discriminant with $(\frac{-D}{N}) = -1$. Then

$$L(f,1)L(f \otimes \varepsilon_D, 1) = \frac{(f,f)}{\sqrt{D}} \frac{m_D^2}{\langle e_f, e_f \rangle} .$$

Proof. We will use Corollary 11.6 and show that

$$\langle e_{f,D}, e_{f,D} \rangle = \frac{m_D^2}{\langle e_f, e_f \rangle}$$

By definition: $m_D = \langle e_f, e_D \rangle = \langle e_f, e_{f,D} \rangle$. Hence

$$e_{f,D} = \frac{m_D^2}{\langle e_f, e_f \rangle} e_f \qquad \text{in} \quad (\mathrm{Pic}(X) \otimes \mathbb{R})^f .$$

The formula in Proposition 13.5 is due to Waldspurger [11], and gives the variation of the special values with the discriminant $-D$. If $-D$ is fundamental and $D \equiv 0(N)$ the correct formula is:

$$L(f,1)L(f \otimes \varepsilon_D, 1) = \frac{(f,f)}{\sqrt{D}} \frac{2 \cdot m_D^2}{\langle e_f, e_f \rangle} .$$

We will not prove this here; it follows from methods similar to those in §8-10.

Corollary 13.6. *The rank of the subgroup spanned by the* t *distinct theta series* g_i *in* M^* *is equal to the number of eigenforms* f *(including the weight 2 Eisenstein series) with* $L(f,1) \neq 0$.

Proof. Since the subgroup spanned is stable under $\mathbb{T}^*$, it suffices to determine which eigenforms f satisfy $g(e_f) \neq 0$. By Proposition 13.5 this will be true if $L(f,1)L(f \otimes \varepsilon_D, 1) \neq 0$ for a fundamental discriminant $-D$ with $(\frac{-D}{N}) = -1$. But Waldspurger has shown [12] that it is always possible to choose D so that $(\frac{-D}{N}) = -1$ and $L(f \otimes \varepsilon_D, 1) \neq 0$, the condition reduces to $L(f,1) \neq 0$.

One case where the rank is less than t is when $N = 389$. Here $t = 22$ and the rank is 21; there is an eigenform with rational Fourier coefficients with $\mathrm{ord}_{s=1} L(f,s) = 2$. It would be interesting to determine the linear relation on the theta-series explicitly in this case.

We end this section by explicitly computing an example, in level $N = 11$. The unique normalized cusp form f was given in (6.4) and corresponds to the elliptic curve $X_0(11)$. We have $e_f = e_2 - e_1$ and $\langle e_f, e_f \rangle = 5$. If $K = \mathbb{Q}(\sqrt{-D})$, and 11 is inert in K, we obtain

$$L(X_0(11)/K) = \frac{(f,f)}{5\sqrt{D}} m_D^2 \tag{13.7}$$

where m_D is the Dth coefficient of $g(e_f) = g_2 - g_1$. If 11 is ramified in K, the above formula holds with m_D^2 replaced by $2m_D^2$.

We recall that the maximal orders in B are distinguished by $w_1 = 2$ and $w_2 = 3$; we find after an explicit description of R_1 and R_2 that

$$g_1 = \frac{1}{2} \sum_{x \equiv y \pmod 2} q^{x^2+11y^2+11z^2} = \frac{1}{2} + q^4 + q^{11} + 2q^{12} + 2q^{15} + q^{16} + \cdots$$

$$g_2 = \frac{1}{2} \sum_{\substack{x \equiv y \pmod 3 \\ y \equiv z \pmod 2}} q^{(x^2+11y^2+33z^2)/3} = \frac{1}{2} + q^3 + q^{12} + 3q^{15} + 3q^{16} + \cdots .$$

Here is a table of the first few coefficients of the eigenform $g(e_f)$:

D	m_D
3	1
4	-1
11	-1
12	-1
15	1
16	2
20	1

D	m_D
23	-1
27	-1
31	-1
36	0
44	1
47	0
48	0

D	m_D
55	1
56	2
59	-1
60	-3
64	-2
67	3
71	1

Since $2g(e_f) \equiv 6G \pmod{5M^*}$, where G is the normalized Eisenstein series defined in (12.11), we obtain the congruences

$$m_D \equiv 3H_{11}(D) \pmod{5} \tag{13.8}$$

for all $D > 0$. Using (13.7) this gives congruences for the special values, which are due to Mazur [6].

In the next section, we shall see that if $-D$ is the discriminant of an imaginary quadratic field K and $m_D \neq 0$, then the conjecture of Birch and Swinnerton-Dyer predicts that

$$m_D^2 = [Ш(X_0(11)/K)] \ . \tag{13.9}$$

In particular, the integer m_D should always annihilate $Ш(X_0(11)/K))$.

14. Elliptic curves with prime conductor

We now consider the special case where the normalized eigenform f has integral Fourier coefficients. Then f corresponds to an isogeny class of elliptic curves $\{E\}$ over $\mathbb{Q}$ with conductor N which appear as quotients of the modular curve $X_0(N)$. The L-series of E over $\mathbb{Q}$ is equal to $L(f,s)$.

In the isogeny class $\{E\}$ there is a distinguished curve E_0, called the strong Weil curve, where the covering

$$(14.1) \qquad \pi : X_0(N) \to E$$

has minimal degree. In this case, the induced map on homology $\pi_* : H_1(X_0(N)(\mathbb{C}),\mathbb{Z}) \to H_1(E_0(\mathbb{C}),\mathbb{Z})$ is surjective, so the induced map of Jacobians has a connected kernel (which is an abelian variety).

$$(14.2) \qquad 0 \to A \to J_0(N) \to E_0 \to 0 .$$

For any curve in the isogeny class, we let ω be a Néron differential on E, $\Delta = \Delta(\omega)$ the minimal discriminant, and t be the order of the finite group $E(\mathbb{Q})_{tor}$. We assume that the parametrization given by (14.1) has minimal degree for E, and adjust its sign so that $\pi^*(\omega) = c \cdot f(q)\frac{dq}{q}$ with $c > 0$. We will assume Manin's conjecture that $c_0 = 1$ for the strong Weil curve E_0 in each isogeny class. (Raynaud has recently proved that $c_0 = 1$ or 2).

The strong Weil curve E_0 is the unique curve in its $\mathbb{Q}$-isogeny class and $t_0 = 1$, except in the following cases [7].

N	curves (see [1, pg. 82-84])	c	t	Δ	deg π
11	$E_0 = J_0(11)$	1	5	-11^5	1
	E_0/μ_5	5	5	-11	5
	$E_0/(\mathbb{Z}/5)$	1	1	-11	5
17	$E_0 = J_0(17)$	1	4	-17^4	1
	E_0/μ_2	2	4	17^2	2
	E_0/μ_4	4	4	17	4
	$E_0/(\mathbb{Z}/4)$	2	2	17	4
19	$E_0 = J_0(19)$	1	3	-19^3	1
	E_0/μ_3	3	3	-19	3
	$E_0/(\mathbb{Z}/3)$	1	1	-19	3
37	$E_0 = J_0(37)^{W=-1}$	1	3	37^3	2
	E_0/μ_3	3	3	37	6
	$E_0/(\mathbb{Z}/3)$	1	1	37	6
$64+u^2$	$E_0: y^2 = x^3-2ux^2+Nx$	1	2	$-N^2$	?
(if π exists)	$E_0/\mu_2: y^2 = x^3+ux^2-16x$	2	2	N	2?

In particular, we have $c \le t \le 5$.

When E_0 is the unique curve in its $\mathbb{Q}$-isogeny class, it is reasonable to conjecture that $\Delta = \pm N$; some evidence for this conjecture is given in [5, §9]. When combined with the information in the previous table, this leads to the slightly more general conjecture that

(14.3) $$t \stackrel{?}{=} c \cdot \mathrm{ord}_N(\Delta) .$$

Now let e_f be an element in the f-isotypical component of Pic(X) which is not divisible by any integer $n > 1$; then e_f is well-determined up to sign and Proposition 13.5 gives the identity

(14.4) $$L(f,1)L(f\otimes\varepsilon_D,1) = \frac{(f,f)}{\sqrt{D}} \frac{m_D^2}{\langle e_f, e_f\rangle} ,$$

where m_D is the Dth coefficient of the form $g(e_f)$ in M^* . The left hand side of this identity is equal to the L-function of E over the field $K = \mathbb{Q}(\sqrt{-D})$ where N is inert. If $L(E/K,1) \neq 0$, then the conjecture of Birch and Swinnerton-Dyer predicts that the rank of E(K) is equal to zero and that

(14.5) $$L(E/K,1) \stackrel{?}{=} \frac{\int_{E(\mathbb{C})} \omega\wedge\bar{\omega}}{\sqrt{D}} \frac{\mathrm{ord}_N(\Delta)}{t^2} [Ш_D] ,$$

where $Ш_D$ is the (conjecturally finite) Tate-Shafarevitch group of E over K .

Since $c^2(f,f) = \deg \pi \cdot \int_{E(\mathbb{C})} \omega\wedge\bar{\omega}$, and Mestre and Oesterlé have recently shown that the identity

$$(14.6) \qquad \langle e_f, e_f \rangle \overset{?}{\doteq} \deg \pi \cdot \mathrm{ord}_N(\Delta)$$

follows from conjecture (14.3), we are led to the following.

Conjecture 14.7. If $m_D \neq 0$ *then* $E(K)$ *is finite of order* t *and* $Ш_D$ *is finite of order* m_D^2.

15. Bibliography

1. Birch, B. J. and Kuyk, W., Modular functions of One Variable IV. Springer Lecture Notes 476 (1975).

2. Eichler, M., Zur Zahlentheorie der Quaternion-Algebren. Crelle J. 195 (1955), 127-151.

3. Gross, B. and Zagier, D., Heegner points and derivatives of L-series. Inv. Math. (1986).

4. Kohnen, W., Newforms of half-integral weight. Crelle J. 333 (1982), 32-72.

5. Kramer, K. and Brumer, A., The rank of elliptic curves. Duke Math J. 44 (1977), 715-743.

6. Mazur, B., On the arithmetic of special values of L-functions. Inv. math. 55 (1979), 207-240.

7. Miyakawa, I., Elliptic curves of prime conductor with $\mathbb{Q}$-rational points of finite order. Osaka J. math. 10 (1973), 309-323.

8. Ohta, M., Theta series mod p . J. Fac. Sci. Tokyo 28 (1981), 679-686.

9. Pizer, A., An algorithm for computing modular forms on $\Gamma_0(N)$. J. Algebra 64 (1980), 340-390.

10. Serre, J-P., Trees, Springer-Verlag (1980).

11. Waldspurger, J-L., Sur les coefficients de Fourier des formes modulaires de poids demi-entier. J. Math. pures et appl. 60 (1981), 375-484.

12. Waldspurger, J-L., Correspondance de Shimura et Quaternions.

13. Zagier, D., The Rankin-Selberg method for automorphic functions which are not of rapid decay. J. Fac. Sci. Tokyo 28 (1982), 415-438.

BENEDICT H. GROSS
DEPARTMENT OF MATHEMATICS
HARVARD UNIVERSITY
CAMBRIDGE, MA 02138, USA.

Canadian Mathematical Society
Conference Proceedings
Volume 7 (1987)

THE EUCLIDEAN ALGORITHM FOR S-INTEGERS

Rajiv Gupta*, M. Ram Murty* and V. Kumar Murty*

1. INTRODUCTION. Let $m, n \in \mathbb{Z}$, $n \neq 0$. Following Euclid, there are $q, r \in \mathbb{Z}$ such that $m = qn + r$ with $r = 0$ or $0 < |r| < |n|$. The analogous assertion for number fields has received considerable attention over the years. To formulate it precisely, let K be a number field and S a finite set of places of K containing the infinite places S_∞. Let $\mathcal{O}_S$ denote the ring of S-integers of K (i.e., elements x of K with $\mathrm{ord}_{\mathfrak{P}} x \geq 0$ for all primes $\mathfrak{P}$ of K not in S).

DEFINITION. $\mathcal{O}_S$ is <u>Euclidean</u> if there is a function

$$\psi : \mathcal{O}_S - \{0\} \longrightarrow \mathbb{Z} \geq 0$$

such that for all $\alpha, \beta \in \mathcal{O}_S$, $\beta \neq 0$, there are $q, r \in \mathcal{O}_S$ satisfying

$$\alpha = q\beta + r$$

and $r = 0$ or $\psi(r) < \psi(\beta)$.

Such a function ψ will be called a <u>Euclidean algorithm</u> for $\mathcal{O}_S$. It is well known that if $\mathcal{O}_S$ is Euclidean, then it is a principal ideal domain. In general, the converse is false. For example, if we take $\mathcal{O} = \mathbf{Z}[\frac{1}{2}(1 + \sqrt{-19})]$ which is the ring of integers in $K = \mathbb{Q}(\sqrt{-19})$, then $\mathcal{O}$ is a principal ideal domain. (One easily checks this from the Minkowski bound [9, p. 119] and the fact that 2 and 3 remain prime in K). If ψ is a Euclidean algorithm for $\mathcal{O}$, take $\alpha_0 \in \mathcal{O}$ satisfying

$$\psi(\alpha_0) = \min \psi(\alpha)$$

where the minimum is over all $\alpha \in \mathcal{O}$ with $\psi(\alpha) \neq 0$. Thus, every residue class mod α_0 is represented by 0 or an element $\alpha \in \mathcal{O}$ with $\psi(\alpha) = 0$. Since $\psi(\alpha) = 0$ implies that α is a unit, we deduce that $N(\alpha_0) \leq 3$. By the fact stated above about 2 and 3, we get a contradiction. Note that in this example $|S| = |S_\infty| = 1$.

* Research partially supported by NSERC grants A8910, U0237 and U0373.

On the other hand, Lenstra [10] and Weinberger [19] have shown that if $\mathcal{O}_S$ is a principal ideal domain and $|S| \geq 2$, then the Riemann Hypothesis for all Dedekind zeta functions implies that $\mathcal{O}_S$ is Euclidean. Our main result here is to show that the assumption of the Riemann Hypothesis may be removed if we insist on a larger set S. More precisely, let us define

$$g \overset{df}{=} \mathrm{GCD}\,\{(N_{k/\mathbb{Q}}(\mathcal{P}) - 1) : \mathcal{P} \in S - S_\infty\} .$$

THEOREM. Suppose that K is Galois over $\mathbb{Q}$ and that

a) $|S| \geq \max(5, 2[K:\mathbb{Q}]-3)$

b) K has a real embedding or $\zeta_g \in K$.

If $\mathcal{O}_S$ is a principal ideal domain, then it is Euclidean.

The idea of the proof is to combine a method used by the first two authors [4] in their work on the Artin primitive root conjecture with a result of the second two authors [15] on the Bombieri-Vinogradov theorem. We would like to heartily thank Lenstra for suggesting that the methods of [4] be developed in this direction.

The study of Euclidean rings of integers before the work of Lenstra and Weinberger proceeded along quite different lines, and in the next section we briefly review the history of the problem and recall the known results. In section 3, we give a technical lemma which is essential in our arguments. In section 4, we give the proof of the Theorem. In section 5, we return to the Artin primitive root conjecture and give a slight refinement of [4].*

Finally, we remark that where $\mathcal{O}_S$ is proved Euclidean by the Theorem, the algorithm provided is <u>not</u> the norm function. These are the first such examples.

2. A BRIEF HISTORY. The problem of deciding which rings $\mathcal{O}_S$ are Euclidean has been studied extensively, but mostly in the case $S = S_\infty$ and ψ the norm function. We give a brief survey of known results, referring the reader to Samuel [18], Lenstra [10,11] and Narkiewicz [16] for further references. When $S = S_\infty$, we shall write $\mathcal{O}_K$ (or simply $\mathcal{O}$) for $\mathcal{O}_S$.

If $[K:\mathbb{Q}] = 2$, there are two cases to consider. If $K = Q(\sqrt{-d})$ is imaginary quadratic, $\mathcal{O}$ has class number 1 for

$$d = 1, 2, 3, 7, 11, 19, 43, 67, 163.$$

Of these, $\mathcal{O}$ is Euclidean only for the first five, and in this case, the norm serves as an algorithm (cf. [18, Prop. 13]). If $K = \mathbb{Q}(\sqrt{d})$ is real quadratic,

* After this paper was written, we learned that R. Heath-Brown has made further refinements. See also M. Ram Murty and S. Srinivasan, "Some remarks on Artin's conjecture", to appear.

it is conjectured that $\mathcal{O}$ has class number 1 for infinitely many d. Heilbronn [6] showed, however, that the norm can be a Euclidean algorithm only for finitely many d. It is classical (cf. [5, ch. 14]) that the norm <u>is</u> an algorithm for the sixteen values

$$d = 2,3,5,6,7,11,13,17,19,21,29,33,37,41,57,73 \quad .$$

Combining work of Hua, Davenport, Inkeri and Chatland, it is known that the above list is complete (see [16] for further references). Thus, $\mathbb{Z}[\sqrt{14}]$, for example, is not Euclidean for the norm.

Write $n_K = [K:\mathbb{Q}] = r+2s$, where r is the number of real embeddings and s is the number of non-conjugate complex embeddings of K. For $\mathcal{O}$ to be Euclidean for the norm means that given any $\alpha \in K^*$, there exists $\beta \in \mathcal{O}$ so that $N_{K/\mathbb{Q}}(\alpha - \beta) < 1$. The norm $N_{K/\mathbb{Q}}$ is a homogeneous form of degree n_K in n_K variables. Given a general homogeneous form (over $\mathbb{R}$) f of degree n in n variables, we define for each point $P \in \mathbb{R}^n$,

$$M(f,P) = \inf |f(Q)|$$

where the infimum is over all $Q \in \mathbb{R}^n$ so that $Q-P \in \mathbb{Z}^n$. We also set

$$M(f) = \sup M(f,P)$$

over all $P \in \mathbb{R}^n$. Then, $M(f)$ is called the <u>inhomogeneous minimum</u> of f. When f is the norm form, $\mathcal{O}$ is Euclidean if and only if $M(f) < 1$.

There is an extensive literature on estimates for $M(f)$, mostly in the case $n=2$. Cassels [1] showed that if $n \leq 4$ and f can be factored into linear forms with the number of factors (counting conjugate factors only once) at most 2, then

$$M(f) > \frac{|\Delta|}{\gamma}$$

where Δ is the discriminant of f, and $\gamma = 420$ if $n=3$, $\gamma = 5300$ if $n=4$. Applying this to f, the norm form, it follows that if $r+s \leq 2$, there are only finitely many K for which $\mathcal{O}$ is Euclidean for the norm.

Another technique for determining whether the norm is a Euclidean algorithm was developed by Lenstra [11]. He defines a constant

$$M = M(K) = \sup\{m : \text{there exist } \alpha_1,\dots,\alpha_m \in \mathcal{O} \text{ such that } \alpha_i - \alpha_j \text{ is a unit for } 1 \leq i < j \leq m\} \quad .$$

If we have the bound

$$M > n!\,n^{-n}(4/\pi)^s |d_K|^{\frac{1}{2}}$$

where $n = n_K$ and d_K is the discriminant of K. then $\mathcal{O}$ is Euclidean for the norm. The proof uses ideas from packing theory. Let $U \subseteq \mathbb{R}^n$ be a bounded measurable set with positive Lebesgue measure $\mu(U)$. If $\{a_i\}$ is a sequence of points in $\mathbb{R}^n$ which are "sufficiently well distributed", and $\mu(U)$ is sufficiently large then the system $\{U+a_i\}$ is not a packing of U in the sense that there are i,j with $i \neq j$ and $(U+a_i) \cap (U+a_j) \neq \emptyset$ (see [11] for a precise statement). Lenstra uses

$$U = \{(x_j) \in \mathbb{R}^n \times \mathbb{C}^s : \sum_{j=1}^{r} |x_j| + 2 \sum_{j=r+1}^{r+s} |x_j| < \frac{1}{2} n_K\} .$$

This set has the property that for $u = (x_j)$, $v = (y_j) \in U$,

$$N(u-v) \stackrel{df}{=} \prod_{j=1}^{r} |x_j - y_j| \prod_{j=r+1}^{r+s} |x_j - y_j|^2 < 1 .$$

Let $\alpha_1, \ldots, \alpha_M \in \mathcal{O}$ be such that the difference of any two is a unit, and for $x \in K^*$, consider the sequence $\{\alpha_i x + \alpha\}$ as α ranges over all of $\mathcal{O}$ and $1 \leq i \leq M$. The above bound on M is what is needed to guarantee that $(U + \alpha_i x + \alpha) \cap (U + \alpha_j x + \beta) \neq \emptyset$ for some $1 \leq i < j \leq M$ and $\alpha, \beta \in \mathcal{O}$. Then

$$x = \frac{\beta - \alpha}{\alpha_i - \alpha_j} + \frac{u-v}{\alpha_i - \alpha_j}$$

for some $u, v \in U$. As $(\alpha_i - \alpha_j)$ is a unit, $\gamma = (\beta - \alpha)/(\alpha_i - \alpha_j) \in \mathcal{O}$ and $N(x - \gamma) < 1$ by the choice of U. Using this method, Lenstra produces 132 examples of Euclidean fields (i.e., K for which $\mathcal{O}$ is Euclidean for the norm) with $4 \leq n_K \leq 8$.

Using a variety of methods, several authors have studied the case of cyclotomic fields $K = \mathbb{Q}(\zeta_m)$. It is known that $\mathbb{Q}(\zeta_m)$ has class number 1 for precisely the following thirty values of m (cf. Masley and Montgomery [13]):

$$m = 1,3,4,5,7,8,9,11,12,13,15,16,17,19,20,$$
$$21,24,25,27,28,32,33,35,36,40,44,45,48,60,84.$$

It is known that $\mathbb{Z}[\zeta_m]$ is Euclidean for the following values:

m	Reason
1, 3, 4	$\phi(m) \leq 2$
5, 7	Lenstra [11, §3.1]
8	Lakein [8]
9, 11, 12, 15	Lenstra (loc. cit.)
16, 20, 24	Ojala [17] .

Finally, we mention that variants of the Euclidean condition have been studied by Cooke [2] and applied by Cooke and Weinberger [3] to study the existence of division chains in number fields.

We return to consider the general case of $\mathcal{O}_S$. Define a sequence of sets

$$E_{-1} = \{0\}$$
$$E_0 = \{0\} \cup \mathcal{O}_S^*$$
$$n \geq 1 \qquad E_n = \{\alpha \in \mathcal{O}_X : \text{every residue class mod } \alpha\mathcal{O}_S \text{ has a representative in } E_{n-1}\} \cup E_{n-1} .$$

Then, Motzkin [14] proved that $\mathcal{O}_S$ is Euclidean if and only if

$$\mathcal{O}_S = \bigcup_{n \geq -1} E_n .$$

Moreover, when $\mathcal{O}_S$ is Euclidean, the function ψ given by

$$\psi(E_n - E_{n-1}) = n \quad (n \geq 0)$$

is a Euclidean algorithm. Motzkin also introduced the notion of a minimal algorithm. Precisely, if $\{\psi_\alpha\}$ is a family of algorithms, with α ranging over some indexing set A, then

$$\psi_A = \min_{\alpha \in A} \psi_\alpha$$

is also an algorithm. In particular, there is a "smallest" algorithm ψ_{min} (say). As Samuel [18] showed, this algorithm satisfies

$$\psi_{min}(xy) \geq \psi_{min}(x) + \psi_{min}(y)$$

if $xy \neq 0$. (Notice that in particular, $\psi_{min}(xy) \geq \psi_{min}(x)$ if $y \neq 0$. In earlier works, this was sometimes imposed as an axiom to be satisfied by any algorithm. That the class of rings which are Euclidean is the same whether we include this axiom or not seems to be due to Veldkamp (cf. [16, p. 114])). It follows that for any $x \in \mathcal{O}_S - \{0\}$,

$$\psi_{min}(x) \geq \sum_{\mathfrak{P} \notin S} \text{ord}_{\mathfrak{P}}(x) \cdot \psi_{min}(\pi)$$

where in the sum, π is a generator of $\mathfrak{P}$. Note that by the semi-additivity, $\psi_{min}(\pi)$ does not depend on the choice of generator π. Define

$$n = \begin{cases} 1 & \text{if } \mathcal{O}_S^* \longrightarrow (\mathcal{O}_S/\mathfrak{P})^* \text{ is surjective} \\ 2 & \text{otherwise ,} \end{cases}$$

and set

$$\theta(x) = \sum_{\mathfrak{P} \notin S} \text{ord}_{\mathfrak{P}}(x) \cdot n_{\mathfrak{P}} .$$

Then certainly, $\psi_{min} \geq \theta$. Samuel studied this function and was led to observe that the condition $n_{\mathfrak{P}} = 1$ was similar in spirit to Artin's primitive root conjecture.

This was formalized by Weinberger [19] and in a refined form by Lenstra [10], who showed the following.

THEOREM. Suppose that $|S| \geq 2$ and all Dedekind zeta functions satisfy the Riemann Hypothesis. Then $\mathcal{O}_S$ is a principal ideal domain if and only if

$\mathcal{O}_S$ is Euclidean. When $\mathcal{O}_S$ is Euclidean, θ is the minimal algorithm.

As was observed in [10] and [19], the difficulty in proving that θ is an algorithm lies entirely at $\beta \in \mathcal{O}_S$ with $\theta(\beta) = 2$. Indeed, let $\alpha, \beta \in \mathcal{O}_S$, $\beta \neq 0$. We want to produce $q, r \in \mathcal{O}_S$ such that $\alpha = q\beta + r$ and $r = 0$ or $\theta(r) < \theta(\beta)$. By the additivity of θ, we may suppose that $(\alpha, \beta) = \mathcal{O}_S$. If $\theta(\beta) = 0$, then $\beta \in \mathcal{O}_S^*$ so $r = 0$. If $\theta(\beta) = 1$, then from the definition of θ, (β) is a prime ideal and $\mathcal{O}_S^*$ surjects on $\mathcal{O}_S/(\beta)$. Thus, we may choose $r = 0$ or an r with $\theta(r) = 0 < \theta(\beta)$. If $\theta(\beta) \geq 3$, then by Dirichlet's theorem on arithmetic progressions in number fields, every coprime residue class $\bmod(\beta)$ contains a prime element π (say). Then $\theta(\pi) \leq 2 < \theta(\beta)$.

When $\theta(\beta) = 2$, we need $r \in \mathcal{O}_S$, $r \equiv \alpha (\bmod \beta)$ and $r \in \mathcal{O}_S^*$ or (r) is a prime ideal and $n_{(r)} = 1$. Let us write

$$\varphi : \mathcal{O}_S^* \longrightarrow (\mathcal{O}_S/\beta\,\mathcal{O}_S)^*$$

for the natural reduction map, and consider the S-ray class field F/K of conductor dividing $\beta\,\mathcal{O}_S$. Thus F is the maximal abelian extension of K whose conductor divides $\beta\,\mathcal{O}_S$ and in which every prime in S splits completely. Moreover, if $K \subseteq M \subseteq F$ and M/K is everywhere unramified, then $M = K$. Such an extension exists and is unique by class field theory, and we have

$$\mathrm{Gal}(F/K) \simeq (\mathcal{O}_S/\beta\,\mathcal{O}_S)^*/\varphi(\mathcal{O}_S^*) .$$

We denote by C the element in $\mathrm{Gal}(F/K)$ corresponding to $\alpha (\bmod \beta)$ under the above isomorphism. The existence of r as above is equivalent to the existence of a prime ideal $\mathfrak{P}$ of K satisfying

$$(\mathfrak{P}, F/K) = C$$
$$n_{\mathfrak{P}} = 1$$

where $(\ ,\ )$ denotes the Artin symbol. We discuss the existence of such primes in section 4. In the next section, we give a technical lemma which is crucial in what follows.

3. A TECHNICAL LEMMA. We will require some notation. Let E_1/E_2 be a Galois extension and D a subset of $\mathrm{Gal}(E_1/E_2)$ stable under conjugation (a conjugacy subset for short). We define

$$d = d(E_1/E_2, D)$$

to be the largest integer such that $\mathbb{Q}(\zeta_d) \subseteq E_1$ and $D|_{\mathbb{Q}(\zeta_d)} = 1$. Let $E/\mathbb{Q}$ be a Galois extension with group G, and C any conjugacy class of G. We define

$$\eta^*(C) = \min_H \max_\omega \frac{|H|}{|H|} \omega(1)$$

where the minimum is over subgroups $H \subseteq G$ satisfying

(i) $H \cap C \neq \emptyset$

(ii) for every non-trivial irreducible character ω of H with $\omega|_{H\cap C} \neq 0$, and every Dirichlet character χ, the Artin L-series $L(s,\omega \otimes \chi)$ is entire,

and the maximum is taken over irreducible non-trivial characters ω of H with $\omega|_{H\cap C} \neq 0$. We also define

$$\eta = \eta(C) = \max(4, 2\eta^*(C)-2) .$$

We can now state the technical lemma.

LEMMA. Let $\epsilon > 0$. There is a constant $\delta_1 > 0$ and $\geq \delta_1 x(\log x)^{-2}$ primes $p \leq x$, satisfying

(i) $(p,E/\mathbb{Q}) \subset C$

(ii) $t \mid (p-1)$, t prime $\Longrightarrow t \mid d(E/\mathbb{Q},C)$ or $t > p^{(1/\eta)-\epsilon}$.

PROOF. This follows from the lower bound sieve method using the twisted Bombieri-Vinogradov theorem of [15, Theorem 7.3].

4. PROOF OF THE THEOREM. We recall our assumptions:

(a) $|S| \geq \max(5, 2n_K-3)$

(b) either K has a real embedding or $\zeta_g \in K$ where $g = \mathrm{GCD}(N_{K/\mathbb{Q}}\mathfrak{P})-1 : \mathfrak{P} \in S-S_\infty)$.

By the discussion of section 2, it suffices to prove the following result Let $\beta \in \mathcal{O}_S$ with $\theta(\beta) = 2$. Let F/K denote the S-ray class field of conductor $\beta\mathcal{O}_S$, and fix $1 \neq C \in \mathrm{Gal}(F/K)$.

PROPOSITION. There are infinitely many primes $\mathfrak{P}$ of K satisfying $(\mathfrak{P},F/K) = C$ and $n_{\mathfrak{P}} = 1$.

The proof will require several lemmas. For any prime ℓ, let L_ℓ denote the Kummer extension of K obtained by adjoining the ℓ-th roots of all elements of $\mathcal{O}_S^*$. (In particular, $\zeta_\ell \in L_\ell$). Let L denote the compositum $F \prod L_\ell$, the product taken over prime divisors ℓ of $d(F/K,C)$, and set

$$C' = \{\sigma \in \mathrm{Gal}(L/K) : \sigma|_F = C \text{ and } \sigma|_{L_\ell} \neq 1 \text{ for } \ell \mid d(F/K,C)\} .$$

By Lenstra [10, pp. 221-223], C' is not empty.

LEMMA 1. $d(L/K,C') = d(F/K,C)$.

PROOF. First, we note that if $\ell \mid d(F/K,C)$, then $\zeta_\ell \in K$. Indeed, $F \supseteq K(\zeta_\ell) \supseteq K$ and so every prime in S splits completely in $K(\zeta_\ell)$. If K has a real embedding, this forces $\ell = 2$ so $\zeta_\ell \in K$. Otherwise, $\ell \mid g$ and so (b) above implies $\zeta_\ell \in K$ in this case also. From this, it follows that L/K is abelian. Now take a prime t dividing $d(L/K,C')$. If $K(\zeta_t)$ is contained in F, then t divides $d(F/K,C)$ also. So suppose $\zeta_t \notin F$. Let $H = \mathrm{Gal}(L/F)$, $J = \mathrm{Gal}(L/K(\zeta_t))$ and for each $\ell \mid d(F/K,C)$, set $H_\ell = \mathrm{Gal}(L/L_\ell)$. Since

$$\sigma_0 H - \bigcup_{\ell \mid d} (\sigma_0 H \cap H_\ell) \subseteq C' \subseteq J \cap \sigma_0 H$$

(for some fixed σ_0 in C'), we have

$$(*) \qquad |H| - \sum_{\ell \mid d} |H \cap H_\ell| \le |J \cap \rho_0 H| \le \tfrac{1}{2}|H| \quad .$$

Now,

$$|H \cap H_\ell| = |\mathrm{Gal}(L/L_\ell F)| = |H| / |\mathrm{Gal}(L_\ell/(L_\ell \cap F))| \quad .$$

Note that L_ℓ/K is of degree a power of ℓ and is ramified at most at primes dividing ℓ and at the primes in S. On the other hand, there are no extensions intermediate between F and K which are everywhere unramified over K. Now, suppose that $L_\ell \cap F \neq K$. Then, there is a prime $\mathcal{L}$ of K dividing ℓ with

$$\mathcal{L} \notin S , \quad n_{\mathcal{L}} = 1 \quad \text{and} \quad \beta \mathcal{O}_S = \mathcal{L}^2 \quad .$$

We use an argument of Lenstra [10, p. 222] to deduce that $F \subseteq L_\ell$. Briefly, since $n_{\mathcal{L}} = 1$, and $\mathrm{Gal}(F/K) \simeq (\mathcal{O}_S/\beta\mathcal{O}_S)^*/\varphi(\mathcal{O}_S^*)$, $[F:K]$ is a power of ℓ and so, by Kummer theory, $F = K(x_1^{1/\ell},\ldots,x_s^{1/\ell})$ for some $x_i \in K^*$, $x_i \in K^{*\ell}$ and $1 \le s \in \mathbf{Z}$. Since F is unramified outside $\mathcal{L}$, we have $\mathrm{ord}_{\mathcal{P}}\, x_i \equiv 0 \pmod{\ell}$ for all $\mathcal{P} \neq \mathcal{L}$. As $\mathcal{O}_S$ is a principal ideal domain, we can adjust the x_i by ℓ-th powers so that $\mathrm{ord}_{\mathcal{P}}\, x_i = 0$ for all $\mathcal{P} \notin S \cup \{\mathcal{L}\}$. By a local discriminant calculation and the fact that the conductor of F/K divides $\mathcal{L}^2$, we can also ensure that $\mathrm{ord}_{\mathcal{L}}\, x_i = 0$. Hence, $x_i \in \mathcal{O}_S^*$ so that $F \subseteq L_\ell$.

Hence, we deduce that in any case,

$$|\mathrm{Gal}(L_\ell/(L_\ell \cap F))| \ge [L_\ell:K]/[F:K] \ge \ell^r/N_{K/\mathbb{Q}}\mathcal{L} \ge \ell^{r-n_K} \quad .$$

Thus, $|H \cap H_\ell| \le |H|/\ell^{r-n_K}$. Using this in (*), we find

$$\frac{1}{2} \le \sum_{\ell \mid d} \frac{1}{\ell^{r-n_K}} \le \sum_{\ell=2}^{\infty} \ell^{n_K-r}$$

$$\le (r-n_K-1)^{-1}$$

which contradicts (a). Thus $\zeta_t \in F$ and t divides $d(F/K,C)$. On the other hand, we clearly have $d(F/K,C) \mid d(L/K,C')$. This proves the lemma.

Now let $\tilde{L}$ denote the normal closure of L over $\mathbb{Q}$. Let C'' denote the inverse image of C' in $\mathrm{Gal}(\tilde{L}/K)$ and let $C^* = \bigcup g^{-1}C''g$ where g ranges over $\mathrm{Gal}(\tilde{L}/\mathbb{Q})$.

LEMMA 2. $d(\tilde{L}/\mathbb{Q},C^*) = d(L/K,C')$.

PROOF. Clearly, $d(\tilde{L}/\mathbb{Q},C^*) = d(L/K,C')$. Now suppose that t is a prime divisor of $d(\tilde{L}/K,C'')$. Then $C'|_{K(\zeta_t)\cap L} = C''|_{K(\zeta_t)\cap L} = 1$. If $K(\zeta_t) \neq K(\zeta_t)\cap L$, then there exist primes $\mathcal{P}$ of K such that $\mathcal{P}$ does not split completely in $K(\zeta_t)$ and $(\mathcal{P},L/K) \subset C'$. The latter is equivalent to the condition $(\mathcal{P},\tilde{L}/K) \subset C''$. This contradicts $t \mid d(\tilde{L}/K,C'')$. Hence, $K(\zeta_t)$ is contained in L and $C'|_{K(\zeta_t)} = 1$. Thus, t divides $d(L/K,C')$. Conversely, it is easily seen that $d(L/K,C')$ divides $d(\tilde{L}/K,C'')$. This proves the lemma.

LEMMA 3. Let $n = \max(4, 2n_K - 4)$. There is a constant $\delta_2 > 0$ and $\geq \delta_2 x/(\log x)^2$ primes $\mathcal{P}$ of K satisfying

(i) $N_{K/\mathbb{Q}}(\mathcal{P}) = p \leq x$

(ii) $(\mathcal{P},F/K) = C$

(iii) $(\mathcal{P},L_\ell/K) \neq 1$ for each ℓ dividing $d(F/K,C)$

(iv) $t \mid (p-1)$, t prime $\Longrightarrow t \mid d(F/K,C)$ or $t > p^{(1/n)-\epsilon}$.

PROOF. For any prime p which is unramified in $\tilde{L}$ that satisfies $(p,\tilde{L}/\mathbb{Q}) \subset C^*$, take a prime $\mathcal{B} \mid p$ in $\tilde{L}$. Then, for some g in $\mathrm{Gal}(\tilde{L}/\mathbb{Q})$, $(\mathcal{B}^g,\tilde{L}/\mathbb{Q})$ lies in C'' and hence in $\mathrm{Gal}(\tilde{L}/K)$. If $\mathcal{P}$ is a prime of K below $\mathcal{B}^g$, then $\mathcal{P}$ is of degree 1 and $(\mathcal{P},L/K) \subset C'$. Thus, such a $\mathcal{P}$ satisfies (i)-(iii).

Now, we recall that L/K is abelian, by our assumption (b). Also, as $K/\mathbb{Q}$ is Galois, it follows that $\tilde{L}$ is also abelian over K. Moreover, the characteristic function of C^* can be written as a linear combination of inductions of (abelian) characters of $\mathrm{Gal}(\tilde{L}/K)$. In the notation of section 3, $\eta^*(D) \leq n_K$ for any conjugacy class D in C^*, and $\eta(D) \leq n$ (as we see by taking $H = \mathrm{Gal}(\tilde{L}/K)$). Now, the lemma of section 3 shows that there are $\gg x/(\log x)^2$ primes $p < x$ satisfying $(p,\tilde{L}/\mathbb{Q}) \subset C^*$ and

$$t \mid (p-1),\ t \text{ prime} \Longrightarrow t \mid d(\tilde{L}/\mathbb{Q},C^*) \text{ or } t > p^{(1/n)-\epsilon}.$$

By Lemmas 1 and 2, $d(\tilde{L}/\mathbb{Q},C^*) = d(F/K,C)$. This proves the lemma.

For any prime $\mathcal{P}$ of K, we denote by $k(\mathcal{P})$ the residue field mod $\mathcal{P}$. If $\mathcal{P} \notin S$, we also define the map $\varphi_{\mathcal{P}} : \mathcal{O}_S^* \longrightarrow k(\mathcal{P})^*$ which is reduction mod $\mathcal{P}$.

LEMMA 4. Let $s = |S| - 1$. Then, the number of primes $\mathcal{P}$ of K for which $|\varphi_{\mathcal{P}}(\mathcal{O}_S^*)| < y$ is $\ll y^{(s+1)/s}$.

PROOF. This is proved in exactly the same way as Lemma 2 or [4]. We have only to note that the rank of $\mathcal{O}_S^*$ is s and that if $\beta \in K$, then $\beta^b - 1$ has $\underline{O}(b)$ prime divisors (counted with multiplicity). Here, the implied constant depends on β and on $[K:\mathbb{Q}]$.

PROOF OF THE PROPOSITION. We consider the primes $\mathcal{P}$ of K satisfying conditions (i)-(iv) of Lemma 3. For such a prime, we have that either $|\varphi_{\mathcal{P}}(\mathcal{O}_S^*)| < p^{(n-1)/n+\epsilon}$ or all prime divisors of the index $[k(\mathcal{P})^* : \varphi_{\mathcal{P}}(\mathcal{O}_S^*)]$ divide $d(F/K,C)$. By Lemma 4, the number of $\mathcal{P}$ with $N_{K/\mathbb{Q}}\mathcal{P} \leq x$ satisfying the first condition is $\ll x^{1-\epsilon}$ as $s \geq n$. Thus, we can discard those primes. Also, a prime ℓ divides the index $[k(\mathcal{P})^* : \varphi_{\mathcal{P}}(\mathcal{O}_S^*)]$ if and only if $\mathcal{P}$ splits completely in L_ℓ. But this contradicts (iii) of Lemma 3. Hence, the index must be 1 and this proves the result.

5. PRIMITIVE ROOTS. Let q,r,s be three distinct integers such that $\omega(q)$, $\omega(r)$ and $\omega(s)$ are odd and such that the group $\Gamma = \{q^a r^b s^c : a,b,c \in \mathbb{Z}\}$ has rank 3. (Here ω denotes the total number of prime divisors.) By the method of proof of Lemma 3, we find that there are at least $\gg x/(\log x)^2$ primes $p \leq x$ satisfying

(i) $\left(\frac{q}{p}\right) = \left(\frac{r}{p}\right) = \left(\frac{s}{p}\right) = -1$

(ii) $\varphi_p(\Gamma) = \mathbb{F}_p^*$

(iii) $t | (p-1)$, t prime $\Longrightarrow t = 2$ or $t > p^{\frac{1}{4}+\epsilon}$.

We refine the result of [4].

THEOREM. There is a constant $\delta_1 > 0$ such that one of the numbers of the set

$$\{q, r, s, q^2 r, qr^2, q^2 s, qs^2\}$$

is a primitive root for $> \delta_1 x/(\log x)^2$ of the primes p x.

PROOF. Let T denote the set of primes p satisfying (i)-(iii). Let $\langle q \rangle$ denote the subgroup of $\mathbb{F}_p^*$ generated by q, and let d denote its order. We note that

$$\#\{p<x : p \text{ prime}, \ d<y\} \le \Sigma\,\omega(q^b-1) \le \Sigma\, b\log q \ll y^2 .$$

The implied constant of course depends on q. If we choose $y = x^{1/2}/(\log x)^{1+\epsilon}$, we see that by throwing out $o(x/(\log x)^2)$ of the $p < x$, we may assume that $d > p^{1/2}/(\log p)^{1+\epsilon}$ for all $p \in T$. We know that d is even by (ii). This and (i) implies that

$$\omega(d) \le \omega(p-1) < \omega(d) + 1 .$$

If $\omega(p-1) = \omega(d)$, then $\langle q\rangle = \mathbb{F}_p^*$, so suppose $p-1 = dt$, t prime. Write d' for the order of the subgroup $\langle q,r\rangle$ of $\mathbb{F}_p^*$ generated by q and r. Write d'' for the order of $\langle q,s\rangle$. Then $d',d'' \equiv 0 \pmod d$, i.e., $d',d'' = d$ or dt. If either of d' or d'' is equal to dt, this means $\langle q,r\rangle$ or $\langle q,s\rangle = \mathbb{F}_p^*$. If $d' = d'' = d$, this means $\langle q\rangle = \langle q,r,s\rangle = \mathbb{F}_p^*$ contradicting $p-1 = dt$. Suppose now that $\langle q,r\rangle = \mathbb{F}_p^*$. Then, one of $\{r, q^2r, qr^2\}$ is a primitive root. Indeed, let g be a primitive root mod p. Say $q = g^a$, $r = g^b \pmod p$. Because of (ii), the elements of $S = \{b, 2a+b, a+2b\}$ are all odd. Also, $(p-1, ab) = 1$ and $t|a$, so t does not divide any element of S. Moreover the vectors $(1,0),(1,2)$ and $(2,1)$ are pairwise linearly independent modulo any prime larger than 3. Hence, any other odd prime divisor t' of $(p-1)$ divides at most one element of S, for p sufficiently large. We conclude that at least one element of S is relatively prime to $p-1$, as required.

REMARKS. 1) The lower bound $x/(\log x)^2$ can be improved to $x(\log\log\log x)/(\log x)^2$ by allowing $p-1$ to have prime divisors $< \log\log p$. To ensure that these primes do not appear in the index $[\mathbb{F}_p^* : \varphi_p(\Gamma)]$, we have to replace (ii) by the condition that p does not split completely in any of the Kummer extensions $\mathbb{Q}(\zeta_\ell, q^{1/\ell})$ for $\ell < \log\log x$. This number is estimated using the effective Chebotarev density theorem.

2) The proof also shows that for infinitely many primes p, two of q,r,s are sufficient to generate $\mathbb{F}_p^*$.

3) It is possible to produce a set of 7 elements directly by the methods of [4]. Indeed, by the same reasoning, there are at least $\delta_1 x/(\log x)^2$ primes $p \le x$ satisfying (i)-(iii) above. Consider any such prime. If g is a primitive root mod p, then

$$q = g^\alpha, \ r = g^\beta, \ s = g^\gamma \pmod p$$

where $(p-1,\alpha,\beta,\gamma) = 1$ and because of (ii), α,β,γ are all odd. Now if $v = (v_1,v_2,v_3) \in \mathbb{Z}^3$, $v_i \ge 0$, then $(q,r,s)^v q^{v_1} r^{v_2} s^{v_3}$ is a primitive root mod p if and only if $(p-1,\lambda,v) = 1$ where $\lambda = (\alpha,\beta,\gamma)$. Consider

$$\tilde{S} = \{(1,2,0),(2,1,0),(1,0,2),(2,0,1),(0,1,2),(0,2,1),(1,2,2)\} .$$

Using, for example, cross products, it is easy to check that any three elements of S are linearly independent. Thus, if the odd prime divisors of $(p-1)$ are sufficiently large, then for each odd prime t dividing $(p-1)$, at most two of $\{\lambda.v, v \in \tilde{S}\}$ are divisible by t (as $\lambda \neq 0 \pmod t$). As $p-1$ has at most three odd prime divisors, for some $v_0 \in S$, $\lambda.v_0$ is not divisible by any odd prime dividing $p-1$. But $\lambda.v$ is odd for any $v \in S$ and so $(p-1, \lambda.v_0) = 1$ as desired. This proves that one of

$$\{qr^2, q^2r, qs^2, q^2s, rs^2, r^2s, qr^2s^2\}$$

is a primitive root (mod p) for at least $\gg x(\log x)^{-2}$ primes $p \leq x$.

4) Replacing condition (ii) by the condition $(\frac{q}{p}) = -1$ yields the result that one of the set

$$\{qr^2, qs^4, qr^2s^2, qr^4s^2, q^3r^2, q^3s^2, q^3r^2s^2\}$$

is a primitive root (mod p) for $\geq \delta\, x/(\log x)^2$ primes $p \leq x$. This is true for any three integers q, r, s such that q is not a square and q, r, s generate a subgroup of $\mathbb{Q}^*$ of rank 3.

5) The above method has an interesting corollary. One can deduce that almost all elements of the form $q^a r^b s^c$ are primitive roots. To be more precise, let q, r, s be as before. Then there is a $\delta > 0$ such that the number of elements of

$$\{q^a r^b s^c : 0 \leq a,b,c \leq N,\ (a,b,c) \neq (0,0,0) \bmod 2\}$$

which are not primitive roots (mod p) for at least $\delta\, x/(\log x)^2$ of the primes $p \leq x$ is $O(N^2)$.

To see this let $u \in \mathbb{F}_2^3 - \{0\}$, and set

$$S_N^u = \{(a,b,c) \in \mathbf{Z}^3 : 0 \leq a,b,c \leq N,\ (a,b,c) = u \pmod 2\}.$$

By replacing (ii) by the condition $(q/p) = -1$, we can get at least $\delta_1 x/(\log x)^2$ primes $p \leq x$ such that for any seven 3-wise linearly independent elements of S_N^u, at least one gives a primitive root mod p. Let M be the minimum number of 3-wise linearly independent elements $v_1, \ldots, v_M$ of S_N^u such that for each $1 \leq i \leq M$, the number of primes $p \leq x$ for which $(q,r,s)^{v_i}$ is a primitive root (mod p) is $\leq \delta_1 x/(\log x)^2$. Then $M \leq 6$. Suppose $v \in S_N^u$ is such that $(q,r,s)^v$ is a primitive root mod p for $< \delta_1 x/(\log x)^2$ primes $p \leq x$. Then v must be in the span of $\{v_i, v_j\}$ for some i, j with $1 \leq i, j \leq M$, and the number of such v is at most

$$\binom{M}{2}\left[\frac{N+1}{2}\right]^2 \leq 15\left[\frac{N+1}{2}\right]^2 .$$

Letting u range over the seven elements of $\mathbb{F}_2^3 - \{0\}$, we get at most $105(N+1)^2/4$ exceptions as claimed. It is thus clear that there is an infinite variation on the seven numbers formed with q, r, s.

REFERENCES

1. J.W.S. Cassels, The inhomogeneous minimum of binary quadratic, ternary cubic and quaternary quartic forms, Proc. Camb. Phil. Soc. 48 (1952) 72-86, 519-520.
2. G. Cooke, A weakening of the Euclidean property for integral domains and applications to algebraic number theory, I, II, J. Reine und Angew. Math. 282 (1976) 133-156, 283/284 (1976) 71-85.
3. G. Cooke and P. Weinberger, On the construction of division chains in algebraic number fields with applications to SL_2, Comm. Alg. 3 (1975) 481-524.
4. R.Gupta and M. Ram Murty, A remark on Artin's conjecture, Inv. Math. 78 (1984) 127-130.
5. G. H. Hardy and E. M. Wright, An introduction to the theory of numbers, 5th ed., Oxford University Press, Oxford, 1979.
6. H. Heilbronn, On Euclid's algorithm in real quadratic fields, Proc. Camb. Phil. Soc. 34 (1938) 521-526.
7. H. Iwaniec, Rosser's sieve, Acta Arith. 36 (1980) 174-202.
8. R. B. Lakein, Euclid's algorithm in complex quartic fields, Acta Arith. 20 (1972) 393-400.
9. S. Lang, Algebraic Number Theory, Addison-Wesley, Reading, 1970.
10. H.W. Lenstra, Jr., On Artin's conjecture and Euclid's algorithm in global fields, Inv. Math. 42 (1977) 201-224.
11. H.W. Lenstra, Jr., Euclidean number fields of large degree, Inv. Math. 38 (1977) 237-254.
12. H.W. Lenstra, Jr., Euclid's algorithm in cyclotomic fields, J. London Math. Soc. 10 (1975) 457-465.
13. J. Masley and H. Montgomery, Cyclotomic fields with unique factorization, J. Reine und Angew. Math. 286/287 (1976) 248-256.
14. Th. Motzkin, The Euclidean algorithm, Bull. Amer. Math. Soc. 55 (1949) 1142-1146.
15. M. Ram Murty and V. Kumar Murty, A variant of the Bombieri-Vinogradov Theorem, these proceedings.
16. W. Narkiewicz, Elementary and analytic theory of algebraic numbers, Polish Scientific Publishers, Warszawa, 1974.
17. T. Ojala, Euclid's algorithm in the cyclotomic field $\mathbb{Q}(\zeta_{16})$, Math. Comp. 31 (1977) 268-273.
18. P. Samuel, About Euclidean rings, J. Algebra 19 (1971) 282-301.
19. P. Weinberger, On Euclidean rings of algebraic integers, Proc. Symp. Pure Math. 24 (1973) 321-332.

DEPARTMENT OF MATHEMATICS, UNIVERSITY OF BRITISH COLUMBIA, VANCOUVER

DEPARTMENT OF MATHEMATICS, McGILL UNIVERSITY, MONTREAL

DEPARTMENT OF MATHEMATICS, CONCORDIA UNIVERSITY, MONTREAL

Canadian Mathematical Society
Conference Proceedings
Volume 7 (1987)

REAL QUADRATIC FUNCTION FIELDS

David R. Hayes[1]

ABSTRACT. Formulas analogous to those of Zagier for the special values of partial zeta-functions over a real quadratic number field are derived for "real quadratic" function fields. They are proved using analogs of Shintani's sector zeta-functions. As an application, two analogs of a theorem of Hirzebruch are proved.

1. INTRODUCTION. Let $k = \mathbf{F}_q(T)$ be the rational function field over the finite constant field $\mathbf{F}_q$ of characteristic $p \neq 2$, and let ∞ be the unique pole of T. The polynomial ring $A = \mathbf{F}_q[T]$ is the ring of functions in k which are holomorphic away from ∞. Let k_∞ be the completion of k at ∞, and let v_∞ be the normalized valuation on k_∞. As one knows, every element $x \in k_\infty^\times$ with $d = -v_\infty(x)$ has a unique Laurent expansion

$$(1.1) \qquad x = \sum_{r \in \mathbf{Z}} \zeta_r T^{-r} \qquad (\zeta_r \in \mathbf{F}_q)$$

where $\zeta_r = 0$ for $r < -d$ and $\zeta_{-d} \neq 0$. We write

$$(1.2) \qquad \deg(x) = d, \quad \|x\| = q^d \quad \text{and} \quad \operatorname{sgn}(x) = \zeta_{-d} .$$

It is also convenient to define $\|0\| = 0$ and $\operatorname{sgn}(0) = 0$. If $\operatorname{sgn}(x) = 1$, then x is called <u>positive</u>.

Let I be the maximal ideal of the valuation ring at ∞, a compact subset of the local field k_∞ which we view as an analogue of the unit interval. Following Artin [A], we put $\langle 0 \rangle = 0$ and for $x \in k_\infty^\times$ we define

$$(1.3) \qquad \langle x \rangle = \sum_{r \geq 1} \zeta_r T^{-r} \in I .$$

We call $\langle x \rangle$ the <u>fractional part</u> of x, and we call $[x] = x - \langle x \rangle$ the <u>polynomial part</u> of x. For $x \neq 0$, we have

$$(1.4) \qquad [x] = \sum_{m=0}^{d} \zeta_{-m} T^m .$$

1. Supported in part by NSF Grant DMS-8201637.

If K/k is a quadratic extension in which ∞ splits completely, then K is called a <u>real quadratic function field</u> over the pair (k,∞). Let ∞ split into the two K-places ∞_1 and ∞_2. A place of K which is not equal to either ∞_1 or ∞_2 is called <u>finite</u>. The integral closure of A in K is the ring $\mathcal{O}_K$ of functions in K which are holomorphic away from ∞_1 and ∞_2, and the finite K-places correspond one-to-one with the non-zero prime ideals of $\mathcal{O}_K$. Since $p \neq 2$, $K = k(\sqrt{\Delta})$ for a uniquely determined squarefree, positive polynomial $\Delta \in A$ of even degree; and we have $\mathcal{O}_K = A + A\sqrt{\Delta}$. Real quadratic function fields were studied extensively by Artin (see [A]), who found that such fields have much in common with real quadratic number fields. In particular, Artin introduced a continued fraction algorithm in k_∞ based on the "polynomial part" function (1.4) and showed how to use the algorithm to compute a fundamental unit of $\mathcal{O}_K$.

In this paper, we derive formulas analogous to those of Zagier [Z2] for the special values at $s = 1 - n$, $n \geq 1$, of the partial zeta-function associated to a rank 2 A-submodule of K. Zagier applies the Gauss theory of binary quadratic forms to derive his formulas, but they may also be proved using the "sector zeta-functions" of Shintani. We will adopt the Shintani approach in establishing our analogues for K (cf. [Ha]). The techniques lead to an expression for this partial zeta-function as a sum of more elementary functions $E(s)$ (see (7.14) below). Using Artin's continued fraction algorithm, we compute the functions $E(s)$ explicitly as rational functions in q^{-s} and then more or less read off the properties we need. We observe in particular that these functions satisfy functional equations relating $E(s)$ to $E(1 - s)$ (see (7.20) and (7.24) below) and take rational values at $s = 1 - n$, $n \geq 1$, with denominator at most $(q - 1)(q^n - 1)$. These results are collected in Theorems 7.25 and 7.29.

One application of Zagier's evaluations at $s = 0$ is the following remarkable theorem of Hirzebruch (see [Hi] or [Z1]):

THEOREM 1.5. For a rational prime ℓ, let $h(\ell)$ and $h(-\ell)$ denote the respective class numbers of the number fields $\mathbf{Q}(\sqrt{\ell})$ and $\mathbf{Q}(\sqrt{-\ell})$. Suppose that $\ell \equiv 3 \pmod 4$ and that $h(\ell) = 1$. Then

$$(1.6)\qquad h(-\ell) = \frac{1}{3} \cdot \sum_{r=1}^{\partial} (-1)^r a_r$$

where $\{a_1, a_2, \ldots, a_\partial\}$ is a minimal period of partial quotients for the continued fraction expansion

$$\sqrt{\ell} = a_0 + \cfrac{1}{a_1 + \cfrac{1}{\ddots\, a_\partial + \cfrac{1}{a_1 + \ddots}}} .$$

An adaption of Zagier's method yields a similar result for the case $\ell \equiv 1 \pmod 8$ and $h(\ell) = 1$. In this case, one may again express $h(-\ell)$ in the form (1.6) but now $\{a_1, a_2, \ldots, a_\partial\}$ is a minimal period of partial quotients for the continued fraction expansion of $2\sqrt{\ell}$.

In §8 below, we prove two analogues of Theorem 1.5 in function fields as an application of our evaluations of partial zeta-functions. I would like to thank M. Rosen for bringing this theorem of Hirzebruch to my attention and for suggesting that a similar result might be true for function fields. I am grateful to K. Ireland who verified formula (8.10) in many examples using computer programs written by R. Small, thereby detecting an error in my original statement of that formula. Some of these examples are presented at the end of §8. Ireland's computations led him also to notice a relation between the partial quotients of a continued fraction and the character sum

$$\sigma_1 = \sum_{A \in \mathbb{A}} \left[\frac{\Delta}{A}\right] \qquad (A \text{ positive, } \deg(A) = 1)$$

where $[\Delta/A]$ is the Jacobi symbol for the ring $\mathbb{A}$. This relation and others turned out to be consequences of our results. They are stated at the end of §7. We try to explain the combinatorial meaning of these relations in Theorem 7.36.

Sections 2, 3 and 5 below sketch well-known background material. They are included to establish notation and for the convenience of the reader. The partial zeta-functions are defined in §4, and the main computations are carried out in §§6 and 7.

This paper is an expanded version of my remarks at the CMS seminar. I greatly appreciate the efforts made by the organizers of the seminar, and I wish to extend special thanks to the Canadian Mathematical Society for its support.

2. EMBEDDING K IN $k_\infty \times k_\infty$. By completing K respectively at the places ∞_1 and ∞_2 which are split over ∞, we define embeddings e_1 and e_2 of K into k_∞. We fix the square root of Δ in K by requiring that $e_1(\sqrt{\Delta})$ be positive. The map $\underset{\sim}{e} = e_1 \times e_2$ embeds K as a dense subfield of the product k-algebra $k_\infty \times k_\infty$. For any vector $\alpha = (\alpha^{(1)}, \alpha^{(2)}) \in k_\infty \times k_\infty$, we define

(2.1) $\underset{\sim}{T}(\alpha) = \alpha^{(1)} + \alpha^{(2)}$, $\underset{\sim}{N}(\alpha) = \alpha^{(1)}\alpha^{(2)}$ and $\underset{\sim}{S}(\alpha) = \alpha^{(2)}/\alpha^{(1)}$, the function $\underset{\sim}{S}$ being defined only if $\alpha^{(1)} \neq 0$. We call $\underset{\sim}{T}$, $\underset{\sim}{N}$ and $\underset{\sim}{S}$ respectively the <u>trace</u>, <u>norm</u> and <u>slope</u> functions on $k_\infty \times k_\infty$. For $x \in K$, we write $\underset{\sim}{e}(x) = (x^{(1)}, x^{(2)})$; and we abuse language somewhat and write, e.g., $\underset{\sim}{T}(x)$ for $\underset{\sim}{T}(\underset{\sim}{e}(x))$. If $x \in \mathcal{O}_K$, then $\underset{\sim}{T}(x)$ and $\underset{\sim}{N}(x)$ are polynomials in A. Since $\underset{\sim}{e}(K^\times)$ consists of invertible elements in the k-algebra $k_\infty \times k_\infty$, the slope of any non-zero element of K is always defined.

A <u>lattice</u> L in $k_\infty \times k_\infty$ is an A-submodule of $k_\infty \times k_\infty$ of rank 2. This means that L is generated over A by two k_∞-linearly independent vectors $\rho, \tau \in k_\infty \times k_\infty$, and we write

(2.2) $$L = [\rho,\tau] = A\rho + A\tau .$$

We say that the pair of basis vectors (ρ,τ) is <u>oriented</u> if the determinant

(2.3) $$V(L) = \det \begin{bmatrix} \rho^{(1)} & \rho^{(2)} \\ \tau^{(1)} & \tau^{(2)} \end{bmatrix} \in k_\infty$$

is positive. Any pair of basis vectors can be oriented by multiplying one vector of the pair by an appropriately chosen constant. Since any two basis vectors for the A-module L can be mapped to ρ,τ by a matrix in $GL_2(A)$, $V(L)/\mathrm{sgn}(V(L))$ is independent of the choice ρ,τ and is therefore an invariant of the lattice L. We agree in the sequel always to compute $V(L)$ using an oriented pair of basis vectors so that $V(L)$ is itself our invariant.

The quotient space $\Pi(L) = (k_\infty \times k_\infty)/L$ is compact because it is homeomorphic to the standard fundamental domain $D(L) = I\rho + I\tau$. Let us introduce in $k_\infty \times k_\infty$ an additive Haar measure normalized so that

$$\mathit{meas}(I \times I) = 1.$$

The total mass of $\Pi(L)$ in the quotient measure is then

(2.4) $$\mathrm{Vol}(L) = \mathit{meas}(D(L)) = \|V(L)\|$$

(see [W], Chap. 1 §2). If L' is a sublattice of L, then

(2.5) $$\#(L/L') = \frac{\mathrm{Vol}(L')}{\mathrm{Vol}(L)}.$$

by (2.4) and the invariance of the measure.

The following lemma will be useful in §6:

LEMMA 2.6. Suppose ρ and τ are any vectors with $\|\underset{\sim}{S}(\rho)\| < \|\underset{\sim}{S}(\tau)\|$. Then ρ and τ are linearly independent over k_∞; and if $L = [\rho,\tau]$, then $\mathrm{Vol}(L) = \|\rho^{(1)}\tau^{(2)}\|$.

Proof: The determinant in (2.3) has absolute value

$$\begin{aligned}\|\rho^{(1)}\tau^{(2)} - \rho^{(2)}\tau^{(1)}\| &= \|\rho^{(1)}\tau^{(1)}\|\cdot\|\underset{\sim}{S}(\tau) - \underset{\sim}{S}(\rho)\| \\ &= \|\rho^{(1)}\tau^{(1)}\|\cdot\|\underset{\sim}{S}(\tau)\| = \|\rho^{(1)}\tau^{(2)}\| \neq 0\end{aligned}$$

since $\underset{\sim}{S}(\rho)$ is defined and $\|\underset{\sim}{S}(\tau)\| > 0$. □

Let M be a rank 2 A-submodule of K. Since e(K) is dense in $k_\infty \times k_\infty$, any module basis for M maps under $\underset{\sim}{e}$ into k_∞-linearly independent vectors ρ and τ. Thus, $\underset{\sim}{e}(M)$ is the lattice $[\rho,\tau]$ generated over A by these vectors. This fact allows us to attach the analytic invariant $V(\underset{\sim}{e}(M))$ to M. For brevity, we write this invariant as $V(M)$.

3. DISCRIMINANTS AND ORDERS. Let $\mathrm{End}(M) = \{z \in K : zM \subset M\} \subseteq \mathcal{O}_K$ be the ring of operators of the rank 2 A-submodule M of K. We say that two such sum-modules M and N are equivalent (notation: $M \sim N$) if $M = yN$ for some function $y \in K$. Equivalent modules have the same ring of operators, and so in computing End(M) we may assume that $M = A + Ax$, where $Ax^2 + Bx + C = 0$ for a relatively prime triple of polynomials $A,B,C \in A$ such that $B^2 - 4AC$ is not a perfect square. Just as in the theory of quadratic number fields, we see by direct computation that

$$\mathrm{End}(M) = A + AAx, \tag{3.1}$$

which shows in particular that End(M) is also a rank 2 A-module.

DEFINITION 3.2. The <u>discriminant</u> of M is the element $\mathrm{Disc}(M) = V(\mathrm{End}(M)) \in k_\infty^\times$, and the <u>norm</u> of M is the element $NM = V(M)/\mathrm{Disc}(M) \in k_\infty^\times$.

The reader may amuse himself by showing that

$$\mathrm{Disc}(A + Ax) = (B^2 - 4AC)^{1/2} = e_1(B + 2Ax) \tag{3.3}$$

and that

$$N(A + Ax) = \mathrm{sgn}(A)/A\ , \tag{3.4}$$

provided that the triple A,B,C is multiplied by a normalizing constant which makes $e_1(B + 2Ax)$ positive. Equation 3.3 shows that Disc(M) can be pulled back via e_1 to an element of $\mathcal{O}_K$. Thus,

$$\mathrm{Disc}(\mathcal{O}_K) = \mathrm{Disc}(A + A\sqrt{\Delta}) = \sqrt{\Delta}.$$

PROPOSITION 3.5. We have $\mathrm{Disc}(M) = F\sqrt{\Delta}$ for a uniquely determined positive polynomial $F \in A$, and $\mathrm{End}(M) = A + AF\sqrt{\Delta}$.

Proof. We may assume $M = A + Ax$. Since 1,Ax is an A-basis for End(M), it is clear from (3.3) that 1,Disc(M) is also an A-basis for End(M). Now because $\operatorname{End}(M) \subset \mathcal{O}_K$, the matrix of transition M from the basis $1,\sqrt{\Delta}$ to the basis $1,\operatorname{Disc}(M)$ has polynomial entries. Thus, $F = \det(M) = \operatorname{Disc}(M)/\sqrt{\Delta}$ is a positive element of A. □

A subring of $\mathcal{O}_K$ which contains A and has K as its field of fractions is called an <u>order</u> in $\mathcal{O}_K$. The ring $\operatorname{End}(M)$ is an order, and conversely any order is its own ring of operators. Proposition 3.5 shows that the orders in $\mathcal{O}_K$ correspond one-to-one with the monic polynomials $F \in A$. Let $\mathcal{O}_F = A + AF\sqrt{\Delta}$ be the order corresponding to F. The polynomial F is called the <u>conductor</u> of $\mathcal{O}_F$. It is clear from the definitions and Proposition 3.5 that

(3.6) $$\#(\mathcal{O}_K/\mathcal{O}_F) = \|F\| = q^{\deg F}.$$

A <u>fractional ideal</u> of $\mathcal{O}_F$ is any $\mathcal{O}_F$-submodule M of K which has $\mathcal{O}_F$-rank 1 (and hence A-rank 2). The fractional ideal M is <u>proper</u> if $\operatorname{End}(M) = \mathcal{O}_F$, and M is <u>integral</u> if $M \subset \mathcal{O}_F$. From the definitions, if M is both proper and integral, then $NM \in A$ and

(3.7) $$\#(\mathcal{O}_F/M) = \|NM\|.$$

Let I(F) be the set of proper fractional ideals of $\mathcal{O}_F$. We have then

PROPOSITION 3.8. The set of ideals I(F) forms a group under the usual multiplication of ideals.

Proof. Given $M \in I(F)$, let $M^{-1} = \{z \in K : zM \subset \mathcal{O}_F\}$. It is immediate that $(yM)^{-1} = y^{-1}M^{-1}$ for any $y \in K$; and so in computing M^{-1}, we may assume that $M = A + Ax$ as in the first paragraph of this section. For such an M, we find by direct computation that

(3.9) $$M^{-1} = AA + A(B + Ax)$$

and that $MM^{-1} \subset \mathcal{O}_F$ contains the relatively prime triple of polynomials A,B,C and hence also contains 1. We conclude that $MM^{-1} = \mathcal{O}_F$ for all $M \in I(F)$. This fact implies both that I(F) is closed under ideal multiplication and that M^{-1} is indeed an inverse for M in I(F). □

For $x \in K$, let x' denote the conjugate of x over k. If $Ax^2 + Bx + C = 0$ with $x \notin K$, then $x + x' = -B/A$. Thus, (3.9) together with (3.4) shows that

(3.10) $$M^{-1} = \frac{1}{NM} \cdot M'$$

for all proper fractional ideals M of $\mathcal{O}_F$ of the form $M = A + Ax$. From the definitions, $N(yM) = \tilde{N}(y) \cdot NM$, and so (3.10) is in fact valid for all $M \in I(F)$. One consequence of (3.10) is the multiplicative property of the norm function on $I(F)$:

(3.11) $\quad MN \in I(F) \Longrightarrow N(MN) = N(M) \cdot N(N)$.

It is convenient to put $d(M) = \deg(\mathrm{Disc}(M))$. Then we have

(3.12) $\quad d(M) = \deg(F) + \frac{1}{2}\deg(\Delta)$

by Proposition 3.5.

4. PARTIAL ZETA-FUNCTIONS. Given a multiplicative character $\chi:\mathbf{F}_q \to \mathbf{C}$ (so $\chi(0) = 0$), we extend the definition of χ to all $x \in k_\infty$ by defining $\chi(x) = \chi(\mathrm{sgn}(x))$.

DEFINITION 4.1. Two rank 2 A-submodules M and N of K are <u>χ-equivalent</u> (notation: $M \sim N \bmod \chi$) if $M = yN$ for some function $y \in K$ such that $\chi(\underset{\sim}{S}(y)) = 1$.

When χ is the quadratic character of $\mathbf{F}_q$, χ-equivalence is a close analogue to "narrow equivalence" in a real quadratic number field. When χ is the trivial character $\chi = \chi_0$, χ-equivalence is the same as the notion of equivalence introduced in §3.

For a fixed character χ, we define the partial zeta-function $Z_\chi(s,M)$ attached to a given rank 2 A-submodule M of K as follows: Let $\mathrm{End}(M) = \mathcal{O}_F$. Then for $s \in \mathbf{C}$,

(4.2) $$Z_\chi(s,M) = \sum_N \frac{1}{\|NN\|^s} \qquad (N \sim M \bmod \ , N \subseteq \mathcal{O}_F)$$

provided that the series is absolutely convergent. Relative to the order $\mathcal{O}_F$, the zeta-function of M is defined as a summation over the integral ideals in the χ-equivalence class containing M. We write $Z_0(s,M)$ for the partial zeta-function defined using the trivial character $\chi = \chi_0$.

In §7 below, we show that the series in (4.2) converges in the complex half-plane $\mathrm{Re}(s) > 1$ to a rational function of q^{-s}. The analytic continuation of $Z_\chi(s,M)$ to a function meromorphic on the whole complex plane is then immediate, and we can read off (more or less!) the values at the negative integers. For these calculations, we need to rewrite $Z_\chi(s,M)$ formally as a sum over certain lattice points in $k_\infty \times k_\infty$.

Let K_χ be the set of functions $y \in K$ such that $\chi(\underset{\sim}{S}(y)) = 1$. By the unit theorem, we have

(4.3) $$K_\chi \cap \mathcal{O}_F^\times = \mathbf{F}_q^\times \cdot \varepsilon_{F,\chi}^{\mathbf{Z}}$$

where the fundamental unit $\varepsilon_{F,\chi}$ is fixed by requiring that $\varepsilon_{F,\chi}^{(2)}$ be positive and that $\|\underset{\sim}{S}(\varepsilon_{F,\chi})\| > 1$. In the case $\chi = \chi_0$, we write this fundamental unit as just ε_F. For $x,y \in K_\chi$, $N = xM$ is integral $\Longleftrightarrow x \in M^{-1}$; and $xM = yM$ $\Longleftrightarrow y = \varepsilon x$ for some unit $\varepsilon \in K_\chi \cap O_F^\times$. Therefore we have the formal identity

$$(4.4)\qquad (q-1)\cdot Z_\chi(s,M) = \sum_x{}' \frac{1}{\|NxM\|^s} = \|NM\|^{-s}\cdot\sum_x{}' \frac{1}{\|\underset{\sim}{N}x\|^s} \qquad (x \in M^{-1} \cap K_\chi)$$

where the prime on the summation indicates that exactly one element x is to be chosen from each orbit of $\varepsilon_{F,\chi}^{\mathbf{Z}}$ acting on $M^{-1} \cap K_\chi$. Let P be the "slit" plane $(k_\infty \times k_\infty)\backslash(\{0\} \times k_\infty)$. The multiplicative group $K^\times$ acts as a group of transformations on P via

$$(4.5)\qquad x\cdot\alpha = x\cdot(\alpha^{(1)},\alpha^{(2)}) = (x^{(1)}\alpha^{(1)},x^{(2)}\alpha^{(2)})$$

for $x \in K$ and $\alpha \in P$. Let $X_{F,\chi}$ be a fundamental domain for the action of $\varepsilon_{F,\chi}^{\mathbf{Z}}$ on P, and let $L = \underset{\sim}{e}(M^{-1})$. Let δ_χ be the characteristic function of the set of vectors $\alpha \in P$ such that $\chi(\underset{\sim}{S}(\alpha)) = 1$. We can meet all the conditions on the elements x in the summations of (4.4) by insisting that $\delta_\chi(\underset{\sim}{e}(x)) = 1$ and that $\underset{\sim}{e}(x) \in X_{F,\chi} \cap L$. We have then

$$(4.6)\qquad (q-1)\cdot Z_\chi(s,M) = \|NM\|^{-s}\cdot \sum_\lambda \frac{\delta_\chi(\lambda)}{\|\underset{\sim}{N}\lambda\|^s} \qquad (\lambda \in X_{F,\chi} \cap L)\ .$$

For any vectors $\rho,\tau \in P$ such that $\|\underset{\sim}{S}(\rho)\| < \|\underset{\sim}{S}(\tau)\|$. let

$$(4.7)\qquad X(\rho,\tau) = \{\alpha \in P\colon \|\underset{\sim}{S}(\rho)\| \le \|\underset{\sim}{S}(\alpha)\| < \|\underset{\sim}{S}(\tau)\|\ \}.$$

We call $X(\rho,\tau)$ the <u>sector</u> determined by ρ and τ; and the function $Z_\chi(s;\rho,\tau)$ defined formally by

$$(4.8)\qquad (q-1)\cdot Z_\chi(s;\rho,\tau) = \|V([\rho,\tau])\|^s\cdot\sum_\lambda \frac{\delta_\chi(\lambda)}{\|N\lambda\|^s} \quad (\lambda \in X(\rho,\tau) \cap [\rho,\tau])$$

is the <u>sector zeta-function</u> of $X(\rho,\tau)$ relative to χ. One convenient fundamental domain for the action of $\varepsilon_{F,\chi}^{\mathbf{Z}}$ on P is the sector

$$(4.9)\qquad X_{F,\chi}(\rho) = X(\rho,\varepsilon_{F,\chi}\rho)\quad ,$$

where ρ is a fixed vector in P. Suppose we could choose ρ so that ρ and and $\varepsilon_{F,\chi}\rho$ are basis vectors for L. Then we would have $Z_\chi(s,M) = q^{-d(M)s}\cdot Z_\chi(s;\rho,\varepsilon_{F,\chi}\rho)$ because

$$(4.10)\qquad V(M^{-1})/\mathrm{Disc}(M) = N(M^{-1}) = 1/N(M)$$

by Definition 3.2 and (3.11). However this way of writing $Z_\chi(s,M)$ is usually <u>invalid</u> for any choice of $\rho \in L \cap P$ because $[\rho,\varepsilon_{F,\chi}\rho] = L$ is almost never true! In §6, we show how to choose the vector ρ in (4.9) so that $Z_\chi(s,M)$

can be evaluated as $q^{-d(M)s}$ times a sum of sector zeta-functions.

The functions $Z_\chi(s,M)$ are of interest because they provide a way of computing the special values of the Artin L-functions associated to "ring class fields" over K. The remainder of this section is devoted to a brief development of this connection with L-functions. Let $P_{F,\chi}$ be the group of principal $\mathcal{O}_F$-fractional ideals generated by the functions $y \in K_\chi$. The elements of the quotient group $\mathrm{Pic}_\chi(\mathcal{O}_F) = I(F)/P_{F,\chi}$ are the χ-equivalence classes of the rank 2 A-submodules N with $\mathrm{End}(N) = \mathcal{O}_F$. For a finite K-place p not dividing F, let $p_F = p \cap \mathcal{O}_F$, a proper prime ideal in $\mathcal{O}_F$. A standard argument shows that $p \longmapsto p_F$ sets up a bijective correspondence between the finite K-places not dividing F and the non-zero prime ideals of $\mathcal{O}_F$ which do not contain F. Now by class field theory, there is an abelian extension $K_{F,\chi}/K$ characterized by the property that a finite K-place p prime to F splits completely in $K_{F,\chi}/K \iff p_F \in P_{F,\chi}$. If we think of the Artin map $\phi: N \longmapsto \sigma_N$ as defined on the integral ideals N of $\mathcal{O}_F$ which do not contain F via the correspondence $p \longleftrightarrow p_F$, then (by class field theory again) ϕ lifts to an isomorphism

(4.11) $$\mathrm{Pic}_\chi(\mathcal{O}_F) \longrightarrow \mathrm{Gal}(K_{F,\chi}/K) = G_{F,\chi} .$$

We see in particular that $\mathrm{Pic}_\chi(\mathcal{O}_F)$ is a finite group.

A character $\psi \in \hat{G}_{F,\chi}$ is F-primitive if the finite part of its conductor is precisely the divisor of F in K. For such a character, let

(4.12) $$L_\infty(s,\psi) = \prod_p \left(1 - \psi(\sigma_p) \cdot N_p^{-s}\right)^{-1} \qquad (p \text{ finite}, \ p \nmid F)$$

be the Artin L-function of ψ with the Euler factors associated to ∞_1 and ∞_2 removed. We have then via the isomorphism (4.11)

(4.13) $$L_\infty(s,\psi) = \sum_M \psi(M) \cdot Z_\chi(s,M)$$

where M runs through a set of representatives for the classes in $\mathrm{Pic}_\chi(\mathcal{O}_F)$. The identity (4.13) is easy to prove when F = 1 just by multiplying out the product in (4.12) in the usual way. The proof of (4.13) in general requires an understanding of the arithmetic properties of the order $\mathcal{O}_F$. We will not develop these properties in this paper because we need only the case F = 1 for the applications in §8 below.

5. CONTINUED FRACTIONS.

In this section, we sketch the theory of continued fractions for k_∞ which was introduced by Artin in [A]. Let $\mathbf{P}^1 = k_\infty \cup \{\mathcal{H}\}$ be the projective line over k_∞, where $\mathcal{H}$ is the "point at infinity". The action of a matrix

(5.1) $$M = \begin{bmatrix} A & B \\ C & D \end{bmatrix} \in GL_2(A)$$

on $x \in \mathbf{P}^1$ is defined by the Möbius formula

(5.2) $$M \circ x = \frac{Ax + B}{Cx + D}$$

with the usual conventions for $x = \infty$ and $x = -D/C$.

DEFINITION 5.3. For $w \in k_\infty \backslash k$, the sequence $\{w_r : r \geq 1\}$ in $k_\infty \backslash k$ is defined recursively as follows: We put $w_1 = w$, and for $r \geq 1$ we define w_{r+1} by

(5.4) $$w_r = A_r + \frac{\zeta_r}{w_{r+1}} = \begin{bmatrix} A_r & \zeta_r \\ 1 & 0 \end{bmatrix} \circ w_{r+1}$$

where $A_r = [w_r]$ (see 1.4 above) and where $\zeta_r \in \mathbf{F}_q^\times$ is chosen in some systematic way. The procedure (5.4) for producing the sequence $\{w_r\}$ is a <u>continued fraction algorithm</u> in k_∞. The polynomials A_r are called the <u>partial quotients</u> of w.

If we choose each $\zeta_r = 1$ in (5.4), then we say that the continued fraction algorithm is <u>standard</u>. Let $w_r^S = \mu_r w_r$ for $r \geq 1$ be the sequence produced from w by the standard algorithm, and let A_r^S for $r \geq 1$ be the sequence of partial quotients of the standard algorithm. The elements μ_r must satisfy the recurrence $\mu_{r+1} \cdot \mu_r = 1/\zeta_r$, which implies that

(5.5) $$\mu_{2j+1} = \prod_{i=1}^{j} \frac{\zeta_{2i-1}}{\zeta_{2i}} \quad \text{and} \quad \mu_{2j} = \prod_{i=0}^{j-1} \frac{\zeta_{2i}}{\zeta_{2i+1}}$$

for $j \geq 0$ with the conventions that $\mu_0 = \zeta_0 = 1$. It is clear that $A_r^S = \mu_r A_r$ for all $r \geq 1$, a fact which will be useful for the calculations in §8 below.

For $r \geq 0$, let

(5.6) $$M_r = M_r(w) = \prod_{j=1}^{r} \begin{bmatrix} A_j & \zeta_j \\ 1 & 0 \end{bmatrix} \in GL_2(A) .$$

We have then

(5.7) $$w = M_r \circ w_{r+1}$$

which of course we can also write in the familiar expanded form

$$w = A_1 + \cfrac{\zeta_1}{A_2 + \cfrac{\zeta_2}{\ddots \; A_r + \cfrac{\zeta_r}{w_{r+1}}}} \quad .$$

The fractions $P_r/Q_r = M_r \circ \mathcal{H} \in k^\times$ for $r \geq 1$ are called the <u>convergents</u> of w. We may normalize the polynomials $P_r, Q_r \in A$ so that

$$(5.8) \qquad M_r = \begin{bmatrix} P_r & \zeta_r P_{r-1} \\ Q_r & \zeta_r Q_{r-1} \end{bmatrix}$$

for $r \geq 0$, provided that we define $P_0 = 1$, $P_{-1} = 0$, $Q_0 = 0$ and $Q_{-1} = 1$. We have then for $r \geq 0$ the recurrence relations

$$(5.9) \qquad \begin{aligned} P_{r+1} &= A_r P_r + \zeta_r P_{r-1} \\ Q_{r+1} &= A_r Q_r + \zeta_r Q_{r-1} \end{aligned}$$

from which we may deduce for $r \geq 1$

$$(5.10) \qquad \begin{aligned} \|P_r\| &= \|A_1 A_2 \ldots A_r\| \\ \|Q_r\| &= \|A_2 A_3 \ldots A_r\| \quad . \end{aligned}$$

We also see that

$$(5.11) \qquad P_r Q_{r-1} - P_{r-1} Q_r = \det(M_r)/\zeta_r = (-1)^r \zeta_1 \zeta_2 \cdots \zeta_{r-1}$$

which shows that P_r and Q_r are relatively prime polynomials.

For fixed w and any element $u \in k_\infty \backslash k$, we define the sequence $\{u_r : r \geq 1\}$ by the equation $u = M_r(w) \circ u_{r+1}$. If $u = w$, then of course $u_r = w_r$ and so $\|u_r\| > 1$ for all $r \geq 2$. In fact, $u = w$ is the only choice for which this phenomenon occurs.

PROPOSITION 5.12. If $u \neq w$, then $\|u_r\| < 1$ for all sufficiently large values of r.

Proof. We have

$$u_{r+1} = M_r^{-1} \circ u = \zeta_r \cdot \frac{Q_{r-1}u - P_{r-1}}{-Q_r u + P_r} = \left(\frac{-\zeta_r Q_{r-1}}{Q_r}\right) \cdot \frac{u - P_{r-1}/Q_{r-1}}{u - P_r/Q_r} \quad .$$

Now as Artin shows, the sequence of partial quotients P_r/Q_r converges in the k_∞-topology to w. Therefore since $u \neq w$, we have $\|u - P_{r-1}/Q_{r-1}\| = \|u - P_r/Q_r\| \neq 0$ for all large r, which implies $\|u_{r+1}\| = \|Q_{r-1}/Q_r\| = 1/\|A_r\| < 1$ by (5.10). ∎

This proposition will be used in the next section.

6. THE CORNERS OF A LATTICE. The "convexity polygon" of a full lattice in $\mathbf{R} \times \mathbf{R}$ is defined as the convex hull of the part of the lattice which lies in the open first quadrant. When the lattice intersects the co-ordinate axes only at the origin, the vertices of the convexity polygon can be explicitly constructed by means of a continued fraction algorithm (see, e.g., [H], §5). We show in this section that most of the important aspects of the convexity polygon can be carried over to lattices in $k_\infty \times k_\infty$, thereby enabling us to base our calculations in §7 on a fundamental domain of type (4.9) which decomposes naturally into a union of smaller sectors.

Throughout this section, L will be a lattice in $k_\infty \times k_\infty$ which has the following property: $\alpha \in L$, $\alpha \neq (0,0) \Rightarrow \alpha^{(1)} \neq 0$ and $\alpha^{(2)} \neq 0$. Such a lattice is said to be scattered. The lattice $\underset{\sim}{e}(M)$ of a rank 2 A-submodule of K is scattered because $\underset{\sim}{e}(K^\times)$ consists of invertible elements in the k-algebra $k_\infty \times k_\infty$.

DEFINITION 6.1. An ordered pair of basis vectors (ρ,τ) of L is called a corner pair for L if: *i*) $\rho^{(2)}$ and $\tau^{(2)}$ are both positive, and *ii*) $\|\rho^{(1)}\| > \|\tau^{(1)}\|$ and $\|\rho^{(2)}\| < \|\tau^{(2)}\|$. A corner of L is either one of the vectors in a corner pair for L.

For a given corner pair (ρ,τ) for L, we have the following properties:

I) $\|S(\tau)\| > \|S(\rho)\|$.

II) The pair (ρ,τ) is uniquely determined by either one of its component vectors.

III) There are vectors $\gamma,\kappa \in L$ such that (γ,ρ) and (τ,κ) are both corner pairs of L.

IV) There is no corner $\rho_* \neq \rho$ of L such that $\|\underset{\sim}{S}(\rho)\| \leq \|\underset{\sim}{S}(\rho_*)\| < \|\underset{\sim}{S}(\tau)\|$.

Proofs. Property I is immediate from the definitions. For II, suppose (ρ,τ) and (ρ,τ_*) are corner pairs of L. The change of basis from (ρ,τ) to (ρ,τ_*) over A takes the form $\rho = \rho$ and $\tau_* = A\rho + \zeta\tau$ where $A \in A$ and $\zeta \in \mathbf{F}_q^\times$. Since $\|\rho^{(1)}\| > \|\tau_*^{(1)}\|$ and $\|\rho^{(1)}\| > \|\tau^{(1)}\|$. A must be zero; and then $\zeta = 1$ because $\tau_*^{(2)}$ and $\tau^{(2)}$ are both positive. The same sort of argument shows that τ determines ρ.

To prove III, define γ by

$$(6.2) \qquad \zeta\gamma = \rho - A\tau$$

where $A = [\rho^{(1)}/\tau^{(1)}]$ and where $\zeta \in \mathbf{F}_q^\times$ is chosen so that $\gamma^{(2)}$ is positive. Because $\|\rho^{(1)}\| > \|\tau^{(1)}\|$, we have $\|A\| > 1$. Now from $\zeta\gamma/\tau = (\rho/\tau) - A$, we deduce first that $\|\gamma^{(1)}/\tau^{(1)}\| < 1$ and second that $\|\gamma^{(2)}/\tau^{(2)}\| = \|A\| > 1$. Since τ, γ are clearly basis vectors for L, (τ,γ) is indeed a corner pair for L. Similarly, if we define κ by

(6.3) $\qquad \zeta\kappa = \tau - B\rho$

with $B = [\tau^{(2)}/\rho^{(2)}]$ and ζ chosen so that $\kappa^{(2)}$ is positive, then (κ,ρ) will be a corner pair for L.

To prove IV, assume that ρ_* is a corner with $\|\underset{\sim}{S}(\rho)\| \le \|\underset{\sim}{S}(\rho_*)\| < \|\underset{\sim}{S}(\tau)\|$. We will show that $\rho_* = \rho$. By III, there is a vector τ_* such that (ρ_*,τ_*) is a corner pair for L. Invoking Lemma 2.6, we conclude that $L_1 = [\rho,\tau_*]$ and $L_2 = [\rho_*,\tau]$ are sub-lattices of L with $\mathrm{Vol}(L_1) = \|\rho^{(1)}\tau_*^{(2)}\|$ and $\mathrm{Vol}(L_2) = \|\rho_*^{(1)}\tau^{(2)}\|$. By this same lemma

(6.4) $\qquad \mathrm{Vol}(L) = \|\rho^{(1)}\tau^{(2)}\| = \|\rho_*^{(1)}\tau_*^{(2)}\|.$

We have then by (2.5)

$$\#(L/L_1) = \mathrm{Vol}(L_1)/\mathrm{Vol}(L) = \|\rho^{(1)}\tau_*^{(2)}\|/\|\rho^{(1)}\tau^{(2)}\| = \|\tau_*^{(2)}/\tau^{(2)}\|$$

and

$$\#(L/L_2) = \mathrm{Vol}(L_2)/\mathrm{Vol}(L) = \|\rho_*^{(1)}\tau^{(2)}\|/\|\rho_*^{(1)}\tau_*^{(2)}\| = \|\tau^{(2)}/\tau_*^{(2)}\| .$$

These indices are therefore positive integers whose product is 1! It follows that $L_1 = L_2 = L$, that $\|\tau_*^{(2)}\| = \|\tau^{(2)}\|$ and via (6.4) that $\|\rho_*^{(1)}\| = \|\rho^{(1)}\|$.

Now we observe that $\|\rho_*^{(1)}\| = \|\rho^{(1)}\| > \|\tau^{(1)}\|$ and that $\|\rho_*^{(2)}\| < \|\tau_*^{(2)}\| = \|\tau^{(2)}\|$ which shows that (ρ_*,τ) is a corner pair for L. Thus $\rho_* = \rho$ by II. $\square$

Let us assume for the moment that a corner pair (γ_0,γ_1) exists for L. Properties I) - IV) then imply the existence of a doubly infinite sequence $\Gamma = \Gamma(L) = \{\gamma_r : r \in \mathbf{Z}\}$ which is uniquely determined by the initial choice (γ_0,γ_1) and the property that every pair (γ_{r-1},γ_r) of consecutive vectors in Γ is a corner pair for L. Every corner of L occurs in Γ and so we call Γ the <u>sequence of corners</u> of L. Note that the enumeration has been chosen so that the slope function is increasing on Γ:

(6.5) $\qquad i > j \Rightarrow \|\underset{\sim}{S}(\gamma_i)\| > \|\underset{\sim}{S}(\gamma_j)\|$.

Equations (6.2) and (6.3) actually provide a continued fraction algorithm for computing Γ in both directions away from (γ_0,γ_1). Writing (6.2) for the corner pair (γ_{r-1},γ_r), we get

(6.6) $\qquad \zeta_r\gamma_{r+1} = \gamma_{r-1} - A_r\gamma_r$

where $A_r = [\gamma_{r-1}^{(1)}/\gamma_r^{(1)}]$ and $\zeta_r \in \mathbf{F}_q^{\times}$ is chosen so that $\gamma_{r+1}^{(2)}$ is positive. This equation defines A_r and ζ_r for all $r \in \mathbf{Z}$. In its first co-ordinate, (6.6) can be written as

(6.7) $$\frac{\gamma_{r-1}^{(1)}}{\gamma_r^{(1)}} = A_r + \frac{\zeta_r}{\gamma_{r+1}^{(1)}/\gamma_r^{(1)}} .$$

Thus for $r \geq 1$, the polynomials A_r are the partial quotients in the continued fraction expansion (3.4) of $w = \gamma_0^{(1)}/\gamma_1^{(1)}$, and $w_r = \gamma_{r-1}^{(1)}/\gamma_r^{(1)}$. The construction of Γ in the negative direction away from (γ_0,γ_1) can be understood as a continued fraction algorithm in the same way. In fact, if we rewrite (6.6) as $\gamma_{r-1} = \zeta_r\gamma_{r+1} + A_r\gamma_r$, then we see from (6.3) that

(6.8) $$A_r = -\zeta_r \cdot [\gamma_{r+1}^{(2)}/\gamma_r^{(2)}] .$$

In particular, we learn that $\operatorname{sgn}(A_r) = -\zeta_r$.

One corollary of (6.7) is that $\gamma_0^{(1)}/\gamma_1^{(1)}$ is not in k since otherwise, as Artin shows, the algorithm (6.7) would end at some finite stage. Of course, since L is scattered, it is easy to prove that $\alpha^{(1)}/\beta^{(1)}$ and $\alpha^{(2)}/\beta^{(2)} \in k_\infty\backslash k$ for all pairs of distinct vectors $\alpha,\beta \in L$. This fact is used in the following proposition, which shows that $\Gamma(L)$ always exists.

PROPOSITION 6.9. There exists a corner pair for the scattered lattice L.

Proof. Let ρ_*,τ_* be a pair of basis vectors for L, let $w = \rho_*^{(1)}/\tau_*^{(1)}$ and let $u = \rho_*^{(2)}/\tau_*^{(2)}$. For $r \geq 1$, let $M_r = M_r(w) \in GL_2(A)$ be the matrix (5.8) associated to the continued fraction expansion of w, and define u_r by $u = M_r \circ u_{r+1}$. Since ρ_*,τ_* are basis vectors, $u \neq w$; and so by Proposition 5.12, there is an index $r = R \geq 1$ such that $\|w_{R+1}\| > 1$ and $\|u_{R+1}\| < 1$. Put

$$\begin{bmatrix} \mu\rho \\ \zeta\tau \end{bmatrix} = M_R^{-1} \begin{bmatrix} \rho_* \\ \tau_* \end{bmatrix}$$

where $\mu,\zeta \in \mathbf{F}_q$ are chosen so that $\rho^{(2)}$ and $\tau^{(2)}$ are positive. Then ρ,τ are basis vectors for L, and $\mu\rho^{(1)}/\zeta\tau^{(1)} = M_R^{-1} \circ w = w_{R+1}$ while $\mu\rho^{(2)}/\zeta\tau^{(2)} = M_R^{-1} \circ u = u_{R+1}$. Thus, (ρ,τ) is a corner pair for L. □

Consider now the special scattered lattice $L = \underset{\sim}{e}(M^{-1})$, where M is a rank 2 A-submodule of K. We can use the sequence of corners of L to write

$Z_\chi(s,M)$ as $q^{-d(M)s}$ times a finite sum of sector zeta-functions of type (4.8). The group of units of the ring $O_F = \mathrm{End}(M)$ acts on L via (4.5), and the subgroup $\varepsilon_{F,\chi}^{\mathbb{Z}}$ acts on the corners of L. Since $\|\underset{\sim}{S}(\varepsilon_{F,\chi})\| > 1$, the action of $\varepsilon_{F,\chi}$ increases the slope of each corner. Therefore there is an integer $\partial = \partial(M^{-1},\chi) > 0$ such that

$$\varepsilon_{F,\chi}\gamma_r = \gamma_{r+\partial} \tag{6.10}$$

for all $r \in \mathbf{Z}$. This equation has profound implications for the continued fraction expansion (6.7):

$$w_{r+\partial} = \frac{\varepsilon_{F,\chi}^{(1)}\gamma_{r-1}^{(1)}}{\varepsilon_{F,\chi}^{(1)}\gamma_r^{(1)}} = \frac{\gamma_{r-1}^{(1)}}{\gamma_r^{(1)}} = w_r \tag{6.11}$$

which in turn implies

$$A_{r+\partial} = A_r \tag{6.12}$$

for all $r \in \mathbf{Z}$. Thus the continued fraction expansion of $\gamma_0^{(1)}/\gamma_1^{(1)}$ is <u>purely periodic</u> with period ∂.

PROPOSITION 6.13. The invariant $\partial_0 = \partial(M^{-1},\chi_0)$ for the case $\chi = \chi_0$ is the minimal period for the continued fraction algorithm (6.7).

Proof. Let $n > 0$ be an integer such that $w_{n+r} = w_r$ for all $r \in \mathbb{Z}$. We observe from the equation $w = M_n \circ w_{n+1}$ that the matrix $M_n \in GL_2(A)$ has an eigenvalue $\varepsilon \in K$ such that $\varepsilon^{(1)} = \gamma_n^{(1)}/\gamma_0^{(1)} = \gamma_{n+1}^{(1)}/\gamma_1^{(1)}$. We must have then $\varepsilon\gamma_0 = \gamma_n$ and $\varepsilon\gamma_1 = \gamma_{n+1}$; and these equations imply that $\varepsilon \in O_F^\times$ because both γ_0,γ_1 and γ_n,γ_{n+1} are pairs of basis vectors for $L = \underset{\sim}{e}(M^{-1})$. We see also that $\varepsilon^{(2)} = \gamma_n^{(2)}/\gamma_0^{(2)}$ is positive and that

$$\|\underset{\sim}{S}(\varepsilon)\| = \frac{\|\underset{\sim}{S}(\gamma_n)\|}{\|\underset{\sim}{S}(\gamma_0)\|} > 1.$$

By (4.3) $\varepsilon = \varepsilon_F^m \Rightarrow n = m\partial_0$ for some integer $m \geq 1$. □

Equation (6.12) implies that the set of integers $\deg(A_r)$ is bounded. In fact, we can write down an explicit upper bound:

PROPOSITION 6.14. For all $r \in \mathbb{Z}$, $\deg(A_r) \leq d(M)$.

Proof. Since γ_{r-1},γ_r are basis vectors for $L = \underset{\sim}{e}(M^{-1})$

$$||A_r|| = \left|\left|\frac{\gamma_{r-1}^{(1)}}{\gamma_r^{(1)}}\right|\right| = \left|\left|\frac{\gamma_{r-1}^{(1)}\cdot\gamma_r^{(2)}}{\underset{\sim}{N}\gamma_r}\right|\right| = \frac{||V(M^{-1})||}{||\underset{\sim}{N}\gamma_r||}$$

by Lemma 2.6. If $\gamma_r = \underset{\sim}{e}(x)$ for $x \in M^{-1}$, then $xM \subset O_F$ and so

$$1 \leq ||N(xM)|| = ||\underset{\sim}{N}\gamma_r|| \cdot ||NM|| \Rightarrow ||A_r|| \leq ||V(M^{-1})||\cdot||NM|| .$$

Now $||A_r|| \leq ||\mathrm{Disc}(M)||$ follows by (4.10). □

Let us specify the corner γ_0 as the vector ρ on which the fundamental domain $X_{F,\chi}(\rho)$ of §4 is based. The fundamental domain then decomposes naturally as a disjoint union:

$$X_{F,\chi}(\gamma_0) = X(\gamma_0,\varepsilon_{F,\chi}\gamma_0) = \bigcup_{r=1}^{\partial} X(\gamma_{r-1},\gamma_r)$$

of smaller sectors. The sector $X(\gamma_{r-1},\gamma_r)$ has the very convenient property that the vectors γ_{r-1},γ_r which define the sector also generate the lattice L. Therefore by (4.10) we have the formal identity

$$(6.15)\qquad Z_\chi(s,M) = q^{-d(M)s} \cdot \sum_{r=1}^{\partial} Z_\chi(s;\gamma_{r-1},\gamma_r)$$

where each summand on the right is one of the sector zeta-functions (4.8).

7. $Z_\chi(s,M)$ AS A RATIONAL FUNCTION IN q^{-s}. Let ρ and τ be vectors in P such that both $\rho^{(2)}$ and $\tau^{(2)}$ are positive and $||\rho^{(1)}|| > ||\tau^{(1)}||$ while $||\rho^{(2)}|| < ||\tau^{(2)}||$. Then by Lemma 2.6, (ρ,τ) is a corner pair for the lattice $L = [\rho,\tau]$. If e is the order of the character χ, then

$$\delta_\chi(\alpha) = \frac{1}{e} \cdot \sum_{j=0}^{e-1} \chi^j(\underset{\sim}{S}(\alpha))$$

for all $\alpha \in P$. Therefore, we have

$$(7.1)\qquad (q-1)\cdot Z_\chi(s;\rho,\tau) = \frac{1}{e}\cdot \sum_{j=0}^{e-1} D(s,\chi^j;\rho,\tau)$$

where

$$(7.2)\qquad D(s,\chi;\rho,\tau) = ||V(L)||^s \cdot \sum_{\lambda} \frac{\chi(\underset{\sim}{S}(\lambda))}{||\underset{\sim}{N}\lambda||^s} \qquad (\lambda \in X(\rho,\tau) \cap [\rho,\tau]) .$$

We show in this section that the series in (7.2) converges for $\mathrm{Re}(s) > 1$ to a rational function in q^{-s}. We may then evaluate $Z_\chi(s,M)$ as a sum of such rational functions via (7.1) and (6.15).

We first determine which lattice points

$$(7.3) \qquad \lambda = A\rho + B\tau \qquad\qquad (A,B \in A)$$

lie in the sector $X(\rho,\tau)$. Let ℓ_1 and ℓ_2 be the positive integers such that

$$(7.4) \qquad \|\rho^{(1)}/\tau^{(1)}\| = q^{\ell_1} \qquad \text{and} \qquad \|\rho^{(2)}/\tau^{(2)}\| = q^{-\ell_2} .$$

LEMMA 7.5. The lattice point (7.3) lies in $X(\rho,\tau)$ if and only if one of the following four conditions holds:

Case 1: $-\ell_2 < \deg(B) - \deg(A) < \ell_1$.
Case 2: $\deg(A) = \deg(B) + \ell_2$ and $\operatorname{sgn}(A) \neq -\operatorname{sgn}(B)$.
Case 3: $B \neq 0$ and $\deg(A) > \deg(B) + \ell_2$.
Case 4: $B = 0$.

In all four cases, $\operatorname{sgn}(\lambda^{(1)}) = \operatorname{sgn}(A\rho^{(1)})$; and $\operatorname{sgn}(\lambda^{(2)})$ is given by the table

Case	1	2	3	4
$\operatorname{sgn}(\lambda^{(2)})$	$\operatorname{sgn}(B)$	$\operatorname{sgn}(A) + \operatorname{sgn}(B)$	$\operatorname{sgn}(A)$	$\operatorname{sgn}(A)$

Proof. We have

$$\|\underset{\sim}{S}(\rho)\| \leq \|\underset{\sim}{S}(\lambda)\| \iff 1 \leq \|\underset{\sim}{S}(\rho^{-1}\lambda)\|$$

$$\iff \|A + B\tau^{(2)}/\rho^{(2)}\| \geq \|A + B\tau^{(1)}/\rho^{(1)}\| .$$

This inequality is true if $B = 0$; and for $B \neq 0$, it is equivalent to

$$\left\|\frac{A}{B} + \frac{\tau^{(2)}}{\rho^{(2)}}\right\| \geq \left\|\frac{A}{B} + \frac{\tau^{(1)}}{\rho^{(1)}}\right\| .$$

Since $\|\tau^{(2)}/\rho^{(2)}\| > 1 > \|\tau^{(1)}/\rho^{(1)}\|$, this last inequality is valid for all choices of A and $B \neq 0$ except for those pairs such that $\|A/B\| = \|\tau^{(2)}/\rho^{(2)}\|$ and $\operatorname{sgn}(A/B) = -1$. These exceptions are precisely the pairs excluded in Case 2.

Similarly, the inequality $\|\underset{\sim}{S}(\lambda)\| < \|\underset{\sim}{S}(\tau)\|$ is equivalent to $A \neq 0$ and

$$(7.6) \qquad \left\|\frac{B}{A} + \frac{\rho^{(2)}}{\tau^{(2)}}\right\| < \left\|\frac{B}{A} + \frac{\rho^{(1)}}{\tau^{(1)}}\right\| .$$

Since $\|\rho^{(2)}/\tau^{(2)}\| < 1 < \|\rho^{(1)}/\tau^{(1)}\|$,

$$(7.6) \text{ holds} \iff \|B/A\| < \|\rho^{(1)}/\tau^{(1)}\| \iff \deg(B) - \deg(A) < \ell_1 .$$

Thus, both inequalities are valid precisely for those pairs A,B with $\deg(B) - \deg(A) < \ell_1$ except for the pairs excluded in Case 2. These are the pairs that comprise the categories in Cases 1-4.

It remains to prove the assertions about signs. We have $\operatorname{sgn}(\lambda^{(1)}) = \operatorname{sgn}(A\rho^{(1)})$ because $\|A\rho^{(1)}\| > \|B\tau^{(1)}\|$ in all four cases. In Case 1, $\|A\rho^{(2)}\| < \|B\tau^{(2)}\| \Rightarrow \operatorname{sgn}(\lambda^{(2)}) = \operatorname{sgn}(B\tau^{(2)}) = \operatorname{sgn}(B)$; and in Case 2, $\|A\rho^{(2)}\| = \|B\tau^{(2)}\| \Rightarrow \operatorname{sgn}(\lambda^{(2)}) = \operatorname{sgn}(A) + \operatorname{sgn}(B)$ since $\rho^{(2)}$ and $\tau^{(2)}$ are both positive and $\operatorname{sgn}(A) \neq -\operatorname{sgn}(B)$. In cases 3 and 4, $\|A\rho^{(2)}\| > \|B\tau^{(2)}\| \Rightarrow \operatorname{sgn}(\lambda^{(2)}) = \operatorname{sgn}(A)$. □

Our aim now is to sum the series of (7.2) in closed form on the complex half-plane $\operatorname{Re}(s) > 1$. It is convenient to arrange the series according to the cases in Lemma 7.5. We write

$$(7.7) \qquad D(s) = D(s,\chi;\rho,\tau) = D_1(s) + D_2(s) + D_3(s) + D_4(s)$$

where for $1 \leq i \leq 4$, $D_i(s) = D_i(s,\chi;\rho,\tau)$ is $\|V(L)\|^s$ times the sum of $\chi(\underset{\sim}{S}(\lambda))/\|\underset{\sim}{N}\lambda\|^s$ over the lattice points (7.3) which satisfy the condition of Case i. We note that

$$\chi(\underset{\sim}{S}(\lambda)) = \chi(\lambda^{(2)}/\lambda^{(1)}) = \bar{\chi}(\lambda^{(1)}) \cdot \chi(\lambda^{(2)})$$

and

$$\|\underset{\sim}{N}\lambda\| = \|\lambda^{(1)}\| \cdot \|\lambda^{(2)}\| = \|A\rho^{(1)} + B\tau^{(1)}\| \cdot \|A\rho^{(2)} + B\tau^{(2)}\| .$$

Our results will be written as rational functions of the variable $U = q^{-s}$.

CASE 1: In this case, we have

$$\chi(\underset{\sim}{S}(\lambda)) = \bar{\chi}(A\rho^{(1)}) \cdot \chi(B)$$

and

$$\|\underset{\sim}{N}\lambda\| = \|A\rho^{(1)}\| \cdot \|B\tau^{(2)}\| = \|V(L)\| \cdot \|A\| \cdot \|B\|$$

by Lemmas 2.6 and 7.5. Therefore,

$$(7.8) \qquad D_1(S) = \sum_t D_1(t,s) \qquad\qquad (-\ell_2 < t < \ell_1)$$

with

$$D_1(t,s) = \bar{\chi}(\rho^{(1)}) \cdot \sum_{A,B} \frac{\bar{\chi}(A)\cdot\chi(B)}{\|A\|^s\cdot\|B\|^s} \qquad (\deg(B) - \deg(A) = t)$$
$$= \bar{\chi}(\rho^{(1)}) \cdot \sum_{n\geq 0} q^{-ns}\cdot q^{-(n+|t|)s} \sum_{A,B} \bar{\chi}(A)\cdot\chi(B)$$

where the inner summation above is over all pairs of polynomials A,B such that $\deg(A) = n$ and $\deg(B) = n + t$ if $t \geq 0$ and otherwise over all pairs of polynomials A,B with $\deg(B) = n$ and $\deg(A) = n + |t|$. Since each pair of non-zero constants appears as the leading coefficients $(\mathrm{sgn}(A),\mathrm{sgn}(B))$ for exactly $q^{2n+|t|}$ pairs of polynomials (A,B) in the inner summation, that summation is zero unless χ is the trivial character χ_0. When $\chi = \chi_0$, the inner summation equals $(q-1)^2q^{2n+|t|}$, and so

$$D_1(t,s) = (q-1)^2q^{(1-s)|t|} \cdot \sum_{n=0}^{\infty} q^{(2-2s)n}$$
$$= \frac{(q-1)^2(qU)^{|t|}}{1-q^2U^2} .$$

We conclude in particular that the series defining $D_1(t,s)$ and $D_1(s)$ are absolutely convergent for $\mathrm{Re}(s) > 1$. Put $H_1(\chi) = 1$ if $\chi = \chi_0$, and put $H_1(\chi) = 0$ if $\chi \neq \chi_0$. Plugging the above evaluation of $D_1(t,s)$ into (7.8) and summing the finite geometric series, we arrive at the following evaluation:

$$(7.9) \qquad D_1(s) = H_1(\chi)\cdot\frac{(q-1)^2}{1-q^2U^2}\cdot\frac{(qU)^{\ell_1}+(qU)^{\ell_2}-qU-1}{(qU)-1}$$

CASE 2: In this case,

$$\chi(\underset{\sim}{S}(\lambda)) = \bar{\chi}(A\rho^{(1)}) \cdot \chi(\mathrm{sgn}(A)+\mathrm{sgn}(B))$$

and

$$\|\underset{\sim}{N}\lambda\| = \|A\rho^{(1)}\| \cdot \|B\tau^{(2)}\| = \|V(L)\|\cdot\|A\|\cdot\|B\|$$

by Lemmas 2.6 and 7.5. Therefore,

$$D_2(s) = \bar{\chi}(\rho^{(1)}) \cdot \sum_{A,B} \frac{\bar{\chi}(A)\cdot\chi(\mathrm{sgn}(A)+\mathrm{sgn}(B))}{\|A\|^s\|B\|^s} \qquad (\deg(A) = \deg(B) + \ell_2)$$
$$= \bar{\chi}(\rho^{(1)}) \cdot \sum_{n\geq 0} q^{-(n+\ell_2)s}\cdot q^{-ns}\cdot q^{2n+\ell_2} \cdot \sum_{\mu,\zeta\neq 0} \bar{\chi}(\mu)\cdot\chi(\mu+\zeta)$$

since for every pair (μ,ζ) of non-zero constants, there are $q^{2n+\ell_2}$ pairs of polynomials (A,B) with $\mathrm{sgn}(A) = \mu$, $\mathrm{sgn}(B) = \zeta$, $\deg(B) = n$ and $\deg(A) = n + \ell_2$. The inner summation equals

$$\sum_{\mu,\zeta\neq 0} \bar{\chi}(\mu)\cdot\chi(\mu+\mu\zeta) = \sum_{\mu,\zeta\neq 0} \chi(1+\zeta) = H_2(\chi)$$

where $H_2(\chi) = (q-1)(q-2)$ if $\chi = \chi_0$, and otherwise $H_2(\chi) = 1-q$. We have then

$$D_2(s) = \bar{\chi}(\rho^{(1)}) \cdot H_2(\chi) q^{\ell_2(1-s)} \cdot \sum_{n \geq 0} q^{(2-2s)n} \quad .$$

We observe that $D_2(s)$ is absolutely convergent for $\mathrm{Re}(s) > 1$ and that

$$(7.10) \qquad D_2(s) = \bar{\chi}(\rho^{(1)}) \cdot \frac{H_2(\chi) \cdot (qU)^{\ell_2}}{1 - q^2U^2}$$

in that half-plane.

CASE 3: In this case,

$$\chi(\underset{\sim}{S}(\lambda)) = \bar{\chi}(A\rho^{(1)}) \cdot \chi(A) = \bar{\chi}(\rho^{(1)})$$

and

$$\|\underset{\sim}{N}\lambda\| = \|A\rho^{(1)}\| \cdot \|A\rho^{(2)}\| = \|\rho^{(1)}\tau^{(2)}\| \cdot \|\rho^{(2)}/\tau^{(2)}\| \cdot \|A\|^2$$
$$= q^{-\ell_2} \cdot \|V(L)\| \cdot \|A\|^2$$

by Lemmas 2.6 and 7.5. Therefore

$$(7.11) \qquad D_3(s) = \bar{\chi}(\rho^{(1)}) \; q^{s\ell_2} \cdot \sum_{A,B} \frac{1}{\|A\|^{2s}} \qquad (\deg(A) > \deg(B) + \ell_2) \; .$$

If we write $\deg(A) = n$ and $\deg(B) = m$, the summation in (7.11) can be written as

$$\sum_{n > m+\ell_2} (q-1)^2 q^{n+m} \cdot q^{-2ns}$$

$$= (q-1)^2 \cdot \sum_{m \geq 0} q^m \cdot \sum_{n > m+\ell_2} q^{(1-2s)n} = (q-1)^2 \cdot \sum_{m \geq 0} q^m \; \frac{(qU^2)^{m+\ell_2+1}}{1 - qU^2}$$

$$= \frac{(q-1)^2 (qU^2)^{\ell_2+1}}{1 - qU^2} \cdot \sum_{m \geq 0} (q^2U^2)^m = \frac{(q-1)^2(qU^2)^{\ell_2}}{1 - qU^2} \cdot \frac{qU^2}{1 - q^2U^2}$$

for $\mathrm{Re}(s) > 1$. Since $q^{s\ell_2} = U^{-\ell_2}$, we get the following evaluation for $D_3(s)$:

$$(7.12) \qquad D_3(s) = \bar{\chi}(\rho^{(1)}) \cdot \frac{(q-1)^2(qU)^{\ell_2}}{1 - qU^2} \cdot \frac{qU^2}{1 - q^2U^2} \quad .$$

CASE 4: In this case,

$$\chi(\underset{\sim}{S}(\lambda)) = \bar{\chi}(\rho^{(1)})$$

and

$$\|\underset{\sim}{N}\lambda\| = q^{-\ell_2} \cdot \|V(L)\| \cdot \|A\|^2$$

just as in Case 3. Therefore,

$$D_4(s) = \bar{\chi}(\rho^{(1)}) \cdot q^{s\ell_2} \cdot \sum_A \frac{1}{\|A\|^{2s}} = \bar{\chi}(\rho^{(1)}) \cdot \frac{1}{U^{\ell_2}} \cdot \sum_{n\geq 0} (q-1)q^n \cdot q^{-2ns} .$$

The series over n is absolutely convergent for $Re(s) > 1/2$, and so

(7.13) $$D_4(s) = \bar{\chi}(\rho^{(1)}) \cdot \frac{q-1}{U^{\ell_2}(1 - qU^2)}$$

in that half-plane.

In order to apply these results to the evaluation of $Z_\chi(s,M)$, we take $L = \underset{\sim}{e}(M^{-1})$; and we write (6.15) via (7.1) as

(7.14) $$(q-1) \cdot Z_\chi(s,M) = q^{-d(M)s} \cdot \frac{1}{e} \cdot \sum_{j=0}^{e-1} E_j(s,\chi,M)$$

where

(7.15) $$E_j(s,\chi,M) = \sum_{r=1}^{\partial} D(s,\chi^j,\gamma_{r-1},\gamma_r) ,$$

$\Gamma = \{\gamma_r : r \in \mathbb{Z}\}$ being the sequence of corners of $\underset{\sim}{e}(M^{-1})$. For $\rho = \gamma_{r-1}$ and $\tau = \gamma_r$, we write $\ell_1 = \ell_1(r)$ and $\ell_2 = \ell_2(r)$; and then we see from (6.7) and (6.8) that

(7.16) $$\ell_1(r) = \deg(A_r) = \ell_2(r+1)$$

for all $r \in \mathbb{Z}$. The polynomial

(7.17) $$P_j(U,\chi,M) = \sum_{r=1}^{\partial} \chi^j(\gamma_{r-1}^{(1)}) \cdot (qU)^{\ell_2(r)} = \sum_{r=1}^{\partial} \chi^j(\gamma_r^{(1)}) \cdot (qU)^{\ell_1(r)}$$

will play an important role in our evaluation of $Z_\chi(s,M)$.

Let us first evaluate $E_j(s,\chi,M)$ when $0 < j \leq e$. After a little algebra, we find from the results in Cases 1-4 above that

$$D(s,\chi^j;\gamma_{r-1},\gamma_r) = D_2(s) + D_3(s) + D_4(s)$$
$$= \frac{(q-1) \cdot \chi^j(\gamma_{r-1}^{(1)})}{1 - qU^2} \cdot \left[U^{-\ell_2(r)} - (qU)^{\ell_2(r)} \right] .$$

Thus by (7.15) and (7.17)

$$(7.18)\qquad E_j(s,\chi,M) = \frac{q-1}{1-qU^2}\cdot[P_j(1/qU,\chi,M) - P_j(U,\chi,M)] .$$

Let $\mathbf{Q}(\chi)$ be the cyclotomic number field obtained by adjoining all the values χ to $\mathbf{Q}$. As an immediate consequence of (7.18), we have

PROPOSITION 7.19. For $0 < j \leq e$, $E_j(s,\chi,M)/(q-1)$ is a Laurent polynomial in U and $1/U$ with coefficients from the ring of integers of $\mathbf{Q}(\chi)$. As a function of s, $E_j(s,\chi,M)$ is entire and satisfies the functional equation

$$(7.20)\qquad E_j(1-s,\chi,M) = q^{1-2s}\cdot E_j(s,\chi,M) .$$

When $j = 0$, it is convenient to sum the result from each of the Cases 1-4 above separately over r before adding them. After some algebra, one finds that

$$(7.21)\qquad \frac{(1-qU)^2}{q-1}\cdot E_0(s,\chi,M) = \left[\partial(q-1) - \frac{q(U-1)^2\cdot P_0(U,\chi,M)}{1-qU^2} + \frac{(1-qU)^2\cdot P_0(1/qU,\chi,M)}{1-qU^2}\right]$$

which enables us to prove

PROPOSITION 7.22. The function $(1-q^{1-s})\cdot E_0(s,\chi,M)/(q-1)$ is a Laurent polynomial in U and $1/U$ with coefficients in $\mathbf{Z}$. As a function of s, $E_0(s,\chi,M)$ is holomorphic in the whole complex plane except for simple poles at the points $s = 1 + 2\pi in/\log(q)$, $n \in \mathbf{Z}$. The residue at each of these poles is

$$(7.23)\qquad \frac{(q-1)^2}{\log(q)}\cdot\deg(\varepsilon^{(2)}_{F,\chi}) .$$

Further, $E_0(s,\chi,M)$ satisfies the functional equation

$$(7.24)\qquad (1-q^s)^2\cdot E_0(1-s,\chi,M) = (1-q^{1-s})^2\cdot E_0(s,\chi,M) .$$

Proof: The right hand side of (7.21) is a Laurent polynomial in U and $1/U$ with coefficients in $\mathbf{Z}$ which takes the value zero at $U = 1/q$. Thus $E_0(s,\chi,M)/(q-1)$ can be written as a rational function of U with denominator $1-qU$. The residue of this rational function at $U = 1/q$ is

$$\frac{-1}{q}\cdot\lim_{U\to 1/q}\frac{\partial(q-1)(1-qU^2) - q(U-1)^2\, P_0(U,\chi,M)}{(1-qU)(1-qU^2)}$$

$$= \frac{1-q}{q}\cdot\sum_{r=1}^{\partial}\ell_2(r)$$

by L'Hopital. Now by (6.8) and (6.10)

$$\sum_{r=1}^{\partial} \ell_2(r) = \sum_{r=1}^{\partial} [\deg(\gamma_r^{(2)}) - \deg(\gamma_{r-1}^{(2)})] = \deg\left[\frac{\gamma_\partial^{(2)}}{\gamma_0^{(2)}}\right] = \deg(\varepsilon_{F,\chi}^{(2)}) \neq 0 .$$

Further, the residue of $E_0(s,\chi,M)/(q-1)$ at $s = 1$ is $-q/\log(q)$ times its residue as a function of U at $1/q$. Thus, (7.23) gives the correct residue for the pole at $s = 1$. Since $E_0(s,\chi,M)$ is a rational function in q^{-s}, it is periodic with primitive period $2\pi i/\log(q)$; and therefore the points $s = 1 + 2\pi in/\log(q)$, $n \in \mathbf{Z}$, are also poles of $E_0(s,\chi,M)$ having the same residue as the pole at $s = 1$. The functional equation (7.24) is immediate from the observation that the right hand side of (7.21) is invariant under the substitution $U \leftarrow 1/qU$. □

We are now in a position to describe the analytic properties of the partial zeta-function $Z_\chi(s,M)$.

THEOREM 7.25. Let M be a rank A-submodule of K with conductor F. The series in (4.6) converges absolutely in the complex half-plane $\mathrm{Re}(s) > 1$. The partial zeta-function $Z_\chi(s,M)$ is a rational function in q^{-s} which is holomorphic in the whole complex plane except for first order poles at the points $s = 1 + 2\pi in/\log(q)$, $n \in \mathbf{Z}$. The residue at each of these poles is

$$(7.26) \qquad \frac{(q-1)/e}{\log(q)} \cdot \frac{\deg(\varepsilon_{F,\chi}^{(2)})}{\|\mathrm{Disc}(M)\|} = \frac{(q-1)/e}{\log(q)} \cdot \frac{\deg(\varepsilon_{F,\chi}^{(2)})}{\|F\sqrt{\Delta}\|}$$

where e is the order of the character χ.

Proof: The convergence of the series in (4.6) and its rationality in q^{-s} follow from (7.14) and the computations in Cases 1-4 above. The residue at the indicated poles is an immediate consequence of (7.14) and Propositions 7.19 and 7.22. □

We can also use (7.14) to describe the values $Z_\chi(1-n,M)$ for integers $n \geq 1$. By (7.16) and Proposition 6.14, the polynomial $P_j(U,\chi,M)$ has degree $d(M)$ or less in U. It follows then from (7.18) and Proposition 7.19 that for $j > 0$

$$(7.27) \qquad q^{-d(M)s} \sum_{j=1}^{e-1} E_j(s,\chi,M)/(q-1)$$

is a polynomial in $\mathbf{Z}[U]$. Specializing s to $1-n$ is equivalent to substituting $U = q^{n-1}$; and so the values (7.27) at $s = 1-n$, $n \in \mathbf{Z}$, $n \geq 1$ belong to $\mathbf{Z}$.

The same reasoning based on (7.21) and Proposition 7.22 shows that

$$q^{-d(M)s}\cdot(1-q^{1-s})\cdot E_0(s,\chi,M)/(q-1) \tag{7.28}$$

is a polynomial in $\mathbb{Z}[U]$. Therefore the values of (7.28) at $s=1-n$, $n\in\mathbb{Z}$, $n\geq 1$ are themselves in $\mathbb{Z}$.

Substituting these results in (7.14) to evaluate $Z_\chi(1-n,M)$, we obtain

THEOREM 7.29. For integers $n\geq 1$,

$$Z_\chi(1-n,M) = \frac{a_\chi}{e(q^n-1)} \tag{7.30}$$

where $a_\chi\in\mathbb{Z}$.

In case $\chi=\chi_0$, we can use our results to study the zeta-function $Z_F(s)$ of the order $\mathcal{O}_F$. Assume $\chi=\chi_0$, and let C run through a system of representatives for the ideal classes in $\mathrm{Pic}(\mathcal{O}_F)$. Then from the definitions

$$Z_F(s) = \sum_C Z_0(s,C) = \sum_C\sum_{i\geq 0} N_i(C)\cdot U^i = \sum_{i\geq 0} c_i(\mathcal{O}_F)\cdot U^i \tag{7.31}$$

where $N_i(C)$ is the number of $\mathcal{O}_F$-integral ideals of degree i which belong to the ideal class of C and where $c_i(\mathcal{O}_F)$ is the total number of $\mathcal{O}_F$-invertible integral ideals of degree i. When $\chi=\chi_0$, (7.14) with $M=C$ simplifies to

$$Z_0(s,C) = q^{-ms}\cdot E_0(s,\chi_0,C)/(q-1) \tag{7.32}$$

where $m=d(\mathcal{O}_F)=\deg(F)+\deg(\Delta)/2$ by (3.12). The analytic properties of $Z_F(s)$ including its functional equation are an immediate consequence of these last two identities together with Proposition 7.22. Going further, let us multiply (7.32) by $1-qU^2$ and compute $U^m\cdot(1-qU^2)\cdot E_0(s,\chi_0,C)/(q-1)$ modulo U^m from (7.21). Substituting the results in (7.32), we obtain the following congruence of formal power series:

$$N_0(C)+N_1(C)\cdot U+\sum_{i\geq 2}(N_i(C)-qN_{i-2}(C))\cdot U^i \equiv U^m\cdot P_0(1/qU,\chi_0,C) \pmod{U^m}. \tag{7.33}$$

Now by (7.17), for each representative ideal C

$$U^m\cdot P_0(1/qU,\chi_0,C) = \sum_{j=1}^{m} b_j(C)\cdot U^{m-j}$$

where for $1\leq j\leq m$, $b_j(C)$ is the number of polynomials of degree j in a minimal period of partial quotients for the continued fraction algorithm (6.7) associated to $\tilde{e}(C^{-1})$. Using this last result in (7.32) and equating like powers of U, we find that

$$N_0(C) = b_m(C),\quad N_1(C) = b_{m-1}(C) \tag{7.34}$$

and for $2\leq i<m$

$$N_i(C) - qN_{i-2}(C) = b_{m-i}(C)\ . \tag{7.35}$$

One can understand these equations in the following way. Given an ideal $B \in I(F)$, call $x \in B$ a corner of B if $\underset{\sim}{e}(x)$ is a corner of the lattice $\underset{\sim}{e}(B)$. From the proof of Proposition 6.14, for $0 \leq i < m$ and every $r \in \mathbb{Z}$

$$\deg(A_r) = m - i \iff \|N(x_r C)\| = q^i$$

where x_r is the corner $\underset{\sim}{e}^{-1}(\gamma_r)$ of C^{-1}. Thus, the corners x of C^{-1} define integral ideals xC of degree less than m which lie in the ideal class of C; and the number of such ideals of given degree i is just $b_{m-i}(C)$. We learn from the second equation in (7.34) that all integral ideals of degree 1 in the class of C can be generated in this fashion. However, equation (7.35) shows that precisely $qN_{i-2}(C)$ integral ideals of degree $2 \leq i < m$ in the class of C are not generated by the corners of $\underset{\sim}{e}(C^{-1})$.

An integral ideal $B \in I(F)$ is called impure if it is divisible by the principal $\mathcal{O}_F$-ideal generated by a non-constant polynomial in A. Otherwise B is pure. We can use this notion to characterize the ideals generated by corners.

THEOREM 7.36. The set $\mathrm{Cor}(C) = \{xC\colon x \text{ a corner of } C^{-1}\}$ is precisely the set of pure $\mathcal{O}_F$-ideals of degree less than m in the class of C.

Proof: For $i \in \mathbb{Z}$, let $P(i)$ and $I(i)$ be respectively the number of pure and impure integral ideals of degree i in $I(F)$. We will prove the following three statements simultaneously by induction on $0 \leq i < m$.

1) $B \in \mathrm{Cor}(C)$, $\deg(B) = i \Rightarrow B$ is pure.
2) $I(i) = \sum_{j \geq 1} q^j \cdot P(i-2j) = qP(i-2) + qI(i-2) = qN_{i-2}$.
3) The pure ideals of degree i in the class of C all belong to $\mathrm{Cor}(C)$.

The degree of $A\mathcal{O}_F$ for $A \in A$ is $2\deg(A)$ which shows that there cannot be any impure ideals of degrees 0 and 1. Therefore 1) and 2) above are trivially true for $i = 0$ and $i = 1$, and 3) follows immediately for these values of i by (7.34). Proceeding by induction, we take $2 \leq i_0 < m$, and we assume the validity 1), 2) and 3) for all $0 \leq i < i_0$.

To prove 1) for $i = i_0$, we assume that $B = xC = AB_p$ where A is a positive polynomial in A and B_p is pure of degree $i \leq i_0$. If $i < i_0$, then $B_p = yC$ for a corner y of C^{-1} by the induction hypotheses; and we have $y = Ax\varepsilon$ for an $\mathcal{O}_F$-unit ε such that $\varepsilon^{(2)}$ is positive. Since such units act on the corners of C^{-1}, y and εx are corners of C^{-1} which differ by a multiplier in A and which therefore have the same slope. By property IV in §6, this forces $y = \varepsilon x$ which contradicts $i < i_0$.

To prove 2) for $i = i_0$, we note that any impure ideal B of degree i_0 has the form $B = AB_p$ where B_p is pure of degree $i_0 - 2j$ and where $A \in A$ is a positive polynomial of degree $j > 0$. We claim that $AB_p = A'B'_p \Rightarrow A = A'$ and $B_p = B'_p$. In fact if x and x' are corners of C^{-1} such that $B_p = xC$ and $B'_p = x'C$, then appealing to property IV of corners as in the above argument, we see that $x = \varepsilon x'$ for some $\mathcal{O}_F$-unit ε. But this forces $B_p = B'_p$ and $A = A'$. Because there are exactly q^j positive polynomials A of degree j, this uniqueness of decomposition enables us to count the number of impure polynomials of degree i_0 by the series in 2) above with $i = i_0$.

To prove 3) for $i = i_0$, we observe from 1), 2) and (7.35) that $\mathrm{Cor}(C)$ contains only pure ideals of degree i_0 and that the total number of pure ideals of degree i_0 in the class of C is exactly the number of ideals of degree i_0 in $\mathrm{Cor}(C)$.

Now 1) and 3) for $0 \leq i < m$ imply the theorem. □

In case $F = 1$, Artin [A] has computed the coefficients $c_i(\mathcal{O}_K)$ in (7.31) in terms of the character sums

$$(7.37) \qquad \sigma_i = \sum_{a \in A} \left[\frac{\Delta}{A}\right] \quad (A \text{ positive, } \deg(A) = i)$$

where $[\Delta/A]$ is the Jacobi symbol for the ring A. His result can be stated as

$$(7.38) \qquad \sigma_i = c_i(\mathcal{O}_K) - qc_{i-1}(\mathcal{O}_K)$$

for $1 \leq i < 2m$ where $2m = \deg(\Delta)$. Using this equation with (7.34) and (7.35) to eliminate the $c_i(\mathcal{O}_K)$, we obtain

$$(7.39) \qquad \sigma_i + q = \sum_C b_{m-1}(C) \quad \text{and} \quad \sigma_i - q\sigma_{i-2} = \sum_C b_{m-i}(C) - q\sum_C b_{m-i+1}(C)$$

for $2 \leq i < m$. Thus, half of the character sums (7.37) can be computed by counting the number of partial quotients of given degree in the continued fraction expansions associated to the ideal classes in $\mathrm{Pic}(\mathcal{O}_K)$. The relationship between b_{m-1} and the character sum σ_1 was first noticed by K. Ireland.

8. CLASS NUMBER FORMULAS. Throughout this section, the character χ is specialized to be the quadratic character of $\mathbf{F}_q$. We will show that the special values

$$Z_0(\pi i/\log(q), \mathcal{O}_K) \qquad \text{and} \qquad Z_\chi(0, \mathcal{O}_K)$$

can be used to prove function field analogs of Theorem 1.5. We begin by noting from the functional equation (7.24) that $E_0(0,\chi,M) = 0$. Therefore (7.14) with $U = 1$ together with (7.16), (7.17) and (7.18) imply that

$$(8.1) \qquad Z_\chi(0,M) = \frac{1}{2} \cdot \sum_{r=1}^{\partial} \chi(\gamma_r^{(1)}) \cdot \frac{\|A_r\| - 1}{q - 1} .$$

We can rewrite (8.1) in terms of the partial quotients $A_r^s (r \geq 1)$ of the standard continued fraction expansion $\gamma_0^{(1)}/\gamma_1^{(1)}$. Since $\chi(\zeta^{-1}) = \chi(\zeta)$ for all $\zeta \in \mathbb{F}_q^\times$, (5.5) implies that

$$(8.2) \qquad \chi(A_r^s) = \chi(\mu_r A_r) = \left(\prod_{j=1}^{r-1} \chi(\zeta_j)\right) \cdot \chi(A_r) = \chi(-1) \cdot \prod_{j=1}^{r} \chi(\zeta_j)$$

for $r \geq 1$ as $\mathrm{sgn}(A_r) = -\zeta_r$. Now also

$$\chi(\gamma_r^{(1)}) = \chi(\gamma_0^{(1)}) \cdot \prod_{j=1}^{r} \chi(\gamma_{j-1}^{(1)}/\gamma_j^{(1)}) = \chi(\gamma_0^{(1)}) \cdot \prod_{j=1}^{r} \chi(A_j)$$
$$= \chi(\gamma_0^{(1)}) \cdot \chi(-1)^r \cdot \prod_{j=1}^{r} \chi(\zeta_j) = \chi(\gamma_0^{(1)}) \cdot \chi(-1)^{r-1} \cdot \chi(A_r^s)$$

by (8.2). Plugging this value into (8.1), we get

$$(8.3) \qquad Z_\chi(0,M) = \frac{\chi(\gamma_0^{(1)})}{2} \cdot \sum_{r=1}^{\partial} \chi\left((-1)^{r-1} A_r^s\right) \cdot \frac{\|A_r^s\| - 1}{q-1}$$

since $\|A_r^s\| = \|\mu_r A_r\| = \|A_r\|$.

The value of ∂ in (8.3) is an appropriate multiple of the minimal period ∂_0 of the continued fraction algorithm (6.7). Because $\gamma_{r-1}^{(2)}/\gamma_r^{(2)}$ is positive for all $r \in \mathbb{Z}$, we can compute the value of ∂_0 from the standard continued fraction expansion of $\gamma_0^{(1)}/\gamma_1^{(1)}$ as follows: ∂_0 is the least integer $n > 0$ such that

$$\frac{w_{r+n}^s}{w_r^s} \in \mathbb{F}_q^\times$$

for all $r \geq 1$. When we speak below of the <u>minimal period</u> of a standard algorithm, we always understand it to be defined in this sense.

Our next task is to derive an expression for $Z_0(s,M)$ at $s = \pi i/\log(q)$ $\Leftrightarrow U = -1$. Let

$$S_1 = \sum_{r=1}^{\partial} (-q)^{\ell_1(r)} = \sum_{r=1}^{\partial} (-q)^{\ell_2(r)}$$

and let

$$S_2 = \sum_{r=1}^{\partial} (-1)^{\ell_1(r)} = \sum_{r=1}^{\partial} (-1)^{\ell_2(r)} .$$

Summing $D(s,\chi_0;\gamma_{r-1},\gamma_r) = D_1(s) + D_2(s) + D_3(s) + D_4(s)$ from $r = 1$ to $r = \partial$, we find by (7.9), (7.10), (7.12) and (7.13) with $U = -1$ and (6.15) that

$$(-1)^{d(M)}(q-1)\cdot Z_0(\pi i/\log(q),M) = \frac{(q-1)^2}{1-q^2}\cdot\frac{2S_1+\partial(q-1)}{-(q+1)} + \frac{(q-1)(q-2)S_1}{1-q^2} +$$

$$+ \frac{(q-1)^2 S_1}{1-q}\cdot\frac{q}{1-q^2} + \frac{q-1}{1-q}\cdot S_2$$

$$= \frac{4q}{(q+1)^2}\cdot S_1 + \partial\left(\frac{q-1}{q+1}\right)^2 - S_2 = \frac{4q}{(q+1)^2}\cdot(S_1-S_2) + (\partial-S_2)\left(\frac{q-1}{q+1}\right)^2$$

which we can write as

$$(8.4)\qquad (q+1)\cdot Z_0(\pi i/\log(q),M) =$$

$$= (-1)^{d(M)}\left[\frac{4q}{q+1}\sum_{r=1}^{\partial}(-1)^{\deg(A_r^S)}\cdot\frac{\|A_r^S\|-1}{q-1} + 2W\left(\frac{q-1}{q+1}\right)\right]$$

where W is the number of polynomials of odd degree in a minimal period $A_1^S, A_2^S, \ldots, A_\partial^S$ of partial quotients for the standard continued fraction expansion of $\gamma_0^{(1)}/\gamma_1^{(1)}$.

We write $Z_N(s)$ for the zeta-function of a finite extension N/k; and for each place P of N we write $Z_{N,P}(s)$ for $Z_N(s)$ with the Euler factor corresponding to P removed. We write $Z_{K,\infty}(s)$ for the zeta-function of K with the Euler factors corresponding to ∞_1 and ∞_2 removed.

LEMMA 8.5. Let P be a place of some finite extension field N of k. Then

$$(8.6)\qquad Z_{N,P}(0) = -\frac{h(P)}{w}$$

where w is the number of roots of unity in N and $h(P)$ is the class number of the ring of functions in N which are holomorphic away from P.

Proof: Let m be the degree of the field of constants in N over $\mathbb{F}_q$, and let dm be the degree of P over $\mathbb{F}_q$. Then we have

$$Z_{N,P}(s) = (1-q^{-dms})\cdot Z_N(s) = \frac{(1-q^{-dms})p(q^{-s})}{(1-q^{-ms})(1-q^{m-ms})}$$

where $P(U)$ is a polynomial in $U = q^{-s}$ such that $P(1) = h$ is the class number of N. Therefore

$$Z_{N,P}(0) = \frac{d\cdot P(1)}{1-q^m} = -\frac{dh}{w} .$$

Now by a standard calculation (see [R]), $dh = h(P)$. □

Suppose now that $K = k(\sqrt{P})$, where P is a positive irreducible of even degree. Choose $\zeta \in \mathbb{F}_q^\times$ with $\chi(\zeta) = -1$. The field $E = k(\sqrt{\zeta P})$ is called <u>imaginary quadratic</u> over (k,∞) because ∞ remains prime in E/k. We write $h(P)$ for the class number of $\mathcal{O}_K$, and we write $h(\zeta P)$ for the class number of the integral closure of A in E. Let $k_2 = k(\sqrt{\zeta})$ be the constant field extension of k of degree 2. We write "∞" also for the unique place in each of E and k_2 which lies over ∞. Consider the Hasse diagram

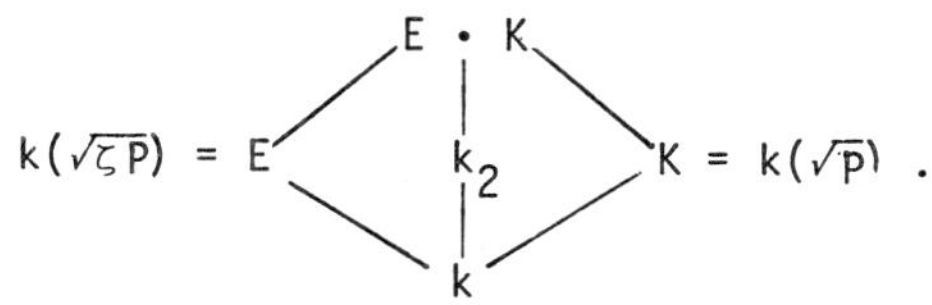

The Galois groups of EK/K, E/k and k_2/k are each cyclic of order 2, and so each group has a unique non-trivial character which we call respectively ψ_1, ψ_2 and ψ_3. For $i = 1,2,3$, let $L_\infty(s,\psi_i)$ be the Artin L-function of the character ψ_i with the Euler factors associated to the places over ∞ removed. Since ψ_1 induces $\psi_2 + \psi_3$ as a character on the Galois group of EK/k, we have

$$(8.7) \qquad L_\infty(s,\psi_1) = L_\infty(s,\psi_2 + \psi_3) = L_\infty(s,\psi_2)\cdot L_\infty(s,\psi_3) .$$

Because

$$L_\infty(s,\psi_2) = Z_{E,\infty}(s)/Z_{k,\infty}(s) \quad \text{and} \quad L_\infty(s,\psi_3) = Z_{k_2,\infty}(s)/Z_{k,\infty}(s),$$

we can calculate the values of these L-functions at $s = 0$ using Lemma 8.5. We obtain

$$L_\infty(0,\psi_2) = \frac{-h(\zeta P)/(q-1)}{-1/(q-1)} = h(\zeta P)$$

and

$$L_\infty(0,\psi_3) = \frac{-1/(q^2-1)}{-1/(q-1)} = 1/(q+1).$$

Therefore (8.7) implies

$$(8.8) \qquad h(\zeta P) = (q+1)\cdot L_\infty(0,\psi_1).$$

THEOREM 8.9. Assume that $h(P) = 1$, and let $D = [\sqrt{P}]$. Then

$$(8.10)\qquad h(\zeta P) = (-1)^{\frac{1}{2}\deg(P)} \cdot \left[\frac{4q}{q+1} \sum_{r=1}^{\partial} (-1)^{\deg(A_r^S)} \cdot \frac{\|A_r^S\|-1}{q-1} + 2W\left(\frac{q-1}{q+1}\right)\right]$$

where W is the number of polynomials of odd degree in the minimal period $A_1^S, A_2^S, \ldots, A_\partial^S$ of partial quotients for the standard continued fraction expansion of $\sqrt{P} + D$.

Proof: Since EK/K is a constant field extension, the Artin L-function of ψ_1 is the twist $Z_K(s+\pi i/\log(q))$ of $Z_K(s)$. Therefore

$$(8.11)\qquad L_\infty(s,\psi_1) = Z_{K,\infty}(s+\pi i/\log(q)) = Z_0(s+\pi i/\log(q), \mathcal{O}_K)$$

from the definitions because $\mathcal{O}_K$ has class number one by hypothesis.

Put $\gamma_0 = \underset{\sim}{e}(\sqrt{P} + D)$, and put $\gamma_1 = (1,1)$. If $\gamma_0^{(2)}$ is positive, then (γ_0, γ_1) is a corner pair for the lattice $\underset{\sim}{e}(\mathcal{O}_K)$ as one easily checks. Evaluating $L_\infty(0,\psi_1)$ by (8.11) and (8.4) with $M = \mathcal{O}_K$, we obtain (8.10) via (8.8) and (3.12). Proposition 6.13 shows that ∂ is indeed the minimal period for the continued fraction expansion of $\gamma_0^{(1)}/\gamma_1^{(1)} = \sqrt{P} + D$.

If $\gamma_0^{(2)}$ is not positive, then $(\zeta\gamma_0, \gamma_1)$ is a corner pair for $\underset{\sim}{e}(\mathcal{O}_K)$ for a suitably chosen constant ζ. We conclude that (8.10) holds with each A_r^S replaced by itself times a constant. However, such a replacement does not effect the value of the sum. □

One can construct a somewhat more elementary proof of this theorem from the class number formula for $h(\zeta P)$ which was derived by Artin ([A], §18,(11a)). Artin's formula expresses $h(\zeta P)$ in terms of the character sums (7.38). The relations (7.39) written for the case $h(P) = 1$ together with the functional equation for $Z_{K,\infty}(s)$ enable one to replace the character sums σ_ν for $1 \leq \nu \leq 2m - 1$ in Artin's formula by the integers $b_{m-i}(\mathcal{O}_K)$ for $1 \leq i \leq m$. One obtains (8.10).

We can also prove an analogue of Theorem 1.5 for another type of "imaginary" quadratic function field. Let $P \in A$ be a positive irreducible of odd degree, and put $E = k(\sqrt{-P})$. Now ∞ ramifies in E/k, but E still deserves to be called imaginary quadratic because there is a unique E-place above ∞. We will pair E with the real quadratic function field $K = k(\sqrt{TP})$. Let $R = k(\sqrt{-T})$, a field in which ∞ also ramifies. Again we write "∞" for the places in E or R which lie over ∞; and we write $h(TP)$ for the class number of $\mathcal{O}_K$ and $h(-P)$ for the class number of the integral closure of A in E. Consider now the Hasse diagram

(8.12)

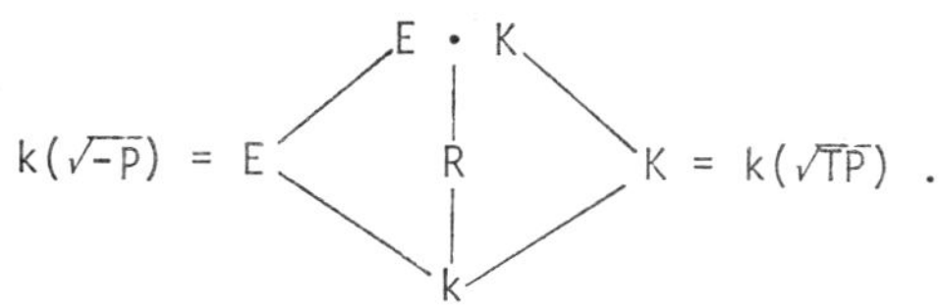

Once again we have non-trivial characters ψ_1, ψ_2 and ψ_3 of the respective Galois groups of EK/K, E/k and R/k, and (8.7) is valid for the associated Artin L-functions with the Euler factors over ∞ removed. Using Lemma 8.5, we compute

$$L_\infty(0,\psi_2) = h(-P) \quad \text{and} \quad L_\infty(0,\psi_3) = 1,$$

and so (8.7) yields

$$h(-P) = L_\infty(0,\psi_1)\ . \tag{8.13}$$

THEOREM 8.14. Assume that $h(TP) = 1$, and let $D = [\sqrt{TP}]$. Then

$$h(-P) = \chi(2)\cdot \sum_{r=1}^{\partial} \chi\left((-1)^{r-1}A_r^S\right)\cdot \frac{\|A_r^S\|-1}{q-1} \tag{8.15}$$

where $A_1^S, A_2^S, \ldots, A_\partial^S$ is the minimal period of partial quotients for the standard continued fraction expansion of $\sqrt{TP} + D$.

Proof: We first identify EK as the ring class field over K of conductor $F = 1$ associated to χ-equivalence. Let $\mathbb{A}_K$ be the adèle ring of K, and let $\Phi:\mathbb{A}_K^\times \longrightarrow \mathrm{Gal}(EK/K)$ be the reciprocity law map. Since $p \neq 2$, the ramification in EK/k over T and P is absorbed by K/k. Therefore EK/K is unramified over every finite place of K. We conclude that $\Phi(V_f) = 1$, where V_f is the subgroup of $\mathbb{A}_K^\times$ consisting of those idèles which have a unit component at every finite place of K.

Let V_∞ be the subgroup of $\mathbb{A}_K^\times$ consisting of the idèles $\underset{\sim}{a}$ which have component 1 at every finite place and satisfy $\chi(\underset{\sim}{a}(\infty_1)) = \chi(\underset{\sim}{a}(\infty_2)) = 1$. The local extensions of EK/K at ∞_1 and ∞_2 are both isomorphic to $k_\infty(\sqrt{-T})/k_\infty$. Since the ramification is tame in this local extension and since T is manifestly a local norm, we conclude that $\Phi(V_\infty) = 1$.

We have now shown that $\ker(\Phi)$ contains the subgroup $V = K^\times\cdot V_f\cdot V_\infty$. Now $h(TP) = 1$ implies that this subgroup has index 2 in $\mathbb{A}_K^\times$, and so V must be the subgroup attached to EK/K by class field theory. Which ideals P of $\mathcal{O}_K$ split completely in EK/K? Let $\underset{\sim}{\pi}_P$ be the idèle which has component 1 at every place of K other than P and which has a component at P which is a local uniformizer. Now by definition of V

$$P \text{ splits in } EK/K \iff \underset{\sim}{\pi}_P \in V \iff P = x\mathcal{O}_K \text{ with } \chi(\underset{\sim}{S}(x)) = 1.$$

Thus EK is indeed our ring class field.

Fix an element $y \in K$ such that $\chi(\underset{\sim}{S}(y)) = -1$. The group $\mathrm{Pic}_\chi(\mathcal{O}_K)$ of §4 contains two classes which are represented by the fractional ideals $\mathcal{O}_K$ and $y\mathcal{O}_K$. Invoking (4.13), we conclude that

$$Z_{K,\infty}(s) = Z_\chi(s,\mathcal{O}_K) + Z_\chi(s,y\mathcal{O}_K) \quad \text{and} \quad L_\infty(s,\psi_1) = Z_\chi(s,\mathcal{O}_K) - Z_\chi(s,y\mathcal{O}_K) .$$

Now since $Z_{K,\infty}(s)$ is missing two Euler factors, $Z_{K,\infty}(0) = 0$ which implies via (8.13) and the above equations with $s = 0$ that

$$(8.16) \qquad h(-P) = 2 \cdot Z_\chi(0,\mathcal{O}_K) .$$

Put $\gamma_0 = \underset{\sim}{e}(\sqrt{TP} + D)$ and $\gamma_1 = (1,1)$. If $\gamma_0^{(2)}$ is positive, then (γ_0,γ_1) is a corner pair for the lattice $\underset{\sim}{e}(\mathcal{O}_K)$ with $\mathrm{sgn}(\gamma_0^{(1)}) = 2$. Now (8.16) and (8.3) with this choice of corner pair and $M = \mathcal{O}_K$ yields formula (8.14). In this case, the unit $\varepsilon_{1,\chi}$ must equal the fundamental unit ε_1 of $\mathcal{O}_K$ because otherwise $\mathrm{Pic}_\chi(\mathcal{O}_K)$ would be trivial (cf. [A], p. 198). Therefore, ∂ is the minimal period for the continued fraction expansion of $\gamma_0^{(1)}/\gamma_1^{(1)} = \sqrt{TP} + D$ by Proposition 6.13.

If $\gamma_0^{(2)}$ is not positive, then $(\zeta\gamma_0,\gamma_1)$ is a corner pair for $\underset{\sim}{e}(\mathcal{O}_K)$ for a suitably chosen constant ζ. Arguing as above, we get a formula for h(-P) in which the contribution of ζ cancels out, yielding (8.15). □

As a first example illustrating Theorem 8.9, we take $q = 3$ and $P = T^4 + T^2 + T + 1$. A minimal period of partial quotients in the standard continued fraction expansion of $\sqrt{P} + T^2 + 2$ is given by the following table:

A_1	A_2	A_3	A_4	A_5
$2T^2 + 1$	$2T$	$T + 1$	$T + 1$	$2T$

Thus $\partial = 5$ in this example, $b_1 = 4$ and $b_2 = 1$. According to the tables in Artin ([A], p. 233), $h(P) = 1$ and $h(-P) = 4$. Computing the second class number by (8.10), we find

$$h(-P) = \frac{12}{4} \cdot (-4 \cdot 1 + 1 \cdot \frac{9-1}{3-1}) + 8 \cdot \frac{2}{4} = 4.$$

As a first example illustrating Theorem 8.14, we take $q = 3$ and $P = T^3 - T - 1$. A minimal period of partial quotients in the standard continued fraction expansion of $\sqrt{TP} + T^2 + 1$ is given by

A_1	A_2	A_3	A_4
$2T^2 + 2$	$T + 2$	$2T$	$T + 2$

Thus $\partial = 4$ in this example, $b_1 = 3$ and $b_2 = 1$. According to the tables in Artin (loc. cit.), $h(TP) = 1$ and $h(-P) = 7$. Computing the second class number by (8.15), we find

$$h(-P) = \chi(2) \cdot [\chi(2) \cdot \frac{9-1}{3-1} + \chi(-1) + \chi(2) + \chi(-1)] = 7.$$

Computer generated examples illustrating Theorem 8.9 with $\deg(P) = 2m > 5$ are available from the work of Ireland and Small. In the examples listed below, $h(P)$ and $h(\zeta P)$ were computed using the formulas derived by Artin. Since the continued fractions involved in these examples have large minimal periods, we list only the integers $b_j = b_j(\mathcal{O}_K)$ for $1 \leq j \leq m$. Formula (8.10) can then be verified in the form

$$h(\zeta P) = (-1)^m \cdot \left[\frac{4q}{q+1} \cdot \sum_{j=1}^{m} (-1)^j \cdot b_j \cdot \frac{q^j - 1}{q-1} + 2W\left(\frac{q-1}{q+1}\right)\right]$$

where

$$W = \sum_i b_{2i-1} \qquad (1 \leq i \leq (m+1)/2).$$

Example 1. Take $q = 3$ and $P = T^6 + T + 2$. Then $h(P) = 1$ and $h(\zeta P) = 26$. From the standard continued fraction for $\sqrt{P} + D$, one obtains $b_1 = 6$, $b_2 = 2$ and $b_3 = 1$.

Example 2. Take $q = 5$ and $P = T^6 + T^2 + 3T + 2$. Then $h(P) = 1$ and $h(\zeta P) = 42$. From the standard continued fraction for $\sqrt{P} + D$, one obtains $b_1 = 30$, $b_2 = 6$ and $b_3 = 1$.

Example 3. Take $q = 3$ and $P = T^8 + T^5 + 2T^4 + T^3 + 2T^2 + T + 2$. Then $h(P) = 1$ and $h(\zeta P) = 48$. From the standard continued fraction for $\sqrt{P} + D$, one obtains $b_1 = 44$, $b_2 = 14$, $b_3 = 4$ and $b_4 = 1$.

Example 4. Take $q = 3$ and $P = T^{10} + T^8 + T^7 + 2T^6 + T^5 + 2T^4 + T^3 + T^2 + T + 2$. Then $h(P) = 1$ and $h(\zeta P) = 314$. From the standard continued fraction for $\sqrt{P} + D$, one obtains $b_1 = 64$, $b_2 = 20$, $b_3 = 8$, $b_4 = 2$ and $b_5 = 1$.

Finally, as the referee has observed, (8.10) and (8.15) both reduce nicely modulo q. This is also true for (7.30) and (7.39).

BIBLIOGRAPHY

[A] E. Artin, "Quadratische Körper im Gebiete der höheren Kongruenzen I, II, Math. Zeitschrift, 19 (1924) p. 153-246.

[H] D. Hayes, "Brumer elements over a real quadratic base field," preprint.

[Hi] H. Hirzebruch, "Hilbert modular surfaces," L'Enseignement Mathématique, 19 (1973) p. 183-281.

[R] M. Rosen, "S-units and S-class group in algebraic function fields," Journal of Algebra, 26 (1973), p. 98-108.

[W] A. Weil, Basic Number Theory, Springer-Verlag, New York(1967).

[Z1] D. Zagier, "Nombres de classes et fractions continues," Journées arithmétique de Bordeaux, Astérisque 24-25 (1975) p. 81-97.

[Z2] D. Zagier, "Valeurs des fontions zêta des corps quadratique réel aux entiers négatifs," Journées arithmétique de Caen, Astérisque 41-42 (1977) 135-151.

DEPARTMENT OF MATHEMATICS AND STATISTICS
UNIVERSITY OF MASSACHUSETTS
AMHERST, MA 01003

Current Address:
Department of Mathematics and Statistics
University of Massachusetts
Amherst, MA 01003

Canadian Mathematical Society
Conference Proceedings
Volume 7 (1987)

IWASAWA MODULES FOR $\mathbb{Z}_p^d$-EXTENSIONS OF NUMBER FIELDS

John Minardi

Abstract. This brief introduction to the theory of $\mathbb{Z}_p^d$-extensions will focus on a generalization of a question considered by Greenberg [6] in the case $d = 1$. A curious feature of this study is the occasional intervention of p-adic independence hypotheses which extend those conjectured by Leopoldt.

I. $\mathbb{Z}_p$-Extensions.

Fix a prime number p and let k be a finite extension of $\mathbb{Q}$. Write $[k : \mathbb{Q}] = r_1 + 2r_2$ with r_1 and $2r_2$ the number of real and complex embeddings of k. Suppose that K is a $\mathbb{Z}_p$-extension of k, so $K = \bigcup_{n\geq 0} k_n$ with $k_n \subseteq k_{n+1}$ and $\mathrm{Gal}(k_n/k) \cong \mathbb{Z}/p^n\mathbb{Z}$. Denote by L_n the p-Hilbert class field of k_n and write $L_K = \bigcup_{n\geq 0} L_n$. Via the Artin map, $\mathrm{Gal}(L_n/k_n) \cong A_n$, where A_n is the p-primary part of the class group of k_n. Then

$$Y_K = \mathrm{Gal}(L_K/K) = \varprojlim \mathrm{Gal}(L_n/k_n) \cong \varprojlim A_n,$$

the inverse limit on the right being taken with respect to the maps induced by the norm maps on ideals. The "Iwasawa module" Y_K is a $\mathbb{Z}_p$-module on which $\Gamma = \mathrm{Gal}(K/k) \cong \mathbb{Z}_p$ acts by inner automorphisms — it can be shown that Y_K thus becomes a finitely generated torsion module for $\Lambda = \mathbb{Z}_p[[\Gamma]] = \varprojlim \mathbb{Z}_p[\Gamma/\Gamma^{p^n}]$. Now for any topological generator γ of Γ, the map sending $\gamma - 1 \mapsto S$ extends to a $\mathbb{Z}_p$-algebra isomorphism $\Lambda \xrightarrow{\sim} \mathbb{Z}_p[[S]]$. A structure theorem (for modules over Noetherian integrally closed domains) provides a map from Y_K to a unique module of the form $\bigoplus_{i=1}^{t} \Lambda/(f_i^{e_i})$ (where $f_i \in \Lambda$ is irreducible) having finite kernel and cokernel. Having chosen an isomorphism $\Lambda \xrightarrow{\sim} \mathbb{Z}_p[[S]]$, one calls $f_{Y_K} = \prod_i f_i^{e_i}$ a characteristic power series for Y_K.

Let μ_{p^∞} denote the p-power roots of unity. Then $\mathrm{Gal}(\mathbb{Q}(\mu_{p^\infty})/\mathbb{Q}) \cong \mathbb{Z}/(p-1)\mathbb{Z} \oplus \mathbb{Z}_p$, so there is a unique ("cyclotomic") $\mathbb{Z}_p$ extension $\mathbb{Q}_c$ of $\mathbb{Q}$ contained in $\mathbb{Q}(\mu_{p^\infty})$. In this way a finite extension k of $\mathbb{Q}$ always has at least one $\mathbb{Z}_p$-extension, namely $k_c = k \cdot \mathbb{Q}_c$.

1980 *Mathematics Subject Classification* 12A70.

Iwasawa obtained in [7] the following asymptotic class number formulas by studying the module structure of Y_K: given $K = \bigcup_{n\geq 0} k_n$ as above, there exist $\lambda, \mu, \nu \in \mathbb{Z}$ such that $v_p(|A_n|) = \lambda n + \mu p^n + \nu$ when n is sufficiently large. ($v_p(\cdot)$ means p-adic valuation and $|\cdot|$ means cardinality.) The exact power of p dividing f_{Y_K} in Λ is p^μ and $\lambda = \dim_{\mathbb{Q}_p}(Y_K \otimes_{\mathbb{Z}_p} \mathbb{Q}_p)$ = Weierstrass degree of f_{Y_K}. It is known that λ or μ can be large, though in the case that $K = k_c$, there are no examples known in which $\mu \neq 0$.

When k is totally real, it seems that Y_{k_c} is always finite; that is, $\lambda_c = \mu_c = 0$. This is the question considered by Greenberg in [6] — no example to the contrary has yet been found. Further, Greenberg showed [5] that if Y_{k_c} is finite whenever k is a totally real abelian extension of $\mathbb{Q}$, then Iwasawa's "main conjecture" for abelian basefields would follow. This conjecture has been proved by Mazur and Wiles, but the finiteness question remains unsettled.

One more aspect of the story can be mentioned here. When K/k is a $\mathbb{Z}_p$-extension, Y_K is finite if and only if every ideal of the integer ring $\mathcal{O}_{k_n}$ representing a class in A_n generates a principal ideal (or "capitulates") in $\mathcal{O}_{k_m}$ for some $m > n$ (cf. [6], p. 267). The phenomenon of capitulation, which appears already in the Principal Ideal Theorem of class field theory, is still not well understood, and remains essentially a mystery in this setting.

II. $\mathbb{Z}_p^d$-Extensions.

Now assume that $\Gamma = \mathrm{Gal}(K/k)$ is isomorphic to the direct product of $d \geq 1$ copies of the additive group $\mathbb{Z}_p$, and define k_n, A_n, L_n, L_K, Y_K in the analogous ways. Note that $\mathrm{Gal}(k_n/k) \cong (\mathbb{Z}/p^n\mathbb{Z})^d$. If $\{\gamma_1, \ldots, \gamma_d\}$ is a $\mathbb{Z}_p$-basis for Γ, then the map $\gamma_i - 1 \mapsto T_i$, $i = 1, \ldots, d$, extends to a $\mathbb{Z}_p$-algebra isomorphism $\mathbb{Z}_p[[\Gamma]] \xrightarrow{\sim} \mathbb{Z}_p[[T_1, \ldots, T_d]]$, so that $\Lambda = \mathbb{Z}_p[[\Gamma]]$ is seen to be a complete regular local domain of dimension $d+1$. In particular, it is a unique factorization domain.

It can be shown (cf. [4]) that Y_K is a finitely generated torsion Λ-module, the action of Γ again being given by inner automorphisms. Cuoco and Monsky have shown in [3], by exploiting the module structure of Y_K, that there exist nonnegative integers m, ℓ such that $v_p(|A_n|) = (mp^n + \ell n + O(1))p^{(d-1)n}$ as $n \to \infty$. They have used this approach to study the growth of class numbers in families of $\mathbb{Z}_p$-subextensions of K/k.

A finitely generated torsion Λ-module M is called pseudo-null (written $M \sim 0$) if M has two relatively prime annihilators in Λ. When $d = 1$ this is equivalent to the finiteness of M. As before there is a map from M to a unique module of the form $\bigoplus_{i=1}^t \Lambda/(f_i^{e_i})$ ($f_i \in \Lambda$ irreducible) having pseudo-null kernel and cokernel. Call $f_M = \prod_i f_i^{e_i}$ a characteristic power series for M; it is uniquely determined up to a unit in Λ. (Note that if K/k is a $\mathbb{Z}_p^d$-extension, then so is K/F whenever $k \subseteq F \subseteq K$, $[F : k] < \infty$. It can readily be seen that $Y_K \sim 0$ as a Λ-module if and only if $Y_K \sim 0$ as a $\mathbb{Z}_p[[\mathrm{Gal}(K/F)]]$-module. This is certainly clear when $d = 1$.) For $d \geq 2$, it can happen that $p | f_{Y_K}$ (v. [2]), but it is not known whether one can always choose $f_{Y_k} = p^\ell$ for some $\ell \geq 0$.

Now denote by $\tilde{k}$ the composite of all $\mathbb{Z}_p$-extensions of k. ¿From class field theory it is seen that $\mathrm{Gal}(\tilde{k}/k) \cong \mathbb{Z}_p^d$, with $r_2 + 1 \leq d \leq [k : \mathbb{Q}]$. Leopoldt's conjecture for (k, p) is that $r_2 + 1 = d$. When k is totally real, then $\tilde{k} = k_c$ under Leopoldt's conjecture (which holds for $k/\mathbb{Q}$ abelian), so a generalization of Greenberg's question would be "Is $Y_{\tilde{k}}$ pseudo-null?"

What follows are some answers to this question in the case that $k = \mathbb{Q}(\sqrt{-D})$ is imaginary quadratic, so that $\mathrm{Gal}(\tilde{k}/k) \cong \mathbb{Z}_p^2$, and $\Lambda \cong \mathbb{Z}_p[[S, T]]$.

Theorem 1. *If $p \nmid h_k$, then $Y_{\tilde{k}} \sim 0$.*

Remarks. (i) In the proof, one shows by Galois theoretic arguments that $Y_{\tilde{k}}/TY_{\tilde{k}}$ is finite, whence it is seen to follow by Nakayama's lemma that $Y_{\tilde{k}} \sim 0$. Here $T = \tau - 1$, where $\tau \in \mathrm{Gal}(\tilde{k}/k)$ lies in an inertia group for a prime of k over p and is not a pth-power in $\mathrm{Gal}(\tilde{k}/k)$.

(ii) If p does not split in $k/\mathbb{Q}$, then $p \nmid h_k$ implies $Y_{\tilde{k}} = 0$. When $p > 2$ and p splits in $k/\mathbb{Q}$. then $Y_{\tilde{k}} = 0$ if and only if $\lambda_c > 1$, which happens exactly when $\mathfrak{p}^{h_k}$ is generated by an element which is not a pth-power in the completion of k at $\bar{\mathfrak{p}}$. Here $\mathfrak{p}, \bar{\mathfrak{p}}$ denote the primes of k lying over p. $Y_{\tilde{k}}$ is not finite unless $Y_{\tilde{k}} = 0$.

(iii) $Y_{\tilde{k}} \sim 0$ implies that $Y_K \sim 0$ whenever $K \supseteq \tilde{k}$ is a $\mathbb{Z}_p^d$-extension of a field F with $k \subseteq F \subseteq \tilde{k}$ and $[F : k] < \infty$. Note that Y_K is taken to be a module over $\mathbb{Z}_p[[\mathrm{Gal}(K/F)]]$. In particular, $Y_{\tilde{F}} \sim 0$ for such F.

Theorem 2. *Let $\mathfrak{p}$ be a prime of k over p and suppose that F/k is a finite abelian extension in which only one prime $\mathfrak{P}$ lies over $\mathfrak{p}$. If the p-primary part of the class group of F is generated by the class of some power of $\mathfrak{P}$, then $Y_{F\tilde{k}} \sim 0$.*

Remarks. (i) Theorem 1 is a special case of Theorem 2. If $\mathrm{Gal}(F/k)$ has exponent dividing $p - 1$, then there is an eigenspace version of Theorem 2.

(ii) It seems very likely that this theorem, taken with an analytic class number formula, can be used in special cases to verify the "2-variable main conjecture" for elliptic curves with complex multiplication.

(iii) When $k = F$, Theorem 2 provides many examples for which $Y_{\tilde{k}} \sim 0$ while $p | h_k$. When $p = 3$, the first five cases are $D = 23, 26, 29, 53, 59$.

Now consider the case when p does not split in $k/\mathbb{Q}$ and $p | h_k$.

Case A. Assume $L_0 \subseteq \tilde{k}$.

Example. $k = \mathbb{Q}(\sqrt{-31})$; $p = 3$ is inert in $k/\mathbb{Q}$. Here $h_k = 3$ so L_0/k is a cyclic unramified extension of degree 3. For brevity, denote L_0 by H. Since $H \subseteq \tilde{k}$, there are 3 primes over 3 in $\tilde{k}$.

On the one hand, one can show that either $Y_{\tilde{k}} \cong \Lambda/(F)$ with F irreducible or $Y_{\tilde{k}} \cong \Lambda/\mathfrak{A}$ for some ideal $\mathfrak{A}$ containing two relatively prime elements, in which case $Y_{\tilde{k}} \sim 0$. Thus, if not pseudo-null, $Y_{\tilde{k}}$ contains no nonzero pseudo-null submodule.

On the other hand, since $\tilde{k}/k$ is locally (at any prime over 3) a maximal abelian totally ramified extension, and since 3 splits in H/k, $\tilde{H}/\tilde{k}$ is an unramified $\mathbb{Z}_3^2$-extension. (Leopoldt's conjecture is known to hold for H.) Let $\mathfrak{p}_1, \mathfrak{p}_2, \mathfrak{p}_3$ denote the primes of H lying over 3. What does $\tilde{H}/\tilde{k}$ look like locally? It turns out that this extension is infinite locally (at any $\mathfrak{p}_i$) if and only if the 3-units $E_3(H)$ of H are "3-adically independent". To be more precise, let $\varphi_i : H \hookrightarrow H_i$ be the canonical embedding into the completion of H at $\mathfrak{p}_i$, and let U_i denote the local units at $\mathfrak{p}_i$ which are congruent to 1 modulo $\mathfrak{p}_i$, $i = 1, 2, 3$. Define $\varphi : E_3(H) \to U_1 \times U_2 \times U_3$ by

$$\varphi(e) = \left(\varphi_1(e \cdot 3^{-v_1(e)})/\omega_1(\varphi_1(e \cdot 3^{-v_1(e)})), \ldots, \varphi_3(e \cdot 3^{-v_3(e)})/\omega_3(\varphi_3(e \cdot 3^{-v_3(e)}))\right),$$

where $v_i(\cdot)$ denotes the $\mathfrak{p}_i$-adic valuation and ω_i is the Teichmüller character for $\mathfrak{p}_i$. Then $\tilde{H}/\tilde{k}$ is infinite locally over the primes $\mathfrak{p}_1, \mathfrak{p}_2, \mathfrak{p}_3$ precisely when the $\mathbb{Z}_3$-submodule generated by $\varphi(E_3(H)) < U_1 \times U_2 \times U_3$ has rank 4. This condition is contained in a general conjecture regarding p-units (v. [8]) which is a consequence of Schanuel's conjecture.

In this case, one can verify that $\varphi(E_3(H))$ does span a submodule of rank 4, whence $\tilde{H}/\tilde{k}$ is locally infinite. It follows that the sum D of the decomposition groups in $Y_{\tilde{k}} = \mathrm{Gal}(L_{\tilde{k}}/\tilde{k})$ at the primes over 3 is a nonzero Λ-submodule which, as a $\mathbb{Z}_3$-module, is a quotient of $\mathbb{Z}_3^3$. Thus $D \sim 0$, so by the observation above, $Y_{\tilde{k}} \sim 0$.

Remarks. (i) Actually $D \cong \mathbb{Z}_3^3$, while its image in $\mathrm{Gal}(\tilde{H}/\tilde{k})$ is isomorphic to $\mathbb{Z}_3^2$.

(ii) It can be shown using a combination of module theory, arithmetic, and Galois theory that there are $\mathbb{Z}_3$-subextensions of $\tilde{k}/k$ in which the classes of k capitulate only in subfields of arbitrarily high degree over k. While it is not an unexpected result, this seems to be the first example in which such delays in capitulation in $\mathbb{Z}_p$-extensions are known to occur.

(iii) Under this conjecture regarding p-units — call it $(*, p)$ — one can formulate general statements about pseudo-nullity, for example:

Theorem 3. *Let p be a prime not splitting in the imaginary quadratic field k. Suppose $p | h_k, p^2 \nmid h_k$, $L_0 \subseteq \tilde{k}$, and $(*, p)$ holds for L_0. Then $Y_{\tilde{k}} \sim 0$.*

Case B. Here it is assumed that $L_0 \cap \tilde{k} = k$.

Example. $k = \mathbb{Q}(\sqrt{-331})$; $p = 3$ is inert in $k/\mathbb{Q}$. Here again $h_k = 3$. Write $L_0 = H$.

Once more it happens that either $Y_{\tilde{k}} \cong \Lambda/(F)$ with $F \in \Lambda$ irreducible or $Y_{\tilde{k}} \cong \Lambda/\mathfrak{A}$ with $\mathfrak{A}$ containing two relatively prime elements. In this case there is only one prime over 3 in $\tilde{k}$, and the corresponding decomposition group D in $Y_{\tilde{k}}$ must be a quotient of $\mathbb{Z}/3\mathbb{Z}$. Since D is a Λ-submodule of $Y_{\tilde{k}}$, if $D = \mathbb{Z}/3\mathbb{Z}$, then $Y_{\tilde{k}} \sim 0$, by the same observation as in Case A. Further, one can show that the converse holds, so the problem is to "locate" $\mathcal{D}$. Again the extension $\tilde{H}/\tilde{k}$ is unramified, but this time it is nonabelian. A natural hope would be that $\tilde{H} \cap L_{\tilde{k}}/\tilde{k}$ is nontrivial locally over 3, although there is no reason to expect that this sufficient condition for the the pseudo-nullity of $Y_{\tilde{k}}$ is also necessary. In any event, a calculation completed recently revealed that $\tilde{H} \cap L_{\tilde{k}}/\tilde{k}$ is locally trivial, squashing this hope; the question is unresolved.

Remark. (i) It follows by the work of Cuoco [2] that $p \nmid f_{Y_{\tilde{k}}}$, so if $Y_{\tilde{k}} \not\sim 0$, this would be a unique example.

(ii) More recently, a similar calculation has shown that $Y_{\tilde{k}} \sim 0$ when $k = \mathbb{Q}(\sqrt{-298})$ and $p = 3$. In this case $(h_k = 3, L_0 \cap \tilde{k} = k$, 3 is inert in $k/\mathbb{Q})$, $\tilde{H} \cap L_{\tilde{k}}/\tilde{k}$ actually is nontrivial locally, so $\mathcal{D} \cong \mathbb{Z}/3\mathbb{Z}$. In no case has it been verified that $Y_{\tilde{k}} \not\sim 0$.

(iii) In both Case A and Case B it can be shown that $Y_{\tilde{k}} \sim 0$ if and only if Y_{k_∞} is finite for some $\mathbb{Z}_3$-subextension k_∞ of $\tilde{k}/k$ which does not contain H. This in turn occurs if and only if the ideals of k capitulate in some finite level of k_∞. A more general type of statement, parallel to the one for $\mathbb{Z}_p$-extensions, can be made using the results of Billot [1]: whenever $k/\mathbb{Q}$ is imaginary quadratic, $Y_{\tilde{k}} \sim 0$ if and only if for each $n \geq 0$ the classes of A_n capitulate in k_m for some $m \geq n$. Once again the phenomenon of capitulation is tied to pseudo-nullity.

III. Summary.

The uncited results above are taken from the author's University of Washington Ph.D. dissertation. It is hoped that these notes give some flavor of the elements of the theory of $\mathbb{Z}_p^d$-extensions — class number formulas, pseudo-nullity, capitulation, etc. At this point there is a number of basic open questions. In particular, as mentioned above, it is not known whether f_{Y_K} always divides p^ℓ for some $\ell \geq 0$ when K/k is a $\mathbb{Z}_p^d$-extension, $d \geq 2$. When $Y_K \sim 0$, the class number formula of Cuoco and Monsky reduces to "$v_p(|A_n|) = O(1)p^{(d-1)n}$ as $n \to \infty$". If $Y_{\tilde{k}}$ is often pseudo-null, then, it would make sense to refine the formula; this work has been begun by Monsky [9]. One would also like to relate the module structures of Y_K and $Y_{K'}$ when K'/k is a $\mathbb{Z}_p^{d'}$-subextension of K/k. A host of questions arises when one considers also the maximal unramified-outside-of-p pro-p-abelian extension M_K of K, and the relation between $X_K = \mathrm{Gal}(M_K/K)$ and Y_K. Finally, it would be of great interest to know the exact significance of pseudo-nullity assertions for "main conjectures," but unfortunately even a more precise formulation of this query lies outside the scope of these notes; see for example the final chapter of the author's dissertation.

BIBLIOGRAPHY

1. Billot, P., "Groupe de Selmer d'une courbe elliptique et théorie d'Iwasawa," *Thèse* (1984), Université de Paris-sud.
2. Cuoco, A., "The Growth of Iwasawa Invariants in a Family," *Comp. Math.*, **41**(1980), 415–437.
3. Cuoco, A., and Monsky, P., "Class Numbers in $\mathbb{Z}_p^d$-Extensions," *Math. Ann.*, **255**(1981), 235–258.
4. Greenberg, R., "The Iwasawa Invariants of Γ-Extensions of a Fixed Number Field," *Amer. J. Math.*, **95**(1973), 204–214.
5. Greenberg, R., "On p-Adic L-Functions and Cyclotomic Fields," *Nagoya Math. J.*, **56**(1974), 61–77.
6. Greenberg, R., "On the Iwasawa Invariants of Totally Real Number Fields," *Amer. J. Math.*, **98**(1976), 263–284.
7. Iwasawa, K., "On Γ-Extensions of Algebraic Number Fields," *Bull. Amer. Math. Soc.*, **65**(1959), 183–226.
8. Jaulent, J.-F., "Sur l'Independance ℓ-Adique de Nombres Algebriques," preprint.
9. Monsky, P., "Class Numbers in $\mathbb{Z}_p^d$-Extensions II," preprint.

DEPARTMENT OF MATHEMATICS
UNIVERSITY OF WASHINGTON
SEATTLE, WASHINGTON 98195

Canadian Mathematical Society
Conference Proceedings
Volume 7 (1987)

A VARIANT OF THE BOMBIERI-VINOGRADOV THEOREM

M. Ram Murty[1] and V. Kumar Murty[2]

INTRODUCTION. Let a,q, be positive integers with $1 \leq a < q$, $(a,q) = 1$. Denote by $\pi(x,q,a)$ the number of primes $p \leq x$, satisfying $p \equiv a \pmod q$. We know from the classical result of Dirichlet [3, ch. 22] that

$$\pi(x,q,a) \sim \frac{1}{\phi(q)} \pi(x)$$

where ϕ is Euler's totient function. Estimates of the size of the error term

$$\pi(x,q,a) - \frac{1}{\phi(q)} \pi(x)$$

are of great importance in applications. It is known that the Riemann Hypothesis for all Dirichlet L-functions implies that

$$\pi(x,q,a) = \frac{1}{\phi(q)} \pi(x) + O(x^{\frac{1}{2}} (\log qx))$$

The celebrated theorem of Bombieri [1] and Vinogradov [30] shows unconditionally that this estimate holds on the average. It states that for any $A > 0$, there is a $B = B(A) > 0$ so that

$$\sum_{q \leq Q} \max_{y \leq x} \max_{(a,q)=1} \left|\pi(y,q,a) - \frac{1}{\phi(q)} \pi(y)\right| \ll \frac{x}{(\log x)^A} \tag{0.1}$$

where $Q = x^{\frac{1}{2}} (\log x)^{-B}$ and we write $f \ll g$ to mean $|f/g|$ is bounded.

The purpose of this paper is to prove a variant of this theorem where a non-abelian splitting condition is introduced. More precisely, let K be a number field. Suppose that it is Galois over $\mathbb{Q}$ with group $G = \mathrm{Gal}(K/\mathbb{Q})$. Let C be a conjugacy class in G. With a and q as above, denote by $\pi_C(x,q,a)$ the number of primes $p \leq x$ which are unramified in K, which satisfy $(p, K/\mathbb{Q}) = C$, and $p \equiv a \pmod q$. (Here, $(p,K/\mathbb{Q})$ is the Artin symbol of p in G). From the

1 Research partially supported by NSERC grant U0237
2 Research partially supported by NSERC grant U0373

Chebotarev density theorem [12], we know that

$$\pi_C(x,q,a) \sim \delta(C,q,a)\ \pi(x)$$

for some density $\delta(C,q,a) \geq 0$. Let ζ_q denote a primitive q-th root of unity. If K and $\mathbb{Q}(\zeta_q)$ are disjoint, then

$$\delta(C,q,a) = \frac{|C|}{|G|} \cdot \frac{1}{\phi(q)} \tag{0.2}$$

(For any finite set S, we write $|S|$ for its cardinality). Our aim is to prove, for any $A > 0$, an estimate of the form

$$\sum_{q \leq Q} \max_{(a,q)=1} \max_{y \leq x} |\pi_C(y,q,a) - \delta(C,q,a)\ \pi(y)| \ll \frac{x}{(\log x)^A} \tag{0.3}$$

with $Q = x^{\alpha - \epsilon}$, $\epsilon > 0$ and the sum is over q such that $K \cap \mathbb{Q}(\zeta_q) = \mathbb{Q}$. Here, α will depend on G and C and satisfy

$$\alpha \geq \min\left(\frac{2}{|G|}, \frac{1}{2}\right)$$

For a precise description of α, see §7. (We remark that with more care, we can even take $Q = x^{\alpha}(\log x)^{-B}$, $B = B(A) > 0$). If we assume (AC) that the L-functions attached to all abelian twists of the non-trivial irreducible characters of G are entire, then we can prove (0.3) with a larger value of α. Indeed, set

$$\delta = \max_{\chi \neq 1} |\chi(1) - 2|$$

where the maximum is over the irreducible characters of G. Then, assuming (AC), we can take $\alpha = \min(\frac{1}{\delta}, \frac{1}{2})$. In particular, if $\delta \leq 2$, we have (0.3) with $\alpha = \frac{1}{2}$. The groups G which satisfy $\delta \leq 2$ can be classified (using results of the type [5, theorem 24.6]).

Our motivation for studying estimates of the form (0.3) comes from the observation that non-abelian analogues of the Bombieri-Vinogradov theorem would have interesting arithmetical consequences. For example, let $\pi_q(x)$ denote the number of primes $p \leq x$ which split completely in the Kummer extension $L_q = \mathbb{Q}(\zeta_q, \sqrt[q]{2})$. It is well-known [19] that an estimate of the form

$$\sum_{q \leq Q} |\pi_q(x) - \frac{1}{q(q-1)}\ \pi(x)| \ll \frac{x}{(\log x)^2}$$

(with Q about $x^{1/2}$) would suffice to imply Artin's primitive root conjecture.

Let f be a cusp form of weight $k \geq 2$ for the congruence subgroup $\Gamma_0(N)$, $N \geq 1$. Suppose that f is an eigenform for the Hecke operators T_p,

($p \nmid N$, p prime). Write $f = \sum_{n \geq 1} a_n e^{2\pi i n z}$ for the Fourier expansions at $i\infty$. For each prime ℓ, let $\pi_\ell(x)$ denote the number of primes $p \leq x$ satisfying $a_p \equiv 0 \pmod{\ell}$. It is well-known [20] that there is a Galois extension $K_\ell/\mathbb{Q}$ such that the condition $a_p \equiv 0 \pmod{\ell}$ is equivalent to the condition that p has a certain splitting type in K . An estimate of the form

$$\sum_{\ell \leq Q} |\pi_\ell(x) - \frac{1}{\ell}\pi(x)| << \frac{x}{(\log x)^2}$$

with Q a power of exp((log x)/(log log x)) would suffice to imply the normal order (and even the statistical distribution) of the number of prime divisors of a_p and a_n. It would also imply lower bounds for a_p valid for a set of p of density 1 (cf. [20], [21]).

The need for an estimate of the precise form (0.3) arose in the problem of determining which rings of S-integers in a number field K possess a Euclidean algorithm. The connection between the two problems is explained in detail in [9]. For other variants of the Bombieri-Vinogradov theorem, see Huxley [10], Wilson [32] and Motohashi [17].

Besides the original argument of Bombieri [1], there are proofs of (0.1) given by Gallagher [6] and Vaughan [31]. Moreover, much work has been done recently by Bombieri, Iwaniec and Friedlander [3] to prove (0.1) with a larger value of Q. Our approach to (0.3) will be an adaptation of the methods of Gallagher [6] and Ramachandra [23]. The paper is self-contained, and in general, we attempt to make explicit the dependence of constants on the number field K. The first three sections contain various preliminaries. The proof of the main theorem is given in sections 4-6, and the last section contains an application to the problem of the least prime whose Artin symbol lies in a given conjugacy class.

Table of Contents

Notation If K is a number field, n_K denotes the degree $[K:\mathbb{Q}]$ of $K/\mathbb{Q}$ and d_K denotes the absolute value of the discriminant of $K/\mathbb{Q}$.

1. Character Sums

1.1 Let G be a finite group and C a conjugacy class of G. Let $\delta_C : G \to \{0,1\}$ denote the characteristic function of C. In terms of characters, we have

$$\delta_C = \frac{|C|}{|G|} \sum_{\eta} \bar{\eta}(g_C)\eta$$

where g_C is any element of C, and the sum is over all irreducible characters η of G. Let H be any subgroup of G such that $H \cap C \neq \phi$. Take $g_C \in H \cap C$ and let C_H denote the conjugacy class of g_C in H. Then

$$\delta_C = \lambda \operatorname{Ind}_H^G \delta_{C_H}$$

where $\delta_{C_H} : H \to \{0,1\}$ is the characteristic function of C_H and $\lambda = |C| \cdot |H| / |G| \cdot |C_H|$. Now, let χ be any character of G. Then by Mackey's induction theorem [24, p. 57] we have

$$\delta_C \otimes \chi = \lambda \operatorname{Ind}_H^G (\delta_{C_H} \otimes \chi|_H) \tag{1.1.1}$$

1.2 Suppose G = Gal(L/F) with $L \supseteq F$, where L, F are number fields. Let k be a non-negative integer, and let ξ be a class function. Define

$$\psi_k(L/F, \xi, x) = \frac{1}{k!} \sum_{Nv^m \leq x} (\log Nv)\left(\log \frac{x}{Nv^m}\right)^k \xi(\sigma_v^m) \tag{1.2.1}$$

where the sum is over powers of places v of F unramified in L, σ_v denotes a Frobenius element at v and N denotes $\mathrm{Norm}_{F/\mathbb{Q}}$. It is convenient to include ramified primes also in this sum. This is done by extending ξ in the usual way. Let v be a place of F and w a place of L above v. Let D_w and I_w denote the decomposition and inertia group at w (respectively). Then $\sigma_w \in D_w/I_w$. We set

$$\xi(\sigma_v^m) = \frac{1}{|I_w|} \sum \xi(g)$$

where the sum is over all elements $g \in D_w$ whose image in D_w/I_w is σ_w^m. A different choice of w conjugates D_w and I_w and thus leaves the above sum unchanged. With this, we let $\tilde{\psi}_k(L/F, \xi, x)$ denote the sum in (1.2.1), now taken over all v^m, v a place of F and $Nv^m \leq x$. We have

$$\psi_k(L/F,\xi,x) = \tilde{\psi}_k(L/F,\xi,x) + O\left(\|\xi\| \frac{(\log x)^{k+1}}{k!} \left\{\frac{1}{|G|} \log d_L + n_F x^{\frac{1}{2}}\right\}\right) \tag{1.2.2}$$

where $\|\xi\| = \sup_{g \in G} |\xi(g)|$. The proof is almost the same as in

Serre [25, Proposition 7] and is omitted.

1.3 We can write the class function ξ as a linear combination

$$\xi = \sum a_\eta \eta$$

where η ranges over the irreducible characters of G and $a_\eta \in \mathbb{C}$. Let $L(s,\eta)$ denote the Artin L-series attached to η. Then

$$\tilde{\psi}_k(L/F, \xi, x) = \frac{1}{2\pi i} \sum_\eta a_\eta \int_{(2)} \frac{L'}{L}(s,\eta) \frac{x^s}{s^{k+1}} ds$$

where the integration is on the line Re(s) = 2.

1.4 Let K be a Galois extension of $\mathbb{Q}$, and $1 \leq a, q \in \mathbb{Z}$, $a < q$, $(a,q) = 1$. We use the above observations with $L = K(\zeta_q)$, where ζ_q is a primitive q-th root of 1, and $F = \mathbb{Q}$. Suppose that $K \cap \mathbb{Q}(\zeta_q) = \mathbb{Q}$. Then we know that $G = \mathrm{Gal}(K(\zeta_q)/\mathbb{Q})$ has a splitting

$$G \simeq \mathrm{Gal}(K/\mathbb{Q}) \times \mathrm{Gal}(\mathbb{Q}(\zeta_q)/\mathbb{Q}) .$$

Every element g of G can thus be written as (g_1, g_2) in an obvious way. Every conjugacy class of G is of the form (C_1, C_2) where C_1 is a class of $\mathrm{Gal}(K/\mathbb{Q})$ and C_2 a class of $\mathrm{Gal}(\mathbb{Q}(\zeta_q)/\mathbb{Q})$. Fix an identification $\mathrm{Gal}(\mathbb{Q}(\zeta_q)/\mathbb{Q}) \simeq (\mathbb{Z}/q)^x$. We use the same letter to denote the image of a under $\mathbb{Z} \to \mathbb{Z}/q$. Let $\delta_{a,q}: G \to \{0,1\}$ be defined by $\delta_{a,q}(g) = \begin{cases} 1 \text{ if } g_2 = a \\ 0 \text{ otherwise} \end{cases}$.

Fix a conjugacy class C in $\mathrm{Gal}(K/\mathbb{Q})$ and let $\xi = \xi(C,a,q) = \delta_C \otimes \delta_{a,q}$. Thus ξ is the characteristic function of the class $(C,\{a\})$. Explicitly, we have

$$\psi_k(K(\zeta_q)/\mathbb{Q},\xi,x) = \frac{1}{k!} \sum_{\substack{p^m \leq x \\ p \nmid d_K \\ p^m \equiv a(q) \\ \sigma_{p^m} \in C}} (\log p)(\log(\frac{x}{p^m}))^k$$

Our aim is to show that given $A > 0$, there is a suitable choice of Q_1 so that

$$(1.4.1) \qquad \sum_{q \leq Q_1}' \max_{(a,q)=1} \max_{y \leq x} |\psi_0(K(\zeta_q)/\mathbb{Q}, \xi(C,a,q),y) - \delta(C,a,q)y| \ll \frac{x}{(\log x)^A}$$

The prime on the summation indicates that we range only over those q satisfying $K \cap \mathbb{Q}(\zeta_q) = \mathbb{Q}$.

1.5 We transform the sum of (1.4.1) in several ways. The first observation is that it suffices to prove an estimate with ψ_0 replaced by ψ_k for some large k. We sketch the argument as it is similar to Gallagher [6]. Write $\psi_k(x)$ for $\psi_k(K(\zeta_q)/\mathbb{Q}, \xi(C,a,q),x)$ for simplicity. Define $\tilde{\psi}_k(x)$ in a similar way. Then, we have

$$\psi_{k+1}(x) = \int_1^x \psi_k(t) \cdot \frac{dt}{t}$$

and for any $\alpha > 0$, we have by the mean value theorem,

$$\frac{1}{\alpha}\int_{e^{-\alpha}x}^{x} \psi_k(t)\frac{dt}{t} \le \psi_k(x) \le \frac{1}{\alpha}\int_x^{e^{\alpha}x} \psi_k(t)\frac{dt}{t} .$$

Thus

$$\max_{y\le x} |\psi_k(y) - \delta(C,a,q)y| \ll \frac{1}{\alpha} \max_{y\le e^{\alpha}x} |\psi_{k+1}(y) - \delta(C,a,q)y| + \alpha\delta(C,a,q)x .$$

Choosing $\alpha = (\log x)^{-A/2}$, we deduce that

$$\sum_{q\le Q_1}{}' \max_{(a,q)=1} \max_{y\le x} | \psi_k(y) - \delta(C,a,q)y|$$

$$\ll (\log x)^{A/2} \sum_{q\le Q_1}{}' \max_{(a,q)=1} \max_{y\le e^{\alpha}x} |\psi_{k+1}(y) - \delta(C,a,q)y| + \frac{x\log Q_1}{(\log x)^{A/2}}$$

Moreover, (1.2.2) shows that replacing ψ with $\tilde{\psi}$ on the right introduces an additional contribution of

$$\ll \frac{(\log x)^{k+2+\frac{A}{2}}}{(k+1)!} Q_1 \{\frac{1}{n_K} \log d_K + \log q + x^{\frac{1}{2}}\}$$

1.6 We have a decomposition

$$\tilde{\psi}_k(x) = \tilde{\psi}_k(K(\zeta_q)/\mathbb{Q}, \xi(C,a,q),x) = \frac{1}{\phi(q)} \sum\bar{\chi}(a)\ \psi_k(K(\zeta_q)/\mathbb{Q}, \delta_C \otimes \chi, x)$$

where the sum is over charcters χ of $\mathrm{Gal}(\mathbb{Q}(\zeta_q)/\mathbb{Q})$. The trivial character contributes a term

$$\frac{1}{\phi(q)} \tilde{\psi}_k(K(\zeta_q)/\mathbb{Q}, \delta_C, x) = \frac{1}{\phi(q)} \tilde{\psi}_k(K/\mathbb{Q}, \delta_C, x)$$

From the effective Chebotarev density theorem [12], we have

$$\tilde{\psi}_k(K/\mathbb{Q}, \delta_C, x) = \frac{|C|}{n_K} x + O(x^\beta \frac{1}{k!}(\log x)^k)$$

$$+ O(x \frac{1}{k!}(\log x)^k \exp(-c\, n_K^{-\frac{1}{2}} (\log x)^{\frac{1}{2}}))$$

where c is a positive absolute constant and β is a possible Siegel zero of $K/\mathbb{Q}$. (The implied constants are absolute). Using the bound

$$\beta < \max(1 - \frac{1}{4 \log d_K}, 1 - \frac{c_1}{d_K^{1/n_K}})$$

of Stark [28] we deduce that for $\log Q_1 \ll \log x$,

(1.6.1)
$$\sum_{q \le Q_1}' \max_{(a,q)=1} \max_{y \le x} |\tilde{\psi}_k(y) - \delta(C,a,q)y|$$

$$= \sum_{q \le Q_1} \frac{1}{\phi(q)} \sum_{\chi \neq 1} \max_{y \le x} |\tilde{\psi}_k(K(\zeta_q)/\mathbb{Q}, \delta_C \otimes \chi, y)|$$

$$+ O(\frac{1}{k!} x(\log x)^{k+1} \exp(-cn_K^{-\frac{1}{2}} (\log x)^{\frac{1}{2}}))$$

(Here the implied constant depends on the field K).

It is convenient to include only primitive characters $\chi \pmod q$ (ie. characters which do not factor through $Gal(\mathbb{Q}(\zeta_{q_1})/\mathbb{Q})$ for some proper divisor q_1 of q). This is easily done by observing that if $\chi \pmod q$ is induced by character $\chi_1 \pmod{q_1}$ with $q_1 | q$, $q_1 \neq q$, then

$$|\tilde{\psi}_k(K(\zeta_q)/\mathbb{Q}, \delta_C \otimes \chi, x) - \tilde{\psi}_k(K(\zeta_q)/\mathbb{Q}, \delta_C \otimes \chi_1, x)|$$

$$\le \frac{1}{k!} \sum_{p^m | q/q_1} (\log p)(\log \frac{x}{p^m})^k \ll \frac{1}{k!} (\log x)^k \log(\frac{q}{q_1})$$

Moreover, we have

$$\sum_{\substack{q_1 | q \\ q \le Q_1}} \frac{1}{\phi(q)} \ll \frac{\log Q_1}{\phi(q_1)}$$

Using $\log Q_1 \ll \log x$, we deduce that the first term on the right hand side of (1.6.1) is

(1.6.2) $$\ll (\log x) \sum_{1<q\le Q_1}' \max_{y\le x} \frac{1}{\phi(q)} \sum_{\chi}^{*} |\tilde{\psi}_k(K(\zeta_q)/\mathbb{Q}, \delta_C \otimes \chi, y)| + O(\frac{1}{k!} Q_1 (\log x)^{k+1})$$

where the asterisk on the sum indicates that we only include primitive χ (mod q).

1.7 Now let H' be a subgroup of $G_1 = \mathrm{Gal}(K/\mathbb{Q})$ with $H' \cap C \neq \phi$. Set $H = H' \times \mathrm{Gal}(\mathbb{Q}(\zeta_q)/\mathbb{Q})$. Using (1.1.1), (1.3) and the invariance of L-series under induction, we deduce that in the above sum, we may replace

$$\tilde{\psi}_k(K(\zeta_q)/\mathbb{Q}, \delta_C \otimes \chi, y) \quad \text{by} \quad \lambda \cdot \tilde{\psi}_k(K(\zeta_q)/M, \delta_{C_{H'}} \otimes \chi|_H, y)$$

where $\lambda = |C|\cdot|H'| / |C_{H'}|\cdot|G_1|$ and M is the subfield of $K(\zeta_q)$ fixed by H. (Note that M is also the subfield of K fixed by H' and so does not depend on q). Moreover, we have

$$\tilde{\psi}_k(K(\zeta_q)/M, \delta_{C_{H'}} \otimes \chi|_H, y) = \frac{|C_{H'}|}{|H'|} \sum_{\omega} \bar{\omega}(g_C)\, \tilde{\psi}_k(K(\zeta_q)/M, \omega \otimes \chi, y)$$

where ω ranges over the irreducible characters of H' and $g_C \in H' \cap C$. We deduce from these observations that

(1.7.1) $$\sum_{1<q\le Q_1}' \frac{1}{\phi(q)} \max_{y\le x} \sum_{\chi}^{*} |\tilde{\psi}_k(K(\zeta_q)/\mathbb{Q}, \delta_C \otimes \chi, y)|$$

$$\le \frac{|C|}{|G_1|} |H'| \max_{\omega} \left\{ \sum_{1<q\le Q_1}' \frac{1}{\phi(q)} \max_{y\le x} \sum_{\chi}^{*} |\tilde{\psi}_k(K(\zeta_q)/M, \omega \otimes \chi, y)| \right\}$$

Finally, decomposing the interval $(1,Q_1]$ into $O(\log Q_1)$ intervals of the form $(\frac{Q}{2}, Q]$, we find that the expression in large parentheses is

$$\ll (\log Q_1) \max_{Q\le Q_1} \max_{y\le x} \frac{\log\log Q}{Q} \sum_{1<q\le Q} \sum_{\chi}^{*} |\tilde{\psi}_k(K(\zeta_q)/M, \omega \otimes \chi, y)| .$$

Summarizing the discussion of the previous paragraphs, we have proved the following.

(1.8) Proposition

$$\sum_{q\le Q_1}' \max_{(a,q)=1} \max_{y\le x} |\psi_k(K(\zeta_q)/\mathbb{Q}, \xi(C,a,q), y) - \delta(C,a,q)y|$$

$$\ll \frac{|C|}{|G_1|} |H'| \ell^2 \cdot \ell_2 \max_{\omega} \max_{Q\le Q_1} \max_{y\le x} \frac{1}{Q} \sum_{1<q\le Q} \sum_{\chi}^{*} |\tilde{\psi}_k(K(\zeta_q)/M, \omega \otimes \chi, y)| + E$$

where $\ell = \log x$, $\ell_2 = \log\log x$ and

$$E \ll \frac{1}{k!}(\log x)^{k+1} x \exp(-cn_K^{-\frac{1}{2}}(\log x)^{\frac{1}{2}}) + \frac{1}{k!}(\log x)^{k+1} Q_1$$

$$+ \frac{1}{k!}(\log x)^{k+1} Q_1 \{\frac{1}{n_K}\log d_K + \log q + x^{\frac{1}{2}}\} .$$

We recall that

$$\tilde{\psi}_k(L(\zeta_q)/M, \omega \otimes \chi, x) = \frac{1}{k!} \sum_{Nv^m \leq x} (\log Nv)(\log \frac{x}{Nv^m})^k \omega(\sigma_v^m)\chi(Nv^m)$$

where v runs over primes of M.

1.9 We observe that if μ and ν are two irreducible characters of a finite Galois group $G = \mathrm{Gal}(L/F)$, and F_μ, F_ν are their Artin conductors then the Artin conductor $F_{\mu\otimes\nu}$ of $\mu \otimes \nu$ satisfies $F_{\mu\otimes\nu} \mid F_\mu^{\nu(1)} F_\nu^{\mu(1)}$ (cf. for example, Martinet [16, p. 80]). This fact will be used repeatedly.

2. Phragmén-Lindelöf Theorem

2.1 We write $s = \sigma + it$. Let $f(s)$ be a function regular in a vertical strip $c \leq \sigma \leq d$ and satisfying in this strip a growth condition

$$|f(s)| \ll e^{|s|^\delta} \tag{2.1.1}$$

for a positive constant δ. Suppose that there are positive constants C, D, α, β and a constant Q satisfying

$$\begin{aligned} |f(c+it)| &\leq C|Q+c+it|^\alpha \\ |f(d+it)| &\leq D|Q+d+it|^\beta \end{aligned} \tag{2.1.2}$$

The Phragmén-Lindelöf theorem gives an estimate for $f(s)$ when $c \leq \sigma \leq d$. We shall need it in the following sharp form given by Rademacher [22].

2.2 Proposition For $c \leq \sigma \leq d$, and f satisfying (2.1.1) and (2.1.2), we have

$$|f(s)| \leq (C|Q+s|^\alpha)^{\frac{d-\sigma}{d-c}} (D|Q+s|^\beta)^{\frac{\sigma-c}{d-c}}$$

2.3 We apply this in the study of a general class of Dirichlet series. Let $L(s)$ be a Dirichlet series satisfying the following properties. First, we require

(i) $L(s) = \prod_p L_p(s)$ for $\sigma > 1$, where L_p is a polynomial in p^{-s}. The product is over all finite primes. Denote by m_p the degree of L_p in p^{-s}.

(ii) There is a positive integer $d = d(L)$ with $m_p \leq d$ for all p and $m_p = d$ for all but finitely many p.

Let us write

$$L_p(s) = \prod_{i=1}^{m_p} (1 - \pi_i\, p^{-s})^{-1}$$

where $\pi_i = \pi_{i,p} \in \mathbb{C}$ and $|\pi_i| = 1$. Define

$$L_p^{v}(s) = \prod_{i=1}^{m_p} (1 - \bar{\pi}_i\, p^{-s})^{-1}$$

and set $L^{v}(s) = \prod_p L_p^{v}(s)$. Let A be a positive real number and a,b non-negative integers with $a + b \leq d$. Set

$$\Lambda(s) = A^{s/2} (\pi^{-s/2}\, \Gamma(\tfrac{s}{2}))^{a} (\pi^{-(s+1)/2}\, \Gamma(\tfrac{s+1}{2}))^{b}\, L(s)$$

$$\Lambda^{v}(s) = A^{s/2} (\pi^{-s/2}\, \Gamma(\tfrac{s}{2}))^{a} (\pi^{-(s+1)/2}\, \Gamma(\tfrac{s+1}{2}))^{b}\, L^{v}(s) .$$

Suppose that

(iii) $\Lambda(s)$ and $\Lambda^{v}(s)$ have an analytic continuation to the entire complex plane except possibly for a pole at $s = 0$ or 1.

(iv) $\Lambda(s) = w\, \Lambda^{v}(1 - s)$ with $w \in \mathbb{C}$ $|w| = 1$.

<u>2.4 Proposition</u> Under the assumtions (i) - (iv) above, we have for $0 \leq \sigma \leq 1$,

$$|L(\sigma + it)| \leq (A(|t| + 2)^{d})^{\frac{1-\sigma}{2}} (\log[A(|t| + 2)^{d}])^{d}$$

<u>PROOF</u> From (i), we see that for $\epsilon > 0$,

$$|L(1 + \epsilon + it)| \leq \zeta(1 + \epsilon)^{d}$$

where ζ denotes the usual Riemann zeta function. By (iv) and Stirling's formula,

$$|L(-\epsilon + it)| \leq A^{\frac{1}{2}+\epsilon} (|t| + 2)^{d(\frac{1}{2}+\epsilon)} \zeta(1 + \epsilon)^{d}$$

By Proposition (2.2), we have

$$|L(\sigma + it)| \leq \zeta(1 + \epsilon)^{d} (A(|t| + 2)^{d})^{\frac{1-\sigma+\epsilon}{2}}$$

valid for $-\frac{1}{2} \leq -\epsilon \leq \sigma \leq 1+\epsilon \leq \frac{3}{2}$. Choose $\epsilon = (\log[A(|t| + 2)^{d}])^{-1}$.

The result follows on noting that $\zeta(1+\epsilon) \ll \epsilon^{-1}$.

We note two interesting consequences of Proposition (2.4). These will not be needed in the remainder of the paper.

2.5 Let f be a holomorphic cusp form of integral weight k for a congruence subgroup of $SL_2(\mathbb{Z})$. Write L(s,f) for its associated L-series. For any Dirichlet character χ, $f \otimes \chi$ is the twist of f by χ. The Shimura correspondence [26] attaches to f a cusp form F of weight $\frac{1}{2}(k+1)$ with the following property. Write

$$F(z) = \sum_{n=1}^{\infty} C(n)e^{2\pi inz}$$

for the Fourier expansion at ∞. There is a constant Ω such that for any fundamental discriminant D of a quadratic field.

$$C(|D|)^2 = \Omega|D|^{\frac{k-1}{2}} L(\tfrac{k}{2}, f \otimes \chi)$$

where $\chi(n) = (\frac{D}{n})$ is the Kronecker symbol. The analog of the Ramanujan conjecture (cf. the discussion in Goldfeld-Hoffstein-Patterson [8, p. 154]) is

$$C(|D|) \ll_{\epsilon} |D|^{\frac{k-1}{4}+\epsilon}$$

for every $\epsilon > 0$. We see from Proposition (2.4) that

(2.5.1) $$C(|D|) \ll |D|^{\frac{k}{4}} (\log |D|)^2$$

This is slightly sharper than the estimate $|D|^{\frac{k}{4}+\epsilon}$ stated in [8, p. 154]) Iwaniec [11] has recently obtained a significant improvement of (2.5.1) where the exponent $\frac{k}{4}$ is replaced by $\frac{k}{4} - \frac{1}{28}$.

2.6 The second application is to the residue of the Dedekind zeta function. Let $L/\mathbb{Q}$ be a number field. Landau [14] showed that there is a constant $C > 0$ so that

(2.6.1) $$\operatorname*{res}_{s=1} \zeta_L(s) \leq C^{n_L} (\log d_L)^{n_L - 1}$$

Siegel [27] obtained sharp values for C. We are interested in the power of $\log d_L$. Suppose there is a subfield $K \subseteq L$ satisfying

(i) L/K is Galois

(ii) for every non-trivial character of Gal(L/K), the associated Artin L-series is entire.

(For example, K = L satisfies these conditions).

Then

$$\frac{\operatorname*{res}_{s=1} \zeta_L(s)}{\operatorname*{res}_{s=1} \zeta_K(s)} \leq \prod_{\chi} (\log A_\chi + \chi(1) \log 2)^{\chi(1)}$$

where $A_\chi = d_K^{\chi(1)} \operatorname{Norm}_{L/Q}(F_\chi)$, and χ ranges over the non-trivial irreducible characters of Gal(L/K). By the conductor-discriminant formula,

$$\log(d_L/d_K) = \sum_{\chi} \chi(1) \log A_\chi$$

Thus, we get (using the arithmetic mean-geometric mean inequality and (2.6.1)),

$$\operatorname*{res}_{s=1} \zeta_L(s) \leq C^{n_K} (\log d_K)^{n_K - 1} 2^n \prod_{\chi} (\log A_\chi)^{\chi(1)}$$

$$\leq C^{n_K} \cdot 2^n (\log d_K)^{n_K - 1} \left(\frac{\log(d_L/d_K)}{n}\right)^n$$

where $n = \sum \chi(1)$. This is an improvement over (2.6.1) whenever K is a proper subfield of L. In particular, if $L/\mathbb{Q}$ is Galois, we deduce that for some absolute constant $C > 0$.

$$\operatorname*{res}_{s=1} \zeta_L(s) \leq C^{n_L} \cdot (\log d_L)^{\frac{n_L}{2}+1}$$

by taking a subfield K of L such that L/K is cyclic of prime order.

2.7 Remark It would be of interest to know whether the power of the logarithm in Proposition (2.4) can be reduced.

3. Zero-free regions

3.1 Let L/F be a Galois extension of number fields and χ be an abelian character of $G = \mathrm{Gal}(L/F)$. We shall need a zero-free region for the Artin L-series $L(s,\chi)$. Let F_χ denote the Artin conductor of χ and set $A_\chi = d_F N_{F/Q}(F_\chi)$.

3.2 Lemma For $\sigma > 1$,

$$-\mathrm{Re}\, \frac{\zeta_F'}{\zeta_F}(s) < \mathrm{Re}\left(\frac{1}{s} + \frac{1}{s-1}\right) + \frac{1}{2} \log\left(\frac{d_F}{2^{2r_2} \pi^{n_F}}\right) + \frac{r_1}{2} \mathrm{Re}\frac{\Gamma'}{\Gamma}\left(\frac{s}{2}\right) + r_2 \mathrm{Re}\frac{\Gamma'}{\Gamma}(s)$$

where $n_F = r_1 + 2r_2$ and r_1 is the number of real embeddings of F.

This is part of Lemma 3 in Stark [28].

3.3 Lemma For $\sigma > 1$, and non-trivial χ,

$$-\mathrm{Re}\,\frac{L'}{L}(s,\chi) < C_1\mathcal{L} - \sum_{\rho}\mathrm{Re}\left(\frac{1}{s-\rho}\right)$$

where $C_1 > 0$ is an absolute constant, the sum is over all zeroes ρ of $L(s,\chi)$ with $0 < \mathrm{Re}\,\rho < 1$, and $\mathcal{L} = \frac{1}{2}\log A_\chi + n_F(|t| + 2)$.

Proof We just sketch the proof. Set

$$\Lambda(s,\chi) = A_\chi^{s/2}\,\mathbf{\Gamma}(s)\,L(s,\chi)$$

where $\mathbf{\Gamma}(s)$ is a product of n_F Γ-factors. We have the functional equation

$$\Lambda(s,\chi) = w\,\Lambda(1-s,\bar{\chi})$$

with $w \in \mathbb{C}$, $|w| = 1$, and also the Hadamard factorization

$$\Lambda(s,\chi) = e^{a+bs}\prod_{\rho}\left(1-\frac{s}{\rho}\right)e^{s/\rho}$$

where $a = a(\chi)$, $b = b(\chi)$ are complex numbers and the product is over all zeroes ρ of $\Lambda(s,\chi)$ (equivalently, over all zeroes ρ of $L(s,\chi)$ with $0 < \mathrm{Re}\,\rho < 1$). By logarithmically differentiating both sides and using $\mathrm{Re}\,b = -\sum_{\rho}\mathrm{Re}(1/\rho)$, we find

$$-\mathrm{Re}\,\frac{L'}{L}(s,\chi) = \frac{1}{2}\log A_\chi - \sum\mathrm{Re}\left(\frac{1}{s-\rho}\right) + \mathrm{Re}\,\frac{\mathbf{\Gamma}'}{\mathbf{\Gamma}}(s)$$

For $\sigma > 1$,

$$\mathrm{Re}\,\frac{\mathbf{\Gamma}'}{\mathbf{\Gamma}}(s) \ll n_F\log(|t|+2)$$

by Stirling's formula. This proves the lemma.

3.4 Proposition There is an absolute constant $C > 0$ such that $L(s,\chi)$ has at most one zero in the region

$$1 - \frac{C}{\mathcal{L}} \le \sigma \le 1$$

If this zero β exists, then it is real and simple and χ is a character of order 1 or 2.

Proof For $\sigma > 1$, we have for non-trivial χ,

$$(3.4.1)\quad 3\left(-\frac{L'}{L}(\sigma,\chi_0)\right) + 4\left(-\mathrm{Re}\,\frac{L'}{L}(\sigma+it,\chi)\right) + \left(-\mathrm{Re}\,\frac{L'}{L}(\sigma+2it,\chi^2)\right) \ge 0$$

where χ_0 is the principal character mod F_χ. Then,

$$-\frac{L'}{L}(\sigma,\chi_0) = \sum_{(\mathfrak{p},F_\chi)=1}(\log N\mathfrak{p})(N\mathfrak{p})^{-ns} \le -\frac{\zeta_F'}{\zeta_F}(\sigma)$$

and by lemma (3.2), we deduce that for $\sigma = 1 + \frac{C_2}{\mathcal{L}}$,

$$-\frac{L'}{L}(\sigma,\chi_0) < \frac{1}{\sigma - 1} + C_3\mathcal{L} \tag{3.4.2}$$

where C_2, C_3 are positive absolute constants. Next, for $\sigma > 1$,

$$\mathrm{Re}\left(\frac{1}{s-\rho}\right) = \frac{\sigma - \beta}{|s-\rho|^2} \geq 0$$

and so, Lemma (3.3) implies that for any zero ρ of $L(s,\chi)$ with $0 < \mathrm{Re}\,\rho < 1$,

$$-\mathrm{Re}\,\frac{L'}{L}(s,\chi) < C_1\mathcal{L} - \mathrm{Re}\left(\frac{1}{s-\rho}\right) \tag{3.4.3}$$

Suppose that χ^2 is not the principal character. Let χ_1 be the primitve character inducing χ^2 mod F_χ. Then

$$\left|\frac{L'}{L}(s,\chi^2) - \frac{L'}{L}(s,\chi_1)\right| \leq \sum_{\gamma | F_\chi} \frac{(\log N\gamma)(N\gamma)^{-\sigma}}{(1 - (N\gamma)^{-\sigma})} \leq \mathcal{L}$$

combining this with (3.4.3),

$$-\mathrm{Re}\,\frac{L'}{L}(s,\chi^2) < C_4\mathcal{L} \tag{3.4.4}$$

Using (3.4.2)-(3.4.4) in (3.4.1), we deduce that if $\rho = \beta + i\gamma$, $\sigma = 1 + \frac{C_2}{\mathcal{L}}$ and $t = \gamma$, then

$$\frac{3}{\sigma - 1} - \frac{4}{\sigma - \beta} + C_5\mathcal{L} \geq 0$$

ie

$$\beta \leq 1 + \frac{C_2}{\mathcal{L}} - \frac{4\,C_2}{(3 + C_2C_5)\mathcal{L}}$$

Choosing $C_2 < 1/C_5$ shows that for an absolute $C > 0$,

$$\beta \leq 1 - \frac{C}{\mathcal{L}}$$

If χ^2 is principal, Lemma (3.2) implies that

$$-\mathrm{Re}\,\frac{L'}{L}(s,\chi^2) < \mathrm{Re}\left(\frac{1}{\sigma - 1 + it}\right) + C_3$$

We deduce that $\beta < 1 - \frac{C}{\mathcal{L}}$ if $|\gamma| \geq C_6/\mathcal{L}$ for some $C_6 > 0$. The usual arguments show that if $|\gamma| < C_6/\mathcal{L}$, then $\gamma = 0$, and that there is at most one such zero and that it is simple (cf. [4, pp. 92-93]).

3.5 Finally, we need an analogue of Siegel's theorem giving a bound for the possible real zero in Proposition (3.4). We adapt a method of Stark [28]. Thus, let χ be a real character of G of conductor F_χ. Let N/F be the extension defined by it. We suppose that χ is not the trivial character, so that N is quadratic over F. As before, we write $A = A_\chi = d_F(NF_\chi)$. We may view χ as a quadratic Hecke character of F.

3.6 The Brauer-Siegel theorem implies that as we range over a set of quadratic Hecke characters χ of F with $NF_\chi \to \infty$, we have

$$L(1,\chi) > \frac{C(\varepsilon)}{C_1^{n_F} (\log d_F)^{n_F}} \cdot \frac{1}{A_\chi^\varepsilon} \tag{3.6.1}$$

where $\varepsilon > 0$ is arbitrary, $C_1 > 0$ is an absolute constant, and $C(\varepsilon) > 0$ depends only on ε. Indeed, if we let R denote the residue of $\zeta_N(s)$ at $s = 1$ and r the residue of $\zeta_F(s)$ at $s = 1$, then $R = L(1,\chi)r$. The Brauer-Siegel theorem states that for any $\varepsilon > 0$,

$$R > C(\varepsilon)/d_N^\varepsilon .$$

Also, we have by (2.6.1) that

$$r \le (C_1 \log d_F)^{n_F} .$$

Since $d_N = d_F^2 NF_\chi$, we conclude that (3.6.1) holds. This implies that the exceptional zero of Proposition (3.4) satisfies

$$\beta \le 1 - \frac{c(\varepsilon)}{(C_1 \log d_F)^{n_F} A_\chi^\varepsilon} .$$

This can be refined using the following result of Stark [28, Lemma 10].

<u>3.7 Lemma</u> Suppose there is a sequence of fields

$$Q = M_0 \subset M_1 \subset M_2 \subset \dots \subset M_t = M$$

such that for $1 \le i \le t$, M_i/M_{i-1} is normal. Suppose there is a real β in the range

$$1 - \frac{1}{16 \log d_M} \le \beta < 1$$

such that $\zeta_M(\beta) = 0$. Then there is a quadratic field $S \subseteq M$ with $\zeta_S(\beta) = 0$. We shall use this to prove the following.

3.8 Proposition Let $\varepsilon > 0$ and χ as in (3.5). Let J denote the normal closure of F over $\mathbb{Q}$. Then

$$\beta \leq \max\left(1 - \frac{C(\varepsilon)}{(d_J^2 A)^{\varepsilon}} , 1 - \frac{1}{16 n_J \log(d_J^2 A)} \right)$$

Proof We may assume that β satisfies

$$1 - \frac{1}{16 n_J \log(d_J^2 A)} \leq \beta < 1 . \tag{3.8.1}$$

Then, the composition JN is Galois over N and so, by a well-known result of Aramata-Brauer, $\zeta_{JN}(\beta) = 0$. Moreover,

$$d_{JN} \leq d_J^{2n_F} d_N^{n_J} \leq (d_J^2 A)^{n_J} \tag{3.8.2}$$

and so (3.8.1) implies that

$$1 - \frac{1}{16 \log d_{JN}} \leq \beta < 1 .$$

Finally, $JN \supseteq J \supseteq \mathbb{Q}$ is a normal tower, and so, by Lemma (3.7), there is a quadratic subfield S of JN with $\zeta_S(\beta) = 0$. By the classical Siegel theorem (see for example [4, section 21] or Goldfeld [7]) there is a constant $C(\varepsilon) > 0$ so that

$$\beta \leq 1 - \frac{C(\varepsilon)}{d_S^{\varepsilon}}$$

Now,

$$d_{JN} \geq d_S^{n_{JN}/2} \geq d_S^{n_J/2} .$$

Hence,

$$\beta \leq 1 - \frac{C(\varepsilon)}{d_{JN}^{2\varepsilon/n_J}}$$

Combining this with (3.8.2) and replacing ε by $\varepsilon/2$ proves the result.

4. Gallagher's Method

4.1 Suppose we have a Dirichlet series

$$L(s) = \sum a_n n^{-s}$$

for $\sigma = \mathrm{Re}(s) > 1$. For any Dirichlet character χ, set

$$L(s,\chi) = \sum a_n \chi(n) n^{-s}$$

(We are thus departing from the notation used in earlier sections for Artin L-functions). Suppose that all the $L(s,\chi)$ satisfy the hypotheses (i)-(iv) of (2.3). Write A for the conductor of $L(s)$ and A_χ for the conductor of $L(s,\chi)$. We assume that

(4.1.1) $$A_\chi \ll Aq^d$$

if q = conductor of χ. Let us also write

$$\frac{1}{L(s,\chi)} = \sum b_n \chi(n) n^{-s}$$

$$-\frac{L'}{L}(s,\chi) = \sum \Lambda(n)\, c_n \chi(n) n^{-s}$$

for $\sigma > 1$. Let $z > 0$ be a parameter to be specified later. For any Dirichlet character χ, define

$$F_z(s,\chi) = \sum_{n \le z} \Lambda(n)\,\chi(n)\,C_n n^{-s}$$

$$G_z(s,\chi) = \sum_{n > z} \Lambda(n)\,\chi(n)\,C_n n^{-s}$$

$$M_z(s,\chi) = \sum_{n \le z} b_n\,\chi(n)\,n^{-s}$$

We use an identity of Gallagher as modified by Bombieri [2]

$$(4.1.1) \qquad -\frac{L'}{L}(s,\chi) = G_z(1-LM_z) + F_z(1-LM_z) - L'M_z$$

4.2 Let k be a positive integer and define for any $C > 1$,

$$\psi_k(x,\chi) = \frac{1}{2\pi i}\int_{(C)} -\frac{L'}{L}(s,\chi)\,\frac{x^s}{s^{k+1}}\,ds$$

(Again, this is a slight departure from the notation of section 1). Our aim in the next few paragraphs will be to obtain estimates for

$$\sum_{1<q\le Q}\ \sum_{\chi}{}^{*}\ |\psi_k(x,\chi)|$$

Here, the inner sum ranges over primitive characters $\chi \pmod q$.
We have from (4.1.1)

$$(4.2.1) \qquad \psi_k(x,\chi) = \frac{1}{2\pi i}\int_{(C)} G_z(1-LM_z)\,\frac{x^s}{s^{k+1}}\,ds \ +$$

$$+ \frac{1}{2\pi i}\int_{(C)} F_z(1-LM_z)\,\frac{x^s}{s^{k+1}}\,ds - \frac{1}{2\pi i}\int_{(C)} L'M_z\,\frac{x^s}{s^{k+1}}\,ds$$

4.3 Since F_z and M_z are Dirichlet polynomials and L, L' are analytic for all s, we can move the line of integration in the second and third terms of the above expression, into the critical strip. Using the inequality $2|ab| \le |a|^2 + |b|^2$ repeatedly, and taking $C = 1 + \frac{1}{\log x}$, $k > \frac{d}{4}$, we obtain

$$(4.3.1) \qquad \sum_{1<q\le Q}\ \sum_{\chi}{}^{*}\ |\psi_k(x,\chi)| \ll x \sum_{q\le Q}\ \sum_{\chi}{}^{*}\int_{(C)} (|G_z|^2 + |1-LM_z|^2)\,\frac{|ds|}{|s|^{k+1}}$$

$$+ x^{\frac{1}{2}} \sum_{q\le Q}\ \sum_{\chi}{}^{*}\int_{(\frac{1}{2})} (1 + |F_z|^2 + |M_z|^2 + |F_zM_z|^2 + |L|^2 + |L'|^2)\,\frac{|ds|}{|s|^{k+1}}$$

Most of these terms can be handled by the large sieve inequality of Gallagher. We briefly review how this is done.

4.4 Lemma (Gallagher) If $\sum|A_n| < \infty$ and $T \geq 1$, we have

$$\sum_{1<q\leq Q} \sum_{\chi}^{*} \int_{-T}^{T} \Big|\sum_{n=1}^{\infty} A_n \chi(n) n^{it}\Big|^2 dt \ll \sum_{n=1}^{\infty} |A_n|^2 (n + Q^2 T)$$

(see for example, Bombieri [2, p. 30]).

4.5 Using this, we find that

$$\sum_{1<q\leq Q} \sum_{\chi}^{*} \int_{C-i}^{C+i} |G_z|^2 \frac{|ds|}{|s|^{k+1}} \ll \sum_{n>z} \frac{\Lambda(n)^2 \, |C_n|^2}{n^{2C}} (n + Q^2)$$

$$\ll d^2 (\log x)^3 (1 + \frac{Q^2}{z})$$

since $|C_n| \leq d$. Moreover, for any integer $j > 1$,

$$\sum_{1<q\leq Q} \sum_{\chi}^{*} \int_{C-(j+1)i}^{C-ji} + \int_{C+ji}^{C+(j+1)i} \ll \frac{1}{j^{k+1}} \sum_{1<q\leq Q} \sum_{\chi}^{*} \int_{C-(j+1)i}^{C+(j+1)i} |G_z|^2 |ds|$$

$$\ll \frac{1}{j^k} d^2 (\log x)^3 (1 + \frac{Q^2}{z})$$

Take $k \geq 2$, and sum over j to deduce that

$$\text{(4.5.1)} \qquad \sum_{1<q\leq Q} \sum_{\chi}^{*} \int_{(C)} |G_z|^2 \frac{|ds|}{|s|^{k+1}} \ll d^2 (\log x)^3 (1 + \frac{Q^2}{z})$$

Next,

$$1 - LM_z = - \sum_{n>z} \sum_{\substack{e|n \\ e\leq z}} b_e a_{\frac{n}{e}} \chi(n) n^{-s}$$

It is easy to see that $b_{p^\alpha} = 0$ if $\alpha > d$ and for $\alpha \leq d$, $|b_{p^\alpha}| \leq \binom{d}{\alpha}$. Also, $|a_n| \leq \tau_d(n)$ where $\tau_j(n)$ represents the number of ways of writing n as an unordered product of j integers. Therefore,

$$\Big| \sum_{\substack{e|n \\ e\leq z}} b_e a_{n/e} \Big| \leq \tau_{d+1}(n) \, \tau(n)^d$$

where $\tau(n) = \tau_2(n)$ is the usual divisor function. Hence,

(4.5.2)
$$\sum_{1<q\le Q}\ \sum_{\chi}^{*} \int_{(C)} |1-LM_z|^2 \frac{|ds|}{|s|^{k+1}} \ll \sum_{n>z} \frac{\tau_{d+1}(n)^2\ \tau(n)^{2d}}{n^{2C}} (n+Q^2)$$

$$\ll (\log x)^{(d+1)^2 2^{2d}} \left((d+1)^2 2^{2d}\right)!\ (1+\frac{Q^2}{z})$$

The other sums are handled similarly, and we obtain

$$\sum_{1<q\le Q}\ \sum_{\chi}^{*} \int_{(\frac{1}{2})} (1+|F_z|^2+|M_z|^2+|F_zM_z|^2)\frac{|ds|}{|s|^{k+1}}$$

$$\ll 2^{2d} d^2 \sum_{n\le z^2} \frac{(\log n)^2\ \tau(n)^4}{n}(n+Q^2)$$

(4.5.3)
$$\ll 2^{2d} d^2\ (Q^2+z^2)(\log z)^{17}$$

All of the implied constants are absolute. It remains only to treat

(4.5.4)
$$\sum_{1<q\le Q}\ \sum_{\chi}^{*} \int_{(\frac{1}{2})} (|L|^2+|L'|^2)\frac{|ds|}{|s|^{k+1}}$$

At this point, we depart from the method used in the proof of the classical Bombieri-Vinogradov theorem since we do not have an approximate functional equation for L in which the dependence on the field and conductor is explicit. In principle, it should be possible to derive such an estimate, but in fact it is more convenient to use a method of Ramachandra to estimate (4.5.4) more directly. This is discussed in the next section. (The referee points out that it may also be possible to generalize Vaughan's proof [31] which does not use the approximate functional equation).

5. Mean Value Estimates

5.1 We use a method of Ramachandra [23]. Write

$$L(s,\chi) = \theta(s,\chi)\ L^{v}(1-s,\bar{\chi})$$

where

$$\theta(s,\chi) = A_{\chi}^{\frac{1}{2}-s}\ \gamma_{\chi}(1-s)/\gamma_{\chi}(s)$$

and

$$\gamma_{\chi}(s) = \left(\pi^{-\frac{s}{2}}\Gamma(\tfrac{s}{2})\right)^{a_{\chi}} \left(\pi^{-(s+1)/2}\Gamma(\tfrac{s+1}{2})\right)^{b_{\chi}}$$

Here, a_χ and b_χ are non-negative integers satisfying $a_\chi + b_\chi = d$. We deduce from Stirling's formula that

$$\theta(s,\chi) \ll (A_\chi(|t|+2)^d)^{\frac{1}{2}-\sigma} \tag{5.1.1}$$

for $-\frac{3}{4} \le \sigma < 1$, as $|t| \to \infty$.

5.2 For any $U > 0$, we have

$$L(s,\chi) = \sum_{n=1}^{\infty} a_n \chi(n) e^{-n/U} n^{-s} - \frac{1}{2\pi i} \int_{(C_1)} L(s+w,\chi) U^w \Gamma(w) dw \tag{5.2.1}$$

where $C_1 = -\frac{1}{2} - \frac{1}{\log V}$. Here, U and V are parameters to be specified. For $\mathrm{Re}(s+w) < 0$, we can write

$$L(s+w,\chi) = \theta(s+w,\chi) \sum_{n=1}^{\infty} \bar{a}_n \bar{\chi}(n) n^{s+w-1}$$

We split the sum into two parts corresponding to $n > U$ and $n \le U$. Substituting into the integral in (5.2.1) we get

$$L(s,\chi) = \sum_{n=1}^{\infty} a_n \chi(n) e^{-n/U} n^{-s} - \frac{1}{2\pi i} \int_{(C_1)} \theta(s+w,\chi) \sum_{n>U} \bar{a}_n \overline{\chi(n)} n^{s+w-1} U^w \Gamma(w) dw$$

$$- \frac{1}{2\pi i} \int_{(C_1)} \theta(s+w,\chi) \sum_{n \le U} a_n \chi(n) n^{s+w-1} U^w \Gamma(w) dw$$

5.3 We move the second integral to the line $C_2 = -\frac{1}{\log V}$ and apply the Cauchy-Schwartz inequality to both sides. Using (5.1.1), we deduce that

$$\int_{-T}^{T} |L(\tfrac{1}{2}+it,\chi)|^2 dt \ll \int_{-T}^{T} \Big| \sum_{n=1}^{\infty} a_n \chi(n) e^{-n/U} n^{-\frac{1}{2}+it} \Big|^2 dt \tag{5.3.1}$$

$$+ \left(\frac{A_\chi T^d}{U}\right)^{-2C_1} \int_{-\infty}^{\infty} \int_{-T}^{T} \Big| \sum_{n>U} \bar{a}_n \overline{\chi(n)} n^{-1-\frac{1}{\log V}+i(y+t)} \Big|^2 \Gamma(C_1+iy)^2 dt\, dy$$

$$+ \left(\frac{A_\chi T^d}{U}\right)^{-2C_2} \int_{-\infty}^{\infty} \int_{-T}^{T} \Big| \sum_{n \le U} \bar{a}_n \overline{\chi(n)} n^{-\frac{1}{2}-\frac{1}{\log V}+i(y+t)} \Big|^2 \Gamma(C_2+iy)^2 dt\, dy$$

Summing both sides over primitive characters $\chi \pmod q$ and $q \le Q$, we can write the right hand side of (5.3.1) as $\sum_1 + \sum_2 + \sum_3$. From Lemma 4.4, we see that

$$\sum_1 \ll \sum_{n=1}^{\infty} |a_n|^2 e^{-2n/U} n^{-1}(n+Q^2T) \ll (U+Q^2T)(\log U)^{d^2}.$$

We choose

$$V = U = (A Q^d T^d)^{\frac{1}{2}}$$

Then,

$$\sum_2 \ll (A Q^d T^d)^{\frac{1}{2}} \Big(\sum_{n>U} \frac{|a_n|^2}{n^{2+(2/\log V)}} (n+Q^2 T) \Big)$$

$$\ll (A Q^d T^d)^{\frac{1}{2}} (1 + \frac{Q^2 T}{U})(\log U)^{d^2} = (U + Q^2 T)(\log U)^{d^2}$$

Similarly,

$$\sum_3 \ll \sum_{n \le U} |a_n|^2 n^{-1-\frac{2}{\log V}} (n + Q^2 T)$$

$$\ll (U + Q^2 T)(\log U)^{d^2}$$

Summarizing, we have

(5.3.2) $$\sum_{q \le Q} \sum_{\chi}{}^{*} \int_{-T}^{T} |L(\tfrac{1}{2}+it,\chi)|^2 \, dt \ll (U + Q^2 T)(\log U)^{d^2}.$$

<u>5.4</u> Proposition (2.4) implies that

$$\int_T^{\infty} |L(\tfrac{1}{2}+it,\chi)|^2 \frac{dt}{t^{k+1}} \ll \int_T^{\infty} (A Q^d t^d)^{\frac{1}{2}} (\log A Q^d t^d)^{2d} \frac{dt}{t^{k+1}}$$

$$\ll \frac{(A Q^d)^{\frac{1}{2}} d^{2d} (\log A Q^d)^{2d} (\log T)^{2d}}{T^{k-\frac{1}{2}d}}.$$

Summing over primitive $\chi \pmod q$ and $q \le Q$ yields

$$\sum_{q \le Q} \sum_{\chi}{}^{*} \int_{T}^{\infty} |L(\tfrac{1}{2}+it,\chi)|^2 \frac{dt}{t^{k+1}} << \frac{(d \log A)^{2d} A^{\frac{1}{2}} \cdot Q^{\frac{d}{2}+2} (\log Q)^{2d} (\log T)^{2d}}{T^{k-\frac{1}{2}d}}$$

Let $\frac{1}{4} > \in > 0$. We choose $T = Q^{\in/d}$ and $k = \frac{d}{2}(1+\frac{d}{\in})$. Then, we get

$$\sum_{q \le Q} \sum_{\chi}{}^{*} \int_{T}^{\infty} |L(\tfrac{1}{2}+it,\chi)|^2 \frac{dt}{t^{k+1}} << Q^2 (\log Q)^{4d} \quad (\in d \log A)^{2d} A^{\frac{1}{2}}$$

The same estimate holds if we change the range of integration on the left to $(-\infty, -T)$. Summarizing, we have proved that

$$(5.4.1) \qquad \sum_{q \le Q} \sum_{\chi}{}^{*} \int_{(\frac{1}{2})} |L(s,\chi)|^2 \frac{|ds|}{|s|^{k+1}} <<$$

$$<< (\log A)^{d^2} (\log Q)^{d\max(4,d)} (d+\in)^{d^2} \{ Q^{2+\frac{\in}{d}} + A^{\frac{1}{2}} Q^{\frac{d+\in}{2}} + Q^2 A^{\frac{1}{2}} \}$$

5.5 An estimate for the the term in (4.5.4) involving $L'(s,\chi)$ is obtained in a similar fashion. We begin with

$$L(s,\chi) = \sum_{n=1}^{\infty} a_n \chi(n) e^{-n/U} n^{-s} - \frac{1}{2\pi i} \int_{(C_1)} \theta(s+w,\chi) L^{v}(1-s-w,\bar{\chi}) U^{w} \Gamma(w) dw$$

and differentiate with respect to s. We obtain

$$L'(s,\chi) = -\sum a_n \chi(n) e^{-n/U} (\log n) n^{-s} - \frac{1}{2\pi i} \int_{(C_1)} \theta'(s+w,\chi) L^{v}(1-s-w,\bar{\chi}) U^{w} \Gamma(w) dw$$

$$+ \frac{1}{2\pi i} \int_{(C_1)} \theta(s+w,\chi) L^{v\,'}(1-s-w,\bar{\chi}) U^{w} \Gamma(w) dw$$

We then proceed as before with the last integral, decomposing the Dirichlet series for $L^{v\,'}$ at U. There is essentially no change in the previous calculation. For the first integral, we note that

$$\theta'(s+w,\chi) = \theta(s+w,\chi) \frac{\theta'}{\theta}(s+w,\chi) .$$

From Stirling's formula,

$$\frac{\theta'}{\theta}(s+w,\chi) << d(\log(|t+y|+2) + \frac{1}{|s+w|})$$

where $s = \frac{1}{2} + it$, $w = C_1 + iy$. Thus the second integral is a magnification by a factor of $O(d \log U)$ of the integral in (5.2.1). Finally, the Phragmén-Lindelöf estimate for $L'(s,\chi)$ is a magnification of the estimate for $L(s,\chi)$ by a factor of $(\log(A(|t| + 2)^d))^d$. Putting all these observations together, we deduce that

(5.5.1)
$$\sum_{q \le Q} \sum_{\chi}{}^{*} \int_{(\frac{1}{2})} |L'(s,\chi)|^2 \frac{|ds|}{|s|^{k+1}} <<$$

$$<< (\log A)^{d(d+2)} (\log Q)^{d(d+2)} (d+\epsilon)^{d(d+2)} \{ Q^{2+\frac{\epsilon}{d}} + A^{\frac{1}{2}} Q^{\frac{d+\epsilon}{2}} + Q^2 A^{\frac{1}{2}} \}$$

5.6 We put together all the estimates of the previous two sections. Using (4.2.1), (4.3.1), (4.5.1), (4.5.3), (5.4.1), (5.5.1) we see that

$$\sum_{q \le Q} \sum_{\chi}{}^{*} |\psi_k(x,\chi)| << ((d+1)^2 2^{2d})!\; x(\log x)^{(d+1)^2 2^{2d}} (1 + \frac{Q^2}{z})$$

$$+ x^{\frac{1}{2}} 2^{2d} d^2 (Q^2 + z^2)(\log z)^{17} + (d+\epsilon)^{d(d+2)} x^{\frac{1}{2}} ((\log A)(\log Q))^{d(d+2)} \cdot$$

$$\cdot \{ Q^{2+\frac{\epsilon}{d}} + A^{\frac{1}{2}} Q^{\frac{d+\epsilon}{2}} + Q^2 A^{\frac{1}{2}} \} .$$

We chose $z = Q(\log x)^{\gamma}$ with $\gamma > (d+1)^2 2^{2d} + D$. Let $\ell = \log x$ and suppose that

(5.6.1)
$$\ell^{\gamma} \le Q \le \min(x^{\frac{1}{2}} \ell^{-(2\gamma+17+D)}, x^{\frac{1}{2}-\epsilon}, x^{\frac{1}{d-2}-\epsilon})$$

Then, we deduce that

(5.6.2)
$$\frac{1}{Q} \sum_{q \le Q} \sum_{\chi}{}^{*} |\psi_k(x,\chi)| << C_d\, x(\log x)^{-D}$$

where

$$C_d = \max((d+1)^2 2^{2d})!\,, (d+\epsilon)^{d(d+2)} A^{\frac{1}{2}} (\log A)^{d(d+2)}) .$$

We remark that this estimate is valid for any Dirichlet series $L(s)$ satisfying the conditions at the beginning of (4.1). Notice that if $d \le 4$, we can take $Q = x^{\frac{1}{2}-\epsilon}$.

6. Estimates for the Initial Range

To take care of the range $Q \le (\log x)^{\gamma}$, we must specialize to the following case. Let L/F be an abelian extension of number fields, with group G. Take for L(s) and Artin L-series of G. Then L(s) satisfies the hypotheses of §4 with $d = n_F = [F : \mathbb{Q}]$. Thus, under the condition (5.6.1), the estimate (5.6.2) holds. We shall need the following estimate.

6.2 Lemma Suppose that L is normal over $\mathbb{Q}$. Let $1 < q$, N be positive integers with $A_\chi << A_q^d \le (\log x)^N$. Then, there are positive constants $C_1 = C_1(N)$ and

$$C_2 = C_2(L,N) = \min(d_L^{-1/N}, (32 n_L \log d_L)^{-1})$$

so that for any primitive Dirichlet character $\chi \pmod q$,

$$|\psi_k(x,\chi)| << (\log A_\chi)\, x \exp(-C_1 C_2 (\log x)^{\frac{1}{2}})$$

Proof This is by the traditional method (see [12] or [18]). We have for $2 \le T \le x$,

$$\psi_k(x,\chi) = - \sum_{|\gamma| < T} \frac{x^\rho}{\rho^{k+1}} + O\left(\frac{dx(\log x)^2}{T}\right) .$$

Denote by $N(t,\chi)$ the number of zeroes ρ of $L(s,\chi)$ with $|t - \operatorname{Im}\rho| \le 1$ and $0 < \operatorname{Re}\rho < 1$. We have

$$N(t,\chi) << \log A_\chi + d \log(|t| + 5) .$$

Hence,

$$- \sum_{|\gamma| < T} \frac{1}{\rho^{k+1}} << \sum_{j < T} \frac{1}{j^{k+1}} N(j,\chi) << \log A_\chi$$

$$<< \log A + d \log q$$

Now using Propositions (3.4) and (3.8) and choosing

$$T = \exp\left(\frac{1}{d}\left((\log x)^{\frac{1}{2}} - \frac{1}{2} \log A_\chi\right)\right)$$

and $\varepsilon = 1/2N$, the result follows.

6.3 Let $\gamma > 0$ and $U = (\log x)^\gamma$. If $q \le U$, then Lemma (6.2) implies that for some $c = c(d,\gamma)$, and for $x >> e^A$, and any $D > 0$,

(6.3.1) $$\frac{1}{U}\sum_{q\le U}\sum_{\chi}^{*}|\psi_k(x,\chi)| \ll U\,x\exp\left(-c(\log x)^{\frac{1}{2}}\right) \ll x(\log x)^{-D}$$

7. Completion of the Proof

7.1 We now return to the notation and context of section 1. Thus $K/\mathbb{Q}$ is a Galois extension with group G_1, and C is a conjugacy class in G_1. Let H' be an abelian subgroup of G_1 such that $H' \cap C \neq \phi$. Let M be the subfield of K left fixed by H'. Then, for any q, $K(\zeta_q)/M$ is abelian. Let ω be an irreducible character of H'. We appeal to the results of sections 4-6 with $L(s)$ the Artin L-series $L(s,\omega)$. Let $\in > 0$ and $D > 0$. We deduce from (5.6.2) and (6.3.1) that for $d = [M:\mathbb{Q}]$ and any

$$Q \le \min\left(x^{\frac{1}{2}-\in},\ x^{\frac{1}{d-2}-\in}\right) \overset{df}{=} Q_1$$

we have

$$\frac{1}{Q}\sum_{q\le Q}\sum_{\chi}^{*}|\psi_k(K(\zeta_q)/M,\ \omega\otimes\chi,\ x)| \ll x(\log x)^{-D}$$

where the implied constant depends on K, $\in$ and D. Notice that this dependence can be made explicit for Q larger than a power of $(\log x)$, but not for small values of Q.

Now, proposition (1.8) implies that

$$\sum_{q\le Q_1}{}' \max_{(a,q)=1}\max_{y\le x}|\psi_k(K(\zeta_q)/\mathbb{Q},\ \xi(C,a,q),y) - \delta(C,a,q)y| \ll x(\log x)^{3-D}$$

By the reductions of section 1, we deduce that

$$\sum_{q\le Q_1}{}' \max_{(a,q)=1}\max_{y\le x}|\psi_0(K(\zeta_q)/\mathbb{Q},\ \xi(C,a,q),y) - \delta(C,a,q)y| \ll x(\log x)^{-D}$$

7.2 Let H be the largest abelian subgroup of G_1 with $H \cap C \neq \phi$. Let $d = [G_1 : H]$. Set

$$\eta = \begin{cases} d - 2 & \text{if} \quad d \geq 4 \\ 2 & \text{if} \quad d \leq 4 \end{cases}$$

Take an element $g \in C$ and let H_1 be the subgroup generated by g. Then $d \leq [G_1 : H_1] \leq \frac{1}{2}|G_1|$. We can now state the main result.

7.3 Theorem Let $Q = x^{\frac{1}{\eta}-\epsilon}$. Then for any $A > 0$,

$$\sum_{q \leq Q}{}' \max_{(a,q)=1} \max_{y \leq x} \left| \pi_C(y,q,a) - \frac{|C|}{|G|\phi(q)} \pi(y) \right| \ll \frac{x}{(\log x)^A}.$$

The implied constant depends on K, ϵ and A.

7.4 In general, it is possible to replace η by a larger value as follows. Let

$$d^* = \min_H \max_\omega [G_1 : H]\, \omega(1)$$

where the minimum is over all subgroups H satisfying

(i) $H \cap C \neq \phi$

(ii) for every irreducible character ω of H and any non-trivial Dirichlet character χ, the Artin L-series $L(s, \omega \otimes \chi)$ is entire. The maximum is over irreducible characters of H. Then, in the def$^{\underline{n}}$ of η we can replace d with d^*.

8. The Least Prime in a Conjugacy Class

8.1 As before, let $K/\mathbb{Q}$ be a Galois extension and C a conjugacy class in $G = \mathrm{Gal}(K/\mathbb{Q})$. For $0 < a$, $q \in \mathbb{Z}$, $(a,q) = 1$, denote by $P_C(q,a)$ the least prime $p \equiv a \pmod q$, p unramified in K and $(p, K/\mathbb{Q}) = C$. Define

$$F_C(q) = \max_{(a,q)=1} P_C(q,a)$$

From the work of Lagarias-Odlyzko-Montgomery [13] it is known that there is an absolute constant $L > 0$ so that

$$F_C(q) \ll d_{K(\zeta_q)}^L$$

Moreover, assuming the Generalized Riemann Hypothesis, we have for every $\epsilon > 0$,

$$F_C(q) \ll_\epsilon (\log d_{K(\zeta_q)})^{2+\epsilon}$$

Notice that $d_{K(\zeta_q)}$ can be as large as $d_K^{\phi(q)} \, q^{\phi(q)[K:\mathbb{Q}]}$. Let η be as in (7.2). As an application of the main theorem, we shall verify the following.

8.2 Proposition For any $\epsilon > 0$, $F_C(q) \leq q^{\eta+\epsilon}$ with the possible exception of a set of q of density 0. The exceptional set depends on K, C and ϵ.

Proof For each Q > 0, set

$$S_Q = \{ \tfrac{1}{2}Q \leq q \leq Q : F_C(q) > q^{\eta+\epsilon} \}$$

Let $x = (\frac{1}{2}Q)^{\eta+\epsilon}$. Then, for $q \in S_Q$, $\pi_C(x,q,a) = 0$ for some a(mod q). From Theorem (7.3) we deduce that

$$\frac{\log\log Q}{Q} |S_Q| \cdot \frac{|C|}{|G|} \operatorname{Li} x \ll \sum_{q \in S_Q} \frac{|C|}{|G|} \cdot \frac{\operatorname{Li} x}{\phi(q)} \ll \frac{x}{(\log x)^4}$$

Thus,

$$|S_Q| \ll \frac{|G|}{|C|} \frac{Q}{(\log\log Q)(\log Q)^3}$$

We therefore get

$$\{q \leq Q : F_C(q) > q^{\eta+\epsilon}\} \ll \frac{|G|}{|C|} \cdot \frac{Q}{(\log Q)^2} = o(Q)$$

Remark The problem of estimating F(q) seems to have first been considered by Turán [29] who showed (assuming the Lindelöf hypothesis) that $F(q) \ll q^{4+\epsilon}$ for a set of q of density 1. When $K = \mathbb{Q}$, the proposition gives $F(q) \ll q^{2+\epsilon}$ for a set of q of density 1. This can be further improved by utilising the recent work of Bombieri-Iwaniec-Friedlander [3].

References

1. E. Bombieri, On the Large Sieve, Mathematika, 12(1965), 201-225.

2. E. Bombieri, Le Grand Crible dans la Theorie Analytique des Nombres, Astérisque No. 18, Soc. Math. France, 1974.

3. E. Bombieri, H. Iwaniec and J. Friedlander, Primes in Arithmetic Progressions to Large Moduli, Acta Mathematica, 1986.

4. H. Davenport, Multiplicative Number Theory, second edition, Springer-Verlag, New York, 1980.

5. W. Feit, Characters of Finite Groups, Benjamin, New York, 1967.

6. P.X. Gallagher, Bombieri's Mean Value Theorem, Mathematika, 15(1968), 1-6.

7. D. Goldfeld, A Simple Proof of Siegel's Theorem, Proc. Nat. Acad. Sci. U.S.A., 71(1974)1055.

8. D. Goldfeld, J. Hoffstein and S. Patterson, On Automorphic Forms of Half-integral Weight with Applications to Elliptic Curves, in Number Theory Related to Fermat's Last Theorem, pp. 153-193, ed. N. Koblitz, Birkhauser, Boston, 1982.

9. R. Gupta, M. Ram Murty and V. Kumar Murty, The Euclidean Algorithm for S-integers, these proceedings.

10. M. Huxley, The Large Sieve Inequality for Algebraic Number Fields, III. Zero Density Results, J. London Math. Soc. (2)3(1971) 233-240.

11. H. Iwaniec, Fourier Coefficients of Modular Forms of Half-Integral Weight, to appear.

12. J. Lagarias and A. Odlyzko, Effective versions of the Chebotarev Density Theorem, pp. 409-464 in: Algebraic number fields, ed. A. Frohlich, Academic Press, New York, 1977.

13. J. Lagarias, A. Odlyzko and H. Montgomery, A Bound for the Least Prime Ideal in the Chebotarev Density Theorem, Inv. Math., 54(1979) 271-296.

14. E. Landau, Nachr. Akad. Wiss. Göttingen, (1918) 78-97, 478-488.

15. S. Lang, Algebraic Number Theory, Addison-Wesley, Reading, 1970.

16. J. Martinet, Character Theory and Artin L-functions, pp. 1-87 in: Algebraic number fields, ed. A. Fröhlich, Academic Press, New York, 1977

17. Y. Motohashi, An Induction Principle for the generalization of Bombieri's prime number theorem, Proc. Japan Academy, 52(1976) 273-275.

18. V. Kumar Murty, Explicit Formulae and the Lang-Trotter Conjecture, in Proc. Edmonton number theory conference, ed. M.V. Subbarao, Rocky Mountain J. Math. 15(1985) 535-551.

19. M. Ram Murty, On Artin's Conjecture, J. Number Theory, 16(1983) 147-168.

20. M. Ram Murty and V. Kumar Murty, Prime Divisors of Fourier Coefficients of modular forms, Duke Math. J. 51(1984) 57-76.

21. M. Ram Murty and V. Kumar Murty, An Analogue of the Erdös-Kac theorem for Fourier coefficients of modular forms, Indian J. Pure Appl. Math. 15(1984) 1090-1101.

22. H. Rademacher, On the Phragmén-Lindelöf Theorem and Some Applications, Math. Z. 72(1959) 192-204.

23. K. Ramachandra, A Simple Proof of the Mean Fourth Power Estimate for $\zeta(1/2 + it)$ and $L(1/2 + it,\chi)$, Ann. Scuola Norm. Sup. Pisa, 1(1974) 81-97.

24. J.P. Serre, Linear Representations of Finite Groups, Springer-Verlag, New York, 1977.

25. J.P. Serre, Quelques Applications du Théorème de Densité de Chebotarev, Publ. Math. IHES, 54(1982) 123-201.

26. G. Shimura, On Modular Forms of Half-Integral Weight, Ann. Math., 97(1973) 440-481.

27. C.L. Siegel, Abschätzung von Einheiten, Nachr. Akad. Wiss. Gottingen, (1969)71-86.

28. H.M. Stark, Some Effective Cases of the Brauer-Siegel Theorem, Inv. Math., 23(1974) 135-152.

29. P. Turán, On Some Recent Results in the Analytical Theory of Numbers, pp. 359-374. in Institute on number theory, ed. D.J. Lewis, Proc. Symp. Pure Math. 20, AMS, Providence, 1971.

30. A.I. Vinogradov, On the Density Conjecture for Dirichlet L-series, Izv. Akad. Nauk SSSR Ser. Mat., 29(1965) 903-934.

31. R.C. Vaughan, An Elementary method in Prime Number Theory, Acta Arithmetica, 27(1980) 111-115.

32. R. Wilson, The Large Sieve in Algebraic Number Fields, Mathematika, 16(1969) 189-204.

M. Ram Murty: Dept. of Mathematics, McGill University, Montreal, Canada

V. Kumar Murty: Dept. of Mathematics, Concordia University, Montreal, Canada

Canadian Mathematical Society
Conference Proceedings
Volume 7 (1987)

ANOTHER PROOF OF THE KRONECKER LIMIT FORMULAE

Tobias Orloff

For complex parameters $z = x + iy \in h$, $w, s \in \mathbb{C}$ define

$$E(z,w;s) = \sum_{c,d} y^s |cz+d+w|^{-2s}$$

where the sum is over all pairs of integers (c,d) such that $cz+d+w \neq 0$. The series converges for $\mathrm{Re}(s) > 1$ and has a meromorphic continuation to the whole s-plane. Moreover,

1) $$E(z,w;0) = \begin{cases} -1 & \text{if } w \in \mathbb{Z}z + \mathbb{Z} \\ 0 & \text{if } w \notin \mathbb{Z}z + \mathbb{Z} \end{cases}$$

and

2) $$E(z, w;s) = \frac{\pi}{s-1} + O(1) \quad \text{near } s = 1.$$

The classical first and second Kronecker limit formulae determine the value of $E'(z,w;0) = \frac{\partial}{\partial s} E(z,w;s)|_{s=0}$: Writing $e(z) = e^{2\pi iz}$ let

3) $$\eta(z) = e(z/24) \sum_{-\infty}^{\infty} (-1)^n e(\tfrac{1}{2}n(3n + 1)z)$$

and

4) $$\theta_1(w,z) = \sum_{-\infty}^{\infty} e(\tfrac{1}{2}(n + \tfrac{1}{2})^2 z + (n + \tfrac{1}{2})(w - \tfrac{1}{2})).$$

Then

5) $$E'(z,w;0) = \begin{cases} -\log(4\pi^2 |\eta(z)|^4 \cdot y) & \text{if } w \in \mathbb{Z}z + \mathbb{Z} \\ -\log\left|\frac{\theta_1(w,z)}{\eta(z)} \exp(\frac{iw(w-\overline{w})}{z-\overline{z}})\right|^2 & \text{if } w \notin \mathbb{Z}z + \mathbb{Z}. \end{cases}$$

Proofs of these formulae generally use the following product expansions for $\eta(z)$ and $\theta_1(w,z)$ to compute their logarithms, and then compare this with an independent calculation of $E'(z,w;0)$:

$$\eta(z) = e(z/24) \prod_{n \geq 1} (1-e(nz))$$

$$\theta_1(w,z) = 2e(z/8)\sin \pi w \prod_{n \geq 1} (1-e(w)e(nz))(1-e(-w)e(nz))(1-e(nz)).$$

From this point of view 5) looks rather miraculous. Our purpose here is to show that 5) also follows from more "theoretical" considerations; in particular we will not use the above product formulae. For other proofs see [B], [M], [Si], and [St].

THE FIRST LIMIT FORMULA

Write $D = y^2 \frac{\partial}{\partial z} \frac{\partial}{\partial \bar{z}}$, the $Sl_2(R)$ invariant Laplacian on h. Since $Dy^s = \frac{1}{4}s(s-1)y^s$ we obtain for $E(z;s) = E(z,0;s)$

$$DE(z;s) = \frac{1}{4}s(s-1)E(z;s).$$

Applying D to $E'(z;0) = \frac{\partial}{\partial s} E(z;s)\big|_{s=0}$ we thus obtain

$$DE'(z;0) = -\frac{1}{4} E(z;0) = \frac{1}{4}.$$

Therefore $E'(z;0) + \log y$ is harmonic. Since it is also invariant under $z \to z+1$ we may write

6) $$E'(z;0) = -\log y + B + Cy + \sum_n a_n e(nz) + \sum_n b_n e(-n\bar{z})$$

where $B, C, a_n, b_n \in C$ and the sums are over non-zero $n \in Z$. Since $E(z;s)$ is slowly increasing as $y \to \infty$ we have $a_n = b_n = 0$ for $n < 0$. Moreover $E(-\bar{z};s) = E(z;s)$ so $a_n = b_n$. Write

7) $$f(z) = \frac{1}{2} iCz - \sum_{n \geq 1} a_n e(nz).$$

Following Siegel [Si] we have $\exp(E'(-1/z;0)) = \exp(E'(z;0))$ and so

$$|z|^2 y^{-1}\exp(-f(-1/z))\exp(-f(1/\bar{z})) = y^{-1}\exp(-f(z))\exp(-f(-\bar{z})).$$

Separating holomorphic and anti-holomorphic pieces we see that

$$\exp(f(-1/z)) = A\cdot z\cdot \exp(f(z))$$

for some constant $A \in C$. Putting $z = i$ we obtain $A = -i$. To determine the value of C in 6) and 7) write $\rho = -\frac{1}{2} + \frac{1}{2} i \sqrt{3}$. Then $\rho + 1 = -1/\rho$, so $f(\rho + 1) = f(-1/\rho) = f(\rho) + \log(-i\rho)$, where the branch of the log is chosen so $\log(1) = 0$. On the other hand $f(z+1) = \frac{1}{2} iC + f(z)$ so $\frac{1}{2} iC = \log(-i\rho) = \frac{\pi i}{6}$. Write $\Delta(z) = \exp(12f(z))$. Then $\Delta(-1/z) = z^{12}\Delta(z)$, $\Delta(z+1) = \Delta(z)$, and $\Delta(z) \to 0$ as $y \to \infty$. Thus

$\Delta(z) = \eta(z)^{24}$ is the unique normalized cusp form of weight 12 for $Sl_2(Z)$. So

$$f(z) = \log \eta(z)^2$$

and

$$E'(z;0) = B - \log(y\ |\eta(z)|^4).$$

The constant B may be determined by computing the constant term in the Fourier expansion of $E(z;s)$. A simple computation shows this is $2y^s\ \zeta(2s) + 2y^{1-s}\ \zeta(2s-1)\ \pi^{1/2}\ \Gamma(s - \frac{1}{2})\ \Gamma(s)^{-1}$, with $\zeta(s)$ the Riemann zeta function. Hence $B = 4\ \zeta'(0) = -2\log(2\pi)$.

THE SECOND LIMIT FORMULA

To derive the second limit formula we first consider $E'(z,w;0)$ in the neighborhood $U = \{r+sz\ |-1 < r,\ s < 1\}$ of $w = 0$. For $w \in U - \{0\}$

$$E(z,w;s) = y^s|w|^{-2s} + \sum_{c,d} y^s|cz+d+w|^{-2s},$$

where the sum is over integers $(c,d) \neq (0,0)$. Hence $E'(z,w;0) + \log|w|^2$ is regular at $w = 0$. Now $\frac{\partial}{\partial w}\frac{\partial}{\partial \bar{w}} E(z,w;s) = s^2y^{-1}E(z,w;s+1)$. Therefore by 2)

$$\frac{\partial}{\partial w}\frac{\partial}{\partial \bar{w}} E'(z,w;0) = \pi\ y^{-1}.$$

Thus we may write

8) $$E'(z,w;0) = -2\log|w| - \frac{\pi(w-\bar{w})^2}{2y} + k(w,z)$$

where $k(w,z)$ is regular in $U \times h$ and harmonic in w. (This line of reasoning was essentially noted by Stark [St].) This implies

9) $$k(w,z) = g(w,z) + h(\bar{w}, -\bar{z})$$

with functions g and h on $U \times h$ holomorphic in w. Since $E(z,w;s) = E(-\bar{z},-\bar{w};s)$ we have from 8) and 9)

$$g(w,z) + h(\bar{w}, -\bar{z}) = g(\bar{w}, -\bar{z}) + h(w,z).$$

Hence

10) $$g(w,z) - h(w,z) = g(\bar{w}, -\bar{z}) - h(\bar{w}, -\bar{z}) = C(z)$$

for some function $C(z)$. Note $C(z) = C(-\bar{z})$. Replacing $g(w,z)$ by $f(w,z) = g(w,z) - \frac{1}{2}C(z)$ we have from 8), 9), and 10)

11) $$E'(z,w;0) = -\log|w|^2 - \frac{\pi(w-\bar{w})^2}{2y} + f(w,z) + f(\bar{w}, -\bar{z}).$$

Note $f(w,z)$ is still holomorphic in w. Now

$$\frac{\partial^3}{\partial w^3}\frac{\partial}{\partial \bar z} E(z,w;s) = -s^2(s+1)(s+2)\sum_{c,d} \frac{y^{s-1}(cz+d+\bar w)}{|cz+d+w|^{2s+2}(cz+d+w)^2}.$$

The sum on the right hand side is convergent at $s=0$, hence $\frac{\partial^3}{\partial w^3}\frac{\partial}{\partial \bar z} E'(z,w;0) = 0$. Therefore by 11)

$$\frac{\partial}{\partial w^3}\frac{\partial}{\partial \bar z} f(w,z) = 0 .$$

Since $f(w,z)$ is holomorphic in w it follows that

12) $$f(w,z) = F(w,z) + w^2 a(z) + wb(z) + c(z)$$

where $F(w,z)$ is holomorphic in w <u>and</u> z. Now $\frac{\partial}{\partial w}\frac{\partial}{\partial \bar z} E(z,w;s) = -\frac{\partial}{\partial w}\frac{\partial}{\partial \bar z} E(z,-w;s)$. Therefore from 11) and 12) $\frac{\partial}{\partial \bar z} b(z) = 0$, and so $wb(z)$ may be incorporated into $F(w,z)$. Further $\frac{\partial}{\partial w}\frac{\partial}{\partial \bar z} E(w,z;s) = \frac{\partial}{\partial w}\frac{\partial}{\partial \bar z} E(z,w+1;s)$. Therefore from 11) and 12)

$$\frac{\partial}{\partial w}\frac{\partial}{\partial \bar z} f(w,z) = \frac{\partial}{\partial w}\frac{\partial}{\partial \bar z} f(w+1,z),$$

so $2w\frac{\partial}{\partial \bar z} a(z) = 2(w+1)\frac{\partial}{\partial \bar z} a(z)$. Hence $\frac{\partial}{\partial \bar z} a(z) = 0$, and we may incorporate $w^2 a(z)$ into $F(w,z)$. (Note that U contains points w and $w+1$.)

As for $c(z)$, we have for $w \in \mathbb{R}$

$$\frac{\partial}{\partial z}\frac{\partial}{\partial \bar z} E(z,w;s) = \frac{1}{4}s(s-1)y^{-2}E(z,w;s).$$

By 1) this implies $\frac{\partial}{\partial z}\frac{\partial}{\partial \bar z} E'(z,w;0) = 0$ for $w \in \mathbb{R}-\mathbb{Z}$. Hence from 11) and 12) $c(z) + c(-\bar z)$ is harmonic, so we may write

$$c(z) + c(-\bar z) = d(z) + d(-\bar z)$$

with a holomorphic function $d(z)$. Replacing $f(w,z)$ by $G(w,z) = -F(w,z) - d(z)$ we have

13) $$E'(z,w;0) = -\log|w|^2 - \frac{\pi(w-\bar w)^2}{2y} - G(w,z) - \overline{G(\bar w,-\bar z)}.$$

Note that $G(w,z)$ is holomorphic in w <u>and</u> z in $U \times h$. Now define

$$H_0(w,z) = we^{G(w,z)}.$$

Then $H_0(w,z)$ is holomorphic in $(w,z) \in U \times h$ and has only a simple zero at $w = 0$. Further

$$\exp(-E'(z,w;\, 0)) = \exp\Big(\frac{\pi(w-\bar{w})^2}{2y}\Big)\, H_0(w,z) \cdot \overline{H_0(\bar{w}, -\bar{z})}.$$

Similarly, for any $\lambda \in \mathbb{Z}z + \mathbb{Z}$ there is a function $H_\lambda(w,z)$ holomorphic in $(\lambda + U) \times h$ with only a simple pole at $w = \lambda$ such that

$$\exp(-E'(z,w;0)) = \exp\Big(\frac{\pi(w-\bar{w})^2}{2y}\Big)\, H_\lambda(w,z) \cdot \overline{H_\lambda(\bar{w}, -\bar{z})}.$$

Comparing these formulae on the overlaps we see that $H_\lambda(w,z) = \pm H_\mu(w,z)$ if $(\lambda + U) \cap (\mu + U) \neq \emptyset$. Since $\mathbb{C}$ is simply connected it follows that there is a global homomorphic function $H(w,z)$ on $\mathbb{C} \times h$ such that

$$14)\qquad \exp(-E'(z,w;0)) = \exp\Big(\frac{\pi(w-\bar{w})^2}{2y}\Big)\cdot H(w,z)\cdot \overline{H(\bar{w}, -\bar{z})}.$$

Further $H(w,z)$ has zeroes only at $w \in \mathbb{Z}z + \mathbb{Z}$, and they are simple.

Now, since $E(z,w+1;s) = E(z,w;s)$ we obtain from 14)

$$H(w+1)\overline{H(\bar{w}+1,-\bar{z})} = H(w,z)\overline{H(\bar{w},-\bar{z})}.$$

Separating holomorphic and anti-holomorphic elements we see that

$$15)\qquad H(w+1,z) = \pm H(w,z).$$

Further, $E(z,w+z;s) = E(z,w;s)$ so

$$e((w-\bar{w}) + \tfrac{1}{2}(z-\bar{z}))\, H(w+z,z)\overline{H(\bar{w}+\bar{z},-\bar{z})} = H(w,z)\overline{H(\bar{w},-\bar{z})}.$$

Hence

$$16)\qquad H(w+z,z) = ce(-w- \tfrac{1}{2}z)\, H(w,z)$$

with a constant $c \in \mathbb{C}$. Since $E'(z,w;0)$ is real-valued it follows that $|c| = 1$. Now, the function $\theta_1(w,z)$ given by 4) satisfies

$$\theta_1(w+1,z) = -\theta_1(w,z)$$
$$\theta_1(w+z,z) = -e(-w- \tfrac{1}{2}z)\, \theta_1(w,z)$$

and has zeroes at $w \in \mathbb{Z}z + \mathbb{Z}$. It follows easily that $\theta_1(w,z)/H(w,z)$ is independent of w. (Note also from the theory of theta functions that the space of holomorphic functions of w satisfying 15) and 16) (for a fixed choice of $\pm$ and c) is one-dimensional.) Now, from the first Kronecker limit formula we have

$$\lim_{w \to 0} (E'(z,w;0) - \log y\, |w|^{-2}) = -\log(4\pi^2\, |\eta(z)|^4\, y).$$

Hence from 13) $G(0,z) = \log(2\pi\, \eta(z)^2)$. We claim

$$H(w,z) = we^{G(w,z)} = \eta(z)^{-1}\,\theta_1(w,z).$$

To derive this formula write $H(w,z) = M(z)\theta_1(w,z)$. Then letting " ' " denote " $\frac{\partial}{\partial w}$ " we have

$$H'(0,z) = M(z)\,\theta_1'(0,z) = e^{G(0,z)} = 2\pi\,\eta(z)^2. \tag{17}$$

From 4)

$$\theta_1'(0,z) = 2\pi i \sum_{-\infty}^{\infty} (n+\tfrac{1}{2})\, e(\tfrac{1}{2}(n+\tfrac{1}{2})^2 z - \tfrac{1}{2}(n+\tfrac{1}{2})).$$

By inspection $\theta_1'(0,z+1) = e(1/8)\,\theta_1'(0,z)$. By Poisson summation $\theta_1'(0,-1/z) = (z/i)^{3/2}\,\theta_1'(0,z)$. Hence $\theta_1'(0,z)^8 = a\cdot\Delta(z) = a\,\eta(z)^{24}$ for some $a \in C$. Comparing leading coefficients in the Fourier expansion we obtain

$$\theta_1'(0,z) = 2\pi\,\eta(z)^3.$$

Hence from 17) $M(z) = \eta(z)^{-1}$. This completes the proof of the second Kronecker limit formula.

BIBLIOGRAPHY

[B] Berndt, B.C.: Identities involving the coefficients of a class of Dirichlet series. Trans. Amer. Math. Soc., 160 (1971), 157-167.

[M] Motohashi, Y.: A new proof of the Kronecker limit formula, Proc. Japan Acad. 44 (1968), 614-616.

[Si] Siegel, C.L. : Lectures on Advanced Analytic Number Theory, Tata Institute of Fundamental Research Studies in Mathematics 9. Tata Institute of Fundamental Research, Bombay, 1980.

[St] Stark, H. : L-functions at s=1. IV. Advances in Math. 35 (1980), no. 3, 197-235.

Canadian Mathematical Society
Conference Proceedings
Volume 7 (1987)

INDEPENDANCE ET GROUPES ALGEBRIQUES

Patrice Philippon

RESUME. Dans ce texte nous présentons quelques résultats nouveaux d'indépendance algébrique autour du théorème dit de Lindemann-Weierstrass. L'aspect quantitatif est mentionné et on décrit l'outil principal de ce progrès:un nouveau Lemme de zéros. Aussi on montre par divers problèmes que le sujet est loin d'être épuisé.

Il y a environ un siècle était démontré le premier théorème d'indépendance algébrique, maintenant connu sous le nom de:

THEOREME DE LINDEMANN-WEIERSTRASS . (voir 3 et 8) Soient $\alpha_1,\ldots,\alpha_n$ des nombres algébriques $\mathbb{Q}$ -linéairement indépendants, alors les nombres $e^{\alpha_1},\ldots,e^{\alpha_n}$ sont algébriquement indépendants sur $\overline{\mathbb{Q}}$.

Dans cette note nous voulons décrire quelques résultats nouveaux étendant ce théorème. Le cadre approprié pour de telles généralisations semble être celui des groupes algébriques commutatifs, pour donner une direction à notre discussion formulons une conjecture assez générale dans ce cadre.

CONJECTURE. Soit G un groupe algébrique commutatif défini sur $\overline{\mathbb{Q}}$ et tel que $\mathrm{Hom}(\mathbb{C}_a,G)=0$. Soit $\underline{\alpha}\in TG(\overline{\mathbb{Q}})$, appelons G' le plus petit sous-groupe algébrique de G contenant $\exp_G(\underline{\alpha})$. La $\overline{\mathbb{Q}}$ -adhérence de Zariski de $\{\exp_G(\underline{\alpha})\}$ dans G est-elle une composante irréductible de G'? (i.e. A-t-on $\mathrm{degtr}_{\overline{\mathbb{Q}}}\,\overline{\mathbb{Q}}(\exp_G(\underline{\alpha})) = \dim G'$?)

N.B. On suppose ici que l'on a choisi une fois pour toute un plongement de $\overline{\mathbb{Q}}$ dans $\mathbb{C}$.

Cette conjecture semble assez difficile à aborder et les progrès récents dans cette direction font toujours jouer un rôle prépondérant aux endomorphismes du groupe G considéré (cf.6). Plus précisement, on considère un sous-groupe de type fini Γ de $\mathrm{End}G$ et l'on note V le plus petit sous-espace vectoriel de TG, défini sur $\overline{\mathbb{Q}}$, et contenant l'orbite Λ de $\underline{\alpha}$ sous $d_0\Gamma=:\{d_0\gamma\ ;\gamma\in\Gamma\}$($d_0\gamma$ désigne l'application linéaire tangente à l'origine de l'endomorphisme γ).

Les résultats s'expriment alors naturellement en fonction du coefficient de répartition de Λ dans V introduit par M.Waldschmidt,

$$\mu(\Lambda,V) = \min \{ \frac{\mathrm{rang}_{\mathbb{Z}}\Lambda/\Lambda\cap W}{\dim V/W} \}$$

où le minimum est pris sur l'ensemble des sous-espaces vectoriels W de V distincts de V.

Posons encore $\rho=1$ si G est linéaire (i.e. si $G=\mathbb{G}_m^d$, dans le présent contexte) et $\rho=2$ sinon. Nous noterons X la $\overline{\mathbb{Q}}$-adhérence de Zariski de $\{\exp_G(\underline{\alpha})\}$ dans G, et nous pouvons énoncer:

THEOREME 1. $\dim X \geq \frac{\mu(\Lambda,V)}{\rho}$.

Si $\rho=1$ on est dans la situation du théorème de Lindemann-Weierstrass (voir exemple 1 ci-après), et lorsque $\rho=2$ on se ramène au cas où G est une variété abélienne. La démonstration du théorème est alors donnée au chapître troisième de 6 modulo une conjecture (conjecture 2.15,p.61) qui est maintenant levée grâce à 7. Dans le cas où $\dim V=1$ le théorème est complètement établi dans 6.

Il est peut-être utile d'éclairer le théorème 1 en détaillant ses conséquences les plus significatives.

EXEMPLE 1. On prend $G=\mathbb{G}_m^d$ $(\rho=1)$, $\underline{\alpha}=(\alpha_1,\dots,\alpha_d) \in \overline{\mathbb{Q}}^d$. Si $\alpha_1,\dots,\alpha_d$ sont $\mathbb{Q}$-linéairement indépendants on a $G'=G$. On pose $\Gamma\simeq\mathbb{Z}^d$ et $\Lambda=\mathbb{Z}\alpha_1+\dots+\mathbb{Z}\alpha_d \subset V=\mathbb{C}$, on vérifie aisément que $\mu(\Lambda,V)=d$. Enfin $\exp_G(\underline{\alpha}) = (e^{\alpha_1},\dots,e^{\alpha_d})$ et on retrouve

$$\dim X = \mathrm{degtr}_{\overline{\mathbb{Q}}}\ \overline{\mathbb{Q}}(e^{\alpha_1},\dots,e^{\alpha_d}) = d,$$

le théorème de Lindemann-Weierstrass.

EXEMPLE 2. On prend $G=E^d$ (où E est une courbe elliptique), $\underline{\alpha}=(\alpha_1,\dots,\alpha_d) \in \overline{\mathbb{Q}}^d$. Si $\alpha_1,\dots,\alpha_d$ sont linéairement indépendants sur $\mathrm{End}E$ on a $G'=G$. On pose $\Gamma\simeq(\mathrm{End}E)^d$ et $\Lambda=\mathrm{End}E.\alpha_1+\dots+\mathrm{End}E.\alpha_d \subset V=\mathbb{C}$, on vérifie aisément que $\mu(\Lambda,V)=d$ ou $2d$ selon que $\mathrm{End}E=\mathbb{Z}$ ou $\mathrm{End}E\neq\mathbb{Z}$. Si $\wp$ est la fonction elliptique de Weierstrass associée à E on peut écrire, comme $\rho=2$,

$$\dim X = \mathrm{degtr}_{\overline{\mathbb{Q}}}\ \overline{\mathbb{Q}}(\wp(\alpha_1),\dots,\wp(\alpha_d)) \geq \frac{d}{2} \text{ en général et } =d \text{ dans le cas}$$

$\mathrm{End}E\neq\mathbb{Z}$.

N.B. G.V.Chudnovsky avait établi dans 1 ce dernier résultat lorsque $d=2$ ou 3. Le cas général est traité dans 2.

EXEMPLE 3. On prend $G=A^{\ell}$ (A variété abélienne de dimension g). On sait que $A(\mathbb{C}) \simeq \mathbb{C}^g/\Omega$ où Ω est un réseau de $\mathbb{C}^g$. Le corps des fonctions de A coïncide avec le corps des fonctions méromorphes sur $\mathbb{C}^g$ périodiques par rapport aux éléments de Ω, son dégré de transcendance est g et choisissons une base de transcendance $(f_1,\dots,f_g)$ de ce corps. On considère $\underline{\gamma}_1,\dots,\underline{\gamma}_{\ell} \in \overline{\mathbb{Q}}^g$ et on note V le plus petit sous-espace vectoriel de $\mathbb{C}^g$, défini sur $\overline{\mathbb{Q}}$, contenant les vecteurs $\underline{\gamma}_1,\dots,\underline{\gamma}_{\ell}$. On pose alors $\Lambda=\mathbb{Z}\underline{\gamma}_1+\dots+\mathbb{Z}\underline{\gamma}_{\ell} \subset V$, $\Gamma\simeq\mathbb{Z}^{\ell}$ et on déduit du théorème 1 que:

$$\mathrm{degtr}_{\overline{\mathbb{Q}}}\, \overline{\mathbb{Q}}\,(f_i(\underline{\gamma}_j);\ i=1,\dots,g\ ;\ j=1,\dots,\ell) \geq \mu(\Lambda,V)/2.$$

Si A est de type C.M. on peut prendre $\ell=2g$ et Λ l'orbite d'un vecteur propre $\underline{\gamma}$ sous l'action de EndA, ainsi $\Lambda = (\mathbb{Z}\gamma_1+\dots+\mathbb{Z}\gamma_{2g})\underline{\gamma}$ et $V=\mathbb{C}\underline{\gamma}$ d'où $\mu(\Lambda,V)=2g$. D'autre part les valeurs $f_i(\gamma_j\underline{\gamma})$ s'expriment algébriquement en fonction des $f_i(\underline{\gamma})$, on a donc:

$$\mathrm{degtr}_{\overline{\mathbb{Q}}}\, \overline{\mathbb{Q}}\,(f_1(\underline{\gamma}),\dots,f_g(\underline{\gamma}))=g.$$

En 1932, K.Mahler donnait dans 4 une version quantitative du théorème de Lindemann-Weierstrass. Il est intéressant de noter qu'une semblable version quantitative du théorème 1 est démontrée par E.M.Jabbouri dans 2. Dans le cas C.M. de l'exemple 2 ci-dessus le résultat est en tout point comparable au théorème de Mahler.

THEOREME 2. (E.M.JABBOURI) Soit $\wp$ une fonction elliptique de Weierstrass d'invariants algébriques et admettant des multiplications complexes. Soient $\alpha_1,\dots,\alpha_d$ des nombres algébriques linéairement indépendants sur le corps des multiplications de $\wp$. Pour tout $\varepsilon>0$ il existe un réel $c(\varepsilon,\wp)$ effectivement calculable tel que si $P \in \mathbb{Z}[X_1,\dots,X_d]$ est non nul, on a

$$\log|P(\wp(\alpha_1),\dots,\wp(\alpha_d))| \geq -c\,[\log H(P)+\exp(d^0P^{d+\varepsilon})].d^0P^{d+\varepsilon}.$$

REMARQUES. On note H(P) le maximum des valeurs absolues des coefficients de P. G.Wüstholz a annoncé dans 9,th.6 une version plus optimiste du théorème 2 ne faisant pas intervenir le terme $\exp(d^0P^{d+\varepsilon})$.

Les démonstrations des théorèmes précédents reposent essentiellement sur trois points;

(1). Un mélange des méthodes classiques de Gel'fond et Siegel pour l'indépendance algébrique,

(2). Les lemmes de zéros dans les groupes algébriques commutatifs,

(3). Les techniques d'éliminations projectives.

Dans le reste de cette note nous voulons présenter les résultats de 7 (se rattachant au point (2)) nécessaires à la démonstration du théorème 1, les points

(1) & (3) sont développés dans 6(et aussi dans 2).

LEMMES DE ZEROS

Le problème est de contrôler la répartition des zéros d'hypersurfaces dans un groupe algébrique commutatif. Commençons par le cas le plus simple, lorsque $G=\mathbb{C}_a^d$.

Soit $\Sigma = \mathbb{Z}\sigma_1 + \ldots + \mathbb{Z}\sigma_\ell$ un $\mathbb{Z}$-module de type fini de $\mathbb{C}^d$, on définit pour $S \geq 0$

$$\Sigma_S = \{ m_1\sigma_1 + \ldots + m_d\sigma_d ; |m_i| \leq S\} .$$

On veut minorer le degré D d'un polynôme non identiquement nul $P \in \mathbb{C}[X_1,\ldots,X_d]$ s'annulant sur Σ_S.

REPONSE. (D.W.Masser) $D \geq (\frac{S}{d})^{\mu(\Sigma,\mathbb{C}^d)}$; <u>où</u> $\mu(\Sigma,\mathbb{C}^d)$ <u>est le coefficient de répartition</u> (<u>de Dirichlet</u>) <u>de</u> Σ <u>dans</u> $\mathbb{C}^d$ <u>déjà introduit</u>.

Ce résultat a été étendu aux groupes algébriques commutatifs connexes généraux G par D.W.Masser & G.Wüstholz (voir 5 et 9,§2). Σ est maintenant un sous-groupe de type fini du groupe algébrique G plongé dans un espace projectif $\mathbb{P}_N$ et on définit Σ_S comme ci-dessus.

On veut alors minorer le degré D d'un polynôme homogène $P \in \mathbb{C}[X_0,\ldots,X_N]$ s'annulant sur Σ_S mais pas identiquement sur G.

REPONSE. (D.W.Masser & G.Wüstholz) <u>Il existe un réel</u> C <u>ne dépendant que du plongement de</u> G <u>dans</u> $\mathbb{P}_N$ <u>tel que</u>

$$D \geq C(\frac{S}{d})^{\mu}$$

<u>où ici</u> $\mu = \min \{ \frac{\mathrm{rang}_{\mathbb{Z}}(\Sigma/\Sigma \cap G')}{\dim(G/G')} \}$ <u>le minimum étant pris sur tous les sous-groupes propres</u> G' <u>de</u> G.

N.B. L'exposant μ est l'exposant le plus grand possible dans les énoncés précédents, qui sont donc, en un certain sens optimaux.

Un aspect intéressant des lemmes de zéros dont nous avons besoin est qu'ils doivent tenir compte de l'ordre d'annulation de P le long d'un sous-groupe analytique de G. Plus Précisement (cf. 5 & 7) un sous-groupe analytique (à k paramètres) Φ de G est un homomorphisme $\Phi:\mathbb{C}^k \to G(\mathbb{C})$ défini, injectif et analytique au voisinage de l'origine de $\mathbb{C}^k$. En notant τ_g la translation sur G par g on dit qu'un polynôme homogène $P \in \mathbb{C}[X_o,\ldots,X_N]$ s'annule en g avec un ordre $\geq T$ le long de Φ si et seulement si la fonction $P \circ \tau_g \circ \Phi$, définie et analytique au voisinage de l'origine de $\mathbb{C}^k$, s'annule en 0 avec un ordre $\geq T$.

On veut maintenant minorer le degré d'un polynôme homogène $P \in \mathbb{C}[X_0,\ldots,X_N]$ s'annulant sur Σ_S avec un ordre $\geq T$ le long de Φ mais pas identiquement sur G. On notera $A = \mathrm{im}\Phi$ et l'on supposera que A est Zariski dense dans $G(\mathbb{C})$.

THEOREME 3. $D \geq c\left(\min\{(\frac{S}{d})^{\mathrm{rang}\Sigma/\Sigma\cap G'} \cdot (\frac{T}{d})^{\dim A/A\cap G'}\}\right)^{\frac{1}{\dim G/G'}}$

<u>où le minimum est pris sur l'ensemble des sous-groupes propres</u> G' <u>de</u> G.

Ce théorème se déduit facilement du théorème 2.1 de 7. D.W.Masser & G.Wüstholz avaient déjà établi une version affaiblie du théorème ci-dessus dans le cas k=1 (cf.5) tandis que G.Wüstholz donnait une estimation du même type que celle du théorème 3, mais G' parcourant l'ensemble plus grand de toutes les sous-variétés algébriques de G (cf.9,th2.).

Nous voudrions conclure cette courte présentation des questions d'indépences algébriques "de type Lindemann-Weierstrass" en mentionnant deux problèmes plus spécifiques que la conjecture que nous avons donné au début et qui pourraient se révéler plus abordables.

QUESTION 1. Soient $\alpha_1,\ldots,\alpha_d$ des nombres algébriques et $\wp_1,\ldots,\wp_n$ des fonctions elliptiques de Weierstrass .Notons pour $i=1,\ldots,n$ par ζ_i la fonction zêta associée à $\wp_i (\zeta'_i = -\wp_i)$ et appelons K le corps engendré sur $\overline{\mathbb{Q}}$ par les valeurs de la fonction exponentielle et des fonctions $\wp_1,\ldots,\wp_n,\zeta_1,\ldots,\zeta_n$ aux points $\alpha_1,\ldots,\alpha_d$.

(i.e. $K = \overline{\mathbb{Q}}\,(e^{\alpha_1},\ldots,e^{\alpha_d},\wp_1(\alpha_1),\ldots,\wp_n(\alpha_d),\zeta_1(\alpha_1),\ldots,\zeta_n(\alpha_d))$).

Donner une minoration nouvelle (i.e non contenue dans le théorème 1) du nombre entier $\mathrm{degtr}_{\overline{\mathbb{Q}}}K$.

QUESTION 2. Soient $\wp$ une fonction elliptique de Weierstrass n'admettant pas de multiplication complexe et $\alpha_1,\ldots,\alpha_d$ des nombres algébriques $\mathbb{Q}$-linéairement indépendants. Montrer que les nombres $\wp(\alpha_1),\ldots,\wp(\alpha_d)$ sont algébriquement indépendants.

REFERENCES

1. G.V.Chudnovsky, "Algebraic independance of value of elliptic functions at algebraic points", Inv.Math.61(1980),267-290.

2. E.M.Jabbouri, "Mesures d'indépendances algébriques de valeurs de fonctions elliptiques et abéliennes", Manuscrit.

3. F.Lindemann, "Uber die Zahl π", Math.Ann.20(1882),213-225.

4. K.Mahler, "Zür Approximation der Exponentialfunktionen und des Logarithms", J.für die Reine und Angew.Math.(1932)118-150.

5. D.W.Masser & G.Wüstholz, "Zero estimates on group varieties II", Inv.Math. 80(1985),233-267.

6. P.Philippon, Pour une théorie de l'indépendance algébrique, Thèse de Doctorat d'Etat de l'Université de Paris XI, Orsay(1983).

7. P.Philippon, "Lemmes de zéros dans les groupes algébriques commutatifs", soumis pour publication.

8. K.Weierstrass, "Zu Lindemann Abhandlung, 'Uber die Ludolphsche Zahl'," Sitzgsber.Preuss.Akad.Wiss.(1885)1067-1085.

9. G.Wüstholz, Recent progress in transcendence theory, in "Number theory, Noordwijkerhout,1983", ed.H.Jager,Springer Lecture Notes in Math.1098(1984) 280-296.

CENTRE DE MATHEMATIQUES
Ecole Polytechnique
91128 PALAISEAU Cédex.
FRANCE.

Canadian Mathematical Society
Conference Proceedings
Volume 7 (1987)

ARITHMETIC OF HILBERT-BLUMENTHAL SURFACES

Dinakar Ramakrishnan*

Introduction.

For any real quadratic field F with ring of integers $\mathcal{O}_F$, the Hilbert modular group $SL_2(\mathcal{O}_F)$ acts as a discontinuous group of analytic automorphisms of $\mathfrak{h}^2$, where $\mathfrak{h}$ denotes the Poincaré upper-half plane in $\mathbb{C}$. For every subgroup Γ of finite index in $SL_2(\mathcal{O}_F)$ we can form the quotient complex surface $\Gamma\backslash\mathfrak{h}^2$, which is smooth if Γ is torsion free. There is a natural singular compactification of $\Gamma\backslash\mathfrak{h}^2$ obtained by the addition of a finite set (of cusps) in bijection with $\Gamma\backslash\mathbb{P}^1_F$. In particular, when $\Gamma = SL_2\mathcal{O}_F$ the number of cusps equals the class number h_F of F. Suitable smooth compactifications of the surfaces $\Gamma\backslash\mathfrak{h}^2$ exist, thanks to F. Hirzebruch, and we will call these smooth models, and their adelic variants, Hilbert-Blumenthal surfaces.

The geometry of these surfaces is fascinating. Even by looking only at $\Gamma = SL_2\mathcal{O}_F$ one gets, as F runs the gamut of real quadratic fields, a rich array of complex surfaces of varying degrees of geometric complexity. It may be worthwhile to remark that many, but not all, of the (arithmetical) questions considered in this brief survey, some of them still very much open, are just as interesting, and not really easier, if we restrict ourselves totally to the case: $\Gamma = SL_2\mathcal{O}_F$. Perhaps one could also make the simplifying assumption that $h_F = 1$ and that F contains a unit of norm -1, especially if one finds that the adelic language (which seems to us to be the natural setting) is forbidding.

What makes these surfaces especially interesting to arithmeticians is the fact that they are definable over number fields.

*Partially supported by a grant from the National Science Foundation.

There are canonical models S_Γ of $\Gamma\backslash\mathfrak{h}^2$, defined by Shimura, over suitable number fields k_Γ, this construction having deep relations to class field theory. These S_Γ's parametrize isomorphism classes of polarized abelian surfaces (with suitable level structure) admitting multiplication by $\mathcal{O}_F$. That one can also construct smooth compactifications $\tilde{S}_\Gamma$ over the canonical number fields k_Γ is a consequence of the works of Mumford and of Rapoport. Rapoport's theory actually yields arithmetical models (algebraic spaces) over the rings of integers of k_Γ.

Recall that in the case of modular curves $X_0(N)$, a lot of beautiful arithmetic is tied up with the Galois action on $H^1_{et}(X_0(N)\times\overline{\mathbb{Q}},\mathbb{Q}_\ell) = T_\ell^\vee \otimes_{\mathbb{Z}_\ell}\mathbb{Q}_\ell$, where $T_\ell^\vee$ denotes the $\mathbb{Z}_\ell$-dual of the Tate module T_ℓ of the Jacobian $J_0(N)$, defined as the inverse limit (over n) of the groups of ℓ^n-division points of $J_0(N)$. In the case of Hilbert-Blumenthal surfaces $\tilde{S}_\Gamma$, perhaps the most interesting thing is the representation:

$$\rho_\ell:\ \mathrm{Gal}(\overline{\mathbb{Q}}/k_\Gamma) \to GL(W_\ell), \tag{0.1}$$

where W_ℓ is the finite-dimensional ℓ-adic vector space: $H^2_{et}(\tilde{S}_\Gamma \otimes \overline{\mathbb{Q}},\mathbb{Q}_\ell)$. The main focus of this article will be the structure of the Galois module W_ℓ. We will discuss a very exciting open question raised by T. Oda - also predicted by the Langlands philosophy, which if answered will throw some light on the elliptic curves (over F) attached to Hilbert modular cusp forms of weight two.

For every number field $k \supset k_\Gamma$, the restriction of ρ_ℓ to $\mathrm{Gal}(\overline{\mathbb{Q}}/k)$ defines an L-function: $L(W_\ell/k,s)$ in the sense of Serre. When we throw out a finite set of Euler factors we can identify, for k abelian, this L-function with a (finite) product of some abelian L-functions and certain automorphic L-functions, by a theorem of Brylinski and Labesse. And, thanks to Asai, we have an integral representation for each relevant automorphic L-function, which allows us to deduce meromorphic continuation and holomorphy except possibly at $s = 2$. Of great interest is the order of pole and the residue at $s = 2$.

We are then naturally led to look at: $\mathrm{Pic}(\tilde{S}_\Gamma\times\overline{\mathbb{Q}})$, the Picard group of divisors on $\tilde{S}_\Gamma\times\overline{\mathbb{Q}}$ modulo linear equivalence. There is a natural homomorphism:

$$c_\ell:\ \mathrm{Pic}(\tilde{S}_\Gamma\times\overline{\mathbb{Q}}) \to W_\ell(1), \tag{0.2}$$

where $W_\ell(1)$ denotes the "Tate-twisted" Galois module: $W_\ell \otimes_{\mathbb{Z}} \mathbb{Z}_\ell(1)$, with $\mathbb{Z}_\ell(1)$ denoting $\varprojlim_n \mu_{\ell^n}$. (It is a fact that c_ℓ is injective for Hilbert-Blumenthal surfaces.) We will look at the deep work of Harder, Langlands and Rapoport on the image under c_ℓ of $\mathrm{Pic}(\tilde{S}_\Gamma \times k)$ for abelian number fields k, and the relationship as predicted by Tate's conjecture to the order of pole of $L(W_\ell/k,s)$ at $s = 2$.

Every holomorphic Hilbert modular cusp form f (relative to Γ) of weight 2 contributes to $H^2(\tilde{S}_\Gamma \times \overline{\mathbb{Q}})$. <u>If</u> f corresponds to a cusp form g on a quaternion algebra B over F such that $B \otimes \mathbb{R} = M_2(\mathbb{R}) \times \mathbb{H}$, $\mathbb{H}$ = Hamilton's quaternions, then there is a (Shimura) curve C_B over F such that the L-function of $C_B \times C_B^\theta$, $\mathrm{Aut}(F/\mathbb{Q}) = \{1,\theta\}$, occurs as a factor of the L-function of W_ℓ. We point out some consequences of this fact in §10.

The next thing we will discuss is the establishment of a conjecture of Beilinson, resulting in an interpretation of the residue at $s = 2$ of $L(W_\ell/k,s)$, $k_\Gamma \subset k \subset \mathbb{Q}^{ab}$, in terms of a regulator involving K_1 of $\tilde{S}_\Gamma \times k$. (This is a natural analog of the classical regulator of Dirichlet on $K_1 \mathcal{O}_k = \mathcal{O}_k^*$.) For imaginary quadratic fields k, we will indicate how to obtain a $\mathbb{Q}$-structure of $H^{1,1}(\tilde{S}_\Gamma(\mathbb{C}),\mathbb{C}))$ by combining a suitable subspace of $K_1(\tilde{S}_\Gamma \times k) \otimes \mathbb{Q}$, defined by means of Hirzebruch-Zagier cycles and modular units, with $\mathrm{Pic}(\tilde{S}_\Gamma \times k) \otimes \mathbb{Q}$.

A consequence of the Hilbert modular surfaces being simply connected is the fact that linear and homological equivalences coincide for divisors on $\tilde{S}_\Gamma$. This allows for the possibility of explicit construction of arithmetic Hilbert modular functions f by starting with divisors $D(f)$ which have trivial image in $W_\ell(1)$. One can then study the values of these functions f at CM points z (not lying on $D(f)$), which by the Shimura-Taniyama theory lie in abelian extensions of the reflex CM fields of z. The question which we would like to raise is whether one can put suitable conditions on $D(f)$ to ensure integrality of $f(z)$.

Finally, we touch upon rationality questions about zero cycles, viz. finite formal sums of points, on $\tilde{S}_\Gamma$ (modulo rational equivalence) and discuss an extraordinary consequence of a recent generalization of the Birch-Swinnerton-Dyer conjecture due to Bloch and Beilinson.

TABLE OF CONTENTS

ACKNOWLEDGEMENTS

It is my pleasure to thank R. P. Langlands for explaining to me his joint work with Harder and Rapoport, and to thank J. A. Shalika for sharing and discussing his ideas (which directly motivated the questions raised in 5.33 and 6.5) on the relationship of abelian surfaces to automorphic forms on GSP(4). I have also greatly benefited from conversations/correspondence with D. Blasius, S. Bloch, Ching-Li Chai, P. Deligne, G. Harder, F. Hirzebruch, H. Jacquet, B. Mazur, M. Rapoport, K. Ribet, G. Van der Geer, A. Wiles, D. Zagier, and the participants of this conference.

§1. The classical setting.

1.1. Choose a square-free integer d and consider the real quadratic field $F = \mathbb{Q}(\sqrt{d})$. The ring of integers $\mathcal{O}_F$ of F equals: $\mathbb{Z} + w_d\mathbb{Z}$, where $w_d = \sqrt{d}$ (resp. $(1+\sqrt{d})/2$) if $d \equiv 2,3 \pmod 4$ (resp. $1 \pmod 4$)). Denote by θ the non-trivial automorphism of F so that $\theta(a+b\sqrt{d}) = a - b\sqrt{d}$, for a, b in $\mathbb{Q}$.

1.2. Let Γ be an arithmetic subgroup of SL_2F. In other words, $\Gamma' \overset{\text{def}}{=} \Gamma \cap SL_2(\mathcal{O}_F)$ has finite index in Γ as well as in $SL_2(\mathcal{O}_F)$. By a theorem of Serre one knows that Γ' is actually a congruence subgroup, i.e., Γ' contains, for some ideal $\mathfrak{a} \subset \mathcal{O}_F$, the principal congruence subgroup:

$$(1.3) \qquad \Gamma(\mathfrak{a}) = \{\gamma \in SL_2(\mathcal{O}_F) \mid \gamma - 1 \in \mathfrak{a}M_2(\mathcal{O}_F)\}.$$

There is a natural embedding (with discrete image):

$$(1.4) \qquad \Gamma \hookrightarrow SL_2(F \otimes \mathbb{R}) \simeq SL_2\mathbb{R} \times SL_2\mathbb{R},$$

given by: $\gamma \mapsto (\gamma, \gamma^\theta)$.
Consequently we get a nice action of Γ on $\mathfrak{h}^2$ defined by:

$$(1.5) \qquad \left(\begin{pmatrix} a & b \\ c & d \end{pmatrix}, (z_1, z_2)\right) \mapsto \left(\frac{az_1+b}{cz_1+d}, \frac{a^\theta z_2 + b^\theta}{c^\theta z_2 + d^\theta}\right).$$

1.6. The quotient space $\Gamma\backslash\mathfrak{h}^2$ is a complex surface, which is smooth if Γ is torsion-free. For general Γ this surface admits a finite number of certain simple singularities called "quotient singularities", arising from the points on $\mathfrak{h} \times \mathfrak{h}$ with non-trivial stabilizers in Γ.

1.7. There is a natural (singular) compactification of $\Gamma\backslash\mathfrak{h}^2$, which we will now describe. Identifying $\mathfrak{h}^2$ with the set: $\{z \in F \otimes \mathbb{C} \mid \mathrm{Im}(z) > 0\}$, we see that $\mathbb{P}^1_F = F \cup \{\infty\}$ is a dense subset of the boundary $\partial(\mathfrak{h}^2) = (F \otimes \mathbb{R}) \cup \{\infty\}$. Set:

$$(1.8) \qquad (\mathfrak{h}^2)^* = \mathfrak{h}^2 \cup \mathbb{P}^1_F.$$

We can extend the Γ-action on $\mathfrak{h}^2$ to $(\mathfrak{h}^2)^*$ by letting Γ act on $\mathbb{P}^1_F$ by: $\left(\begin{pmatrix} a & b \\ c & d \end{pmatrix}, (u,v)\right) \mapsto \frac{au+bv}{cu+dv}$. Then $\Gamma\backslash\mathbb{P}^1_F$ is finite, and the quotient $\Gamma\backslash(\mathfrak{h}^2)^*$ is a normal, projective (complex) surface ([HvG]), called the Baily-Borel-Satake compactification of $\Gamma\backslash\mathfrak{h}^2$. It is useful to note that:

(1.9) $$\#(SL_2(\mathcal{O}_F)\backslash\mathbb{P}^1_F) = h_F ,$$

the class number of F.

1.10. As mentioned in the Introduction, the general theory of canonical models of Shimura ([Sh1], [D1]) implies in our (super-special) case that for every Γ there exists a quasi-projective surface S_Γ defined over a (canonical) abelian number field k_Γ such that:

(1.11) $$S_\Gamma(\mathbb{C}) = \Gamma\backslash\mathfrak{h}^2 .$$

One way to get at the canonical model is to start by finding a k_Γ-structure for the Γ-invariant meromorphic functions, viz. Hilbert modular functions relative to Γ, on $\mathfrak{h}^2$. A second, perhaps more natural*, way is to first identify $\Gamma\backslash\mathfrak{h}^2$ as the solution space over $\mathbb{C}$ to a suitable moduli problem, and then to note that this problem may be posed over k_Γ with the resulting solution space being S_Γ.

Let us now review the moduli problem (cf. [HvG], Chap. 7) in the special case where Γ equals:

(1.12) $$\Gamma(\mathcal{O},\mathfrak{a}) \overset{\text{def}}{=} \left\{\begin{pmatrix} a & b \\ c & d\end{pmatrix} \in SL_2(F) \,\middle|\, \begin{matrix} a,d \in \mathcal{O}, \\ c \in \mathfrak{a}\ ,\ b \in \mathfrak{a}^{-1}\end{matrix}\right\} ,$$

where $\mathfrak{a}$ denotes a fractional ideal contained in the inverse different δ^{-1} of F. Recall that δ^{-1} consists of the set of all elements x in F such that $Tr_{F/\mathbb{Q}}(x\mathcal{O}_F) \subset \mathbb{Z}$. If $\mathfrak{a} \supset \mathcal{O}$, then:

(1.13) $$\Gamma(\mathcal{O},\mathfrak{a}) \cap SL_2(\mathcal{O}_F) = \Gamma^0(\mathfrak{a}^{-1}) \overset{\text{def}}{=} \left\{\begin{pmatrix} a & b \\ c & d\end{pmatrix} \in SL_2\,\mathcal{O}_F \,\middle|\, b \in \mathfrak{a}^{-1}\right\}.$$

Clearly $\Gamma^0(\mathfrak{a}^{-1})$ contains the principal congruence subgroup $\Gamma(\mathfrak{a}^{-1})$, and it has finite index in $\Gamma(\mathcal{O},\mathfrak{a})$ as well as in $SL_2(\mathcal{O}_F)$.

---

* This article has modest goals, and it is by no means meant to be a complete summary of the various arithmetical methods in the study of Hilbert modular surfaces. What may be obvious is our bias in favor of methods involving abelian varieties, but what may not be too evident is our respect for all the different approaches to the subject.

Given any point z in $\mathfrak{h}^2 \subset F \otimes \mathbb{C}$, and a fractional ideal $\mathfrak{a} \subset \delta^{-1}$, we get a lattice in $F \otimes \mathbb{C}$ defined by:

(1.14) $$L_z(\mathfrak{a}) = 2\pi i(\mathfrak{a} + z\mathfrak{O}).$$

Set:

(1.15) $$A_z(\mathfrak{a}) = (F \otimes \mathbb{C})/L_z(\mathfrak{a}).$$

Then $A_z(\mathfrak{a})$ is a complex torus of dimension 2. One has the following:

(1.16) (i) The lattice $L_z(\mathfrak{a})$ admits a skew-symmetric, non-degenerate, $\mathbb{Z}$-valued bilinear form E_z $(= E_z(\mathfrak{a}))$;

(ii) $F \otimes \mathbb{C}$ admits a positive-definite Hermitian form H_z such that $\mathrm{Im}(H_z) = E_z$ on $L_z(\mathfrak{a})$, i.e., $A_z(\mathfrak{a})$ is a polarized abelian surface over $\mathbb{C}$ (see [M1]. Chapter 1); and

(iii) There is a representation

$$m: \mathfrak{O} \hookrightarrow \mathrm{End}(A_z(\mathfrak{a})), \text{ sending } 1 \text{ to } \mathrm{Id}.$$

Since $\mathrm{Tr}_{F/\mathbb{Q}}(\mathfrak{a}\mathfrak{O}) \subset \mathbb{Z}$, we can define a bilinear form:

(1.17) $$E: (\mathfrak{a} \oplus \mathfrak{O})^2 \to \mathbb{Z}$$

by setting $E((u_1,v_1),(u_2,v_2)) = \mathrm{Tr}_{F/\mathbb{Q}}(u_1v_2 - u_2v_1)$, for all u_1, u_2 in $\mathfrak{a}$ and v_1, v_2 in $\mathfrak{O}$. Clearly E is skew-symmetric and non-degenerate. (Thus E gives rise to a non-trivial linear form: $\Lambda^2_{\mathbb{Z}}(\mathfrak{a} \oplus \mathfrak{O}) \to \mathbb{Z}$). The desired "polarization" E_z on $L_z(\mathfrak{a})$ is defined by transport of structure. This gives (i).

To obtain (ii) let us define:

(1.18) $$H_z: (F \otimes \mathbb{C})^2 \to \mathbb{C}$$

by: $H_z(w_1,w_2) = \tilde{E}_z(iw_1,w_2) + i\tilde{E}_z(w_1,w_2)$, where $\tilde{E}_z$ denotes the $\mathbb{R}$-linear extension of E_z to $F \otimes \mathbb{C} = L_z \otimes \mathbb{R}$. It is not too difficult to verify that H is hermitian, and even positive-definite.

Finally, since $\mathfrak{a}$ is a fractional ideal, there is a natural (multiplication) of $\mathfrak{O}$ on $\mathfrak{a} \oplus \mathfrak{O}$, and hence on $L_z(\mathfrak{a})$. It is easy to see that the resulting representation $\mathfrak{O} \to \mathrm{End}(A_z(\mathfrak{a}))$ sends 1 to Id and is injective, whence (iii).

Conversely, let $A = \mathbb{C}^2/L$ be a polarized abelian surface over $\mathbb{C}$ with $\mathcal{O} \hookrightarrow \mathrm{End}(A)$. Then L is an $\mathcal{O}$-module of rank 2, which is locally free, i.e., $L \otimes \mathcal{O}_{\mathfrak{p}}$ is free over the local completion $\mathcal{O}_{\mathfrak{p}}$ of $\mathcal{O}$ for every prime $\mathfrak{p} \subset \mathcal{O}$. We can then find a fractional ideal $\mathfrak{a} \subset F$ such that $L \xrightarrow{\sim} \mathfrak{a} \oplus \mathcal{O}$. More simply, if E is the polarization of L, then $\mathfrak{a}$ is the dual (relative to the non-degenerate form given by $\mathrm{Tr}_{F/\mathbb{Q}}$) of:

$$(1.19) \qquad \mathfrak{a}^{\vee} \overset{\text{def}}{=} \{a \in F \mid E_{\mathbb{Q}}(L \times aL) \subset \mathbb{Z}\} \quad ,$$

where $E_{\mathbb{Q}}$ denotes the $\mathbb{Q}$-linear extension of E to $L \otimes \mathbb{Q}$. The isomorphism $L \xrightarrow{\sim} \mathfrak{a} \oplus \mathcal{O}$ leads to isomorphisms $\mathbb{C}^2 \xrightarrow{\sim} F \otimes \mathbb{C}$ <u>and</u> $A \xrightarrow{\sim} A_z$ for some $z \in \mathfrak{h}^2$. The isomorphism class of A determines z modulo the action of $\Gamma(\mathcal{O}, \mathfrak{a})$, whence a point in $\Gamma(\mathcal{O}, \mathfrak{a}) \backslash \mathfrak{h}^2$.

<u>1.20</u>. What happens if $\mathfrak{a}$ is replaced by $\alpha\mathfrak{a}$ for some totally positive integer α? The "moduli spaces" $\Gamma(\mathcal{O}, \mathfrak{a}) \backslash \mathfrak{h}^2$ and $\Gamma(\mathcal{O}, \alpha\mathfrak{a}) \backslash \mathfrak{h}^2$ are analytically isomorphic, but they parametrize abelian surfaces with different polarizations. For more on such matters, and for a discussion of embeddings of $\Gamma(\mathcal{O}, \mathfrak{a}) \backslash \mathfrak{h}^2$ in Siegel modular three-folds, see [HvG].

<u>1.21</u>. One can naturally pose the above moduli problem for $\Gamma(\mathcal{O}, \mathfrak{a})$ already over $\mathbb{Q}$ $(= k_{\Gamma(\mathcal{O}, \mathfrak{a})})$. This results in the construction of the (normal, quasi-projective) surface $S_{\Gamma(\mathcal{O}, \mathfrak{a})}$, the set of whose complex points identifies with $\Gamma(\mathcal{O}, \mathfrak{a}) \backslash \mathfrak{h}^2$.

<u>1.22</u>. For every arithmetic subgroup Γ of $SL_2(F)$, there is a projective, normal surface $\overline{S}_{\Gamma}$ over k_{Γ} (see §2) so that $\overline{S}_{\Gamma}(\mathbb{C}) = \Gamma \backslash (\mathfrak{h}^2)^*$, where $(\mathfrak{h}^2)^* = \mathfrak{h}^2 \cup \mathbb{P}^1_F$. We write S_{Γ}^{∞} for $\overline{S}_{\Gamma} - S_{\Gamma}$. The cusps are all rational over cyclotomic fields. The Galois action on them will be described in the natural (!) adelic setting in §2.

<u>1.23</u>. Let us now briefly review the smooth compactification $\tilde{S}_{\Gamma}(\mathbb{C})$ of $S_{\Gamma}(\mathbb{C})$, due to F. Hirzebruch ([H1], [HvG]) cf. also Ash, Mumford, et al ([AMRT], Chapter 1, §5). Firstly, if Γ acts on $\mathfrak{h}^2$ in such a way that some point has a non-trivial (finite cyclic) stabilizer, then there is a harmless <u>quotient singularity</u> in $S_{\Gamma}(\mathbb{C})$. Such a singularity can be easily resolved by replacing it by a (finite) chain of rational curves. For example, if the stabilizer of a point has order two, then the

point gets blown up into a rational curve of self-intersection -2.

It is more interesting, and important, to understand the nature of the resolution of the cusps. Given any cusp κ in $S_\Gamma^\infty(\mathbb{C})$, we can find some g in $GL_2^+(F)$ such that $g(\kappa) = \infty$. So by replacing Γ by $g\Gamma g^{-1}$, we may assume that $\kappa = \infty$. For every positive real number t, set:

$$W_t = \{z = (z_1, z_2) \in \mathfrak{h}^2 \mid \mathrm{Im}(z_1)\cdot\mathrm{Im}(z_2) > t\}$$

and

$$\Gamma_\infty = \left\{\begin{pmatrix} a & b \\ 0 & a^{-1}\end{pmatrix} \in \Gamma\right\}.$$

Then Γ_∞ stabilizes the cusp ∞, and we have:

$$\Gamma_\infty \backslash W_t \hookrightarrow \Gamma\backslash\mathfrak{h}^2, \quad \text{for } t >> 0.$$

It is easy to see that the stabilizer of ∞ is a semi-direct product:

$$\Gamma_\infty = A \ltimes N,$$

where $N = \{\begin{pmatrix} 1 & x \\ 0 & 1\end{pmatrix} \mid x \in \mathfrak{b}\}$, and $A = \left\{\begin{pmatrix} \varepsilon^n & 0 \\ 0 & \varepsilon^{-n}\end{pmatrix} \middle| n \in \mathbb{Z}\right\}$, for some fractional ideal $\mathfrak{b}$ and a unit ε of infinite order.

What one does now is to add to $N\backslash W_t$ an infinite chain Σ of rational curves, and extend the A-action to Σ so that:

(i) $A\backslash\Sigma$ is finite, and

(ii) $A\backslash(N\backslash W_t \cup \Sigma)$ is smooth and compact. Since $A\backslash(N\backslash W_t) = \Gamma_\infty\backslash W_t$, this will give us what we want.

There is a natural way to do this: Start with a $\mathbb{Z}$-basis (α,β) of $\mathfrak{b}$ such that $\alpha\beta^\theta - \alpha^\theta\beta > 0$. (We can do this.) Look at the continued fraction expansion:

$$\frac{\alpha}{\beta} = a_0 - \frac{1}{a_1} - \frac{1}{a_2} - \cdots \overset{\text{def}}{=} [[a_0, a_1, a_2, \ldots]]$$

By Legendre this continued fraction is periodic (since α/β is real quadratic). Let $(b_0, b_1, \ldots, b_{r-1})$ be a period, with each $b_i \geq 2$. Since α/β is irrational, some b_i must be ≥ 3. Set: $e_0 = 1$, and define for every positive integer $m = nr + \ell$, $1 \leq \ell \leq r$:

$$e_m = (w_{\ell-1}^{-1} e_{\ell-1}) e_r^n ,$$

where

$$w_{\ell-1} = [[b_{\ell-1},\ldots,b_{r-1}]].$$

It is not hard to see that:

$$\beta^{-1}\mathfrak{b} = \mathbb{Z}e_m + \mathbb{Z}e_{m+1}, \ \forall\, m \geq 0.$$

The elements $\{e_m\}$ can be viewed geometrically as the vertices on the convex hull of $N_\beta \cap \mathbb{R}^2_+$, where $N_\beta = \{\begin{pmatrix}1 & \beta^{-1}\mathfrak{b}\\ 0 & 1\end{pmatrix}\}$:

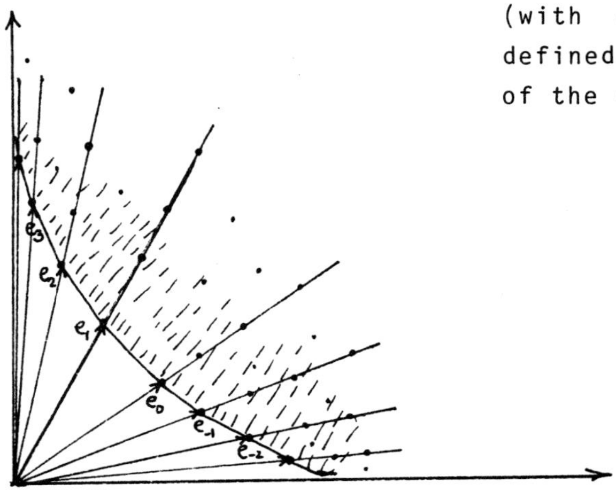

(with e_m for $m < 0$ defined by using periodicity of the continued fraction)

Assume for simplicity that the scaling factor β is 1. One sets:

$$R_m = \mathbb{C}e_m + \mathbb{C}e_{m+1} = \{z_m e_m + w_m e_{m+1}\},$$

$$U_m = \{(z_m, w_m) \in R_m \mid z_m \neq 0\},$$

and

$$V_m = \{(z_m, w_m) \in R_m \mid w_m \neq 0\}.$$

There are analytic isomorphisms:

$$\phi_m : U_{m-1} \xrightarrow{\sim} V_m \ ,$$

given by: $(z_{m-1}, w_{m-1}) = (z_{m-1}^{b_m} w_{m-1}, (z_{m-1})^{-1})$.
Denote by Y the space obtained by gluing $\{R_m\}$ along $\{\phi_m\}$.
Now set:

$$C_m = \text{The closure of } \{w_{m-1} = 0\} \ (= \{z_m = 0\}) \text{ in } Y.$$

The $\Sigma \overset{\text{def}}{=} \bigcup_m C_m$ is an infinite chain of rational curves. Evidently A acts on Σ, and identifies C_m with C_{m+r}. Thus we see:

$\Gamma_\infty \setminus W_t \cup A \setminus \Sigma$ is an r-gon of rational curves

$$\{E_0, E_1, \ldots, E_{r-1}\}$$

such that

$$E_m^2 = -b_m \text{ , and}$$

$$E_m \cdot E_{m\pm 1} = 1 \quad \text{(with the convention: } E_r = E_0\text{)}.$$

That this smooth compactification of S_Γ can be done over k_Γ is a consequence of the work of Rapoport ([Rp], [HLR]). (It can also be deduced from [M2].) We will look at it in the next section.

§2. Adelic formulation.

2.1. It will be convenient for us to pass to GL_2 and work with the $\mathbb{Q}$-algebraic group:

(2.2) $$G = R_{F/\mathbb{Q}} \, GL_2/F,$$

the restriction of scalars (à la Weil) for GL_2 viewed as an F-group. By definition one has, for every $\mathbb{Q}$-algebra R, the identification:

(2.3) $$G(R) = GL_2(F \otimes R).$$

In particular,

(2.4) $$G(\mathbb{Q}) = GL_2(F) \quad \underline{\text{and}} \quad G(\mathbb{R}) \simeq GL_2(\mathbb{R}) \times GL_2(\mathbb{R}).$$

Let us set:

(2.5) $$\mathfrak{X} = F \otimes \mathbb{C} - F \otimes \mathbb{R}$$

Then its component $\mathfrak{X}^0$, say, containing $1 \otimes i$ can be identified with $\mathfrak{h}^2$. The usual action of $SL_2(F \otimes \mathbb{R})$ on $\mathfrak{h}^2$ extends naturally to an action of $G(\mathbb{R})$ on $\mathfrak{X}$. For a natural description of $\mathfrak{X}$ consider the $\mathbb{R}$-algebraic group:

(2.6) $$\mathfrak{S} = R_{\mathbb{C}/\mathbb{R}} \mathbb{G}_m ,$$

and the homomorphism (of $\mathbb{R}$-algebraic groups):

(2.7) $$h: \mathcal{S} \to G_{\mathbb{R}},$$

which is defined on the real points by:

(2.8) $$h(a+ib) = \left(\begin{pmatrix} a & b \\ -b & a \end{pmatrix}, \begin{pmatrix} a & b \\ -b & a \end{pmatrix}\right) \in GL_2\mathbb{R} \times GL_2\mathbb{R},$$

for all $a+ib$ in $\mathbb{C}^*$.

Set:

(2.9) $$L_\infty = \left\{\begin{pmatrix} a & b \\ -b & a \end{pmatrix} \in GL_2(\mathbb{R})\right\} = Z_{GL_2(\mathbb{R})} \cdot O(2) \simeq \mathbb{C}^*,$$

where Z_H denotes for any group H the center of H. Define:

(2.10) $$K_\infty = L_\infty \times L_\infty \subset G(\mathbb{R}).$$

It is easy to see that:

(2.11) K_∞ is the centralizer of $h(\mathbb{C}^*)$ in $G(\mathbb{R})$

<u>and</u>

$$\mathcal{X} = G(\mathbb{R})/K_\infty .$$

Thus $\mathcal{X}$ identifies with the set of $G(\mathbb{R})$-conjugates of $h(\mathbb{C}^*)$.

If we set:

(2.12) $$G' = R_{F/\mathbb{Q}}SL_2/F \quad \underline{\text{and}} \quad T = R_{F/\mathbb{Q}}\mathbb{G}_m F,$$

then we have an exact sequence of $\mathbb{Q}$-algebraic groups:

(2.13) $$1 \longrightarrow G' \longrightarrow G \xrightarrow{\det} T \longrightarrow 1,$$

and $\mathcal{X}^0 = \mathfrak{h}^2 = G'(\mathbb{R})/K'_\infty$, where $K'_\infty = G'(\mathbb{R}) \cap K_\infty$.

<u>2.14</u>. The connected components of $\mathcal{X}$ are in bijection with $\pi_0(T(\mathbb{R})/\det(K_\infty)) = \pi_0((\mathbb{R}^*)^2/(\mathbb{R}^*_+)^2)$. (For any space Y, $\pi_0(Y)$ denotes as usual the set of connected components of Y.)

2.15. For any number field E let $\mathbb{A}_E$ denote the ring of adeles of E. Then class field theory gives us an identification:

(2.16) $$\mathrm{Gal}(E^{ab}/E) \simeq \pi_0(\mathbb{A}_E^*/E^*) .$$

When $E = \mathbb{Q}$ we will simply write $\mathbb{A}$ instead of $\mathbb{A}_{\mathbb{Q}}$. It is useful to recall that:

(2.17) $$\mathbb{A} = \mathbb{R} \times \mathbb{A}_f \, , \; \mathbb{A}_f = \hat{\mathbb{Z}} \otimes \mathbb{Q},$$

and $\hat{\mathbb{Z}} = \varprojlim_n \mathbb{Z}/n\mathbb{Z} \simeq \prod_p \mathbb{Z}_p$.

2.18. For any algebraic group H over $\mathbb{Q}$ we will often write H_f (resp. H_∞) instead of $H(\mathbb{A}_f)$ (resp. $H(\mathbb{R})$), viewed as a locally compact group. We have a direct product decomposition:

$$H(\mathbb{A}) = H_\infty \times H_f .$$

2.19. The theory of canonical models ([Sh1], [D1], [D2]) furnishes in our situation a quasi-compact, separated scheme $S = S(G,h)$ over $\mathbb{Q}$ equipped with a continuous G_f-action such that:

(2.20) $$S(\mathbb{C}) = G(\mathbb{Q})\backslash G(\mathbb{A})/K_\infty \quad (= G(\mathbb{Q})\backslash \mathfrak{X} \times G_f).$$

The continuity of the action of G_f signifies that:

(2.21)(i) For each compact open subgroup K of G_f there is a surface $S_K = S/K$ over $\mathbb{Q}$;

(ii) For K,L: compact open subgroups of G_f and $x \in G_f$ such that $x^{-1}Kx \subset L$ there is a homomorphism $J_{L,K}(x): S_K \to S_L$, with: $J_{M,L}(y)J_{L,K}(x) = J_{M,K}(y,x)$ and $J_{K,K}(x) = \mathrm{Id}$.

(iii) For K,L: compact open with K normal in L, one gets an action of L/K on S_K and $J_{L,K}(e)$ defines an isomorphism of $S_K/(L/K)$ with S_L;

and (iv) $S = \varprojlim_K S_K$ (inverse limit).

2.22. S_K is smooth for K: small enough, and it is in general not geometrically connected. The (possible) disconnectedness is precisely described by the following bijection:

$$(2.23) \qquad \pi_0(S_K(\mathbb{C})) \simeq \pi_0(T(\mathbb{Q})\backslash T(\mathbb{A})/\det(K)).$$

The key points behind this bijection are: (i) the one-to-one correspondence between $\pi_0(G(\mathbb{Q})\backslash G(\mathbb{A}))$ and $\pi_0(T(\mathbb{Q})\backslash T(\mathbb{A}))$, which is a consequence of G' being simply-connected as an algebraic group, and (ii) the connectedness of K_∞.

2.24. We can make (2.23) explicit as follows: By strong approximation theorem ([P]) we can find elements $x_1, x_2, \ldots, x_h$, h depending on K, in G_f such that:

$$(2.25) \qquad G(\mathbb{A}) = \bigcup_{j=1}^{h} G_{\mathbb{Q}}\, x_j\, G_\infty^+ K,$$

with $\{\det x_j \mid 1 \leq j \leq h\}$ forming a complete set of representatives for $F^* \det(G_\infty^+ K)\backslash F_{\mathbb{A}}^*$.

As a consequence we get an isomorphism of complex surfaces

$$(2.26) \qquad S_K(\mathbb{C}) \xrightarrow{\sim} \bigcup_{j=1}^{h} \Gamma_j \backslash \mathfrak{h}^2 ,$$

where $\Gamma_j = G(\mathbb{Q}) \cap x_j G_\infty^+ K x_j^{-1}$.

2.27. A consequence of (2.23) is the fact that when K is the "full" congruence subgroup $G(\hat{\mathbb{Z}})$ the connected components of $S_K(\mathbb{C})$ are in bijection, noting that $T(\mathbb{Q})\backslash T(\mathbb{A})$ equals $F^*\backslash \mathbb{A}_F^*$, with the ideal classes of F.

2.28. For an understanding of the Galois action on the components of S_K it is necessary to use the basic homomorphism

$$(2.29) \qquad r: \mathbb{G}_{m_{\mathbb{C}}} \to \mathfrak{S}_{\mathbb{C}} ,$$

which is, on the $\mathbb{C}$-points, given by: $r(z) = (z,1) \in \mathbb{C}^* \times \mathbb{C}^*$, for all z in $\mathbb{C}^*$. The composite map

$$(2.30) \qquad h\circ r: \mathbb{G}_{m_{\mathbb{C}}} \xrightarrow{r} \mathfrak{S}_{\mathbb{C}} \xrightarrow{h} G_{\mathbb{C}}$$

is a $\mathbb{C}$-homomorphism between $\mathbb{Q}$-groups. The conjugacy class of $h\circ r$ is definable over $\mathbb{Q}$, and it is this fact which furnishes the canonical model $S = S(G,h)$ over $\mathbb{Q}$.

The Shimura reciprocity law is the $\mathbb{Q}$-homomorphism

$$r(G,h):\mathbb{G}_m \to T, \tag{2.31}$$

defined to be the inverse of the composite map $\det\circ h\circ r$, seen explicitly on the $\mathbb{Q}$-points to be the natural morphism of $\mathbb{Q}^*$ into F^*.

Composing the class field theory isomorphism $\mathrm{Gal}(\mathbb{Q}^{ab}/\mathbb{Q}) \simeq \pi_0(\mathbb{A}^*/\mathbb{Q}^*)$ with the map on $\pi_0(\mathbb{A}^*/\mathbb{Q}^*)$ induced by $r(G,h)$ one obtains a morphism

$$\lambda:\mathrm{Gal}(\mathbb{Q}^{ab}/\mathbb{Q}) \to \pi_0(T(\mathbb{Q})\backslash T(\mathbb{A})), \tag{2.32}$$

which describes the Galois action on the components of $S_K(\mathbb{C})$.

2.33. Denote by $S_K^0(\mathbb{C})$ the identity component of $S_K(\mathbb{C})$. Then it is defined over the finite abelian extension $\mathbb{Q}(K)$ of $\mathbb{Q}$ corresponding to the open subgroup of $\mathbb{A}^*/\mathbb{Q}^*$ defined by $r(G,h)^{-1}(\det K \cap r(G,h)(\mathbb{A}_f^*))$.

In particular, when $K = G(\hat{\mathbb{Z}})$, we see that S_K^0, and indeed every component of S_K, is defined over $\mathbb{Q}$. It can be seen that each component is of the form:

$$\Gamma(\mathfrak{O},\mathfrak{a})\backslash \mathfrak{h}^2, \tag{2.34}$$

for a fractional ideal $\mathfrak{a}$ in F (cf. (1.12) for the definition of $\Gamma(\mathfrak{O},\mathfrak{a})$).

2.35. To every point (z,g_f) on $\mathfrak{X}\times G_f$ we can attach a complex torus of dimension 2:

$$A_z = (F\otimes\mathbb{C})/L_z, \tag{2.36}$$

where L_z is the lattice $2\pi i(\delta_F^{-1} + z\mathfrak{O}_F)$. Since where $\Lambda^2_{\mathfrak{O}_F} L_z$ is a projective $\mathfrak{O}_F$-module of rank 1 we get a polarization

$$p:\Lambda^2_{\mathfrak{O}_F} L_z \xrightarrow{\sim} (\det g_f^{-1})\delta_F^{-1}, \tag{2.37}$$

where $(\det g_f)$ signifies the fractional ideal defined by the idele $\det g_f$. Evidently (A_z, p) admits multiplication by $\mathcal{O}_F$.

For every integer $n \geq 1$, set:

(2.38) $$K_n = \left\{ g_f = \begin{pmatrix} a & b \\ c & d \end{pmatrix} \in G_f \;\middle|\; \begin{matrix} a \equiv d \equiv 1 \pmod{n} \\ c \in \delta_F,\ b \in n\delta_F^{-1}, \\ \det g_f \in \mathcal{O}_f^* \end{matrix} \right\},$$

where $\mathcal{O}_f$ denotes $\hat{\mathbb{Z}} \otimes \mathcal{O}_F$. Then one has:

(2.39) S_{K_n} = The moduli space over $\mathbb{Q}$ of isomorphism classes of polarized abelian surfaces A with multiplication by $\mathcal{O}_F$, equipped with the "level structure": $\frac{1}{n}\delta_F^{-1}/\delta_F^{-1} \xrightarrow{\sim} \mathrm{Ker}(A \xrightarrow{.n} A)$.

For more (precise) details on such moduli aspects, see [Dl], [Ka] and [Rp]. See also [Zn].

2.40. For each compact open subgroup K of G_f there is a Baily-Borel-Satake compactification:

(2.41) $$\overline{S}_K = S_K \cup S_K^\infty,$$

which can be embedded as a closed, normal subvariety of a suitable projective space. Here S_K^∞ is a finite set of "cusps". It is a fact that:

(2.42) $$S_K^\infty(\mathbb{C}) = B^+(F)N_f(F)\backslash G_f/K,$$

$$N = \{\begin{pmatrix} 1 & * \\ 0 & 1 \end{pmatrix}\} \subset B = \{\begin{pmatrix} * & * \\ 0 & * \end{pmatrix}\} \subset GL_2,$$

$$B^+(F) = \{b \in B(F) \mid \det b >> 0\}, \text{ and } N_f(F) = N_f \otimes_{\mathbb{Q}} F.$$

Each of the cusps is rational over an abelian extension of $\mathbb{Q}$, and $\mathrm{Aut}\,\mathbb{C}$ acts on the cusps via the quotient group $\mathrm{Gal}\,(\mathbb{Q}^{ab}/\mathbb{Q})$. There is a natural identification

(2.43) $$\mathrm{Gal}(\mathbb{Q}^{ab}/\mathbb{Q}) \simeq D(\hat{\mathbb{Z}}) \quad (\simeq \prod_p \mathbb{Z}_p^*),$$

where D is the algebraic Q-group $\{\begin{pmatrix} * & 0 \\ 0 & 1 \end{pmatrix} \in GL_2\}$, and the $Gal(\mathbb{Q}^{ab}/\mathbb{Q})$-action on $S_K^{\infty}(\mathbb{C})$ corresponds precisely to the left action of $D(\hat{\mathbb{Z}})$ $(= D(\mathbb{Q})^{+}\backslash D_f)$ on $B^{+}(F)N_f(F)\backslash G_f/K$.

2.44. By the work of Rapoport there is a smooth resolution over $\mathbb{Q}$:

$$\pi_K : \tilde{S}_K \to \overline{S}_K \ . \tag{2.45}$$

We write:

$$\tilde{S}_K^{\infty} = \pi_K^{-1}(S_K^{\infty}). \tag{2.46}$$

For a clear exposition of the construction of $\tilde{S}_K$ see [HLR], §1. We would like to close this section with two remarks.

Firstly, every cusp in S_K^{∞} gives rise to a 2-dimensional $\mathbb{Q}$-vector space $H(s)$ containing a lattice $N(s)$, and a cone $C(s)$ of positive elements in the vector space $H(s) \otimes \mathbb{R}$. $N(s)$ is a normal subgroup of the stabilizer $B(s)$, and $B(s)/N(s)$ acts on $H(s)$ preserving $N(s)$ and $C(s)$. A rational polyhedral decomposition of $C(s)$ furnishes, as in §1.23, a polygon $\pi_K^{-1}(s)$ of rational curves resolving the singularity at s. The main problem in doing this over $\mathbb{Q}$ is that one needs to match up, in a compatible way, cone decompositions at every pair of cusps in the same $Gal(\mathbb{Q}^{ab}/\mathbb{Q})$-orbit.

Secondly, Rapoport's work ([Rp]) actually furnishes certain algebraic spaces $\tilde{S}_{K_{\mathbb{Z}}}$, which are models over $\mathbb{Z}$ of $\tilde{S}_K$. Unfortunately there is no published reference that they are schemes over $\mathbb{Z}$. (It is easy to check that they are schemes over a suitable open subset of $\operatorname{Spec} \mathbb{Z}$, cf. [RoT], [HLR]). For an application to p-adic L-functions, see [DR]. See also [Ka].

We would like to point out that by mimicking Chai's thesis ([Ch]) on the integral structure of Siegel varieties, and by making use of theta functions with real multiplication, it can be shown, by exhibiting an ample line bundle, that each $\tilde{S}_K$ admits a model which is actually <u>projective</u>, hence a scheme over $\mathbb{Z}$. (The details will be hopefully worked out elsewhere.) For the principal congruence subgroup K_n (see (2.38)), it can be shown that the model over $\mathbb{Z}$ is regular outside the primes dividing n. To find a model which is regular over all of Spec $\mathbb{Z}$ seems to be a very deep problem. (For the case of modular curves, see [KM].) The only thing we would like to say in this context is that the case of the rational primes p which split in F do not seem to present a problem.

§3. Decomposition of H^2 according to Hecke.

3.1. The ℓ-adic representation of $\mathrm{Gal}(\overline{\mathbb{Q}}/\mathbb{Q})$ that we are interested in is given by the natural Galois action, since $\tilde{S}_K$ is defined over $\mathbb{Q}$, on:

$$(3.2) \qquad W_\ell \stackrel{\mathrm{def}}{=} H^2_{et}(\tilde{S}_K \times_{\mathbb{Q}} \overline{\mathbb{Q}}, \mathbb{Q}_\ell) .$$

This is a finite-dimensional $\mathbb{Q}_\ell$-vector space, and we need to first understand how it can be broken up into a direct sum of indecomposable Galois modules. Set:

$$IW_\ell = \mathrm{Im}(H^2_c(S_K \times_{\mathbb{Q}} \overline{\mathbb{Q}}, \mathbb{Q}_\ell) \to W_\ell)$$

and

$$W_\ell^\infty = \{\alpha \in W_\ell \mid \alpha \text{ has support in } \tilde{S}_K^\infty \times_{\mathbb{Q}} \overline{\mathbb{Q}}\},$$

where H_c^* denotes the compactly supported cohomology. We will call IW_ℓ the "intersection piece" of W_ℓ, and it is the natural analog of the parabolic cohomology of a modular curve.

A fundamental result (see [HLR], §5.3) states that one has a decomposition:

$$(3.4) \qquad W_\ell \simeq IW_\ell \oplus W_\ell^\infty$$

as $\mathrm{Gal}(\overline{\mathbb{Q}}/\mathbb{Q})$-modules.

For a vast generalization of this decomposition theorem see [BBD].

Set:

$$W_B = H^2_B(\tilde{S}_K(\mathbb{C}), \mathbb{Q}),$$

where H_B^* signifies Betti (singular) cohomology. Then there is a natural injection

(3.6) $$j_B : W_B \to W_\ell ,$$

which induces an isomorphism of $W_B \otimes \mathbb{Q}_\ell$ with W_ℓ. Note, however, that there is no $\mathrm{Gal}(\overline{\mathbb{Q}}/\mathbb{Q})$-action on W_B, only a $\mathrm{Gal}(\mathbb{C}/\mathbb{R})$-action. (To get more structure, one usually considers the action of the Mumford-Tate group [DMOS] on W_B.)

We can define IW_B and W_B^∞ as before. Then one has a $\mathrm{Gal}(\mathbb{C}/\mathbb{R})$-decomposition ([HLR], §5.3):

$$W_B \simeq IW_B \oplus W_B^\infty .$$

As is to be expected j_B induces isomorphisms:

(3.7) $$IW_B \otimes \mathbb{Q}_\ell \xrightarrow{\ \sim\ } IW_\ell \quad \text{and} \quad W_B^\infty \otimes \mathbb{Q}_\ell \xrightarrow{\ \sim\ } W_\ell^\infty .$$

The Galois module W_ℓ^∞ has a simple structure. It decomposes as:

(3.8) $$W_\ell^\infty \simeq \bigoplus_{s \in S_K^\infty} (\mathbb{Q}_\ell(-1)^{\oplus n(s)}),$$

where $n(s)$ denotes, for each cusp s, the number of irreducible components of $\pi_K^{-1}(\{s\})$ (which is a polygon of rational curves). By $\mathbb{Q}_\ell(-1)$ we mean as usual the 1-dimensional ℓ-adic Galois module dual to the Tate module $\mathbb{Q}_\ell(1) \overset{\text{def}}{=} \mathbb{Z}_\ell(1) \otimes_{\mathbb{Z}} \mathbb{Q}$, where

$$\mathbb{Z}_\ell(1) = \varprojlim_n \mu_{\ell^n} .$$

Our next objective is to decompose IW_ℓ according to the action of the Hecke algebra.

Pick any x in G_f. Then $L \overset{\text{def}}{=} K \cap xKx^{-1}$ is also a compact open subgroup. The <u>Hecke</u> <u>correspondence</u> T_x is defined by means of the diagram:

(3.10) $$\begin{array}{ccc} & S_L & \\ {}^{R(1)}\swarrow & & \searrow^{R(x)} \\ S_K & \overset{T_x}{\dashrightarrow} & S_K \end{array}$$

where (cf. (2.21)(ii) $R(1) = J_{K,L}(1)$ and $R(x) = J_{K,L}(x)$ (well defined because $x^{-1}Lx \subset K$. We identify $R(1)$ and $R(g)$ with their graphs (algebraic cycles of codimension 2) in $S_L \times S_K$, and by T_x we mean the cycle $R(g) \circ {}^tR(1)$. It is useful to note that $R(x)$ is the map induced by right multiplication by x.

The Hecke correspondence T_x on S_K is defined over $\mathbb{Q}$. It does not in general preserve the connected components of $S_K(\mathbb{C})$. The fact which is of importance to us is that it extends to an algebraic correspondence of $\overline{S}_K$. The action on S_K^∞ can be seen to be the one induced by the right action of G_f (see (2.42)).

3.12. For every field $E \supset \mathbb{Q}$ denote by $\mathcal{H}_E = \mathcal{H}_E^K$ the algebra over E generated by the Hecke correspondences T_x. This algebra is semi-simple, and it acts on $IW_B \otimes E$ as well as on $IW_\ell \otimes E$ (commuting with the Galois action). The map $j_B \otimes E$ is equivariant for this action.

We obtain over $\overline{\mathbb{Q}}_\ell$ an isotypic decomposition:

$$(3.13) \qquad IW_\ell \otimes \overline{\mathbb{Q}}_\ell \xrightarrow{\sim} \bigoplus_\tau \overline{W}'_\ell(\tau),$$

where the (finite) direct sum is over certain irreducible $\mathcal{H}_{\overline{\mathbb{Q}}}$-representations τ. $\overline{W}'_\ell(\tau)$ can be identified with the $\mathcal{H}_{\overline{\mathbb{Q}}_\ell}$-invariants of $(IW_\ell \otimes \overline{\mathbb{Q}}_\ell) \otimes \tau$. Since $\mathcal{H}_{\overline{\mathbb{Q}}_\ell}$ is semi-simple, and since $\mathcal{H}_{\mathbb{Q}}$ preserves W_ℓ, we see that:

(3.14) (i) Every τ is defined over a number field $E(\tau)$.

(ii) $\overline{W}'_\ell(\tau)$ = extension to $E(\tau) \otimes \overline{\mathbb{Q}}_\ell$ of a finite-dimensional $E(\tau) \otimes \mathbb{Q}_\ell$-module $W'_\ell(\tau)$.

Furthermore, there is a canonical non-negative integer $m(\tau,k)$ and an $E(\tau) \otimes \mathbb{Q}_\ell$ module $W_\ell(\tau)$ such that:

$$(3.15) \qquad W'_\ell(\tau) = W_\ell(\tau)^{\oplus m(\tau,k)}.$$

In Deligne's terminology ([D3]) $W_\ell(\tau)$ will signify the ℓ-adic realization of a motive over $\mathbb{Q}$ with coefficients in $E(\tau)$, and $m(\tau,K)$ occurs classically in the study of old forms.

We will be concerned, for each τ, with the representation

(3.16) $$\rho_\ell(\tau)\colon \mathrm{Gal}(\overline{\mathbb{Q}}/\mathbb{Q}) \to GL(W_\ell(\tau))$$

It looks at this point as though the restriction of ρ_ℓ on $W_\ell(\pi)$ is not canonically defined because of the multiplicity $m(\tau,K)$. The best way to see that there is no problem is to take limits over all K and decompose

(3.17) $$IH^2_{et}(\tilde{S} \times \overline{\mathbb{Q}}, \mathbb{Q}_\ell) \overset{\text{def}}{=} \lim_K IH^2_{et}(\tilde{S}_K \otimes \bar{\mathbb{Q}}, \mathbb{Q}_\ell)$$

according to the action of the "finite adelic" Hecke algebra $H^f_{\mathbb{Q}_\ell}$ generated by all the $H^K_{\mathbb{Q}_\ell}$'s. Then there is a "multiplicity one" theorem which says that the components of $IH^2_{et}(\tilde{S} \times \overline{\mathbb{Q}},\mathbb{Q}_\ell)$ are certain representations $\overline{W}_\ell(\tau)_f$'s, each occurring exactly once, such that:

(3.18) (i) $\overline{W}_\ell(\tau)^K_f = \overline{W}'_\ell(\tau)$, $\forall K$

and (ii) $\exists$ compact open $K(\tau)$ s.t.

$$\overline{W}_\ell(\tau)^{K(\tau)}_f \simeq \overline{W}_\ell(\tau).$$

For each place λ of $E(\tau)$ dividing ℓ we get a λ-adic representation $\rho_\lambda(\tau)$ in a natural way, and the knowledge of $\{\rho_\lambda(\tau) \mid \lambda|\ell\}$ completely determines $\rho_\ell(\tau)$.

3.19. By means of the Hecke action on $W_B \otimes \overline{\mathbb{Q}}$, we also get an $E(\tau)$-vector space $W_B(\tau)$ such that $W_B(\tau) \otimes \mathbb{Q}_\ell \simeq W_\ell(\tau)$.

3.20. It may be useful to digress and briefly note that there is no nice action of $H^K_{\mathbb{Q}}$ on all of W_ℓ, just on IW_ℓ, because it is not possible to find a suitable smooth compactification $\tilde{S}_K$ on which all the Hecke correspondences can be simultaneously made to act. The problem is the compatibility of the cone decompositions at all the cusps lying in the same G_f-orbit. One can get around this problem, however, by taking a projective limit over a family of smooth compactifications of S_K.

3.21. What kinds of representation τ of $H^K_{\mathbb{Q}}$ do we have to be concerned with? Well, given any irreducible, admissible representation $\pi = \pi_\infty \otimes \pi_f$ of $G(\mathbb{A})$ $(= G_\infty \times G_f)$ (see [JL] for definition) on $V_\pi = V_{\pi_\infty} \otimes V_{\pi_f}$, then $H^K_{\mathbb{C}}$ acts naturally on $H^0(K, V_{\pi_f}) = V^K_{\pi_f}$. If V_{π_f} is defined over a number field E and if $V^K_{\pi_f} \neq 0$, then we get a candidate for τ. We say that τ is associated to π in this case.

3.22. An irreducible admissible representation $\pi = \pi_\infty \otimes \pi_f$ of $G(\mathbb{A})$ as above will be called cuspidal of weight 2 iff
(i) π is a constituent of the space of cusp forms on $G(\mathbb{A})$, and
(ii) $\pi_\infty \simeq \mathcal{D}_2 \otimes \mathcal{D}_2$, where $\mathcal{D}_2$ denotes the (lowest) discrete series resp. of $GL_2(\mathbb{R})$ with trivial central character of weight two (see [Kn] for a definition). Such a π corresponds to a classical holomorphic Hecke eigenform $h = h(\pi)$ on $\mathfrak{h}^2$ of wt.2 ([Sz]; [G],3.10). The field of definition of π_f is simply the number field E_π generated by the fourier coefficients of $h(\pi)$ ([W]).

The basic result (see [HLR]) is:

(3.23) $W_\ell(\tau) \neq 0 \implies \tau$ is associated to an irreducible admissible representation π of $G(\mathbb{A})$ which is

either one-dimensional (trivial on G^+_∞)

or cuspidal automorphic of weight 2.

We will write $W_\ell(\pi)$ instead of $W_\ell(\tau)$ if there is no confusion.

3.24. First consider the case when π is one-dimensional. Then $\pi = \nu \circ \det$, where ν is an idele class character of F. Since π is trivial on G^+_∞, ν has finite order. Let χ_π denote the restriction of ν to the idele classes of $\mathbb{Q}$. We view χ_π as a Galois character by class field theory. Then (cf. [HLR]):

$$W_\ell(\pi) = \chi_\pi(-1) \oplus \varepsilon_{F/\mathbb{Q}}\, \chi_\pi(-1), \tag{3.25}$$

where $\chi_\pi(-1)$ denotes the Tate-twisted one-dimensional Galois module and $\varepsilon_{F/\mathbb{Q}}$ denotes the quadratic character associated to the extension $F/\mathbb{Q}$.

The most interesting, and non-trivial, case is when $\pi = \pi_\infty \otimes \pi_f$ is cuspidal automorphic of weight 2 with $V^K_{\pi_f} \neq 0$. In this case (cf. [HLR]) one has:

(3.26) $W_\ell(\pi)$ is four-dimensional over $E_\pi \otimes \mathbb{Q}_\ell$.

We will investigate the structure of this Galois representation in the next two sections. Let us close this section, however, indicating why one should expect $W_\ell(\pi)$ to be four-dimensional. It suffices to look at the E_π-rank of $W_B(\pi)$, or equivalently the rank over $E_\pi \otimes \mathbb{C}$ of $W_B(\pi) \otimes \mathbb{C}$. Look at the holomorphic cusp form $h = h(\pi)$ on $\mathfrak{H}^2$ associated to π. Assume for simplicity that F has a unit u of norm -1. Then to h we can associate four (conjugate sets of) differential forms $\omega_h^i = (\omega_h^i(\sigma))_{\sigma\in\mathrm{Hom}(E_\pi,\mathbb{C})}$, $1 \leq i \leq 4$, on $\mathfrak{H}^2$, representing independent classes in $W_B(\pi) \otimes \mathbb{C}$, given by:

(3.27)
$$\omega_h^1(\sigma) = (2\pi i)^2\, h^\sigma(z_1,z_2)dz_1 \wedge dz_2$$
$$\omega_h^2(\sigma) = (2\pi i)^2\, h^\sigma(uz_1,u^\theta\bar{z}_2)dz_1 \wedge d\bar{z}_2$$
$$\omega_h^3(\sigma) = (2\pi i)^2\, h^\sigma(u^\theta\bar{z}_1,uz_2)d\bar{z}_1 \wedge dz_2$$
and
$$\omega_h^4(\sigma) = (2\pi i)^2\, h^\sigma(-\bar{z}_1,-\bar{z}_2)d\bar{z}_1 \wedge d\bar{z}_2 .$$

Recall that in the elliptic modular case every cusp form of weight two gives rise to two independent differentials, one holomorphic and the other anti-holomorphic.

A more functorial way will be to relate $W_B(\pi) \otimes \mathbb{C}$ to the relative (Lie algebra) H^2 of π_∞, which is four-dimensional (see [HLR] and [Cs]).

§4. L-functions.

4.1. Following Serre ([S1]), see also Deligne [D3], we can attach to any compatible system $V. = \{V_\ell\}$ of finite-dimensional ℓ-adic representations of $\mathrm{Gal}(\bar{\mathbb{Q}}/\mathbb{Q})$ an Euler product:

(4.2)
$$L(V.,s) = \prod_p Z_p(V.,p^{-s})^{-1},$$

where

(4.3)
$$Z_p(V.,T) = \det(1-F_pT|V_\ell^{I_p}), \text{ for some } \ell \neq p.$$

Here F_p denotes the geometric Frobenius at p and I_p the inertia subgroup of the decomposition group of p. For this definition to be meaningful we need to know, of course, that the characteristic polynomial of F_p acting on the inertial invariants of V_ℓ is independent of the choice of ℓ ($\neq p$).

4.4. $L(W.,s)$ is, for our $W. = \{W_\ell\}$, simply the H^2-piece of the Hasse-Weil Zeta function of $\tilde{S}_K$.

$\tilde{S}_K$ has good (non-singular) reduction at almost all p. At such a p V_ℓ is unramifield, i.e., $V_\ell^{I_p} = V_\ell$, for every $\ell \neq p$, and by Deligne's proof of the Riemann hypothesis ([D4]) for varieties over finite fields one has:

$$(4.5) \qquad Z_p(W.,T) = \prod_{i=1}^{b_2} (1-\alpha_i T) \in \mathbb{Z}[T] ,$$

with each α_i an algebraic integer (independent of ℓ) of absolute value p. Here b_2 denotes the second Betti number $\dim_Q W$.

A consequence of Deligne's theorem is:

(4.6) The Euler product $\prod_{p:\text{good}} Z_p(W.,p^{-s})^{-1}$ converges absolutely in the right-half plane $\{s \in \mathbb{C} | \mathrm{Re}(s) > 2\}$.

4.7. Serre conjectures that for every p the inverse roots $\{\lambda_\alpha\}$ of F_p on $V_\ell^{I_p}$ satisfy: $|\lambda_\alpha| = p^{j_\alpha/2}$ for some j_α between 0 and 2.

4.8. We can attach an Euler factor at infinity as follows. Let c denote complex conjugation, and let F_∞ denote the "Frobenius at ∞" acting on $\tilde{S}_K(\mathbb{C})$ induced by c. If A is any subring of $\mathbb{C}$ invariant under c, then the "real involution" $F_\infty \otimes c$ acts on $W_B \otimes A$. For any $F_\infty \otimes c$-space X let $X^\pm$ denote the (± 1)-eigenspace for this action. We have Hodge decomposition:

$$(4.9) \qquad W_B \otimes \mathbb{C} = H^{2,0}(W_B \otimes \mathbb{C}) \oplus H^{1,1}(W_B \otimes \mathbb{C}) \oplus H^{0,2}(W_B \otimes \mathbb{C}).$$

(4.10) Set: $H^{1,1}(W_B \otimes \mathbb{R}(j)) = H^{1,1}(W_B \otimes \mathbb{C}) \cap W_B \otimes \mathbb{R}(j)$, $\forall j$, where $\mathbb{R}(j) = (2\pi i)^j \mathbb{R} \subset \mathbb{C}$. Write:

$$h^{p,q} = \dim_{\mathbb{C}} H^{p,q}(W_B \otimes \mathbb{C}),\ h^{1,\pm} = \dim_{\mathbb{R}} H^{1,1}(W_B \otimes \mathbb{R}(1))^{\pm},$$

$$\Gamma_{\mathbb{R}}(s) = \pi^{-s/2}\Gamma(s/2) \quad \underline{\text{and}} \quad \Gamma_{\mathbb{C}}(s) = 2(2\pi)^{-s}\Gamma(s).$$

Then the desired factor at infinity is:

(4.11) $$L_\infty(W.,s) = \Gamma_{\mathbb{R}}(s)^{h^{1,+}} \cdot \Gamma_{\mathbb{R}}(s)^{h^{1,-}} \cdot \Gamma_{\mathbb{C}}(s)^{h^{0,2}}.$$

<u>4.12</u>. Combining the results of Brylinski-Labesse ([BL]) and of Asai [A] (cf. [J] and [HLR], §3, for an adelic formulation), we see that there exists a finite set Q of primes such that:

(i) $L_Q(W.,s) = \prod_{p \notin Q} Z_p(W.,p^{-s})^{-1}$ admits a meromorphic continuation to the whole s-plane with no pole except possibly at the "edge" $s = 2$;

(ii) $L_Q^*(W.,s) \overset{\text{def}}{=} L_Q(W.,s)L_\infty(W.,s)$ admits a functional equation:

$$L_Q^*(W.,s) = \varepsilon_Q(W.,s)L_Q^*(W.,3-s),$$

where the "ε-function" $\varepsilon_Q(W.,s)$ admits a factorization: $\varepsilon_\infty(W.,s)\cdot \prod_{p\notin Q} \varepsilon_p(W.,s)$. (See [D5] for a definition of the ε-factors.)

To understand this in a little bit more detail, let us factor $L(W.,s)$ according to the decomposition of the Galois module W_ℓ. Firstly we have:

<u>4.13</u>. $$L(W.,s) = L(IW.,s)L(W_.^\infty,s)$$

and

$$L(W_.^\infty,s) = \prod_{s\in S_K^\infty} \prod_{j=1}^{n(s)} \zeta(s-1).$$

<u>4.14</u>. Second, $L(IW.,s)$ is completely determined by the collection $\{L(W.(\pi),s) \mid W_\ell(\pi) \neq 0\}$. If <u>$\pi$ is one-dimensional</u> we have, by (3.25):

(4.15) $$L(W.(\pi),s) = L(\chi_\pi, s-1)L(\chi_\pi \cdot \varepsilon_{F/\mathbb{Q}}, s-1),$$

where χ_π is a Dirichlet character of $\mathbb{Q}$.

4.16. From now on let us assume that π is automorphic cuspidal of weight 2.

If E_π denotes the field of definiton of π we view $L(W.(\pi),s) = (L(W.(\pi^\sigma),s))_{\sigma\in\mathrm{Hom}(E_\pi,\mathbb{C})}$ as an $E_\pi \otimes \mathbb{C}$-valued function.

Let us recall the definition of the Langlands dual of G:

(4.17) $${}^LG = (GL_2\mathbb{C} \times GL_2\mathbb{C}) \ltimes \mathrm{Gal}(F/\mathbb{Q}),$$

where $\mathrm{Gal}(F/\mathbb{Q})$ acts on $GL_2\mathbb{C} \times GL_2\mathbb{C}$ by permuting the factors. There is a natural representation

(4.18) $$\rho: {}^LG \to GL(\mathbb{C}^2 \otimes \mathbb{C}^2) \simeq GL_4(\mathbb{C})$$

defined by:

$$\rho(g_1,g_2;1)(v_1 \otimes v_2) = g_1(v_1) \otimes g_2(v_2),$$

and

$$\rho(1,1;\theta)(v_1 \otimes v_2) = v_2 \otimes v_1 ,$$

for all v_1, v_2 in $\mathbb{C}^2$ and g_1, g_2 in $GL_2(\mathbb{C})$.

To the pair (π,ρ) Langlands attaches an Euler product

(4.19) $$L(\pi,\rho,s) = \prod_p L_p(\pi,\rho,s),$$

and an Euler factor $L_\infty(\pi,\rho,s)$ at infinity. By definition,

(4.20) $$L_p(\pi,\rho,s) = \det(1-\rho(A(\pi_p))p^{-s+1})^{-1},$$

where $A(\pi_p)$ is the conjugacy class in ${}^LG$ associated to π_p by functoriality ([L2]), [HLR]).

Let us describe $\rho(A(\pi_p))$ assuming that p is unramified in F and that if $\mathfrak{p}$ is a prime in F dividing p then π is an unramified principal series representation $\pi(\alpha_\mathfrak{p}, \beta_\mathfrak{p})$ (see [HLR]). If p splits in F as a product of two primes $\mathfrak{p}$, $\mathfrak{p}'$, then:

$$(4.21) \qquad \rho(A(\pi_p)) \sim \begin{pmatrix} {}^{\alpha}\mathfrak{p}\,{}^{\alpha}\mathfrak{p}' & & & \\ & {}^{\alpha}\mathfrak{p}'\,{}^{\beta}\mathfrak{p} & & \\ & & {}^{\alpha}\mathfrak{p}\,{}^{\beta}\mathfrak{p}' & \\ & & & {}^{\alpha}\mathfrak{p}'\,{}^{\beta}\mathfrak{p}' \end{pmatrix}$$

And <u>if p remains prime</u> $(= \mathfrak{p})$ in F, then:

$$(4.22) \qquad \rho(A(\pi_p)) \sim \begin{pmatrix} {}^{\alpha}\mathfrak{p} & 0 & 0 & 0 \\ 0 & 0 & {}^{\alpha}\mathfrak{p} & 0 \\ 0 & {}^{\beta}\mathfrak{p} & 0 & 0 \\ 0 & 0 & 0 & {}^{\beta}\mathfrak{p} \end{pmatrix}$$

<u>4.23</u>. It is very important to note that:

$$L(\pi,\rho,s) \neq L(\pi,s),$$

where $L(\pi,s)$ is the <u>standard</u> Jacquet-Langlands L-function attached to π ([JL]). To see this a little bit more intuitively, take F to have class number 1. Write

$$(4.24) \qquad L(\pi,s) = \sum_{\mathfrak{a}}{}' C(\mathfrak{a})N(\mathfrak{a})^{-s},$$

where the sum is over non-zero integral ideals of $\mathcal{O}_F$. Then we have:

$$(4.25) \qquad L(\pi,\rho,s) = \zeta(2s-2) \sum_{m\geq 1} C(m\,\mathcal{O}_F)m^{-s}.$$

The Euler factor at infinity is:

$$(4.26) \qquad L_\infty(\pi,\rho,s) = (2\pi)^{-2s}\Gamma(s)\Gamma(s-1).$$

The main theorem of Brylinski-Labesse ([BL]) is that:

$$(4.27) \qquad L_Q(W.(\pi),s) = L_Q(\pi,\rho,s).$$

We view both sides as an $E_\pi \otimes \mathbb{C}$-valued function. The Euler factors at infinity are also the same.

4.28. From what has been said so far it is not clear why $L_{\mathbb{Q}}(\pi,\rho,s)$ admits a meromorphic continuation to the whole plane. To see this we appeal to a theorem of Asai ([A]) which says that there is an integral representation:

$$L_Q(\pi,\rho,s) = \zeta(2s-2)\cdot\int_{H(\mathbb{Q})Z_H(\mathbb{A})\backslash H(\mathbb{A})}\phi_\pi(h)\mathcal{E}(h,s-1)dh, \tag{4.29}$$

where $H = GL_2$, Z_H = center of H, ϕ_π: a suitable function in the space V_π of π, and $\mathcal{E}(h,s)$ an Eisenstein series on $GL_2(\mathbb{A})$ of weight 0. See [HLR], §3, for details. See also [Sh3], Th.3.11.

The desired meromorphic continuation and functional equation (relating s to $3-s$) now follow from the known analytic properties of Eisenstein series.

Finally, we remark that if ε' is any idele class character of F whose restriction to $\mathbb{A}^*/\mathbb{Q}^*$ corresponds to $\varepsilon_{F/\mathbb{Q}}$, then one has the identity:

$$L(\pi,\rho,s)L(\pi\otimes\varepsilon',\rho,s) = L(\pi\times\pi^\theta,s), \tag{4.30}$$

where the L-function on the right is the Rankin type L-function attached to the pair (π,π^θ), π^θ = the θ-conjugate of π, by Jacquet in [J].

§5. Structure of the Galois representation $(\rho_\ell, W_\ell(\pi))$.

5.1. The object of this section is to understand better the structure of the ℓ-adic representation:

$$\rho_\ell : \mathrm{Gal}(\overline{\mathbb{Q}}/\mathbb{Q}) \to GL(W_\ell(\pi)),$$

associated to a cuspidal automorphic representation π of $G(\mathbb{A})$ $(= GL_2(\mathbb{A}_F))$ of weight two. Here ρ_ℓ denotes more precisely the restriction of ρ_ℓ to the component $W_\ell(\pi)$ of IW_ℓ.

5.2. Recall that $W_\ell(\pi)$ is 4-dimensional over $E_\pi\otimes\mathbb{Q}_\ell$, where E_π denotes the field of definition of π, or equivalently the field generated by the field of Fourier coefficients of the holomorphic cusp form h_π which "generates" π. The principle of functoriality due to Langlands ([L2]) suggests, on the other hand, that every "motivic" automorphic representation of $GL_2(\mathbb{A}_F)$ such as our π should correspond to a compatible system of 2-dimensional ℓ-adic representations of $\mathrm{Gal}(\overline{\mathbb{Q}}/F)$. See also Oda [Od1], where a precise Hodge-theoretic version of this is proposed and analyzed. Specifically one hopes for the truth of the following:

Conjecture 5.3. (i) To every irreducible cuspidal automorphic representation π of $GL_2(\mathbb{A}_F)$ there should correspond a compatible system of irreducible representations

$$\tau_\ell : \mathrm{Gal}(\overline{\mathbb{Q}}/F) \to GL(V_\ell(\pi)),$$

with $V_\ell(\pi)$ being 2-dimensional over $E_\pi \otimes \mathbb{Q}_\ell$, such that:

$$L(\pi,s) = L(V.(\pi),s) \ ;$$

(ii) When π is not associated ([JL]) to a grossen-character of a quartic CM field containing F, $V_\ell(\pi)$ should remain irreducible when restricted to any open subgroup of $\mathrm{Gal}(\overline{\mathbb{Q}}/F)$; and

(iii) There should be an isomorphism as $\mathrm{Gal}(\overline{\mathbb{Q}}/F)$-modules:

$$W_\ell(\pi)\Big|_{\mathrm{Gal}(\overline{\mathbb{Q}}/F)} \xrightarrow{\sim} V_\ell(\pi) \otimes V_\ell(\pi)^\theta .$$

5.4. This conjecture is still very much open, even though, as we will see presently, one has (cf. [HLR]) a reasonable candidate $V_\ell(\pi)$ for each cuspidal π of weight two. The hitch is that one does not see any way of proving the global L-function identity of (i), though one has some local information (cf. (5.24)).

5.5. Let us first see how, given that the hypothetic 2-dimensional $\mathrm{Gal}(\overline{\mathbb{Q}}/F)$-module $V_\ell(\pi)$ exists, how one might go about constructing the 4-dimensional $\mathrm{Gal}(\overline{\mathbb{Q}}/\mathbb{Q})$-module $W_\ell(\pi)$. Intuitively this is clear because giving a homomorphism τ_ℓ of $\mathrm{Gal}(\overline{\mathbb{Q}}/F)$ into $GL_2(E_\pi \otimes \mathbb{Q}_\ell)$ is tantamount to having a homomorphism $\tilde{\tau}_\ell$

of $\mathrm{Gal}(\overline{\mathbb{Q}}/\mathbb{Q})$ into $(GL_2(E_\pi \otimes \mathbb{Q}_\ell) \times GL_2(E_\pi \otimes \mathbb{Q}_\ell)) \rtimes (\mathrm{Gal}(F/\mathbb{Q}))$, and composing $\tilde{\tau}_\ell$ with ρ (see (4.18) for definition) we get a morphism into $GL_4(E_\pi \otimes \mathbb{Q}_\ell)$. Let us make this construction explicit as follows:

We can write:

$$\mathrm{Gal}(\overline{\mathbb{Q}}/\mathbb{Q}) = \mathrm{Gal}(\overline{\mathbb{Q}}/F) \cup \tilde{\theta}\, \mathrm{Gal}(\overline{\mathbb{Q}}/F),$$

where $\tilde{\theta}$ is a lift of the non-trivial automorphism θ of F. Define:

$$(5.6) \qquad \tilde{\tau}_\ell : \mathrm{Gal}(\overline{\mathbb{Q}}/\mathbb{Q}) \to (GL_2(E_\pi \otimes \mathbb{Q}_\ell) \times GL_2(E_\pi \otimes \mathbb{Q}_\ell)) \rtimes \mathrm{Gal}(F/\mathbb{Q})$$

by (for σ in $\mathrm{Gal}(\overline{\mathbb{Q}}/F)$):

$$\tilde{\tau}_\ell(\sigma) = (\tau_\ell(\sigma), \tau_\ell(\sigma); 1)$$

and

$$\tilde{\tau}_\ell(\tilde{\theta}) = (1, \tau_\ell(\tilde{\theta}^2); \theta).$$

Since θ has order 2, $\tilde{\theta}^2$ will lie in $\mathrm{Gal}(\overline{\mathbb{Q}}/F)$ and so $\tau_\ell(\tilde{\theta}^2)$ makes sense. Also we have: $\tilde{\tau}_\ell(\tilde{\theta}^2) = (\tau_\ell(\tilde{\theta}^2), \tau_\ell(\tilde{\theta}^2); 1)$, as it should be. Now set:

$$(5.7) \qquad \rho_\ell = \rho \circ \tilde{\tau}_\ell \, .$$

It is easy to see that the naturality of the construction of ρ_ℓ yields:

$$(5.8) \qquad \rho_\ell \big|_{\mathrm{Gal}(\overline{\mathbb{Q}}/F)} = \tau_\ell \otimes \tau_\ell^\theta \, .$$

It can also be seen that $\Lambda^2(\mathrm{Ind}_{\mathrm{Gal}(\overline{\mathbb{Q}}/F)}^{\mathrm{Gal}(\overline{\mathbb{Q}}/\mathbb{Q})}(\tau_\ell))$ decomposes as a direct sum of ρ_ℓ and $\mathrm{Ind}_{\mathrm{Gal}(\overline{\mathbb{Q}}/F)}^{\mathrm{Gal}(\overline{\mathbb{Q}}/\mathbb{Q})}(\Lambda^2(\tau_\ell))$.

To reverse this process one needs to devise a way to go from a homomorphism of $\mathrm{Gal}(\overline{\mathbb{Q}}/\mathbb{Q})$ into $GL_4(E_\pi \otimes \mathbb{Q}_\ell)$ to a homomorphism of $\mathrm{Gal}(\overline{\mathbb{Q}}/F)$ into $GL_2(E_\pi \otimes \mathbb{Q}_\ell) \times GL_2(E_\pi \otimes \mathbb{Q}_\ell)$. As shown by Harder, Langlands and Rapoport in [HLR], §4, this can be done. The key point is that $GL_2 \times GL_2$ is very close to being an orthogonal group in four variables. With this in mind one is led to show:

(5.9) There is a natural non-degenerate quadratic form q on $W_\ell(\pi)$ such that the image of ρ_ℓ is contained in $GO(W_\ell(\pi),q)$.

By $GO(W_\ell(\pi),q)$ we mean the group of orthogonal similitudes on $W_\ell(\pi)$ relative to q.

The existence of q comes from the fact that there is a natural pairing

(5.10) $$\mathcal{B}_\ell : H^2_{et}(\tilde{S}_K \times \overline{\mathbb{Q}}, \mathbb{Q}_\ell) \times H^2_{et}(\tilde{S}_K \times \overline{\mathbb{Q}}, \mathbb{Q}_\ell) \to H^4_{et}(\tilde{S}_K \times \overline{\mathbb{Q}}, \mathbb{Q}_\ell),$$

defined by taking cup-products of pairs of (ℓ-adic) cohomology classes ([Mi]). This bilinear form is symmetric and non-degenerate. It is the ℓ-adic analog of the more familiar pairing in Betti (singular) cohomology corresponding, by Poincaré duality, to the intersection pairing on homology.

5.11. Since one has isomorphisms: $H^4_{et}(\tilde{S}_K \times \overline{\mathbb{Q}}, \mathbb{Q}_\ell) \simeq H^4_B(\tilde{S}_K(\mathbb{C}), \mathbb{Z}) \otimes \mathbb{Q}_\ell \simeq H_{0,B}(\tilde{S}_K(\mathbb{C}), \mathbb{Z}) \otimes \mathbb{Q}_\ell$, we see that if $\tilde{S}_K(\mathbb{C})$ were connected the $\mathbb{Q}_\ell$-vector space $H^4_{et}(\tilde{S}_K \times \overline{\mathbb{Q}}, \mathbb{Q}_\ell)$ would be one-dimensional. This is not in general the case. To get around this consider first the restriction of $\mathcal{B}_\ell$ to $IW_\ell \times IW_\ell$, and note that this (restricted) pairing is equivariant under the action of Hecke correspondences. It is a fact that $H^4(\tilde{S}_K \times \overline{\mathbb{Q}}, \mathbb{Q}_\ell)$ is the same as the intersection cohomology group $IH^4_{et}(\overline{S}_K \times \overline{\mathbb{Q}}, \mathbb{Q}_\ell)$ (of Goresky-MacPherson and Deligne [BBD]). So the Hecke algebra acts nicely on this group (since it depends only on $\overline{S}_K$ and not on $\tilde{S}_K$). Taking isotypic components according to the action of the Hecke algebra we get:

(5.12) $$\mathcal{B}_\ell(\pi) : W_\ell(\pi) \times W_\ell(\pi) \to Y_\ell(\pi),$$

where $Y_\ell(\pi)$ denotes the isotypic component, corresponding to π, of $IH^4_{et}(\overline{S}_K \times \overline{\mathbb{Q}}, \mathbb{Q}_\ell)$.

We next claim that $Y_\ell(\pi)$ is one-dimensional. This is because otherwise it would result in a bigger-than-one-dimensional space of G_f-invariant bilinear forms on the space of π_f, and this would violate Schur's lemma (since π_f is irreducible). Now using the fact that the contragredient $\pi^\vee$ satisfies:

(5.13) $$\pi^\vee \simeq \pi \otimes \omega_\pi^{-1} ,$$

where ω_π denotes the central character, one deduces that the character by which the Hecke algebra acts on the one-dimensional space $Y_\ell(\pi)$ is associated to the one-dimensional G_f-module $\omega_{\pi_f} \circ \det$. What is needed, however, is to determine how $\mathrm{Gal}(\overline{\mathbb{Q}}/\mathbb{Q})$ acts on $Y_\ell(\pi)$. Before doing it note that, if we view ω_π as a character of $\mathrm{Gal}(\overline{\mathbb{Q}}/F)$, we can compose it with the transfer map from $\mathbb{Q}$ to F to get a character χ_π of $\mathrm{Gal}(\overline{\mathbb{Q}}/\mathbb{Q})$, which is analogous to the way we saw how to construct ρ_ℓ from τ_ℓ. It can be checked that the idèle class character of $\mathbb{Q}$ corresponding to χ_π, again denoted by the same symbol by abuse of notation, satisfies:

$$(5.14) \qquad \chi_\pi = \omega_\pi \Big|_{\mathbb{A}_{\mathbb{Q}}^*} .$$

Now one appeals to the description of the way $\mathrm{Gal}(\overline{\mathbb{Q}}/\mathbb{Q})$ acts (via its abelian quotient) on $\pi_0(S_K \times \overline{\mathbb{Q}})$ (see (2.32)), and deduces an identification as Galois modules:

$$(5.15) \qquad Y_\ell(\pi) = \chi_\pi(-2).$$

The desired quadratic form q is the one associated to $\mathcal{B}_\ell(\pi)$. This proves (5.9).

Note that:

$$(5.16) \qquad GO(W_\ell(\pi),q) = \{g \in GL(W_\ell(\pi)) \,|\, q(gx,gy) = \mu(g)q(x,y)\}.$$

The "multiplier" μ is actually a homomorphism of $GO(W_\ell(\pi),q)$ into the multiplicative group such that μ^2 differs from the determinant by ± 1. Identifying $\mathrm{Gal}(F/\mathbb{Q})$ with $\{\pm 1\}$ one gets (as in [HLR], §4) a commutative diagram with exact rows:

$$(5.17) \qquad \begin{array}{ccccccccc} 1 & \longrightarrow & SGO(W_\ell(\pi),q) & \longrightarrow & GO(W_\ell(\pi),q) & \xrightarrow{\nu} & \{\pm 1\} & \longrightarrow & 1 \\ & & \uparrow \rho_\ell^F & & \uparrow \rho_\ell & & \| & & \\ 1 & \longrightarrow & \mathrm{Gal}(\overline{\mathbb{Q}}/F) & \longrightarrow & \mathrm{Gal}(\overline{\mathbb{Q}}/\mathbb{Q}) & \longrightarrow & \mathrm{Gal}(F/\mathbb{Q}) & \rightarrow & 1 \end{array}$$

Here ν denotes the "similitude norm" map: $g \mapsto \mu(g)^2/\det(g)$, and

$$(5.18) \qquad \rho_\ell^F : \mathrm{Gal}(\overline{\mathbb{Q}}/F) \to \mathrm{SGO}(W_\ell(\pi), q)$$

the restriction of ρ_ℓ to $\mathrm{Gal}(\overline{\mathbb{Q}}/F)$. "SGO" stands for the group of orthogonal similitudes of similitude norm 1. It is bigger than the special orthogonal group SO.

5.19. For each place λ dividing ℓ we get a λ-adic representation ρ_λ^F. As always, the knowledge of $\{\rho_\lambda^F \mid \lambda \mid \ell\}$ determines ρ_ℓ^F. Let us now view ρ_λ^F as the representation on a four-dimensional vector space over $\overline{E}_\lambda$, the algebraic closure (viewed as a subfield of $\mathbb{C}$) of the completion of E_π at λ.

It is possible to determine an isomorphism of $\overline{W}_\lambda(\pi)$ onto $M_2(\overline{E}_\lambda)$ such that the following diagram commutes:

$$(5.20) \qquad \begin{array}{ccc} & \overline{E}_\lambda^* & \\ {\scriptstyle \det \times \det} \nearrow & & \nwarrow {\scriptstyle \mu} \\ GL_2(\overline{E}_\lambda) \times GL_2(\overline{E}_\lambda) & \xrightarrow{\beta} & \mathrm{SGO}(\overline{W}_\lambda(\pi)) \end{array},$$

where β denotes the map: $(A,B) \mapsto \{M \to AM^tB\}$. From this we deduce a short exact sequence:

$$(5.21) \qquad 1 \longrightarrow \overline{E}_\lambda^* \xrightarrow{j} GL_2(\overline{E}_\lambda) \times GL_2(\overline{E}_\lambda) \longrightarrow \mathrm{SGO}(\overline{W}_\lambda(\pi)) \longrightarrow 1,$$

where j denotes the map: $x \mapsto (x, x^{-1})$. This gives us in turn a long exact cohomology sequence (for the <u>trivial</u> action of $\mathrm{Gal}(\overline{\mathbb{Q}}/F)$):

$$(5.22) \qquad \ldots \to \mathrm{Hom}(\mathrm{Gal}(\overline{\mathbb{Q}}/F), \overline{E}_\lambda^*) \to \mathrm{Hom}(\mathrm{Gal}(\overline{\mathbb{Q}}/F), GL_2(\overline{E}_\lambda) \times GL_2(\overline{E}_\lambda)) \to$$
$$\mathrm{Hom}(\mathrm{Gal}(\overline{\mathbb{Q}}/F), \mathrm{SGO}(\overline{W}_\lambda(\pi))) \to H^2(\mathrm{Gal}(\overline{\mathbb{Q}}/F), \overline{E}_\lambda^*) \to \ldots$$

By a theorem of Tate (see Serre [S2], §6) one knows that $H^2(\mathrm{Gal}(\overline{\mathbb{Q}}/F), \overline{E}_\lambda^*) = 0$. This shows (for each $\lambda \mid \ell$):

(5.23) (i) ρ_λ^F can be lifted to a homomorphism $(\tau_\lambda, \tau'_\lambda)$ of $\mathrm{Gal}(\overline{\mathbb{Q}}/F)$ into $GL_2(\overline{E}_\lambda) \times GL_2(\overline{E}_\lambda)$.

(ii) If $(\tau_\lambda, \tau'_\lambda)$ and $(\tau''_\lambda, \tau'''_\lambda)$ are two different liftings, then there exists a character $\beta_\lambda : \mathrm{Gal}(\overline{\mathbb{Q}}/F) \to \overline{E}_\lambda^*$ such that $\tau''_\lambda = \tau_\lambda \otimes \beta_\lambda$ and $\tau'''_\lambda = \tau'_\lambda \otimes \beta_\lambda^{-1}$.

It can be seen that:

(5.24) $\exists$ lifting $(\tau_\lambda, \tau'_\lambda)$ of ρ_λ^F such that $\tau'_\lambda = \tau_\lambda^\theta$.

It is natural to hope for this choice of lifting that the ℓ-adic representation $(\tau_\ell, V_\ell(\pi))$ determined by $\{\tau_\lambda \mid \lambda | \ell\}$ gives an affirmative answer to the conjecture (5.3). But the best that can be proven (so far) is only that:

(5.25) (i) $\rho_\ell^F \simeq \tau_\ell \otimes \tau_\ell^\theta$;

(ii) τ_ℓ is irreducible, and if π is not of type CM, τ_ℓ remains irreducible when restricted to any open subgroup;

and (iii) At almost all places v of F,

$$L_v(V_\ell(\pi)^{\delta_v}, s) = L_v(\pi, s),$$

where δ_v is either 1 or θ.

<u>5.26</u>. We say that π is of <u>type CM</u> iff π is associated to an algebraic Hecke character Ω (necessarily of weight 1) ([JL]) of a quartic CM field L containing F. In this case the representation τ_ℓ sought can be explicitly constructed by setting:

$$\tau_\ell = \mathrm{Ind}_{\mathrm{Gal}(\overline{\mathbb{Q}}/L)}^{\mathrm{Gal}(\overline{\mathbb{Q}}/F)} (\Omega) , \tag{5.27}$$

identifying Ω with the ℓ-adic Galois character of $\overline{\mathbb{Q}}$ over L defined by it. The fact that π is cuspidal implies that Ω is not fixed by $\mathrm{Aut}(L/F)$. This in turn implies that τ_ℓ is irreducible as a 2-dimensional $\mathrm{Gal}(\overline{\mathbb{Q}}/F)$-module. It is, however, clear (in this case) that the restriction of τ_ℓ to $\mathrm{Gal}(\overline{\mathbb{Q}}/L)$ is reducible.

5.28. Part (i) of (5.25) gives the Galois module analog of the L-function identity (4.30).

5.29. To prove part (ii) one exploits an idea of Ribet ([Ri1], Th. 2.3). Suppose τ_ℓ is reducible. Then its semi-simplication τ_ℓ^{ss} should be a sum of two ℓ-adic characters ν_1 and ν_2. If χ_ℓ denotes the ℓ-adic cyclotomic character, using a theorem of Serre and Lang ([S4], Ch.III), and the fact that F is totally real, we can write each ν_i as: $\chi_\ell^{r_i}\beta_i$, where χ_ℓ denotes the ℓ-adic cyclotomic character and each β_i a character of finite order. One checks that $\nu_1\nu_2 = \chi_\ell^{-1}\mu$ with the consequence: $r_1 = r_2 = -1/2$.

Now look at the exterior square representation of ρ_ℓ^F. Then:

$$(5.30)\qquad \Lambda^2(\rho_\ell^F) \simeq \mathrm{sym}^2(\tau_\ell) \otimes \Lambda^2(\tau_\ell^\theta) \oplus \mathrm{sym}^2(\tau_\ell^\theta) \otimes \Lambda^2(\tau_\ell).$$

Then by using: $\tau_\ell^{ss} \approx \nu_1 \oplus \nu_2$ we can write $L(\Lambda^2(\rho_\ell^F),s)$ as a product of 4 abelian L-functions. Furthermore, replacing τ_ℓ by $\tau_\ell \otimes \det \tau_\ell^{-1}$, which is harmless, we can show, using the explicit nature of ν_1 and ν_2, that $L(\Lambda^2(\rho_\ell^F),s)$ has a pole of order 2 at the edge of the critical strip. But on the other hand, we have:

$$(5.31)\qquad L(\Lambda^2(\rho_\ell^F),s) = L(\mathrm{sym}^2(\pi),s)L(\mathrm{sym}^2(\pi^\theta),s) .$$

By the results of [GJ] and [Sd] on the sym^2 L-functions of cuspidal representations of $GL_2(\mathbb{A}_F)$, we see that the order of pole of the right hand side of (5.31) at the edge of the critical strip cannot be 2. This gives the desired contradiction.

5.32. At almost all places v of F, given the Hecke eigenvalue c(v) of π, one can check only that c(v) is the Frobenius eigenvalue at v of either τ_ℓ or τ_ℓ^θ. This is very nice, but it does not suffice to prove conjecture 5.3, because it is apriori possible for, say, half the eigenvalues c(v) to be associated to τ_ℓ and the other half to τ_ℓ^θ. This perhaps points up the limitations of using Hilbert modular surfaces. However, if π is essentially equivalent to π^θ (more on it in §6 and §9), then τ_ℓ can be identified with the twist by a character of the restriction to $\mathrm{Gal}(\overline{\mathbb{Q}}/F)$ of a 2-dimensional representation of $\mathrm{Gal}(\overline{\mathbb{Q}}/\mathbb{Q})$.

5.33. Let me conclude this section by suggesting an alternate approach which, given a $\pi \not\approx \pi^\theta$ such that ω_π is invariant under θ, one can perhaps use in the quest for the "right motive" $\{\tau_\ell, \tau_B\}$. Since π is cuspidal and not equivalent to π^θ, one can lift π (cf.[JPSS], [V]) to a cuspidal automorphic representation Π of $GSP_4(\mathbb{A})$ such that:

$$(5.34) \qquad L(\pi,s) = L(\Pi,s),$$

with Π having a Whittaker model. Explicit computations show that the L-packet at infinity contains, however, two elements $\{\Pi_\infty, \Pi'_\infty\}$ with Π'_∞ a holomorphic limit of discrete series representation (corresponding to a holomorphic Siegel modular form of weight two). Appealing to a certain conjecture of Arthur ([Ar]) we see that (at least conjecturally) there should be a global automorphic cuspidal representation Π' of $GSP_4(\mathbb{A})$, having the same L-function as Π, such that the infinity-component of Π' is isomorphic to Π'_∞. We can thus attach to π (classical) holomorphic genus 2 Siegel modular cusp form ϕ of weight 2.* By multiplying ϕ by a suitable Eisenstein series and by thus jacking up the weight, we should get into cohomology. By partially mimicking Deligne-Serre ([DS]), it will be interesting to attach a 4-dimensional ℓ-adic representation σ_ℓ of $\mathrm{Gal}(\overline{\mathbb{Q}}/\mathbb{Q})$ such that:

$$(5.35) \qquad \sigma_\ell \Big|_{\mathrm{Gal}(\overline{\mathbb{Q}}/F)} \xrightarrow{\sim} \tau_\ell \oplus \tau_\ell^\theta .$$

Note that it is $\tau_\ell \oplus \tau_\ell^\theta$ and <u>not</u> $\tau_\ell \otimes \tau_\ell^\theta$ which occurs here!

§6. Elliptic curves over F.

6.1. Given a cuspidal automorphic representation π of $GL_2(\mathbb{A}_F)$ of weight two with E_π = real, we can ask, following Oda [Od1], §20, if the (conjecturally correct) compatible family $\tau. = \{\tau_\ell\}$ of ℓ-adic representations arises as the first ℓ-adic

*The situation is the same for a π coming from GL(2) of an imaginary quadratic field, for suitable infinity-type of π.

cohomology of an abelian variety A_π over F of dimension $[E_\pi:\mathbb{Q}]$, admitting multiplication by E_π. Then we would have:

$$L(\pi,s) = L(A_\pi,s), \tag{6.2}$$

the L-function on the right being attached, as usual, to the $\mathrm{Gal}(\overline{\mathbb{Q}}/F)$-module $H^1_{et}(A_\pi \times_F \overline{\mathbb{Q}},\mathbb{Q}_\ell)$, which is dual to the Tate module of A_π.

6.3. Now suppose that $E_\pi = \mathbb{Q}$. In other words, the Hecke eigenvalues of π are all rational. Then π should naturally give rise to an elliptic curve A_π over F. It is reasonable to hope for the converse as well, and (as in [Od1]) make the following bold

Conjecture 6.4. There is a natural bijection preserving L-functions:

$$\left\{\begin{matrix}\text{Cuspidal automorphic reps. } \pi \\ \text{of weight 2 such that} \\ E_\pi = \mathbb{Q}\end{matrix}\right\} \longleftrightarrow \left\{\begin{matrix}\text{Elliptic curves}/F \\ \text{up to isogeny}\end{matrix}\right\}$$

Needless to say this is the analog over F of the famous conjecture that elliptic curves/$\mathbb{Q}$ are all modular [W1]. It extends in a completely analogous way to any totally real field ([Od1])--see also [L2], [W2].

6.5. Every elliptic curve A_π over F defines, by restriction of scalars, an abelian surface $\tilde{A}_\pi$ over $\mathbb{Q}$. There is a polarization on $\tilde{A}_\pi$ giving rise to a skew-symmetric bilinear form B on $H^1_{et}(\tilde{A}_\pi \times \overline{\mathbb{Q}},\mathbb{Q}_\ell)$. The Galois group of $\overline{\mathbb{Q}}$ over $\mathbb{Q}$ acts as symplectic similitudes on this four-dimensional ℓ-adic vector space (relative to B). Then by the principle of functoriality (due to Langlands [L2]), noting that the L-group of GSP_4 is again GSP_4 ([Bo],[JPSS]), one sees that there should be a corresponding automorphic representation Π of $GSP_4(\mathbb{A})$, cuspidal if A_π is not the base-change of an elliptic curve over $\mathbb{Q}$. This fits in nicely with the remarks at the end of §5.

6.6. In [Od1] and [Od2] Oda has attached--when $E(\pi)$ is totally real--an abelian variety A_π^o. His (transcendental) construction is quite analogous to the way Kuga and Satake [KS] attach abelian varieties to K-3 surfaces. Though geometrically pleasing, it seems to be very difficult in general even to determine if the isogeny class/$\mathbb{C}$ of A_π^o contains a representative defined over $\overline{\mathbb{Q}}$, let alone verify if it is related to the hypothetical A_π. We will come back to this question after we look at Tate's conjecture in §9.

§7. The Picard group.

7.1. In this section we will briefly review certain basic facts and conjectures about divisors on algebraic surfaces. Let X be a smooth, projective surface over $\mathbb{Q}$, and let $\overline{X}$ denote $X \times_{\mathbb{Q}} \overline{\mathbb{Q}}$.

By a divisor on $\overline{X}$ we mean a finite formal sum:

(7.2) $$D = \sum n_C\, C, \quad n_C \in \mathbb{Z},$$

of (closed) irreducible curves on $\overline{X}$. Write:

(7.3) $$\mathrm{Div}(\overline{X}) = \text{The (free abelian) group of divisors on } \overline{X}.$$

We say that a divisor $D = \sum n_C\, C$ is rational over a number field k iff $n_C = n_{C^\sigma}$, for every automorphism σ of $\overline{\mathbb{Q}}$ over k. By prime divisor on X_k we mean a divisor of the form:

(7.4) $$D = \sum_\sigma C^\sigma,$$

where C is a curve defined over a finite extension L of k, with σ running over elements of $\mathrm{Aut}(L/k)$. If we set:

$$\mathrm{Div}(X_k) = \{k\text{-rational divisors on } X\},$$

then it is simply the $\mathrm{Gal}(\overline{\mathbb{Q}}/k)$-invariants of $\mathrm{Div}(\overline{X})$, and is free abelian on the prime divisors on X_k. It is easy to see that every divisor on $\overline{X}$ belongs to $Div(X_k)$ for some number field k.

<u>7.5</u>. Two divisors D, D' on $\overline{X}$ (resp. X_k) are <u>linearly equivalent</u> over $\overline{\mathbb{Q}}$ (resp. k) iff there is a function f in $\overline{\mathbb{Q}}(X)^*$ (resp. $k(X)^*$) such that

$$D - D' = div(f),$$

where $div(f)$ denotes the divisor: $\sum ord_C(f)C$. Define the <u>Picard group</u> by:

$$\text{(7.6)} \qquad Pic(\overline{X}) = Div(\overline{X})/\text{linear equivalence over } \overline{\mathbb{Q}} ,$$

and

$$Pic(X_k) = Div(X_k)/\text{linear equivalence over } k.$$

It is a fact that if two elements D, D' in $Div(X_k)$ are linearly equivalent over $\overline{\mathbb{Q}}$, then they are already linearly equivalent over k. One deduces:

$$\text{(7.7)} \qquad Pic(X_k) = Pic(\overline{X})^{Gal(\overline{\mathbb{Q}}/k)} .$$

A basic fact in ℓ-adic cohomology ([Mi]) is that there is a $Gal(\overline{\mathbb{Q}}/\mathbb{Q})$-equivariant isomorphism:

$$\text{(7.8)} \qquad \beta_\ell : Pic(\overline{X}) \xrightarrow{\sim} H^1_{et}(\overline{X}, \mathbb{G}_m) .$$

Now for every integer $n \geq 1$ look at the Kummer sequence:

$$1 \longrightarrow \mu_{\ell^n} \longrightarrow \mathbb{G}_m \xrightarrow{\cdot \ell^n} \mathbb{G}_m \longrightarrow 1.$$

Taking limits over n we get:

$$\text{(7.9)} \qquad 1 \to \mathbb{Z}_\ell(1) \to \mathbb{G}_m \to \mathbb{G}_m \to 1.$$

This allows us to define the <u>cycle class map</u>:

$$(7.10) \qquad c\ell_\ell : \mathrm{Pic}(\overline{X}) \to H^2_{et}(\overline{X}, \mathbb{Q}_\ell(1))$$

as the composite morphism:

$$\mathrm{Pic}\,\overline{X} \xrightarrow{\beta_\ell} H^1_{et}(\overline{X}, \mathbb{G}_m) \xrightarrow{\partial_\ell} H^2_{et}(\overline{X}, \mathbb{Z}_\ell(1)) \hookrightarrow H^2_{et}(\overline{X}, \mathbb{Q}_\ell(1)),$$

where ∂_ℓ is the coboundary map in the long exact cohomology sequence associated to the ℓ-adic Kummer sequence (7.9).

It is important to note that $c\ell_\ell$ is equivariant for the action of $\mathrm{Gal}(\overline{\mathbb{Q}}/\mathbb{Q})$ so that for every number field k one has:

$$(7.11) \qquad c\ell_\ell(\mathrm{Pic}(X_k)) \subset H^2_{et}(\overline{X}, \mathbb{Q}_\ell(1))^{\mathrm{Gal}(\overline{\mathbb{Q}}/k)} .$$

Conjecture 7.12. (Tate [T1]) For every number field k one has: $c\ell_\ell(\mathrm{Pic}(X_k)) \otimes \mathbb{Q}_\ell = H^2_{et}(\overline{X}, \mathbb{Q}_\ell(1))^{\mathrm{Gal}(\overline{\mathbb{Q}}/k)}$.

7.13. This is one of the deepest conjectures in the arithmetic of surfaces. There is a perfectly analogous conjecture for surfaces over finite fields, and in the case of a surface over $\mathbb{F}_q$ defined by an elliptic curve E over a function field over $\mathbb{F}_q$ Tate's conjecture is intimately related to the Birch-and-Swinnerton-Dyer conjecture for E.

7.14. Denote by $L^{(2)}(X_k,s)$ the global L-function (as in §4.1) attached to the $\mathrm{Gal}(\overline{\mathbb{Q}}/k)$-module $H^2_{et}(\overline{X}, \mathbb{Q}_\ell)$. As a consequence of Deligne's proof of the Riemann hypothesis for varieties over finite fields, one sees that $L^{(2)}(X_k,s)$ converges absolutely, at least after removing a finite number of bad factors, in the right-half plane $\{\mathrm{Re}(s) > 2\}$.

Conjecture 7.15. (Tate [T1])

$$\mathrm{rk}.c\ell_\ell(\mathrm{Pic}(X_k)) = -\mathrm{ord}_{s=2} L^{(2)}(X_k,s)$$

7.16. Note that this is analogous to Brauer's theorem on Artin L-functions of complex representations of Galois groups. Indeed the conjectures 7.12 and 7.15 together imply that the order of pole at $s = 2$ of $L^{(2)}(X_k,s)$ equals the multiplicity of the trivial representation occurring in $H^2_{et}(\overline{X}, \mathbb{Q}_\ell)$ viewed as a $\mathrm{Gal}(\overline{\mathbb{Q}}/k)$-module.

It will serve our purposes adequately to define the Néron-Severi group by setting:

(7.17) $$NS(\overline{X}) = c\ell_{\ell}(\mathrm{Pic}(\overline{X}))$$

(resp. $NS(X_k) = c\ell_{\ell}(\mathrm{Pic}(X_k))$).

7.18. It is more common, and more appropriate, to define $NS(\overline{X})$ as the group of divisors on $\overline{X}$ modulo algebraic equivalence ([M1]). But it is a fact that a divisor D is algebraically equivalent to zero on a surface $\overline{X}$ iff $c\ell_{\ell}([D]) = 0$, for every prime ℓ, where $[D]$ denotes the class of D in $\mathrm{Pic}(\overline{X})$. So we are justified in our definition.

7.19. Let us digress a little to look at the situation over $\mathbb{C}$. We can, of course, define $\mathrm{Div}(X_{\mathbb{C}})$ and $\mathrm{Pic}(X_{\mathbb{C}})$ in the obvious way. Then one has a $\mathrm{Gal}(\mathbb{C}/\mathbb{R})$-equivariant isomorphism ([DMOS]):

(7.20) $$\beta:\mathrm{Pic}(X_{\mathbb{C}}) \xrightarrow{\sim} H^1(X_{\mathbb{C}}, \mathcal{O}^*_{X_{\mathbb{C}}}).$$

Look at the exponent sequence:

(7.21) $$0 \longrightarrow \mathbb{Z}(1) \longrightarrow \mathcal{O}_{X_{\mathbb{C}}} \xrightarrow{\exp} \mathcal{O}^*_{X_{\mathbb{C}}} \longrightarrow 1,$$

where $\mathbb{Z}(1)$ denotes the Tate structure (in Hodge theory): $2\pi i\,\mathbb{Z} \subset \mathbb{C}$.

The cycle class map in the complex analytic setting is:

(7.22) $$c\ell_B:\mathrm{Pic}(X_{\mathbb{C}}) \to H^2_B(X(\mathbb{C}),\mathbb{Q}(1)),$$

defined as the composite morphism:

$$\mathrm{Pic}(X_{\mathbb{C}}) \xrightarrow{\beta} H^1(X_{\mathbb{C}}, \mathcal{O}^*_{X_{\mathbb{C}}}) \xrightarrow{\partial} H^2_B(X(\mathbb{C}),\mathbb{Z}(1)) \hookrightarrow H^2_B(X(\mathbb{C}),\mathbb{Q}(1)),$$

where ∂ is the coboundary map in the long exact cohomology sequence associated to (7.21).

For any divisor D, $c\ell_B(D)$ is $(2\pi i)$-times the cohomology class $c\ell_{top}(D)$ one obtains by taking the Poincaré dual of the class of the topological 2-cycle defined by D. $c\ell_{top}(D)$ is also the cohomology class of the d-closed current ([GH]) δ_D of type (1,1), defined to be the linear form on 2-forms given by integration over $D(\mathbb{C})$. This yields:

$$(7.23) \qquad c\ell_B(\mathrm{Pic}(X_{\mathbb{C}})) \subset H^2_B(X(\mathbb{C}),\mathbb{Q}(1)) \cap H^{1,1}(X(\mathbb{C}),\mathbb{C}).$$

A famous result of Lefschetz ([GH]), which led Hodge to his general conjecture, asserts:

$$(7.24) \qquad c\ell_B(\mathrm{Pic}(X_{\mathbb{C}})) \otimes \mathbb{Q} = H^2_B(X(\mathbb{C}),\mathbb{Q}(1)) \cap H^{1,1}(X(\mathbb{C}),\mathbb{C}).$$

We will prove a variant of this theorem, as conjectured by Beilinson [B2], for Hilbert-Blumenthal surfaces in §12.

A standard theorem says that even though the inclusion: $\mathrm{Pic}(\overline{X}) \subset \mathrm{Pic}(X_{\mathbb{C}})$ is often proper, one always has:

$$(7.25) \qquad c\ell_B(\mathrm{Pic}(\overline{X})) = c\ell_B(\mathrm{Pic}(X_{\mathbb{C}})).$$

One has a commutative triangle (for every ℓ):

$$(7.26) \qquad \begin{array}{ccc} & \overset{c\ell_B}{\nearrow} & H^2_B(X(\mathbb{C}),\mathbb{Q}(1)) \\ \mathrm{Pic}(\overline{X}) & & \big\downarrow j_B \\ & \underset{c\ell_\ell}{\searrow} & H^2_{et}(\overline{X},\mathbb{Q}_\ell(1)) \end{array}$$

consequently $NS(\overline{X})$ is also the image of $\mathrm{Pic}(\overline{X})$ under $c\ell_B$. Thus we have, using (7.23):

$$NS(\overline{X}) \otimes \mathbb{Q} \simeq H^2_B(X(\mathbb{C}),\mathbb{Q}(1)) \cap H^{1,1}(X(\mathbb{C}),\mathbb{C})$$

One defines for any field $L \supset \mathbb{Q}$:

$$(7.27) \qquad \mathrm{Pic}^0(X_L) = \mathrm{Ker}(\mathrm{Pic}(X_L) \to NS(X_L)).$$

When X has an L-rational point, $\mathrm{Pic}^0(X_L)$ is an abelian variety. In this case there is an isogeny:

$$(7.29) \qquad \phi: \mathrm{Pic}^0(X_L) \to A\ell b(X_L),$$

where $A\ell b(X_L)$ denotes the Albanese variety ([Lg1]). It is helpful to recall that:

$$A\ell b(X(\mathbb{C})) \simeq (H^{1,0}(X(\mathbb{C}),\mathbb{C}))^*/H_{1,B}(X(\mathbb{C}),\mathbb{Z}) . \tag{7.30}$$

<u>7.31</u>. The dimension q of $H^{1,0}(X(\mathbb{C}),\mathbb{C})$, which is half the first Betti number b_1, is called the <u>irregularity</u> of the surface X.

<u>7.32</u>. The (deep) conjecture of Birch and Swinnerton-Dyer asserts that, over any number field k, the rank of $\mathrm{Pic}^0(X_k)$ is equal to the order of zero at the critical center of $L^{(1)}(X_k,s)$, the L-function attached to $H^1_{et}(\overline{X},\mathbb{Q}_\ell)$.

<u>7.33</u>. When X is a Hilbert-Blumenthal surface $\tilde{S}_K$, however, the irregularity q is zero ([Sv], [HvG]). So (7.29) and (7.30) imply:

$$\mathrm{Pic}^0(\tilde{S}_K \times \overline{\mathbb{Q}}) = 0. \tag{7.34}$$

<u>7.35</u>. Actually, when K is a principal congruence subgroup, [Sv] says that $\tilde{S}_K(\mathbb{C})$ is even simply connected. For general K's there may be some torsion in the fundamental group, but in any case $q = 0$.

<u>7.36</u>. Consequently the equality $\mathrm{Pic}(\tilde{S}_K \times \overline{\mathbb{Q}}) = \mathrm{NS}(\tilde{S}_K \times \overline{\mathbb{Q}})$ holds for Hilbert-Blumenthal surfaces $\tilde{S}_K$. So the Tate's conjectures (7.12) and (7.15) (see §9 and §10) and the question of interpreting the residue at the "Tate point" $s = 2$ of $L^{(2)}(\tilde{S}_K,s)$ (see §11 and §12), are the most significant arithmetical problems for divisors on $\tilde{S}_K$. For a discussion of very interesting geometric problems on canonical divisors, etc., see [HvG] and [H2].

§8. Hirzebruch-Zagier cycles.

8.1. Every Hilbert-Blumenthal surface admits a "diagonal" modular curve. By translating this curve around by Hecke correspondences one gets many non-trivial divisors on the surface, called the Hirzebruch-Zagier cycles.

8.2. Let us first motivate things by looking at these cycles in the classical setting (cf. §1). Let Γ be an arithmetic subgroup of $SL_2(F)$, and let $\delta: \mathfrak{h} \hookrightarrow \mathfrak{h}^2$ denote the diagonal embedding. Then the image of $\delta(\mathfrak{h})$ in $\Gamma\backslash\mathfrak{h}^2$, called the "diagonal" curve, can be identified, for Γ small enough, with the modular curve $\Delta\backslash\mathfrak{h}$, where $\Delta = \Gamma \cap SL_2(\mathbb{Q})$. Now pick any element g of $GL_2^+(F)$ and look at the image of $g\delta(\mathfrak{h})$ in $\Gamma\backslash\mathfrak{h}^2$, which corresponds to the modular curve $\Delta_g\backslash\mathfrak{h}$, where $\Delta_g = g^{-1}\Gamma g \cap SL_2(\mathbb{Q})$. This process gives a whole slew of "Hirzebruch-Zagier" divisors, as we vary g.

It is easy to see that if $g = \begin{pmatrix}\alpha & \beta\\ \gamma & \delta\end{pmatrix} \in GL_2^+(F)$, then the complex curve $g\delta(\mathfrak{h})$ is simply the set of points (z_1,z_2) in $\mathfrak{h}^2$ satisfying:

$$(8.3)\qquad z_2 = \frac{(\beta^\theta\gamma-\alpha^\theta\delta)z_1 + (\beta\alpha^\theta-\beta^\theta\alpha)}{(\delta^\theta\gamma-\gamma^\theta\delta)z_1 + (\gamma^\theta\beta-\delta^\theta\alpha)}\ .$$

In other words $g\delta(\mathfrak{h})$ is the intersection of $\mathfrak{h}^2$ with the complex points of an affine curve defined by a quadratic equation with coefficients in F. If $F = \mathbb{Q}(\sqrt{D})$ then one can start with the data (a,b,λ,N) with $a,b \in \mathbb{Z}$, $\lambda \in \mathfrak{O}$, and N a positive integer, and look for a g satisfying:

$$(8.4)\qquad a\sqrt{D} = (\delta^\theta\gamma-\gamma^\theta\delta),\ b\sqrt{D} = (\beta^\theta\alpha-\beta\alpha^\theta),$$

$$\lambda = (\gamma^\theta\beta-\delta^\theta\alpha),\ \text{and}\quad N = \lambda\lambda^\theta+abD.$$

Then $g\delta(\mathfrak{h})$ identifies with the set of (z_1,z_2) in $\mathfrak{h}^2$ satisfying:

$$(8.5)\qquad a\sqrt{D}\, z_1z_2 + \lambda z_2 - \lambda^\theta z_1 + b\sqrt{D} = 0,$$

which is the equation that Hirzebruch and Zagier prefer to work with (cf. [HZ], [HvG], Chap.3). When $\Gamma = SL_2(\mathcal{O})$ they define, for each integer $N \geq 1$, a (reducible) curve T_N in $\tilde{S}_\Gamma(\mathbb{C})$ as the closure of the union of the images in $S_\Gamma(\mathbb{C})$ of the $g\delta(\mathfrak{h})$'s for g's satisfying (8.4). These T_N's are in a natural way the counterparts of the classical Hecke correspondences T_N on $\overline{SL_2(\mathbb{Z})\backslash\mathfrak{h}} \times \overline{SL_2\mathbb{Z}\backslash\mathfrak{h}}$ (see [HvG], Chap.4).

8.6. Now let us construct these (Hirzebruch-Zagier) divisors adelically. First we need to recal some basic facts about modular curves. Set:

$$H = GL_2/\mathbb{Q}\ ,\quad \mathfrak{h}^{\pm} = \mathbb{C} - \mathbb{R}\ , \tag{8.7}$$

and define a homomorphism of $\mathbb{R}$-algebraic groups (see (2.6)):

$$h_0 : \underset{\sim}{S} \to H_{\mathbb{R}}\ , \tag{8.8}$$

defined on the real points by: $h(a+ib) = \begin{pmatrix} a & b \\ -b & a \end{pmatrix} \in GL_2(\mathbb{R})$. Then L_∞ (see (2.9)) is the centralizer of h_0 in $H(\mathbb{R})$ and one has:

$$\mathfrak{h}^{\pm} = H(\mathbb{R})/L_\infty\ . \tag{8.9}$$

Let M denote the Shimura "variety" attached to (H, h_0), equipped with a continuous H_f-action (see (2.21)). Then M is simply the universal modular curve ([DRp]), defined over $\mathbb{Q}$, such that:

$$M(\mathbb{C}) = H(\mathbb{Q})\backslash \mathfrak{h}^{\pm} \times H_f\ . \tag{8.10}$$

8.11. For every compact open subgroup L of H_f the (smooth) curve $M_L \overset{\text{def}}{=} M/L$ parametrizes over $\mathbb{Q}$ the (isomorphism) classes of elliptic curves E with suitable "level structure". In particular, if $L = L(m) \overset{\text{def}}{=} \mathrm{Ker}(H(\hat{\mathbb{Z}}) \to H(\hat{\mathbb{Z}}/m\hat{\mathbb{Z}}))$, then $M_{L(m)}$ is the moduli space/$\mathbb{Q}$ of isomorphism classes of elliptic curves E together with an isomorphism: $(\mathbb{Z}/m\mathbb{Z})^2 \xrightarrow{\sim} \mathrm{Ker}(E \xrightarrow{.m} E)$.

8.12. The Shimura reciprocity law in this case (the analogue of (2.32)) is a little simpler. One gets, for each $L \subset H_f$, a bijection:

$$\mathrm{Gal}(\mathbb{Q}(L)/\mathbb{Q}) \simeq \pi_0(M_L(\mathbb{C})), \tag{8.13}$$

where $\mathbb{Q}(L)$ denotes the abelian extension of $\mathbb{Q}$ defined by the open subgroup $\det(L)$ in $\mathbb{A}^*$. And the multiplication map: $\mathrm{Gal}(\mathbb{Q}(L)/\mathbb{Q}) \times \mathrm{Gal}(\mathbb{Q}(L)/\mathbb{Q}) \to \mathrm{Gal}(\mathbb{Q}(L)/\mathbb{Q})$ induces via (8.13) a transitive action of $\mathrm{Gal}(\mathbb{Q}(L)/\mathbb{Q})$ on the set of connected components of $M_L(\mathbb{C})$. Since M_L is defined over $\mathbb{Q}$, one gets that each of its components is defined (canonically) over $\mathbb{Q}(L)$.

8.14. It is worthwhile to note that when L is the principal congruence subgroup $L(m)$ the connected components of $M_L(\mathbb{C})$ are in bijection with $\mathrm{Gal}(\mathbb{Q}(\mu_m)/\mathbb{Q})$. On the other hand, if $L = L_0(m) \overset{\text{def}}{=} \phi_m^{-1}(\{\begin{pmatrix} * & * \\ 0 & * \end{pmatrix}\})$, where ϕ_m is the reduction map: $H(\hat{\mathbb{Z}}) \to H(\hat{\mathbb{Z}}/m\hat{\mathbb{Z}})$, then $M_L(\mathbb{C})$ is connected.

8.15. There is a natural smooth compactification $\overline{M} = M \cup M^\infty$ over $\mathbb{Q}$ such that:

$$\overline{M} = \varprojlim_{L} \overline{M}_L ,$$

where $\overline{M}_L$ is, for each $L \subset H_f$, a smooth projective curve. One has an identification, analogous to (2.42):

$$M_L^\infty(\mathbb{C}) \simeq B^+(\mathbb{Q})N_f \backslash \mathbb{H}_f/L . \tag{8.16}$$

Again the cusps are all defined over $\mathbb{Q}^{ab}$, and $\mathrm{Aut}\,\mathbb{C}$ acts on $M_L^\infty(\mathbb{C})$ via its quotient $\mathrm{Gal}(\mathbb{Q}^{ab}/\mathbb{Q})$. This action corresponds, via (2.43), to the left action of $D(\hat{\mathbb{Z}})$ on the double coset space in (8.16).

8.17. There is a nice (moduli-theoretic) proper model $\overline{M}_{\mathbb{Z}}$ over $\mathbb{Z}$ for $\overline{M}$ which is regular [KM].

8.18. One has a natural "diagonal" map of $\mathbb{Q}$-algebraic groups:

(8.19) $$\delta : H \hookrightarrow G\ .$$

We will use δ to also denote the corresponding embeddings $M \hookrightarrow S$, $M^{\infty} \hookrightarrow S^{\infty}$ and $\overline{M} \hookrightarrow \overline{S}$.

8.20. The diagonal embedding $M \hookrightarrow S$ can be viewed moduli-theoretically as follows. Every point m in M corresponds to an isomorphism class of an elliptic curve E with "level infinity" structure, i.e., it comes equipped with an isomorphism: $(\hat{\mathbb{Z}})^2 \xrightarrow{\sim} \varprojlim_{m} \operatorname{Ker}(E \xrightarrow{\cdot m} E)$. Set: $A = R_{F/\mathbb{Q}}(E_F)$. Then A is a polarized abelian surface admitting multiplication by $\mathcal{O}_F$, and it has a natural "level infinity" structure. Then $\delta(m)$ may be identified with the point on S defined by the class of A (together with its polarization, endomorphism, and level structures).

Define:

(8.21) $$p_K : \overline{S} \to \overline{S}_K$$

to be the natural projection over $\mathbb{Q}$ such that $p_K(S) = S_K$ and $p_K(S^{\infty}) = S_K^{\infty}$.

8.22. Recall that $\overline{S}$ admits a (right) action ρ_g of G_f. For every g in G_f and compact open $K \subset G_f$ set:

(8.23) $$C_{g,K} = p_K(\rho_g(\delta(M))) \subset S_K\ , \ \underline{\text{and}}$$

$$\overline{C}_{g,K} = p_K(\rho_g(\delta(\overline{M}))) \subset \overline{S}_K\ .$$

Then $\overline{C}_{g,K}$ is a projective algebraic curve over $\mathbb{Q}$, which may at times be singular.

Note that if m_1, m_2 are in $\overline{M}$ such that $m_2 = \rho_k(m_1)$ for some k in $H_f \cap gKg^{-1}$, then one has: $p_K(\rho_g(\delta(m_1))) = p_K(\rho_g(\delta(m_2)))$. Thus one obtains a non-trivial morphism (over $\mathbb{Q}$):

(8.24) $$\lambda_{g,K} : \overline{M}_{H_f \cap gKg^{-1}} \to \overline{C}_{g,K}\ ,$$

8.25. The morphism $\rho_g : S \to S$ does not (of course) descend to a morphism: $S_K \to S_K$, but rather to a self-correspondence of S_K, which is none other than the Hecke correspondence T_g (cf. (3.10)). This shows that we may view each $C_{g,K}$ as the image under the (Hecke) correspondence T_g of $M_{H_f \cap K}$ (the diagonal curve in S_K).

8.26. Clearly the curves $\overline{C}_{g,K}$ are not geometrically connected (in general), and this disconnectedness can be understood in terms of (8.13).

Let us now set:

$$(8.27) \qquad \tilde{C}_{g,K} = \text{The closure of } C_{g,K} \text{ in } \tilde{S}_K .$$

We will call these $\tilde{C}_{g,K}$'s Hirzebruch-Zagier cycles or H-Z cycles for short. (In [HLR] the authors prefer to always view these cycles modulo homological equivalence, and think of them as naturally lying in $W_\ell(1)$. We will not adopt their convention here. This will become important in §12.)

8.28. It is of interest to know where the curves $\overline{C}_{g,K}$ meet. If g and g' are two elements in $\delta(H_f)$, then $\overline{C}_{g,K}$ and $\overline{C}_{g,K}$ meet only at the cusps. But this need not be the case for an arbitrary pair (g,g') of elements in G_f. To answer this question we need to know a little about CM points ([Sh1]) - called special points by Deligne ([D1]) - on S_K.

If L is any quartic CM field containing F, i.e., if L is a purely imaginary quadratic extension of F, then we can define an embedding:

$$(8.29) \qquad \phi_L : L^* \hookrightarrow GL_2(F) .$$

Explicitly, if $L = F + F\gamma$, with $\gamma^2 = -\Delta$, $\Delta \in F$, then we can define ϕ_L by:

$$(8.30) \qquad x + y\gamma \mapsto \begin{pmatrix} x & -\Delta y \\ y & x \end{pmatrix} .$$

Now we may view L^* in the $\mathbb{Q}$-rational points of a torus T_L over $\mathbb{Q}$ (defined by restriction of scalars). Then ϕ_L can be realized as the map on $\mathbb{Q}$-points induced by a morphism: $T_L \hookrightarrow G$ of $\mathbb{Q}$-groups. We can also easily define a homomorphism

(8.31) $$h_L : \mathbb{S} \to T_{L_{\mathbb{R}}}$$

such that $\phi_L \circ h_L = h$ (see (2.7)). This results in an injection of Shimura varieties ([D1]):

(8.32) $$u_L : S(L, h_L) \hookrightarrow S = S(G, h) \ .$$

The fact that such a map u_L is defined for every quartic $\mathbb{C}$M field $L \supset F$ is the true meaning of S being a <u>canonical</u> model.

For any compact open $K \subset G_f$ let us set:

(8.33) $$\Sigma_K^L = p_K(u_L(S(L, h_L))), \text{ and}$$

$$\Sigma_K = \bigcup_L \Sigma_K^L \ .$$

It is not hard to see that, for each L, $\Sigma_K^L(\mathbb{C})$ is <u>finite</u>. The reciprocity law asserts ([Sh1], [D1]) that the points of Σ_K^L are all defined over $(L^\vee)^{ab}$, where $L^\vee$ denotes the reflex $\mathbb{C}$M field of L.

In some ways it is better not to think of the (dense) subset $\Sigma_K(\mathbb{C})$ as a set of "special" points on $S_K(\mathbb{C})$. Rather one should think of S_K as a "special" variety which is somehow built out of the primary object Σ_K.

<u>8.34</u>. In the classical case ([BCS], [Sh2]) of modular curves, the special points correspond to imaginary quadratic fields of $\mathbb{Q}$. They become "super-special" points on S_K under the diagonal embedding. Now the definition of H-Z cycles shows that:

(8.35) $$\left\{\begin{matrix}\text{Intersection points}\\ \text{of H-Z cycles on } \overline{S}_K\end{matrix}\right\} = S_K^\infty \cup \left\{\begin{matrix}\mathbb{C}\text{M-points corresponding}\\ \text{to biquadratic } \mathbb{C}\text{M fields}\\ \text{containing } F\end{matrix}\right\}$$

This set is still dense on $\overline{S}_K(\mathbb{C})$.

§9. Tate cycles and poles of L-functions.

9.1. In this section we will try to understand the essential ideas behind the proof of the following:

Theorem 9.2 (Harder, Langlands and Rapoport [HLR]). Fix an arbitrary compact open subgroup $\underset{\sim}{K}$ of G_f. Then the Tate conjectures 7.12 and 7.15 hold for $\tilde{S}_K$ over any abelian extension of $\mathbb{Q}$.

Strictly speaking, the (joint) authors of this (deep) theorem concentrate on $\overline{S}_K$ and prove the conjectures for IW_ℓ $(= IH^2(\overline{S}_K \times \overline{\mathbb{Q}}, \mathbb{Q}_\ell(1)))$. But recall that one has (cf. (3.4), (3.8) and (4.13)):

$$(9.3) \qquad W_\ell \simeq IW_\ell \oplus W_\ell^\infty, \text{ (as Galois modules)}$$

$$(9.4) \qquad W_\ell^\infty \simeq \bigoplus_{s \in S_K^\infty} (\mathbb{Q}_\ell(-1)^{\oplus n(s)}),$$

and

$$(9.5) \qquad L(W_\cdot^\infty, s) = \prod_{s \in S_K^\infty} \left(\prod_{j=1}^{n(s)} \zeta(s-1) \right),$$

where $n(s)$ denotes, for each s, the number of irreducible components of the polygon $\pi_K^{-1}(\{s\})$ of rational curves. It is now an easy matter to see that:

(9.6) The Tate's conjectures 7.12 and 7.15 hold for $W_\cdot^\infty(1)$ over any number field k.

9.7. Furthermore, the Tate classes in the ℓ-adic representation $W_\ell^\infty(1)$ are all generated by (classes of) divisors on $\tilde{S}_K \times \overline{\mathbb{Q}}$ supported on the cusp resolution $S_K^\infty \times \overline{\mathbb{Q}}$.

9.8. Now one can restrict attention to IW_ℓ. Let us set:

$$\mathcal{T}_\ell = \{\lambda \in IW_\ell(1) \mid \gamma \text{ is fixed by an open subgroup of } \mathrm{Gal}(\overline{\mathbb{Q}}/\mathbb{Q})\},$$

the (ℓ-adic) space of Tate cycles on $\overline{S}_K \times \overline{\mathbb{Q}}$.

Since the Theorem 9.2 concerns only with "abelian" Tate classes, it makes sense to consider:

$$\mathfrak{T}_\ell^{ab} = \mathfrak{T}_\ell^{\mathrm{Gal}(\overline{\mathbb{Q}}/\mathbb{Q}^{ab})} .$$

Clearly $\mathrm{Gal}(\mathbb{Q}^{ab}/\mathbb{Q})$ acts nicely on $\mathfrak{T}_\ell^{ab}$, and one gets an isotypic decomposition:

$$\mathfrak{T}_\ell^{ab} \otimes \overline{\mathbb{Q}}_\ell \simeq \sum_\omega \overline{\mathfrak{T}}_\ell(\omega)$$

where ω runs over characters of $\mathrm{Gal}(\overline{\mathbb{Q}}/\mathbb{Q})$, and $\overline{\mathfrak{T}}_\ell(\omega)$ signifies: $((\mathfrak{T}_\ell^{ab} \otimes \overline{\mathbb{Q}}_\ell) \otimes \omega)^{\mathrm{Gal}(\mathbb{Q}^{ab}/\mathbb{Q})}$.

Since the Galois and Hecke actions on IW_ℓ commute one gets (cf. (3.13), (3.15), (3.18) and (3.28):

(9.9)
$$\mathfrak{T}_\ell^{ab} \otimes \overline{\mathbb{Q}}_\ell \simeq \sum_\omega \sum_{\pi \in \mathcal{V}_K} \overline{\mathfrak{T}}_\ell(\pi;\omega)^{\oplus m(\pi,K)} ,$$

where each $\overline{\mathfrak{T}}_\ell(\pi;\omega)$ denotes $\overline{\mathfrak{T}}_\ell(\omega) \cap \overline{W}_\ell(\pi)(1)$, equipped with a model $\mathfrak{T}_\ell(\pi;\omega)$ over $E_\pi \mathbb{Q}(\omega) \otimes \mathbb{Q}_\ell$. The set $\mathcal{V}_K$ consists of irreducible representations of $G(\mathbb{A})$ as in (3.28).

It is convenient to set: (see (7.10))

(9.10)
$$\mathfrak{Z}_\ell(\pi,\omega) = \mathfrak{T}_\ell(\pi,\omega) \cap c\ell_\ell(\mathrm{Pic}(\tilde{S}_K \times \overline{\mathbb{Q}})) \otimes \mathbb{Q}_\ell .$$

Then Theorem 9.2 will follow from:

<u>Theorem 9.11</u> ([HLR]). For every π in $\mathcal{V}_K$, and for every character ω of $\mathrm{Gal}(\mathbb{Q}^{ab}/\mathbb{Q})$, one has:

(a) $\mathfrak{Z}_\ell(\pi,\omega) = \mathfrak{T}_\ell(\pi,\omega)$

(b) $\mathrm{rk}\ \mathfrak{T}_\ell(\pi,\omega) = -\mathrm{ord}_{s=2} L_Q(W.(\pi) \otimes \omega, s)$

(c) $\mathfrak{Z}_\ell(\pi,\omega)$ is generated by the ω-components of classes of Hirzebruch-Zagier cycles.

It is important to remember that:

(9.12)
$$W_\ell(\pi) \otimes \omega = W_\ell(\pi \otimes \omega'),$$

for any extension ω' to $\mathbb{A}_F^*/F^*$ of ω (viewed as an idele class character of $\mathbb{Q}$). Thus <u>when π is cuspidal</u>, one has by the theorem of Brylinski-Labesse ([BL]), earlier looked at in §4.27,

$$L_Q(W.(\pi) \otimes \omega, s) = L_Q(\pi \otimes \omega', \rho, s), \tag{9.13}$$

for some finite set Q of rational primes.

Recall that for any cuspidal π in $\mathcal{V}_K$ there exists (cf. (5.25)(i)) a representation τ_ℓ of $\mathrm{Gal}(\overline{\mathbb{Q}}/F)$ on a rank 2 $E_\pi \otimes \mathbb{Q}_\ell$-module $V_\ell(\pi)$ such that:

$$W_\ell(\pi) \simeq V_\ell(\pi) \otimes V_\ell(\pi)^\theta. \tag{9.14}$$

This isomorphism allows us to say quite a bit about the Tate cycles in $W_\ell(\pi)$. To begin, note that, since $V_\ell(\pi)$ is irreducible, one has:

$$W_\ell(\pi)(1)^{\mathrm{Gal}(\overline{\mathbb{Q}}/F)} \neq 0 \iff V_\ell(\pi)^\theta \simeq V_\ell(\pi)^\vee . \tag{9.15}$$

<u>9.16</u>. When we do have $V_\ell(\pi)^\theta \simeq V_\ell(\pi)^\vee$, we get an isomorphism: $W_\ell(\pi) \simeq (\mathrm{Sym}^2(V_\ell(\pi)) \otimes \omega_\pi^{-1}) \oplus \mathbb{Q}_\ell(-1)$. Note that $\mathbb{Q}_\ell(-1)$ is acted upon trivially by Hecke, but not trivially by $\mathrm{Gal}(\overline{\mathbb{Q}}/F)$. A very interesting thing is that, since $V_\ell(\pi)^\vee \simeq V_\ell(\pi) \otimes \omega_\pi^{-1}$, $\mathrm{sym}^2(V_\ell(\pi))^\theta \otimes \omega_\pi^{-\theta} \simeq \mathrm{sym}^2(V_\ell(\pi) \otimes \omega_\pi^{-1}) \otimes \omega_\pi^{-\theta} \simeq \mathrm{sym}^2(V_\ell(\pi)) \otimes \omega_\pi^{-1}$; thus the <u>decomposition</u>: $W_\ell(\pi) \simeq (\mathrm{Sym}^2(V_\ell(\pi)) \otimes \omega_\pi^{-1}) \oplus \mathbb{Q}_\ell(-1)$ <u>is actually stable under $\mathrm{Gal}(\overline{\mathbb{Q}}/\mathbb{Q})$</u>. In other words the Tate cycle corresponding to the trivial submodule of $W_\ell(\pi)(1)$ is already defined over $\mathbb{Q}$. Furthermore, still assuming the isomorphism $V_\ell(\pi)^\theta \simeq V_\ell(\pi)^\vee$, if we choose a character (of finite order) μ of $\mathbb{A}_F^*/F^*$ such that $\overline{\mu}^{1-\theta} = \omega_\pi$, then we see that $V_\ell(\pi \otimes \mu)$ is invariant under θ, hence it is the restriction of a representation of $\mathrm{Gal}(\overline{\mathbb{Q}}/\mathbb{Q})$.

Extending the above argument to the case of an arbitrary open subgroup U of the absolute Galois group, remembering that for a non $\mathbb{C}$M cuspidal π the restriction of $V_\ell(\pi)$ to U remains irreducible (5.25), and noting the rather simple nature of $V_\ell(\pi)$ ($\simeq \chi_\pi(-1) \oplus \varepsilon\, \chi_\pi(-1)$, cf. (3.25)) when π is 1-dimensional, one gets the following basic

Proposition 9.17. Let π be in $\mathcal{V}_K$. Then

(a) If π is not of type $\mathbb{C}$M, then all the Tate cycles in $W_\ell(\pi)(1)$ are defined over $\mathbb{Q}^{ab}$;

(b) If π is any cuspidal representation in $_K$, then for every character ω of $Gal(\mathbb{Q}^{ab}/\mathbb{Q})$ one has:

Dim $\mathfrak{T}_\ell(\pi,\omega) \leq 1$ with equality holding iff we can find a character ν' of $\mathbb{A}_F^*/F^*$ such that:

(i) $V_\ell(\pi \otimes \nu'\omega')^\theta \simeq V_\ell(\pi \otimes \nu'\omega')$, where $\omega' = \omega|_{\mathbb{A}_F^*}$,

and (ii) $\det(V_\ell^0(\pi \otimes \nu'\omega')) = \nu\varepsilon_{F/\mathbb{Q}}$,

where $V_\ell^0(\pi \otimes \nu'\omega')$ is an extension (whose existence is guaranteed by (i)) of $V_\ell(\pi \otimes \nu'\omega')$ to a rank 2 $E_\pi \otimes \mathbb{Q}_\ell$-module acted upon by $Gal(\overline{\mathbb{Q}}/\mathbb{Q})$.

9.18. It can be shown that for a $\mathbb{C}$M representation π the number of Tate classes can be 0, 1 or 2. (Use (5.27).) So part of the content of (b) is that even when there are two classes in $\mathfrak{T}_\ell(\pi)$, only one can occur in any given ω-eigenspace.

9.19. On the other hand, if π is a non-CM cuspidal representation, then $sym^2(V_\ell(\pi))$ is strongly irreducible, i.e., it remains irreducible when restricted to any open subgroup of $Gal(\overline{\mathbb{Q}}/F)$; hence π can support at most one Tate class.

9.20. Let us interject now with a relevant digression. Denoting by $W_B(\pi)$ the π-isotypic component of the Hodge analogue IW_B $(\simeq IH_B^2(\tilde{S}_K(\mathbb{C}),\mathbb{Q}))$ of IW_ℓ (see (3.6) and (3.7)), define the space of π-Hodge cycles to be:

$$Hg(\pi) = (W_B(\pi) \cap (W_B(\pi) \otimes \mathbb{C})^{1,1})(1) , \tag{9.21}$$

where $(*)^{1,1}$ denotes the (1,1)-piece of $(*)$, and $W_B(\pi)(1)$ denotes $W_B(\pi) \otimes_{\mathbb{Q}} \mathbb{Q}(1)$. Then by Lefschetz's theorem ([GH]) one has:

$$Hg(\pi) = \mathfrak{Z}(\pi), \tag{9.22}$$

i.e., every Hodge (divisor) class is algebraic, where

$$\mathfrak{Z}(\pi) = Hg(\pi) \cap c\ell_B(Pic(\tilde{S}_K \times \mathbb{C})) \otimes \mathbb{Q} . \tag{9.23}$$

See (7.22) for a definition of $c\ell_B$.

So the Tate's conjecture (Th. 9.11(b)) is equivalent to:

Conjecture 9.24. $\mathfrak{T}_\ell(\pi) = Hg(\pi) \otimes \mathbb{Q}_\ell$, $\forall \pi$ in $\mathcal{V}_K$. Here we have identified $Hg(\pi)$ with its image under the natural ($Gal(\mathbb{C}/\mathbb{R})$-equivariant) embedding $j_B : IW_B(1) \hookrightarrow IW_\ell(1)$ (see (3.7)).

In view of the Remark 9.19, we arrive at the following explicit

Question 9.25. Given any non-$\mathbb{C}$M π in $\mathcal{V}_K$ can we show directly that there is atmost one (up to scalar multiplication) (1,1)-class in $W_B(\pi) \otimes \mathbb{C}$ all of whose periods are in $2\pi i\mathbb{Q}$?

Even this (weaker) question does not have an easy answer. Let us note, however, that Oda had shown in [Od1] that, when $\pi \simeq \pi^\theta$ and $\omega_\pi = 1$ (i.e., when π is a base-change (cf. [L1]) to F of a cuspidal representation π_0 of $GL_2(\mathbb{A}_\mathbb{Q})$), $Hg(\pi)$ is one-dimensional.

9.26. There is one case where the conjecture 9.24 above can be settled immediately, and that is when π is a one-dimensional represenation $\nu_\pi \circ \det$, trivial on G_∞^+. From (3.25) we see that:

$$(9.27) \qquad \dim \mathfrak{T}_\ell(\pi,\omega) \neq 0 \iff \chi_\pi\omega = 1 \text{ or } \varepsilon_{F/\mathbb{Q}},$$

where χ_π is the character of finite order defined by the restriction of ν_π to $\mathbb{A}_\mathbb{Q}^*/\mathbb{Q}^*$. We also have:

$$(9.28) \qquad W_B(\pi) \otimes \mathbb{C} \simeq H^2(\mathfrak{g}_\infty, K_\infty; \mathbb{C}_{\nu_\pi \circ \det}), \quad \mathfrak{g}_\infty = \mathrm{Lie}(G_\infty),$$

which has rank 2 over $E_\pi \otimes \mathbb{C}$ (see [Cs] and [HLR]). The proof of Conjecture 9.24 comes down to a simple computation. Let us just remark in passing that the classes in $W_B(\pi) \otimes \mathbb{C}$ correspond in the classical setting to G_∞-invariant - not just Γ-invariant - forms on $\mathfrak{h}^2$. For example, when $\nu_\pi = 1$, we get a basis consisting of classes defined by the volume forms on the factors:

$$(9.29) \qquad \eta_1 = \frac{dz_1 \wedge d\bar{z}_1}{y_1^2}, \quad \eta_2 = \frac{dz_2 \wedge d\bar{z}_2}{y_2^2},$$

with $y_j = im(z_j)$.

Furthermore we can verify the equality of the number of Tate cycles and the pole of the L-function by exploiting the identity (4.15). We have thus shown how one can prove parts (a) and (b) of Theorem 9.11 for 1-dimensional π's.

9.30. Now let us get back to look at what the Proposition 9.17 implies for a cuspidal automorphic π in $\mathcal{V}_K$. The condition (i) of (b) says that when $\mathfrak{T}_\ell(\pi,\omega)$ has dimension 1, the local components $(\pi \otimes \nu'\omega')_v$ at almost all v must be θ-invariant. (Recall (cf. 5.25) that outside a finite number of v's each $(\pi \otimes \nu'\omega')_v$ corresponds to either $(V_\ell(\pi))_v$ or $(V_\ell(\pi)^\theta)_v$. By strong multiplicity one theorem ([JL]) this means that $\pi \otimes \nu'\omega'$ is globally invariant under θ, and hence is in the image of base-change from $\mathbb{Q}$ to F ([L1]). With this in mind, Harder, Langlands and Rapoport make the following

Definition 9.31. An irreducible π in $\mathcal{V}_K$ is distinguished relative to ω iff the intertwining operator

$$T_\omega : \{\text{space of } \pi\} \to \operatorname{Ind}_{H(\mathbb{A})}^{G(\mathbb{A})}(\omega^{-1}),$$

$$f \mapsto T_\omega f : g \mapsto \int_{H(\mathbb{Q})Z_H(\mathbb{R})\backslash H(\mathbb{A})} \omega(\det h) f(hg)\,dh,$$

is non-zero.

Here dh denotes the measure on $H(\mathbb{Q})Z_H(\mathbb{R})\backslash H(\mathbb{A})$ induced by a Haar measure on $H(\mathbb{A})$. Note that $T_\omega f$ makes sense for a cuspidal π because f has excellent asymptotic behavior ([JL]). And since the volume of $H(\mathbb{Q})Z_H(\mathbb{R})\backslash H(\mathbb{A})$ is finite, T_ω makes sense also for one-dimensional π's.

$\operatorname{Ind}_{H(\mathbb{A})}^{G(\mathbb{A})}(\omega^{-1})$ consists of K-finite functions on $G(\mathbb{A})$ which transform under the left action of $H(\mathbb{A})$ according to the character ω^{-1}. And $G(\mathbb{A})$ acts on these functions by right translation ρ_g. It is evident that the image of T_ω is stable under this action of $G(\mathbb{A})$.

To tie everything together they prove the following key

Proposition 9.32. For every cuspidal π in $\mathcal{V}_K$, and every character ω of $\operatorname{Gal}(\mathbb{Q}^{ab}/\mathbb{Q})$, the following are equivalent:

(a) $\dim \mathfrak{C}_\ell(\pi,\omega) = 1$;

(b) There exist a character ν' of $\mathbb{A}_F^*/F^*$ such that:

(i) $\pi \otimes \nu'\omega'$ is the base-change to F of a cuspidal representation π_0 of $H(\mathbb{A})$ $(= GL_2(\mathbb{A}_\mathbb{Q}))$;

and (ii) $\omega_{\pi_0} = \nu\varepsilon_{F/\mathbb{Q}}$;

(c) π is distinguished relative to ω

(d) $\mathfrak{C}_\ell(\pi,\omega)$ is generated by the (π,ω)-component of an H-Z cycle.

Again, in part (ii) of (b), ω' denotes an extension of ω to $\mathbb{A}_F^*/F^*$.

By the Remark 9.30 we get the equivalence of (a) and (b). The equivalence of (b) and (c) is intuitively clear, but is delicate to prove. For a proof of this non-trivial statement we refer to §2 of [HLR]. For an alternate approach via "relative trace formula" ([JLai]) see the recent Columbia thesis of Ye Yangbo.

To get the equivalence of (c) and (d) we only have to observe that, for every f in π and g in G_f, $T_\omega f(g)$ can be interpreted as the integral of $f\cdot(\omega \circ \det)$ over a homogeneous space of $\rho_g(\delta(H(\mathbb{A})))$. The admissibility of π on the other hand implies that f is fixed by a compact open subgroup of G_f. This shows that $T_\omega f(g)$ is really an integral of f over some H-Z cycle $C_{g,K}(\mathbb{C})$. And the non-triviality of T_ω is equivalent to the (π,ω)-component of some $C_{g,K} \times \overline{\mathbb{Q}}$ being non-trivial. Since $\mathfrak{C}_\ell(\pi,\omega)$ is one-dimensional in this case, it must be generated by the component of such an H-Z cycle. Whence the Proposition 9.32.

To compute the order of pole at $s = 2$ of $L_Q(W_\ell(\pi) \otimes \omega, s)$ we appeal to the adelic version of the integral representation (see §4.28) of Asai which, when used in conjunction with (9.13) and (4.30), yields the following

Proposition 9.33. Let π be cuspidal in $\mathcal{V}_K$. Then for every character ω of $\mathrm{Gal}(\mathbb{Q}^{ab}/\mathbb{Q})$ one has:

$$\left\{\begin{matrix}\pi\text{:distinguished}\\ \text{relative to } \omega\end{matrix}\right\} \Longleftrightarrow \left\{\begin{matrix}L_Q(W_\ell(\pi) \otimes \omega, s) \text{ has a pole}\\ \text{of order 1 at } s = 2\end{matrix}\right\}$$

This finishes the proof of the Theorem 9.11.

9.34. When Hirzebruch and Zagier introduced the (now famous) cycles named after them, they asked ([HZ]), in the case of $\Gamma = SL_2(\mathcal{O}_F)$, $F = \mathbb{Q}(\sqrt{D})$, D a prime $\equiv 1(4)$, whether the whole Picard group was generated by the (trivial) cycles in the cusp resolution and by the H-Z cycles. This conjecture is basically correct, but for general D's, and for $\Gamma = \Gamma(\mathcal{O}, \mathfrak{a})$ (see 1.12), there are Tate classes (see [HLR], §4.7), which are not defined over $\mathbb{Q}^{ab}$. By Proposition 9.17(a) we see that such classes must come from $\mathbb{C}$M forms π. Note that all the H-Z cycles, and the trivial ones, are all defined over cyclotomic fields.

Furthermore, the original definition in [HZ] of H-Z cycles includes, besides the ones we have considered, also the images under Hecke correspondences of Shimura curves/$\mathbb{Q}$. One has an inverse system of Shimura curves over $\mathbb{Q}$ attached to every quaternion algebra B over $\mathbb{Q}$ s.t. $B \otimes \mathbb{R} \simeq M_2(\mathbb{R})$ (see [Sh2], [C]). These curves parametrize polarized abelian surfaces with level structure whose endomorphism algebras/$\mathbb{Q}$ contain B. If such a B splits over F, then the associated curves embed naturally in the Hilbert-Blumenthal surfaces attached to F. The theorem implies in particular, since homological and linear equivalences coincide for H-B surfaces, that:

(9.35) Every Shimura curve/$\mathbb{Q}$ sitting on a Hilbert-Blumenthal surface $\tilde{S}_K$ is linearly equivalent to a H-Z divisor coming from modular curves.

9.36. We close this section by considering an analogous situation. Let B be a quaternion algebra over F such that B is completely split at ∞, i.e.,

$$B_\infty = B \otimes_{\mathbb{Q}} \mathbb{R} \simeq M_2(\mathbb{R}) \times M_2(\mathbb{R}) .$$

Then we have a natural monomorphism h_B of $\mathbb{C}^*$ in $B_\infty^\times$, analogous to (2.7), and so one gets a Shimura variety S^B attached to $(B^\times, h_B)$. For every compact open subgroup K' of $B^*(\mathbb{A}_{F,f})$ one gets this way a surface, which is already projective (!), because $B^\times$ has no unipotents.

It is a very interesting open question to prove the Tate conjectures for such a Shimura (or compact Hilbert-Blumenthal) surface.

Now suppose further that there is a quaternion algebra B_0 over $\mathbb{Q}$ s.t.

(9.37) $$B_0 \otimes_{\mathbb{Q}} \mathbb{R} \simeq M_2(\mathbb{R}), \text{ and } B = B_0 \otimes_{\mathbb{Q}} F.$$

Then in this case, the Shimura curves attached to B_0^* embed in S^B, and one can define natural analogs of H-Z cycles. For every irreducible automorphic representation π' of $B^*(\mathbb{A}_F)$ one can extend the notion of being distinguished. A theorem of Jacquet and Lai ([JLai])asserts that if π' corresponds to some cuspidal π of $GL_2(\mathbb{A}_F)$, the existence of which is proved in [JL] (see also [G]), then:

(9.38) $$\pi'\text{:distinguished} \Longleftrightarrow \pi\text{:distinguished}.$$

Using this Lai ([Lai]) has proven the analogue of Theorem 9.11 for $S_{K'}^B$, for $K' \subset B^*(\mathbb{A}_{F,f})$.

§10. Relations with twisted products of Shimura curves over F.

10.1. Suppose $\pi = \otimes \pi_v$ is a cuspidal automorphic representation of $GL_2(\mathbb{A}_F)$ of weight two such that:

(10.2) At some finite place w, π_w is in the discrete series.

Note that the only discrete series representations of $GL_2(F_w)$ are the special and the supercuspidal representations ([JL]).

Then we may choose a quaternion algebra B over F such that

(10.3) (i) $B \otimes_{\mathbb{Q}} \mathbb{R} \simeq M_2(\mathbb{R}) \times \mathbb{H}$, $\mathbb{H}$:Hamilton's quaternions;

(ii) B_w is ramified; and

(iii) $B_v \simeq M_2(F_v)$, for every finite place $v \neq w$.

Furthermore, by Jacquet-Langlands [JL] we can find an irreducible automorphic represenation π' of $B^*(\mathbb{A}_F)$ such that:

(10.4) $$L(\pi',s) = L(\pi,s).$$

One takes π'_∞ to be: $\mathcal{D}_2 \otimes 1$, and π'_w to be the irreducible finite-dimensional representation of B^*_w (which is the multiplicative group of a Quaternion algebra by (ii)) corresponding to the discrete series representation π_w given by the Local correspondence ([JL]). At every other (finite) place v, one takes: $\pi'_v = \pi_v$.

There is an obvious embedding h_B of $\mathbb{C}^*$ in B^*_∞, and the associated Shimura variety in this case is a projective <u>curve</u> $C = C(B,h_B)$ whose canonical field of definition is the real quadratic field F.

<u>10.5</u>. One can find a compact open subgroup K (resp. K') of $GL_2(\mathbb{A}_{F,f})$ (resp. $B(\mathbb{A}_{F,f})$) such that π (resp. π') contributes to $IH^2(\overline{S}_K)$ (resp. $H^1(C_B)$). And a theorem of Carayol [C], see also [Sh2], says that:

(10.6) To every such π' there exists an ℓ-adic representation $V_\ell(\pi')$ over $E_{\pi'} \otimes \mathbb{Q}_\ell$ of $Gal(\overline{\mathbb{Q}}/F)$ such that $L(V_\ell(\pi'),s) = L(\pi',s)$.

It can be seen that the field of definition $E_{\pi'}$ of π' equals E_π.

Thus if we set:

(10.7) $$V_\ell(\pi) = V_\ell(\pi'),$$

we get a perfect candidate for the elusive 2-dimensional ℓ-adic representation, having the right L-function, attached to π (see Conjecture 5.3).

Then we have

(10.8) $$\alpha_\pi = V_\ell(\pi') \otimes V_\ell(\pi')^\theta \simeq W_\ell(\pi) \text{ as } Gal(\overline{\mathbb{Q}}/F)\text{-modules}$$

and

$$L(V_\ell(\pi') \otimes V_\ell(\pi')^\theta,s) = L(W_\ell(\pi),s).$$

By a <u>twisted product</u> of C we mean the surface:

(10.9) $$X = C \times C^\theta .$$

It is immediate that if K is as in §10.5:

(10.10) $V_\ell(\pi') \otimes V_\ell(\pi')^\theta$ is a component of $H^2_{et}(X_{K'} \times \overline{\mathbb{Q}}, \mathbb{Q}_\ell)$.

Since $W_\ell(\pi)$ is on the other hand a component of $IH^2_{et}(\overline{S}_K \times \overline{\mathbb{Q}}, \mathbb{Q}_\ell)$, the general conjectures of Tate ([T1]) imply the following special.

<u>Conjecture 10.11</u>. Let π be as in §10.1. Then there exists an algebraic correspondence $\gamma: X_{K'} \times \overline{\mathbb{Q}} \longrightarrow \overline{S}_K \times \overline{\mathbb{Q}}$, defined over F, which induces the isomorphism α_π (of (10.8)).

<u>10.12</u>. An immediate consequence of the existence of such a γ will be a proof of Tate's conjecture for π-divisors on $\overline{S}_K \times \mathbb{Q}$, because one knows by Faltings ([F]) the truth of Tate's conjecture for divisors on a product of 2 curves. But the existence of γ will mean much more. This is so because we can start with explicit divisors on $X_{K'} \times \overline{\mathbb{Q}}$, which one can construct as in [MR], by mimicking the works of Ribet [Ri2] and Momose [Mo], and push forward these divisors by γ onto $\overline{S}_K \times \overline{\mathbb{Q}}$. Can one explicitly construct non-Hirzebruch-Zagier cycles this way?

<u>10.13</u>. On the Hodge-theoretic side, by studying the periods, Oda shows in [Od2] how his transcendental abelian varieties A_π tend to be isogenous to the arithmetic abelian varieties $A_{\pi'}$ coming from $H^1(C_{K'})$. More than abelian varieties, what the <u>really</u> interesting thing in this work of his is the coincidence, under suitable hypothesis on π, of the Hodge structures on the π-part of $IH^2_B(\tilde{S}_K)$ and on the π'-part of $H^2_B(X_{K'})$. This fact, when used in conjunction with (10.8) will imply the full Tate conjecture for π-divisors for any such π which comes from a quaternion algebra B, satisfying (10.3).

§11. <u>Regulators</u>.

<u>11.1</u>. Let us first briefly recall the classical regulator map on the group of units $\mathcal{O}_k^*$ of a number field k of degree $n = r_1 + 2r_2$. Set:

(11.2) $$Y_k = \mathbb{Z}^{\mathrm{Hom}(k,\mathbb{C})}.$$

Note that $\mathrm{Gal}(\overline{\mathbb{Q}}/\mathbb{Q})$ acts on Y_k.

The regulator map of Dirichlet is:

$$r: \mathcal{O}_k^* \oplus Y_k^{\mathrm{Gal}(\overline{\mathbb{Q}}/\mathbb{Q})} \to (Y_k \otimes \mathbb{R})^+ \simeq \mathbb{R}^{r_1+r_2} , \tag{11.3}$$

where $r|_{\mathcal{O}_k^*}$ is the map: $u \mapsto \sum_\sigma \log|u|_\sigma \sigma$, and $r|_{Y_k^{\mathrm{Gal}(\overline{\mathbb{Q}}/\mathbb{Q})}}$ is the natural injection.

One has the following

Theorem 11.4 (Dirichlet-Dedekind) ([Sa], [Lg2])

(a) $\mathrm{Im}(r)$ is a lattice in $(Y_k \otimes \mathbb{R})^+$.

(b) The volume of $\mathrm{Im}(r)$, obtained by comparing it to Y_k^+, equals (up to a non-zero rational number) the leading coefficient at $s = 0$ of $\zeta_k(s)$.

(c) $\mathrm{Ker}(r)$ consists of roots of unity in $\mathcal{O}_k^*$.

We prefer this non-standard formalism, which generalizes to surfaces well. Indeed, what this theorem tells us is:

$$\mathrm{rk}.\,\mathcal{O}_k^* + \mathrm{rk}.\,Y_k^{\mathrm{Gal}(\overline{\mathbb{Q}}/\mathbb{Q})} = r_1 + r_2 . \tag{11.5}$$

It is in some sense an accident that we know:

$$Y_k^{\mathrm{Gal}(\overline{\mathbb{Q}}/\mathbb{Q})} \simeq \mathbb{Z} . \tag{11.6}$$

(In general we will have two groups, neither one of which will have an easily determinable rank.) Combining this with the well-known properties of $\zeta_k(s)$ we get:

(11.7) (a) $\mathrm{rk}.\,Y_k^{\mathrm{Gal}(\overline{\mathbb{Q}}/\mathbb{Q})} = 1 = -\mathrm{ord}_{s=1}\, \zeta_k(s)$

and

(b) $\mathrm{rk}.\,\mathcal{O}_k^* = r_1 + r_2 - 1 = \mathrm{ord}_{s=0}\, \zeta_k(s)$.

11.8. Now let X be a smooth projective surface over the number field k. Then Tate's conjectures 7.12 and 7.15 give an analog of 11.7(a), namely:

$$(11.9)\qquad \dim_{\mathbb{Q}_\ell}(H^2_{et}(X \times_k \overline{\mathbb{Q}}, \mathbb{Q}_\ell(1))^{Gal(\overline{\mathbb{Q}}/k)}) = -\mathrm{ord}_{s=2} L^{(2)}(X_k, s).$$

One should also have:

$$(11.10)\qquad NS(X_k) \otimes \mathbb{Q}_\ell \simeq H^2_{et}(X \times_k \overline{\mathbb{Q}}, \mathbb{Q}_\ell(1))^{Gal(\overline{\mathbb{Q}}/k)} .$$

It is easy to see that everything can be formulated over $\mathbb{Q}$ by considering the restriction of scalars: $X_o \overset{def}{=} R_{k/\mathbb{Q}} X/k$. The conjecture becomes:

$$(11.11)\qquad NS(X_o) \otimes \mathbb{Q}_\ell \cong H^2_{et}(X_o \times_{\mathbb{Q}} \overline{\mathbb{Q}}, \mathbb{Q}_\ell(1))^{Gal(\overline{\mathbb{Q}}/\mathbb{Q})} .$$

11.12. One wishes (see [Bl]) that there is a nice "motivic" cohomology theory $H^*_M(X \times_{\mathbb{Q}} \overline{\mathbb{Q}}, \mathbb{Z}(**))$, admitting a Galois action and various realizations such as étale or Betti. Then $H^2_M(X \times_{\mathbb{Q}} \overline{\mathbb{Q}}, \mathbb{Z}(1))$ will be the right analog of Y_k. But unfortunately nobody seems to know how to get at such an arithmetic/motivic object, at least till now.

11.13. In his exciting work ([Bl]) Beilinson also proposes the use of a certain (Chow) group $CH^{2,1}(X_k)$ (to be defined below) in the place of $\mathcal{O}^*_k$, and defines a regulator map

$$(11.14)\qquad r: CH^{2,1}(X_k) \oplus NS(X_k) \to Y^+_{\mathbb{R}}$$

where

$$Y_{\mathbb{R}} = H^{1,1}(X \times_{\mathbb{Q}} \mathbb{C}, \mathbb{C}) \cap H^2_B(X \times_{\mathbb{Q}} \mathbb{C}, \mathbb{R}(1)) .$$

It is not hard to deduce that:

$$(11.15)\qquad H^2_B(X \times_{\mathbb{Q}} \mathbb{C}, \mathbb{R}(1)) \simeq \prod_{\sigma \in Hom(k,\mathbb{C})} H^2_B(X^\sigma(\mathbb{C}), \mathbb{R}(1)) .$$

The restriction of r to $NS(X_k)$ is simply the cycle class map $c\ell_B$ (see (7.22)).

11.16. What is $CH^{2,1}(X_k)$? Simply put, every element β of $CH^{2,1}(X_k)$ is represented (cf. [Q]), by a finite formal sum $\sum_j (C_j, f_j)$, rational over k, where each C_j is a closed irreducible curve on $X_{\bar{k}}$ and f_j a function invertible at the generic point of C_j such that:

(11.17) $\sum_j \operatorname{div}(f_j) = 0$ as a zero-cycle on $X_{\bar{k}}$.

To get a simple example of such an arithmetic/geometric object assume there are two copies C_i of $\mathbb{P}^1$ on $X_{\bar{k}}$ intersecting only at $\{0,\infty\}$. Take f_1 to be the parameter t_1 on C_1 and f_2 to be t_2^{-1}. Then the zero (resp. pole) of f_1 at 0 (resp. ∞) is cancelled by the pole (resp. zero) of f_2 at ∞ (resp. 0). The formal sum $(C_1,f_1) + (C_2,f_2)$ gives rise to a class in $CH^{2,1}(X_{\bar{k}})$.

Given a finite set of curves C_j, we can choose each f_j to be an integer n_j, in which case (11.17) will be automatically satisfied. And $\sum(C_j,n_j)$ can be identified with the divisor $\sum n_jC_j$. This way one gets a map of $\operatorname{Pic}(X_{\bar{k}})\otimes\bar{k}^*$ into $CH^{2,1}(X_{\bar{k}})$.

Note also that the (codimension 2) condition (11.17) implies that each f_j is a unit on the intersection of C_j with the non-singular part of $\bigcup_j C_j$.

11.18. Let $\langle\ ,\ \rangle$ be the self-dual pairing on $H^{1,1}$ defined by:

(11.19) $$\langle\omega_1,\omega_2\rangle = \int_{X(\mathbb{C})} \omega_1\wedge\omega_2 .$$

Then, for each β in $CH^{2,1}(X_k)$, the image $r(\beta)$ in $Y^+_{\mathbb{R}}$ (under the Beilinson regulator r) is completely described by the formula $(\forall\sigma\in\operatorname{Hom}(k,\mathbb{C}))$:

(11.20) $$\langle r(\beta)^\sigma,\omega^\sigma\rangle = \frac{1}{2\pi i}\sum_j \int_{C_j'^\sigma(\mathbb{C})} \log|f_j^\sigma|\cdot\omega^\sigma|_{C_j} ,$$

if $\sum_j (C_j,f_j)$ represents β. By C_j' we mean the $C_j-\operatorname{div}(f_j)$.

It is a fact that $r(\beta)^\sigma$ actually lies in the plus-part of $Y_{\mathbb{R}}$ when σ is real.

For every pair (g,h) of rational functions on the surface $X_{\bar{k}}$, define the Tame symbol of (g,h) at a curve $C\subset X_{\bar{k}}$ by:

(11.21) $$\operatorname{Tame}_C(g,h) = (-1)^{\operatorname{ord}_C(g)\operatorname{ord}_C(h)}\cdot\left(\frac{g^{\operatorname{ord}_C(h)}}{h^{\operatorname{ord}_C(g)}}\right)\Bigg|_C$$

This is the geometric analog of the Norm-residue symbol, and it gives a rational function on C. A basic result ([Q]) is:

(11.22) $\sum_C \operatorname{div}(\mathrm{Tame}_C(g,h)) = 0$ as a zero-cycle on $X_{\bar{k}}$.

We say that two sums: $\sum(C_j^1, f_j^1)$ and $\sum(C_i^2, f_i^2)$ satisfying (11.17) are equivalent if they differ by a linear combination of elements of the form: $\sum_C \operatorname{div}(\mathrm{Tame}_C(g,h))$.

$CH^{2,1}(X_k)$ consists of such equivalence classes, rational over k.

11.23. As a side remark, it may be worthwhile to say that $CH^{2,1}(X_k) \otimes \mathbb{Q}$ is the "weight two" ([So]) subspace of $K_1(X_k) \otimes \mathbb{Q}$, which fact points up the analogy with $\mathcal{O}_k^* \simeq K_1(\mathcal{O}_k)$. In the "motivic" terminology, $CH^{2,1}(X_k) \otimes \mathbb{Q}$ is $H_M^3(X_k, \mathbb{Q}(2))$, while $\mathcal{O}_k^* \otimes \mathbb{Q}$ is $H_M^1(\operatorname{spec} \mathcal{O}_k, \mathbb{Q}(1))$.

The beautiful thing which makes everything work is the following

Proposition 11.24. ([B2]) For every pair (g,h) of rational functions on $X_{\bar{k}}$ one has:

$$r\Big(\sum_C (C, \mathrm{Tame}_C(g,h))\Big) = 0.$$

This is a non-obvious generalization of the well-known product formula for symbols on curves ([S3]). As a consequence we can define r on $CH^{2,1}(X_k)$.

11.25. Let us now take a proper model $X_{\mathcal{O}_k}$ of X over $\mathcal{O}_k$. One can define the "integral part" $CH^{2,1}(X_{\mathcal{O}_k}) \otimes \mathbb{Q}$ of $CH^{2,1}(X_k) \otimes \mathbb{Q}$ ([B1]). (One hopes that this is independent of the choice of the model $X_{\mathcal{O}_k}$.) Let us assume the standard conjectures ([S1]) on the analytic properties of $L^{(2)}(X_k, s)$.

Conjecture 11.26. (Beilinson [B1], §3)

(a) $r(CH^{2,1}(X_{\mathcal{O}_k}) \otimes \mathbb{Q} \oplus NS(X_k) \otimes \mathbb{Q})$ is a $\mathbb{Q}$-lattice of $Y_{\mathbb{R}}^+$.

(b) The volume of this $\mathbb{Q}$-lattice equals, up to a non-zero rational number, the leading coefficient of $L^{(2)}(X_k, s)$ at $s = 1$.

(c) $\mathrm{rk.CH}^{2,1}(X_{\mathcal{O}_k}) = \mathrm{ord}_{s=1} L^{(2)}(X_k,s).$

The explicit definition of the Euler factor at ∞ of $L^{(2)}(X_k,s)$ (see (4.11)), which depends only on the Hodge type, allows us to deduce:

(11.27) $$\dim Y^+_{\mathbb{R}} = \mathrm{ord}_{s=1} L^{(2)}(X_k,s) - \mathrm{ord}_{s=2} L^{(2)}(X_k,s).$$

We thus see the analogy with (11.5), and the compatibility with Tate's conjectures (11.9) and (11.10).

11.28. How does one define the volume of a $\mathbb{Q}$-lattice (= $\mathbb{Q}$-structure) of $Y^+_{\mathbb{R}}$? The key thing is to make use of the short exact sequence:

(11.29)
$$0 \to \sum_{\sigma\in \mathrm{Hom}(k,\mathbb{C})} (F^2H^2_{DR}(X_k)\otimes_{k,\sigma}\mathbb{C})^+ \xrightarrow{p} H^2_B(X\times_{\mathbb{Q}}\mathbb{C},\mathbb{R}(1))^+ \to Y^+_{\mathbb{R}} \to 0,$$

where $H^*_{DR}(X_k)$ denotes the algebraic DeRham cohomology of X_k ([DMOS]), and $F^{\cdot}$ the Hodge filtration (defined algebraically) on H^*_{DR}. The "period map" p is induced, after identifying $H^2_{DR}(X_k)\otimes_{k,\sigma}\mathbb{C}$ with $H^2_B(X(\mathbb{C}),\mathbb{C})$, by the projection morphism: $\mathbb{C} = \mathbb{R}(1)\oplus\mathbb{R}(2) \to \mathbb{R}(1)$.

Note that p does not map the $\mathbb{Q}$-structure $L_{DR} \overset{\mathrm{def}}{=} \sum_\sigma F^2H^2_{DR}(X_k)^+$ into the $\mathbb{Q}$-structure $L_B \overset{\mathrm{def}}{=} H^2_B(X\times_{\mathbb{Q}}\mathbb{C},\mathbb{Q}(1))^+$. But nevertheless we can set:

(11.30) $$L = (\Lambda^{\max} L_B)\otimes(\Lambda^{\max} L^{\vee}_{DR}),$$

which is a $\mathbb{Q}$-structure of $(\Lambda^{\max} Y^+_{\mathbb{R}})$. If $\mathcal{L}$ is any $\mathbb{Q}$-lattice in $Y^+_{\mathbb{R}}$, the (co-) volume of $\mathcal{L}$ is defined by:

(11.31) $$\Lambda^{\max}\mathcal{L} = \mathrm{vol}(\mathcal{L})L.$$

11.32. To conclude let us note that it can happen that $Y^+_{\mathbb{R}} = 0$. Then the period morphism p is an isomorphism, and the point $s = 1$ becomes a critical point in the sense of Deligne

([D3]). The group $CH^{2,1}(X_{\mathcal{O}_k})$ should be torsion in this case. Observe however that L still defines a $\mathbb{Q}$-structure of the one-dimensional real vector space $\Lambda^{max}(Y^+_{\mathbb{R}})$. So we get:

$$\text{vol}(\mathcal{L}) = \det(p), \quad \underline{\text{and}} \tag{11.33}$$

$$L^{(2)}(X_k,1) \sim_{\mathbb{Q}^*} \det(p), \tag{11.34}$$

which is just Deligne's conjecture on critical points ([D3]).

§12. <u>On the residues of poles of L-functions</u>.

<u>12.1</u>. In this section we will briefly discuss an interpretation, as predicted by Beilinson's conjecture 11.26, of the leading coefficient at $s=1$ of the H^2-piece of the L-function of any Hilbert-Blumenthal surface $\tilde{S}_K$ in terms of the periods of certain (1,1)-forms on $\tilde{S}_K(\mathbb{C})$ defined by the regulator r. Since the functional equation for $L_Q^{(2)}(\tilde{S}_K,s)$ (see §4.28) relates s to $3-s$, the point $s = 1$ is the reflex point of the Tate point $s = 2$. Thus the leading coefficient at $s = 1$ is up to certain simple factors the residue at $s = 2$.

<u>Theorem 12.2</u>. ([R1], [R2]) Let K be any compact open subgroup of G_f, and let k be any number field, abelian over $\mathbb{Q}$. Then there is a natural $\mathbb{Q}$-subspace $\mathcal{R}_K(k)$ of $CH^{2,1}(\tilde{S}_K \times k) \otimes \mathbb{Q}$ (see 11.16) such that:

(a) $r(\mathcal{R}_K(k) \oplus NS(\tilde{S}_K \times k) \otimes \mathbb{Q})$ is a $\mathbb{Q}$-lattice of $Y^+_{\mathbb{R}}$.

(b) The volume of $r(\mathcal{R}_K(k) \oplus NS(\tilde{S}_K \times k))$ is a non-zero rational multiple of the leading coefficient at $s = 1$ of $L_Q^{(2)}(\tilde{S}_K \times k,s)$.

Recall that:

$$Y_{\mathbb{R}} = \sum_{\sigma: k \hookrightarrow \mathbb{C}} H^{1,1}(\tilde{S}_K^{\sigma}(\mathbb{C}), \mathbb{R}(1)). \tag{12.3}$$

<u>12.4</u>. Theorem 12.2 makes essential use of the proof (cf. §9) of Tate's conjecture for abelian number fields due to Harder, Langlands and Rapoport, the existence of "modular units"

(see below) on H-Z cycles, the adèlic version (4.29) of Asai's "period integral" representation for the L-function, suitable functoriality properties of Gauss sums and ε-factors ([R3], [D5]), and certain methods derived from Jacquet ([J]) which allow us to prove the surjectivity of $r \otimes \mathbb{R}$. This result is a (non-trivial) analog of Beilinson's proof of the corresponding statement for a product of two modular curves ([B1], §6), extending Bloch's earlier unpublished work on $\mathrm{Jac}(X_0(37))$.

<u>12.5</u>. <u>Definition of</u> $\mathcal{R}_K(k)$ <u>via H-Z-cycles and modular units</u>:

Recall (from §8.18-8.26) that every g in G_f defines an H-Z cycle on $\overline{S}_K$ defined by:

(12.6) $$\overline{C}_{g,K} = p_K(\rho_g(\delta(\overline{M}))),$$

where δ denotes the diagonal embedding of $\overline{M}$ in $\overline{S}$, and p_K the canonical projection: $\overline{S} \to \overline{S}_K$. Furthermore there is a non-trivial morphism (defined over $\mathbb{Q}$):

(12.7) $$\lambda_{g,K}: \overline{M}_L \to \overline{C}_{g,K}, \text{ where } L = H_f \cap gKg^{-1}.$$

By a <u>modular unit</u> on $\overline{M}_L \times \overline{\mathbb{Q}}$ we mean an element of $\mathcal{O}^*_{\overline{M}_L \times \overline{\mathbb{Q}}}$, viz, a function on $\overline{M}_L \times \overline{\mathbb{Q}}$ all of whose zeros and poles lie entirely at the cusps ([KuL], [R2]).

(12.8) By a <u>modular unit on the H-Z cycle</u> $\overline{C}_{g,K} \times \overline{\mathbb{Q}}$ we will mean a function f such that $f \circ \lambda_{g,K}$ is a modular unit on $\overline{M}_L \times \overline{\mathbb{Q}}$.

(12.9) Call a finite formal sum $\sum_j (C_j, f_j)$ <u>K-admissible</u> iff

(i) each C_j is a H-Z cycle $C_{g,K} \times \overline{\mathbb{Q}}$,

(ii) each f_j is a modular unit on C_j,

and (iii) $\sum_j \mathrm{div}(f_j) = 0$ as a zero-cycle on $\overline{S}_K \times \overline{\mathbb{Q}}$.

It makes perfectly good sense to talk about the rationality over a number field k of such a K-admissible sum. It is not hard to see that every K-admissible sum is rational over some cyclotomic field. This is so because the components of H-Z cycles and the cusps on $\overline{M}$ are all rational over $\mathbb{Q}^{ab}$.

Note that the intersection points of the C_j's occurring in any K-admissible sum are rather special. Compare with (8.35).

Now let us look at the situation on the smooth model $\tilde{S}_K$. Denoting, as in (8.27), by $\tilde{C}_{g,K}$ the H-Z cycle in $\tilde{S}_K$ mapping down to $\overline{C}_{g,K}$ under the resolution map $\pi_K:\tilde{S}_K \to \overline{S}_K$, we see that it may happen that $\overline{C}_{g,K}$ and $\overline{C}_{g',K}$ may meet on $\overline{S}_K$, but $\tilde{C}_{g,K}$ and $\tilde{C}_{g',K}$ may not. This poses only a mild problem for us. Indeed, since the components of the cusp resolutions are all rational curves, we claim:

(12.10) Given any K-admissible, k-rational sum $\beta = \sum_j (C_j,f_j)$ on $\overline{S}_K \times k$, we can find a finite formal sum $\sum_i (E_i,\phi_i)$, rational over $\mathfrak{O}_k$, with each E_i a component of the resolution of some cusp and ϕ_i: (an $\mathfrak{O}_k$-integrally defined) function on E_i, such that:

$$\sum_j \operatorname{div}(f_j) + \sum_i \operatorname{div}(\phi_i) = 0.$$

We will call $\tilde{\beta} \overset{\text{def}}{=} \sum_j (C_j,f_j) + \sum_i (E_i,\phi_i)$ a "<u>lift</u>" of β to $\tilde{S}_K \times k$. In general, there are more than one lift, and the difference of two such lifts will be (integrally defined and) supported on $\tilde{S}_K^{\infty}$. Set:

$$(12.12) \qquad \mathcal{R}'_K(k) = \begin{cases} \text{The } \mathbb{Q}\text{-subspace of } CH^{2,1}(\tilde{S}_K \times k) \otimes \mathbb{Q} \\ \text{generated by lifts of K-admissible,} \\ \text{k-rational sums } \sum_j (C_j,f_j). \end{cases}$$

If M is an open subgroup of K, then the natural morphism $p_{K/M}:\tilde{S}_M \to \tilde{S}_K$ (denoted by $J_{K,M}(e)$ in (2.20)(ii)) induces a (transfer) map:

$$(12.13) \qquad p_{K,M_*}:CH^{2,1}(\tilde{S}_M \times k) \otimes \mathbb{Q} \to CH^{2,1}(\tilde{S}_K \times k) \otimes \mathbb{Q}.$$

The space we are after is:

$$(12.14) \qquad \mathcal{R}_K(k) \overset{\text{def}}{=} \begin{cases} \text{The } \mathbb{Q}\text{-subspace of } CH^{2,1}(\tilde{S}_K \times k) \text{ generated} \\ \text{by } \{p_{K,M_*}(\mathcal{R}'_M(k)) \mid M\text{:open in } K\}. \end{cases}$$

12.15. Integrality of $\mathcal{R}_K$: We saw in §2.44 that one does not yet have a regular model over $\mathbb{Z}$ for $\tilde{S}_K$. Our construction of $\mathcal{R}_K$ is rather natural, and it will come from any regular model/$\mathbb{Z}$, (if and) when such a thing is proven to exist. In the meanwhile let us get around this problem by doing the following:

One knows (cf. Katz-Mazur [KM]) that there is a regular, proper model $\overline{M}_{\mathbb{Z}}$ of $\overline{M}$ over $\mathbb{Z}$. Since each Hirzebruch-Zagier cycle is a suitable translate of the modular curve, given a finite union C of such cycles, we can find a nice proper model $C_{\mathbb{Z}}$ over $\mathbb{Z}$. On the other hand, if f is a modular unit, we can analyze, by using the ideas in [B1], §5, when f can be spread out to a function on the integral model.

From these facts, and by using the very simple nature of $\tilde{S}_K^{\infty}$, we can show the following:

Theorem 12.17. ([R1], [R2]) For every open, compact subgroup K of G_f, we can find a (reducible) curve C on $\tilde{S}_K$ with proper model $C_{\mathbb{Z}}$ over $\mathbb{Z}$ such that:

$$\mathcal{R}_K(k) \subset \operatorname{Im}(K_1'(C_{\mathbb{Z}} \times \mathcal{O}_k) \xrightarrow{j_C} K_1(\tilde{S}_K \times k)) \otimes \mathbb{Q},$$

where j_C is the homomorphism induced by the natural (composite) map: $C_{\mathbb{Z}} \to C \hookrightarrow \tilde{S}_K$.

Let us now come back to the proof of the main Theorem 12.2. Note first that the regulator

$$\text{(12.18)} \qquad r: \mathcal{R}_K(k) \oplus NS(\tilde{S}_K \times k) \otimes \mathbb{Q} \to Y_{\mathbb{R}}(k)^+$$

commutes with the Galois action, and we can decompose it accordingly. Theorem 12.2 will now follow from:

Theorem 12.19. For every character ω of $\mathrm{Gal}(\mathbb{Q}^{ab}/\mathbb{Q})$, one has: $r^{\omega}(\mathcal{R}_K(\omega) \oplus NS(\tilde{S}_K,\omega))$ is a $\mathbb{Q}(\omega)$-structure of $Y_{\mathbb{R}}(\omega)^+$ whose volume is, up to a non-zero number in $\mathbb{Q}(\omega)$, the leading coefficient at $s = 1$ of $L^{(2)}(\tilde{S}_K(\omega),s)$.

The meaning of the terms in this theorem are self-evident. Let us take note that in [Od1] T. Oda has a certain "period" integral" representation for the L-function at $s = 1$, but he does not relate it to any regulator. See also [Sh3], Th.3.11(iii).

Corresponding to the decomposition (cf. (3.4))

(12.20) $$H^2(\tilde{S}_K(\omega)) \simeq IH^2(\tilde{S}_K(\omega)) \oplus H^2_{S_K^\infty}(\tilde{S}_K(\omega)),$$

there is a decomposition of $\mathcal{R}_K(\omega)$, $NS(\tilde{S}_K,\omega)$ and also the regulator r^ω.

12.13. The piece of r^ω coming from $\tilde{S}_K^\infty(\omega)$ can be handled by reducing the problem, thanks to (2.44), (3.8) and (4.13), to questions about cyclotomic regulators, which are known.

12.21. We can next decompose the "intersection piece" of r^ω according to the action of the Hecke algebra. This way we get, for every π in $\mathcal{R}_K$, a morphism:

(12.22) $$r^\omega_\pi: \mathcal{R}(\pi,\omega) \oplus NS(\pi,\omega) \to Y_{\mathbb{R}}(\pi,\omega)^+ .$$

12.23. When π is one-dimensional, we can appeal to (3.25) and (4.15), and once again reduce the problem to cyclotomic regulators.

12.24. Now let π be cuspidal, and distinguished relative to ω (cf. 9.31). Then in this case we have, thanks to Theorem 9.32(a):

(12.25) $$\dim NS(\pi,\omega) = 1 = -\mathrm{ord}_{s=2} L_Q(\pi \otimes \omega',\rho,s),$$

where ω' is an extension to $\mathbb{A}_F^*$ of ω (viewed as a character of $\mathbb{A}_{\mathbb{Q}}^*$). Besides, it can be proved ([R2]) that:

(12.26) $$r^\omega_\pi(\mathcal{R}(\pi,\omega)) = 0.$$

Furthermore (see §9.16) we have the "motivic" decomposition

(12.27) $$W_\ell(\pi,\omega) \simeq M_\ell \oplus \omega(-1) \quad \underline{\text{and}} \quad W_B(\pi,\omega) \simeq M_B \oplus \omega(-1),$$

where M_ℓ the ℓ-adic representation attached to the symmetric square of a 2-dimensional $\mathrm{Gal}(\overline{\mathbb{Q}}/\mathbb{Q})$-module.

The regulator breaks up accordingly. The $\omega(-1)$-piece is described by the cycle class map on $NS(\pi,\omega)$, and this part of

Theorem 12.19 is a consequence of the theorem of Harder, Langlands and Rapoport.

On the other hand, the "motive" M is critical in the sense of Deligne ([D3]). Here Beilinson's conjecture is equivalent, as we saw in §11.32, to Deligne's critical value conjecture. Thus, the corresponding part of Theorem 12.19 is a consequence of the proof, due independently to J. Sturm ([St]) and D. Zagier ([Z1]) of Deligne's conjecture for the symmetric square L-function of cusp forms of weight two.

12.28. It suffices now to consider the crucial case when π is cuspidal and not distinguished. In this case one has:

(12.29) $$NS(\pi,\omega) = 0, \quad \underline{\text{and}}$$
$$\text{ord}_{s=1} L_Q(\pi\otimes\omega',\rho,s) = \dim Y_{\mathbb{R}}(\pi,\omega)^+ = 1.$$

Theorem 12.2 will now follow immediately from:

Proposition 12.30. ([R2]) For every non-distinguished, cuspidal π:

$$r_\pi^\omega(\mathcal{R}(\pi,\omega)) = (L_Q'(\pi\otimes\omega',\rho,1))\cdot L_\pi^\omega \subset Y_{\mathbb{R}}(\pi,\omega)^+,$$

where L_π^ω is the $E_\pi\otimes\mathbb{Q}(\omega)$-structure of $Y_{\mathbb{R}}(\pi,\omega)^+$ defined as in §11.28.

12.18. We will now briefly indicate the idea of proof of this proposition. First of all observe that, for any modular unit f, the harmonic function $\log|f|$ can be written, thanks to the Kronecker limit formula ([Rc], [R2]), as the limit of an Eisenstein series on $GL_2(\mathbb{A}_\mathbb{Q})$ as the parameter s tends to 1^+. This fact, when combined with the definitions of the regulator (cf. (11.20)) and of $\mathcal{R}_K$ in terms of H-Z cycles and modular units (cf. §12.5), shows that the periods of the differential forms in $\text{Im}(r_\pi^\omega)$ are simple multiples of integrals of the form:

(12.19) $$\lim_{s\to 1^+} \sum_{g\in I} \int_{M(\mathbb{C})} \mathcal{E}_g(z,x_f;s)\omega(\det x_f)\eta_{\rho_g h}(z,x_f)dx_f,$$

where I is a finite subset of G_f, $\mathcal{E}_g$ an Eisenstein series on $GL_2(\mathbb{A}_\mathbb{Q})$ of weight zero (without singularity at $s = 1$) attached to g, h an $E_\pi \otimes \omega$-rational non-zero automorphic form in the space of $\pi \otimes \omega'$, and $\eta_{\rho_g h}$ a differential form of type (1,1) on $\tilde{S}_K(\mathbb{C})$ attached to $\rho_g h$.

By the adèlic version (4.29) of Asai's integral representation for the L-function, we see that (12.19) is a Rankin-Selberg integral, and is a suitable multiple of $L_Q(\pi^\vee \otimes \overline{\omega}', \rho, 2)$, which is in turn related to $L_Q'(\pi \otimes \omega, \rho, 1)$ by the functional equation involving an ε-factor.

By appealing to functoriality of ε-factors under Galois action ([R3], [D2]), and to the rationality properties of local integrals of Whittaker vectors, we deduce that:

$$r_\pi^\omega(\mathcal{R}(\pi,\omega)) \subset L_Q'(\pi \otimes \omega', \rho, 1) L_\omega^\pi. \tag{12.20}$$

Finally, by suitably adapting certain arguments of Jacquet in [J], we can show that for every non-distinguished π, and character ω, there exists a finite family of Eisenstein series $\{\mathcal{E}_g | g \in I\}$ such that (12-19) is non-zero; whence the identity:

$$r_\pi^\omega(\mathcal{R}(\pi,\omega)) \otimes \mathbb{R} = Y_{\mathbb{R}}(\pi,\omega)^+. \tag{12.21}$$

Since L_ω^π is an $E_\pi \otimes \mathbb{Q}(\omega)$-lattice in $Y_{\mathbb{R}}(\pi,\omega)^+$ we obtain from (12.20) and (12.21):

$$r_\pi^\omega(\mathcal{R}(\pi,\omega)) = L_Q'(\pi \otimes \omega', \rho, 1) L_\omega^\pi . \tag{12.22}$$

This finishes the sketch of our proof of Theorem 12.2. The details will appear in [R2].

<u>12.23</u>. As a consequence of our theorem we can prove a variant, due to Beilinson [B2], of the Hodge conjecture (cf.7.24) for $\tilde{S}_K(\mathbb{C})$. Explicitly we can prove:

<u>Theorem 12.24</u>. ([R1],[R2]) Fix any imaginary quadratic field k. Then every (1,1)-class in $H_B^2(\tilde{S}_K(\mathbb{C}),\mathbb{C})$ is a $\mathbb{C}$-linear combination of classes of H-Z cycles over k, and of images under r of elements in $\mathcal{R}_K(k)$.

§13. Implications for the construction of arithmetic Hilbert modular functions.

13.1. Recall that the irregularity q $(= b_1/2)$ of any Hilbert-Blumenthal surface $\tilde{S}_K$ is zero (cf. §7.33), and that consequently linear and homological equivalences coincide for divisors on such a surface (cf. (7.27)). For principal congruence subgroups K one even has the stronger result that $\tilde{S}_K(\mathbb{C})$ is simply connected, which was proved by Švarčman ([Sv]) by making use of Vasserstein's theorem ([Va]) on the generation of $SL_2(\mathcal{O}_F)$ by elementary matrices. We can check homological triviality of a divisor in either Hodge or ℓ-adic cohomology. We thus get the following useful

Proposition 13.2. Let D be a divisor on $\tilde{S}_K \times \overline{\mathbb{Q}}$ rational over a number field k. Then

$$c\ell_\ell([D]) = 0 \iff \begin{cases} \exists \text{ a function } f \text{ on } \tilde{S}_K \times \overline{\mathbb{Q}}, \text{ rational} \\ \text{over } k, \text{ such that } D = \operatorname{div}(f) \end{cases}$$

13.3. This proposition provides a geometric way to construct arithmetically defined functions on $\tilde{S}_K$, which give rise to arithmetic Hilbert modular functions. Contrast with the usual construction of modular functions by using Eisenstein series, etc. See also Hecke's thesis ([He]) and the related paper by Baily ([Ba]).

13.4. Let us first note that since the cusps have codimension 2 on $\overline{S}_K$, there will be no Hilbert modular functions with zeros and poles only at the cusps. Using the fact that the intersection matrix of a cusp resolution is negative-definite ([HvG]), one can also see that there cannot be any function f on $\tilde{S}_K(\mathbb{C})$ whose divisor is supported entirely on $\tilde{S}_K^\infty(\mathbb{C})$. The moral is that one has to learn to work with functions on $\tilde{S}_K(\mathbb{C})$ whose divisors criss-cross the open surface $S_K(\mathbb{C})$, and try to extract arithmetical information.

13.5. Why do we care? Well, classically the values of elliptic modular units, viz. arithmetical functions on modular curves with zeros and poles only at the cusps, at $\mathbb{C}M$ points w

give rise to units in abelian extensions of the $\mathbb{C}$M field $\mathbb{Q}(w)$. See [KuL], [Rc] and [Sk]. Can one find an analog in the Hilbert-modular case?

13.6. The starting point has to be the fundamental theorem of Shimura ([Sh1]) (see also [He]) asserting that the value (if defined) of an arithmetic Hilbert modular function f at a CM point w (corresponding to a quartic $\mathbb{C}$M field L) is an algebraic number lying in an abelian extension of the reflex $\mathbb{C}$M field $L^{\vee}$. This leads us to pose the following

Problem 13.7. Can one prescribe certain necessary conditions on the nature of the divisors of (arithmetically defined) functions f on $\tilde{S}_K(\mathbb{C})$ which will guarantee that the values of f at a set of $\mathbb{C}$M points (off $\mathrm{div}(f)$) are algebraic integers?

A positive solution, probably very hard to establish, will be the first significant step in the quest for finding explicit units in abelian extensions of quartic $\mathbb{C}$M fields. This will have implications for the conjecture of Stark and Tate ([T2]). Construction and understanding of a good model of $\tilde{S}_K$ over $\mathbb{Z}$ should play a role.

13.8. Since we know well (only) the cusp resolutions and the H-Z cycles, it is natural to ask the following 2 questions:

Question A: Can one find efficient ways to construct (many) homologically trivial divisors D on $\tilde{S}_K(\mathbb{C})$ all of whose components come from H-Z cycles and cusp relations?

Question B: For every homologically trivial divisor D (= $\mathrm{div}(f)$) made up of H-Z cycles and components of cusp resolutions, can we find a set of $\mathbb{C}$M points w (off $\mathrm{div}(f)$) such that each $f(w)$ is integral?

13.9. That a good answer to Question A may exist is seen from the work of Hirzebruch and Zagier on intersection numbers of curves in Hilbert-Blumenthal surfaces ([HZ]). For instance, take $\Gamma = SL_2(\mathcal{O}_F)$, $F = \mathbb{Q}(\sqrt{p})$, $p \equiv 1 \pmod 4$ and T_N: the H-Z divisor (see §8.2) on $\tilde{S}_\Gamma(\mathbb{C})$ defined as the union of curves whose pre-images on $\mathfrak{h}^2$ satisfy:

$$a\sqrt{p}\, z_1 z_2 + \lambda z_1 - \lambda^{\theta} z_2 + b\sqrt{p} = 0, \tag{13.10}$$

for some $a,b \in \mathbb{Z}$, $\lambda \in \mathcal{O}_F$, and $abp + \lambda\lambda^{\theta} = N$. Let $[T_N^c]$ denote the component of the (Betti) homology class $[T_N]$ in the "intersection piece" of H_2. (Thus $[T_N] - [T_N^c]$ will represent a homology class supported on the cusp resolutions.) For any class γ in $H_2(\tilde{S}_\Gamma(\mathbb{C}),\mathbb{Z})$ the intersection number $\langle\gamma,[T_N^c]\rangle$ is well-defined. The main result of [HZ] is:

(13.11) For every class γ in $H_2(\tilde{S}_\Gamma(\mathbb{C}),\mathbb{Z})$ the power series $\sum_{N\geq 1} \langle\gamma,[T_N]\rangle e^{2\pi i N\tau}$, $\tau \in \mathfrak{h}$, is a modular form of weight 2, level p and Nebentypus $(\frac{-}{p})$.

Furthermore, if M is any integer ≥ 1 relatively prime to N, then one has the identity:

$$(13.12)\qquad \langle[T_M],[T_N^c]\rangle = \sum_{\substack{r^2<4NM \\ r^2\equiv 4NM \pmod p}} H\left(\frac{4NM-r^2}{p}\right) + I_p(MN)m$$

where I_p is a suitable arithmetical function, and $H(k)$ is the class number of positive-definite binary quadratic forms of discriminant $-k$.

We will now appeal to a nice argument of Zagier (cf. [Z2], §7) showing how to use (13.11) to build homologically trivial divisors on $\tilde{S}_\Gamma(\mathbb{C})$: Note that the space of holomorphic modular forms of level p, weight 2, and nebentypus $(\frac{-}{p})$ is finite-dimensional ([HZ]), and that this space has a basis consisting of forms with rational coefficients. From this it follows that, for any homology class γ, the power series $\sum_{N\geq 1} \langle\gamma,[T_N^c]\rangle e^{2\pi i N\tau}$ being a modular form of the type above is equivalent to having an infinite number of relations:

$$(13.13)\qquad \sum_{N\geq 1} a_N\langle\gamma,[T_N^c]\rangle = 0, \text{ with } a_N = 0 \text{ for almost all } N.$$

For each $(a_1,a_2,\ldots)$, consider the divisor:

$$(13.14)\qquad D_a = \sum_{N\geq 1} a_N T_N^c ,$$

whose homology class is $\sum_{N\geq 1} a_N[T_N^c]$. Then (13.13) signifies that D_a is homologically trivial. So there is a function f_a such that:

(13.15) $$D_a = \mathrm{div}(f_a).$$

For an explicit example where $p = 101$, $N = 37$, see [Z2].

13.16. We hope to present elsewhere some explicit calculations in a special case, which show that Question B of §13.8 is not purely idle speculation.

§14. Zero cycles over number fields.

14.1. By a zero cycle on a (smooth, projective) surface X we mean a finite formal sum of points on X. A zero cycle Z is said to be rationally equivalent to zero iff we can find a finite formal sum: $\sum_j (C_j, f_j)$ with each C_j: an irreducible curve on X and f_j a function on C_j (invertible at the generic point of C_j), such that:

(14.2) $$Z = \sum_j \mathrm{div}(f_j).$$

As we already saw in §11, each $\mathrm{div}(f_j)$ defines a zero cycle on X with supports in C_j and so (14.2) makes sense. Two zero cycles Z_1, Z_2 are rationally equivalent iff $Z_1 - Z_2$ is rationally equivalent to zero. One defines the Chow group ([Bℓ2]) of zero cycles on X modulo rational equivalence to be:

(14.3) $$CH^2(X) = \{\text{zero cycles on } X\}/\text{rational equivalence}.$$

(The superscript 2 indicates the codimension. Some authors prefer to write $CH_0(X)$ instead.)

If X is defined over k, and if L is a field extension of k, one has:

(14.4) $$\mathrm{Ker}(CH^2(X_k) \to CH^2(X_L)) \text{ is torsion.}$$

There are examples where this kernel is not zero. Contrast this with the situation of $\mathrm{Pic}(X)$ (cf. (7.7)).

14.5. Every zero cycle has a well-defined degree and a rationally trivial cycle has degree zero. Thus the degree map is well-defined on $CH^2(X)$. Set:

(14.6) $$A_0(X) = \mathrm{Ker}(CH^2(X) \xrightarrow{\deg} \mathbb{Z}) .$$

14.7. Let us once and for all take X to be defined over $\mathbb{Q}$. There is an abundance of zero cycles of degree zero modulo rational equivalence on $X_{\mathbb{C}}$. To understand this let us make the definition (for any field $k \supset \mathbb{Q}$):

(14.8) $A_0(X_k)$ is representable $\iff$ $\left\{\begin{array}{l}\exists \text{ curve } C \text{ mapping into} \\ X \text{ such that the induced} \\ \text{morphism } \mathrm{Jac}(C) \to A_0(X_k) \\ \text{is surjective.}\end{array}\right.$

The concrete meaning of this condition is that one can find a (possibly disconnected) curve C on X such that every point on X can be moved off to C up to rational equvialence.

There is a natural surjective morphism over k:

(14.9) $$\alpha_k : A_0(X_k) \to A\ell b(X_k),$$

which can be defined explicitly over $\mathbb{C}$ by the period integral:

(14.10) $$Z \mapsto (\omega \mapsto \int_Z \omega) \in (H^{1,0}(X_{\mathbb{C}}))^*/H_1(X_{\mathbb{C}},\mathbb{Z}) \simeq A\ell b(X_{\mathbb{C}}).$$

A non-obvious fact is:

(14.11) $A_0(X_k)$ is representable $\iff$ α_k is an isomorphism up to torsion.

One has the following

Theorem 14.12. (Mumford [M3]) If $p_g(X) > 0$, then $A_0(X_{\mathbb{C}})$ is not representable.

Here $p_g(X)$ denotes the geometric genus, i.e., $\dim H^{2,0}(X_{\mathbb{C}})$. Bloch extends this theorem to all algebraically closed fields with many parameters. He also conjectures the converse of Mumford's theorem. To get an idea of the enormity of $A_0(X_{\mathbb{C}})$ in the non-representable situation see [Rt1].

14.13. Over a number field k, $A_0(X_k)$ should be reasonable.

Conjecture 14.14. $A_0(X_k)$ is finitely generated over any number field k.

If X is a rational surface, $A_0(X_k)$ is known to be finite ([Co]) for any number field k.

$A_0(X_k)$ is like the Mordell-Weil group. Bloch ([Bℓ1]) and Beilinson ([B1]) have taken this analogy further and have made (independently) the following

Conjecture 14.15. For every number field k, one has:

$$\mathrm{rk}.A_0(X_k) = \mathrm{ord}_{s=2}L^{(3)}(X_k,s).$$

Here $L^{(3)}(X_k,s)$ is the H^3-piece of the Hasse-Weil Zeta function of X_k. The point $s = 2$ is at the center of the critical strip.

Recall (see (7.32)) that the Birch and Swinnerton-Dyer conjecture predicts that:

(14.16) $$\mathrm{rk}.\mathrm{Pic}^0(X)(k) = \mathrm{ord}_{s=1}L^{(1)}(X_k,s).$$

Proposition 14.17. Modulo the Bloch-Beilinson conjecture (14.15) and the Birch-Swinnerton-Dyer conjecture (14.16), one has (for every number field k):

$$\alpha_k: A_0(X_k) \to A\ell b(X)(k) \text{ is an isomorphism mod torsion.}$$

Compare this with Mumford's Theorem 14.12 (over $\mathbb{C}$) and (14.11).

To prove this proposition one observes that Poincaré duality for surfaces in ℓ-adic cohomology (cf. [Mi]) gives:

(14.18) $$\mathrm{ord}_{s=2}L^{(3)}(X_k,s) = \mathrm{ord}_{s=1}L^{(1)}(X_k,s).$$

Combining this with the Conjectures 14.15 and 14.16, we get:

(14.19) $$\mathrm{rk}.A_0(X_k) = \mathrm{rk}.\mathrm{Pic}^0(X)(k).$$

Since $\mathrm{Pic}^0 X_k$ is isogenous over k to $A\ell b(X_k)$, we have:

(14.20) $$\mathrm{rk}.A_0(X_k) = \mathrm{rk}.A\ell b(X)(k).$$

Now the surjectivity of α_k forces the proposition.

14.21. A theorem of Roitman (cf. [Rt2] and [Bℓ2]) says that $\alpha_{\overline{\mathbb{Q}}}$ induces an isomorphism of the torsion subgroup of $A_0(X_{\overline{\mathbb{Q}}})$ onto the torsion subgroup of $A\ell b(X_{\overline{\mathbb{Q}}})$. This fact, when combined with Proposition 14.17, implies (conjecturally) that $A_0(X_{\overline{\mathbb{Q}}})$ is totally indistinguishable from $A\ell b(X_{\overline{\mathbb{Q}}})$, not just mod torsion.

14.22. Now let us take X to be a Hilbert-Blumenthal surface, and first look briefly at the situation over $\mathbb{C}$. The Hecke algebra $H_{\mathbb{C}}^{K}$ acts on $A_0(\tilde{S}_K \times \mathbb{C}) \otimes \mathbb{C}$ as an algebra of correspondences. Using this we can refine Mumford's theorem and establish:

Theorem 14.23. ([R4]) There exist $p_g(\tilde{S}_K)$ components of $A_0(\tilde{S}_K \times \mathbb{C}) \otimes \mathbb{C}$, mutually inequivalent for the action of the Hecke algebra, such that none of these components is representable.

This takes away a little bit of the mystery behind the role of the geometric genus in this context.

14.24. What happens over a number field k? Recall that the irregularity q is zero for Hilbert-Blumenthal surfaces. Since $A\ell b(\tilde{S}_K)$ has dimension q, we must have:

$$A\ell b(\tilde{S}_K) = \mathrm{Pic}^0(\tilde{S}_K) = 0. \tag{14.25}$$

So Prop. 14.16 suggests, in view of Conj. 14.14, that:

Conjecture 14.26. $A_0(S_K \times k)$ is finite for any number field k.

14.27. This conjecture has strong implications for zero cycles of degree 0 made up of special points. If Z is such a zero cycle, necessarily rational over a number field (see §8.28), then there must be arithmetically defined curves C_j and functions f_j (on C_j) such that:

$$Z = \sum_j \mathrm{div}(f_j) \tag{14.28}$$

It will be very interesting to find some examples of this phenomenon, especially in the case of cycles of the form: $P - Q$, with P, Q: special points on a H-Z cycle.

To end let us note that if a zero cycle Z of degree 0 over $\overline{\mathbb{Q}}$ can be written as $\sum \mathrm{div}(f_j)$ as in (14.28), and also as

$\sum \mathrm{div}(f_i')$ for some $\sum_i (C_i', f_i')$, then the finite sum $\sum_j (C_j, f_j) - \sum_i (C_i', f_i')$ represents a class in $CH^{2,1}(\tilde{S}_K \times \overline{\mathbb{Q}})$, the group studied in §11 and §12. Possibly this will provide a way to construct elements in $CH^{2,1}(\tilde{S}_K \times \overline{\mathbb{Q}})$ which are rational over non-abelian number fields.

BIBLIOGRAPHY

[Ar] J. Arthur, "On some problems suggested by the Trace formula", in Lie Group Representations II, Springer Lecture Notes 1041, 1-49 (1983).

[A] T. Asai, On certain Dirichlet series associated with Hilbert modular forms and Rankin's method, Math. Ann., 226, 81-94 (1977).

[AMRT] A. Ash, D. Mumford, M. Rapoport and Tai, Smooth compactification of locally symmetric varieties, Math. Sci. Press, Brookline, MA (1975).

[Ba] W. L. Baily, Jr., "Galois action on Eisenstein series and certain abelian extensions of CM fields", preprint (1983).

[B1] A. A. Beilinson, "Higher regulators and values of L-functions", (English translation), Journal of Soviet Math., 30, no. 2, 2036-2070, July (1985).

[B2] ———————, "Notes on absolute Hodge cohomology", in Applications of Algebraic K-theory to Algebraic Geometry and Number Theory,

[BBD] A. A. Beilinson, J. N. Bernstein and P. Deligne, "Faisceaux pervers", in Analyse et topologie sur les espaces singuliers I, Astérisque 100 (1983).

[Bℓ1] S. Bloch, "Algebraic cycles and values of L-functions", J. Reine Angew. Math. 350, 94-108 (1984).

[Bℓ2] ————, Lectures on Algebraic cycles, Duke University Math. Series IV, Durham, NC (1980).

[BL] J.-L. Brylinski and J.-P. Labesse, "Cohomologie d'intersection et fonctions L de certaines variétés de Shimura", Ann. Scient. Éc. Norm. Sup., 4e série, 17, 361-412 (1984).

[Bo] A. Borel, "Automorphic L-functions", in Automorphic Forms, Representations and L-functions, Proc. Symp. Pure Math., AMS, 33, Part 2 (1979).

[BCS] A. Borel, S. Chowla, C. S. Herz, K. Iwasawa and J.-P. Serre, Seminar on complex multiplication, Springer Lecture Notes, 21 (1966).

[C] H. Carayol, Courbes de Shimura, Formes automorphes et Representations Galoisiennes, Thèse de Doctorat d'Etat, Univ. Paris VII (1984).

[Cs] W. Casselman, "The Hasse-Weil ζ-function of some moduli varieties of dimension greater than one", in Automorphic Forms, Representations and L-functions, Proc. Symp. Pure Math., AMS, 33, Part 2 (1979).

[Ch] C-L. Chai, Compactification of Siegel moduli schemes, Thesis, Harvard University (1984).

[Co] J.-L. Colliot-Thélène, "Hilbert's theorem 90 for K_2 with applications to Chow groups of rational surfaces", Inv. Math., no. 1, 71, 1-20 (1983).

[D1] P. Deligne, "Travaux de Shimura", Sém. Bourbaki, 23e année, no 389 (1970/71).

[D2] ——————, "Variétés de Shimura: Interprétation modulaire et techniques de construction de modèles canoniques", Proc. Symp. Pure Math., AMS, 33, part 2, 247-289 (1979).

[D3] ——————, "Périodes d'intégrales et valeurs de fonction L", Proc. Symp. Pure Math., AMS, 33, part 2, 313-346 (1979).

[D4] ——————, "La Conjecture de Weil I", Publ. Math. IHES, 43, 273-308 (1974).

[D5] ——————, "Les constantes des équations fonctionnelles des fonctions L", Modular functions of one variable II, 501-595, Springer Lecture Notes, 349 (1973).

[DMOS] P. Deligne, J. S. Milne, Ogus and Shih, Hodge cycles, Motives and Shimura varieties, Springer Lecture Notes, 900 (1982).

[DRp] P. Deligne and M. Rapoport, "Les schémas de modules de cowibes elliptiques", Modular Functions of One Variable II, 143-316, Springer Lecture Notes, 349 (1973).

[DRi] P. Deligne and K. Ribet, "Values of abelian L-functions at negative integers over totally real fields", Inv. Math. 59, 227-286 (1980).

[DS] P. Deligne and J.-P. Serre, "Formes modulaires de poids 1", Ann. Scient. Éc. Norm. Sup., 4e serie, 7, 507-530 (1974).

[F] G. Faltings, "Endlichkeitssätze für abelsche varietäten über zahlkörpern", Inv. Math. 73, 349-366 (1983).

[G] S. Gelbart, Automorphic forms on Adèle groups, Annals of Math. Studies 83, Princeton University Press (1975).

[GJ] S. Gelbart and H. Jacquet, "A relation between automorphic forms on GL(2) and GL(3), Ann. Scient. Ec. Norm. Sup., 4° série, 11, 471-542 (1978).

[GH] P. Griffiths and J. Harris, Principles of Algebraic Geometry, John Wiley & Sons, N.Y. (1978).

[HLR] G. Harder, R. P. Langlands and M. Rapoport, "Algebraische zykeln auf Hilbert-Blumenthal-Flächen", Crelles Journal, Band 366, 53-120 (1986).

[He] E. Hecke, "Höhere Modulfunktionen und ihre Anwendung auf die Zahlentheorie", Math. Ann. 71, 1-37 (1912) (in Werke I).

[H1] F. Hirzebruch, "Hilbert Modular surfaces", L'Ens. Math. 71, 183-281 (1973).

[H2] ———————, "The canonical map for certain Hilbert modular surfaces", Proc. Chern Symposium, Springer (1981).

[HvG] F. Hirzebruch and G. van der Geer, Lectures on Hilbert Modular Surfaces, Les Presses de L'Université de Montréal (1981).

[HZ] F. Hirzebruch and D. Zagier, "Intersection numbers of curves on Hilbert modular surfaces and modular forms of Nebentypus, Inv. Math. 36, 57-113 (1976).

[J] H. Jacquet, Automorphic Forms on GL(2)-Part II, Springer Lecture Notes 278 (1972).

[JLai] H. Jacquet and F. Lai, "A relative trace formula", Comp. Math., 54, no. 2, 243-310 (1985).

[JL] H. Jacquet and R. P. Langlands, Automorphic Forms on GL(2), Springer Lecture Notes 114 (1970).

[JPSS] H. Jacquet, I. I. Piatetski-Shapiro and J. A. Shalika, "The converse theorem for GSP(4)", in preparation.

[Ka] N. Katz, "p-adic L-functions for CM fields", Inv. Math. 49, 199-297 (1978).

[KM] N. Katz and B. Mazur, Arithmetic moduli of elliptic curves, Ann. Math. Studies 108, Princeton (1985).

[Kn] A. W. Knapp, "Representations of $GL_2(\mathbb{R})$ and $GL_2(\mathbb{C})$", Proc. Symp. Pure Math., AMS, 33, part 1 (1979).

[KL] D. Kubert and S. Lang, Modular units, Springer Verlag, Grundleheren der Math. Wiss., 244 (1981).

[KS] M. Kuga and I. Satake, "Abelian varieties attached to polarized K-3 surfaces", Math. Ann. 169, 239-242 (1967).

[Lai] F. Lai, "Algebraic cycles on compact Shimura surfaces", Math. Zeitschrift, 189, 593-602 (1985).

[Lg1] S. Lang, Abelian varities, Interscience tracts 7 (1959).

[Lg2] ———, Algebraic Number Theory, Addison-Wesley (1968).

[L1] R. P. Langlands, Base Change for GL(2), Ann. Math. Studies 96, Princeton (1980).

[L2] ———, "Problems in the theory of Automorphic forms", in Modern Analysis and its Applications, Springer Lecture Notes 170, 18-61 (1970).

[Mi] J. S. Milne, Étale cohomology, Ann. Math. Studies 33, Princeton (1980).

[Mo] F. Momose, "On the ℓ-adic representations attached to modular forms", J. Fac. Sci. Univ. Tokyo, Sec. IA, 28, no. 1, 89-109 (1981).

[M1] D. Mumford, Abelian varieties, Tate Institute Lectures; Published by Oxford Univ. Press (1970).

[M2] ———, "Hirzebruch's proportionality theorem in the non-compact case", Inv. Math. 42, 239-272 (1977).

[M3] ———, "Rational equivalence of 0-cycles on surfaces, J. Math. Kyoto Univ. 9, 195-204 (1968).

[MR] V. K. Murty and D. Ramakrishnan, Paper in preparation on Shimura curves and Fermat curves.

[Od1] T. Oda, Periods of Hilbert modular surfaces, Progress in Math., Birkhäuser.

[Od2] ———, "Hodge structures of Shimura varieties attached to the unit groups of Quaternion algebras", in Galois Groups and Their Representations, Adv. Studies in Pure Math. 2, 15-36 (1983).

[P] G. Prasad, "Strong approximation for semi-simple groups over function fields", Ann. Math. (2), 105, 553-572 (1977).

[Q] D. Quillen, "Higher Algebraic K-theory I", Springer Lecture Notes 341, 85-147 (1973).

[Rc] K. Ramachandra, "Some applications of Kronecker's limit formulas", Ann. Math. 80, 104-148 (1964).

[R1] D. Ramakrishnan, "Valeurs de fonctions l des surfaces d'Hilbert-Blumenthal en s = 1", C. R. Acad. Sc. Paris, t.301, Série I, n° 18, 809-812 (1985).

[R2] D. Ramakrishnan, "Periods of integrals arising from K_1 of Hilbert-Blumenthal surfaces", pre-print (1984).

[R3] ——————, "Galois action on ε-factors of GL(n)", pre-print.

[R4] ——————, "Zero cycles on surfaces admitting semi-simple rings of correspondences", pre-print (1985).

[Rp] M. Rapoport, "Compactifications de l'espaces de modules de Hilbert-Blumenthal", Comp. Math. 36 (3), 255-335 (1978).

[Ri1] K. Ribet, "Galois representations attached to eigenforms with Nebentypus", in Modular Functions of One Variable V, Springer Lecture Notes 627, 18-52 (1977).

[Ri2] K. Ribet, "Twists of modular forms and endomorphisms of abelian varieties", Math. Ann. 253, no. 1, 43-62 (1980).

[RoT] J. Rogawski and J. Tunnell, "On Artin L-functions associated to Hilbert modular forms of weight one", Inv. Math. 74, 1-42 (1983).

[Rt1] A. A. Roitman, "Rational equivalence of zero cycles", Math. USSR 18, no. 4, 571-588 (1972).

[Rt2] ——————, "The torsion of the group of 0-cycles modulo rational equivalence", Ann. Math. (2), 111, no.3, 553-569 (1980).

[S1] J.-P. Serre, "Facteurs locaux des fonctions Zeta des variétés algébriques (définitions et conjectures)", Sém. Delange-Pisot-Poitou, exposé 19 (1969/70).

[S2] ——————, "Modular forms of weight one and Galois representations", in Algebraic Number Fields, 193-268, Acad. Press (1977).

[S3] ——————, Groupes algébriques et corps de classes, Hermann, Paris (1959).

[S4] ——————, Abelian ℓ-adic representations and elliptic curves, Benjamin, NY (1968).

[Sa] P. Samuel, Algebraic Theory of Numbers, Hermann, Paris (1970).

[Sd] F. Shahidi, "On certain L-functions", American J. Math. 103, 297-356 (1981).

[Sh1] G. Shimura, "On canonical models of arithmetic quotients of bounded symmetric domains I, II", Ann. Math. (2) 91, 144-222 and 92, 528-549 (1970).

[Sh2] ——————, Introduction to arithmetic theory of automorphic functions, Iwanomi Shoten and Princeton University Press (1971).

[Sh3] G. Shimura, "On certain zeta functions attached to two Hilbert modular forms II: The case of automorphic forms on a quaternion algebra", Ann. Math. 114, 569-607 (1981).

[Sz] H. Shimizu, "On discontinuous groups acting on the product of upper half planes, Ann. Math. 77, 33-71 (1963).

[So] C. Soulé, "Opérations en K-théorie algébrique", Canadian Journal of Math., XXXVII, 488-550 (1985).

[Sk] H. Stark, "Derivatives of L-series at s = 0", in Automorphic Forms, Representation Theory, and Arithmetic, Tata Institute, Bombay; published by Springer Verlag (1981).

[St] J. Sturm, "Special values of zeta functions and Eisenstein series of half integral weight", American J. Math. 102, no. 2, 219-240 (1980).

[Sv] Svarčman, "Simple connectedness of the factor space of the Hilbert modular group", Func. Anal. Appl. 8, 99-100 (1974).

[T1] J. Tate, "Algebraic cycles and poles of zeta-functions", in Arithmetic Algebraic Geometry, O.F.G.Schilling, ed., 93-110, Harper and Row (1965).

[T2] ————, Les conjectures de Stark sur les fonction L d'Artin en s = 0, Progress in Mathematics 47, Birkhäuser (1984).

[Va] Vaserstein, "The group SL_2 of Dedekind rings of arithmetic type", Math. USSR 18, no. 2, 321-332 (1972).

[V] M.-F. Vignéras, "De $GL(2,\mathbb{Q}(\sqrt{d}))$ à $GSP(4,\mathbb{Q})$", C.R.Acad.Sci. Paris 301, Série I, nº 6, 249-252 (1985).

[Wa] J.-L. Waldspurger, "Quelques propriétés arithmétiques de certaines formes automorphes sur GL(2)", Comp. Math. 54, no. 2, 121-171 (1985).

[W1] A. Weil, "Über die Bestimmung Dirichletscher Reihen durch Funktionalgleichungen", Math. Ann. 168, 149-156 (= [1967a] in the Collected Papers).

[W2] ————, "Remarks on Hecke's lemma and its use", in Algebraic Number Theory, Kyoto, 267-274 (= [1977a] in the Collected Papers).

[Zn] Yu. G. Zarhin, "The equations defining the moduli of abelian varieties with endomorphisms from a totally real field", Transactions Moscow Math. Soc., 2, 1-46 (1981).

[Z1] D. Zagier, "Modular forms whose Fourier coefficients involve zeta-functions of quadratic fields", in Modular Functions of One Variable V, Springer Lecture Notes 627, 105-169 (1977).

[Z2] D. Zagier, "Modular points, modular curves, modular surfaces, and modular forms", in Arbeitstagung Bonn Proceedings (1984), 225-246, Springer Lecture Notes 1111.

Addendum:

Recently the Tate's conjecture (7.12) has been proved for Hilbert-Blumenthal surfaces over all number fields. Actually there are two (independent) proofs available, which are very different, the first due to C. Klingenberg, "Die Tate-vermutungen für Hilbert-Blumenthal flächen" (pre-print), and the second due jointly to V. K. Murty and the author, "Period relations and the Tate conjecture for Hilbert modular surfaces" (pre-print). Either way one has a proof for the first time that the Néron-Severi group is not always exhausted by the H-Z cycles and the cusp resolutions.

Dinakar Ramakrishnan
Department of Mathematics
Cornell University
Ithaca, NY 14853

Canadian Mathematical Society
Conference Proceedings
Volume 7 (1987)

ELLIPTIC CURVES AND VALUES OF L-FUNCTIONS

David E. Rohrlich*

Let E be an elliptic curve over $\mathbb{Q}$ with complex multiplication by the ring of integers of an imaginary quadratic field of class number one, and let F be the function field of E over $\mathbb{Q}$. The regulator map of Bloch [2], Deligne, Ramakrishnan, and Beilinson [1] can be viewed in this context as a homomorphism

$$r\colon F^{\times} \otimes F^{\times} \to \mathbb{R}.$$

It is defined by the formula

$$r(f\otimes g) = \frac{\int_{E(\mathbb{C})} \log |f|\ \overline{dg/g} \wedge \omega}{\pi i \int_{E(\mathbb{R})^{\circ}} \omega},$$

where $E(\mathbb{R})^{\circ}$ is the identity component of $E(\mathbb{R})$, with some fixed orientation, and ω is any nonzero regular differential on E. Note that the definition is independent of ω. Let $L(s,E)$ denote the L-function of E over $\mathbb{Q}$, and for a positive integer n let E_n denote the subgroup of points on E of order dividing n. The purpose of this note is to prove the following variant of a theorem of Bloch [2], which, while probably well known, does not appear to be stated in the literature:

<u>THEOREM</u>. Suppose that the divisors of f and g are supported on E_n for some positive integer n. Then

*Research partially supported by an N.S.F. grant

$$r(f\otimes g) = a_{f,g}\, L'(0,E),$$

with $a_{f,g} \in n^{-1}\mathbb{Z}$. Furthermore, there is a choice of n, f, and g for which $a_{f,g} \neq 0$.

It is a pleasure to thank Dinakar Ramakrishnan for generous and illuminating comments, as well as for providing me with a copy of chapter 2 of an unpublished joint work with Bloch (the chapter in question deals with alternative definitions of the regulator map). I am also grateful to Ehud de Shalit for remarks pertaining to formula (1) below, which helped me to see this formula as a special case of a more general fact. Finally, I thank Ray Ross for pointing out an error in an earlier draft of this note.

1. Let $\mathcal{L}$ be a lattice in $\mathbb{C}$. By an elliptic function for $\mathcal{L}$ we mean, as usual, a meromorphic function on $\mathbb{C}$ which is periodic with respect to $\mathcal{L}$. Given z in $\mathbb{C}$ we define $E(z,s)$ by

$$E(z,s) = \sum_{\substack{\beta\in\mathcal{L} \\ z+\beta\neq 0}} |z+\beta|^{-2s}$$

for $\mathrm{Re}\, s > 1$ and by analytic continuation in general. We write ord_z to denote order at z.

LEMMA. If f is an elliptic function for $\mathcal{L}$ different from 0, then

$$(1)\qquad -\frac{1}{\pi i}\int_{\mathbb{C}/\mathcal{L}} \log|f(z)|\,E(z,s)\,dz\wedge d\bar{z} = (s-1)^{-2}\sum_{z\in\mathbb{C}/\mathcal{L}} (\mathrm{ord}_z f)\,E(z,s-1) \qquad (\mathrm{Re}\, s<1).$$

PROOF. The lemma can be viewed as a special case of the fact that Green's function on a compact Riemann surface is the resolvent kernel of the Laplacian (see for example [3], pp. 393-394). Here we shall present a proof which is based on an abelian analogue of the Rankin-Selberg method. We observe at the outset that as a function of z alone, with $s \neq 1$ fixed, the Eisenstein series $E(z,s)$ is periodic with respect to z, continuous on

$\mathbb{C} \setminus \mathcal{B}$, and bounded by a multiple of $1 + |z|^{-2\mathrm{Re}\, s}$ for z near 0. Consequently, the integral on the left-hand side of (1) is absolutely convergent for $\mathrm{Re}\, s < 1$. We shall now assume that $f(0) = 1$, so that $\log |f(z)|$ is $O(|z|)$ for z near 0 and the integral is absolutely convergent in the larger region $\mathrm{Re}\, s < 3/2$. The assumption that $f(0) = 1$ will subsequently be eliminated.

Put

$$I(r) = \frac{1}{2\pi} \int_0^{2\pi} \log |f(re^{i\theta})|\, d\theta \quad (r > 0)$$

and

(2) $$M(s) = \int_0^\infty I(r) r^{1-2s}\, dr.$$

Then

(3) $$M(s) = \frac{1}{2\pi} \int_0^\infty \int_0^{2\pi} \log |f(re^{i\theta})| r^{1-2s}\, d\theta dr$$

$$= \frac{-1}{4\pi i} \int_{\mathbb{C}} \log |f(z)| \; |z|^{-2s}\, dz \wedge d\bar{z}$$

$$= -\frac{1}{4\pi i} \int_{\mathbb{C}/\mathcal{B}} \log |f(z)|\, E(z,s)\, dz \wedge d\bar{z}$$

for $1 < \mathrm{Re}\, s < 3/2$. In particular, the integral (2) defining $M(s)$ is absolutely convergent for $1 < \mathrm{Re}\, s < 3/2$. On the other hand, Jensen's formula gives

(4) $$I(r) = \sum_{|z|<r} (\mathrm{ord}_z f) \log (r/|z|).$$

Since f has neither zero nor pole at 0, we deduce that $I(r) = 0$ if r is small. Hence the integral defining $M(s)$ is absolutely convergent for $\mathrm{Re}\, s > 1$ and gives an analytic continuation of $M(s)$ to this region. Furthermore, for $\mathrm{Re}\, s > 2$ the following calculation is valid:

(5) $$M(s) = \sum_{z \in \mathbb{C}} (\mathrm{ord}_z f) \int_{|z|}^\infty \log (r/|z|)\, r^{1-2s} dr \quad \text{(by (4))}$$

$$= \sum_{z \in \mathbb{C}} (\mathrm{ord}_z f) \; |z|^{2-2s}/(2-2s)^2$$

$$= (2-2s)^{-2} \sum_{z\in\mathbb{C}/\mathcal{L}} \operatorname{ord}_z f \sum_{\beta\in\mathcal{L}} |z+\beta|^{2-2s}$$

$$= (2-2s)^{-2} \sum_{z\in\mathbb{C}/\mathcal{L}} (\operatorname{ord}_z f)\ E(z,s-1).$$

On comparing formulas (3) and (5) we obtain formula (1) in the region $\operatorname{Re} s < 3/2$, under the assumption that $f(0) = 1$.

Next suppose that f is holomorphic and nonzero at 0. Then formula (1) holds when f is replaced by the function $g = f/f(0)$. To prove the lemma for f itself, it suffices to observe that

(6) $$\int_{\mathbb{C}/\mathcal{L}} E(z,s)\, dz \wedge d\bar{z} = 0 \qquad (\operatorname{Re} s < 1).$$

Now we obtain (6) in the region $\operatorname{Re} s < 0$ (and therefore, by analytic continuation, in the region $\operatorname{Re} s < 1$) from term-by-term integration of the functional equation

$$E(z,s) = \frac{\pi^{2s-1}}{\operatorname{area}(\mathcal{L})} \frac{\Gamma(1-s)}{\Gamma(s)} \sum_{\substack{\beta\in\mathcal{L}^{\perp} \\ \beta\neq 0}} |\beta|^{2s-2} e^{2\pi i\langle\beta,\, z\rangle} \qquad (\operatorname{Re} s < 0),$$

where $\langle\ ,\ \rangle$ is the inner product

(7) $$\langle w,z\rangle = \operatorname{Re}(w\bar{z}),$$

$\mathcal{L}^{\perp}$ is the dual of $\mathcal{L}$ with respect to (7), and $\operatorname{area}(\mathcal{L})$ is the area of a period parallelogram of $\mathcal{L}$.

Finally, to complete the proof of the lemma, consider the case where f has a zero or pole at 0. For u small but nonzero, the function $f_u(z) = f(z+u)$ is holomorphic and nonzero at 0, and formula (1) holds with f replaced by f_u. We obtain the formula for f itself on remarking that

$$\sum_{z\in\mathbb{C}/\mathcal{L}} (\operatorname{ord}_z f_u)\ E(z,s-1) = \sum_{z\in\mathbb{C}/\mathcal{L}} (\operatorname{ord}_z f)\ E(z-u,s-1)$$

and that $E(z,s-1)$ is a continuous function of z for $\operatorname{Re} s < 1$.

Now put

$$\mathfrak{g}(z) = \Delta(\mathcal{L})^{1/12}\, e^{-z\eta(z)/2}\, \sigma(z),$$

where η and σ are the Weierstrass eta and sigma functions for $\mathcal{L}$, and the twelfth root of the discriminant is chosen arbitrarily (cf. [4], p.29). By Kronecker's second limit formula,

$$\text{(9)} \qquad \frac{d}{ds} E(z,s)\Big|_{s=0} = -2 \log |\mathfrak{g}(z)| \qquad (z \notin \mathcal{L}).$$

We also have

$$\text{(10)} \qquad \frac{\partial}{\partial \bar{z}} E(z,s) = -s\, E_1(z,s+1),$$

where $\displaystyle E_1(z,s) = \sum_{\substack{\beta\in\mathcal{L} \\ z+\beta\neq 0}} (z+\beta)|z+\beta|^{-2s}$

for $\operatorname{Re} s > 3/2$.

PROPOSITION. (i) If f is an elliptic function for $\mathcal{L}$ different from 0, then

$$\text{(11)} \qquad \frac{1}{\pi i}\int_{\mathbb{C}/\mathcal{L}} \log|f(z)| \log|\mathfrak{g}(z)|\, dz \wedge d\bar{z} =$$

$$\frac{1}{2} \sum_{z\in\mathbb{C}/\mathcal{L}} (\mathrm{ord}_z f)\, \frac{d}{ds} E(z,s)\Big|_{s=-1}$$

(ii) If f and g are elliptic functions for $\mathcal{L}$ different from 0, then

$$\text{(12)} \qquad \frac{1}{\pi i}\int_{\mathbb{C}/\mathcal{L}} \log|f(z)|\, \overline{g'/g(z)}\, dz \wedge d\bar{z} =$$

$$\sum_{z,w\in\mathbb{C}/\mathcal{L}} (\mathrm{ord}_z f)(\mathrm{ord}_w g)\, \frac{d}{ds} E_1(z-w,s)\Big|_{s=0}$$

REMARK. A formula similar to (12) is stated by Beilinson in [1].

PROOF. (i) We differentiate both sides of (1) with respect to s and let $s = 0$, bearing in mind that $E(z,-1) = 0$. On substituting (9) we obtain (11).

(ii) The factorization of an elliptic function in terms of the sigma function shows that we may write

$$g(z) = c.\ \prod_{a\,\in\, A} \mathfrak{g}(z-a)^{\mathrm{ord}_a g},$$

where c is a constant and A is a certain set of representatives in $\mathbb{C}$ for the zeros and poles of g in $\mathbb{C}/\mathcal{L}$. Applying formula (11) to the functions

$$f_{a-u}(z) = f(z+a-u) \qquad (a \in A)$$

where u is an arbitrary complex number, we obtain

$$\frac{1}{\pi i} \int_{\mathbb{C}/\mathcal{L}} \log |f(z)| \, (\log |g(z+u)| - \log |c|) \, dz \wedge d\bar{z}$$
$$= \frac{1}{2} \sum_{\substack{z \in \mathbb{C}/\mathcal{L} \\ a \in A}} (\mathrm{ord}_z f)(\mathrm{ord}_a g) \, \frac{d}{ds} E\,(z-a+u,s)\Big|_{s=-1}$$

by linearity. Now take $2\,\partial/\partial\bar{u}$ of both sides and let $u = 0$, so that

$$\frac{1}{\pi i} \int_{\mathbb{C}/\mathcal{L}} \log |f(z)| \, \overline{g'/g(z)} \, dz \wedge d\bar{z} =$$
$$\sum_{\substack{z \in \mathbb{C}/\mathcal{L} \\ a \in A}} (\mathrm{ord}_z f)(\mathrm{ord}_a g) \, \frac{d}{ds} \left(\frac{\partial}{\partial \bar{z}} E\,(z-a,s)\right)\Big|_{s=-1}$$

After substituting (10) and using the fact that $E_1(z,0) = 0$, we obtain (12).

2. In this section, E denotes an elliptic curve over $\mathbb{R}$ with complex multiplication by the ring of integers $\mathcal{O}$ in an imaginary quadratic field k of class number one. We let D denote the discriminant of k. For $D \neq -4$ we define an invariant sign $(E) = \pm 1$ in the following manner: Let

$$y^2 = 4x^3 - g_2 x - g_3$$

be a Weierstrass equation for E over $\mathbb{R}$. For $D = -3$ put

$$\mathrm{sign}\ (E) = \mathrm{sign\ of}\ g_3$$

and for $D < -4$ put

$$\mathrm{sign}\ (E) = \mathrm{sign\ of}\ g_2/g_3.$$

Then sign (E) depends only on E over $\mathbb{R}$, not on the particular equation chosen. It is a special case of the invariant used to classify elliptic curves with a given j-invariant over a field which is not algebraically closed.

PROPOSITION. Fix an embedding of k in $\mathbb{C}$ and an orientation of $E(\mathbb{R})^\circ$. Then there exists a unique complex analytic homomorphism

$$\theta\colon\ \mathbb{C} \longrightarrow E(\mathbb{C})$$

with the following three properties:

(i) θ is defined over $\mathbb{R}$.

(ii) The kernel of θ is a fractional ideal $\mathfrak{b}$ of $\mathcal{O}$, and $\mathfrak{b} \cap \mathbb{R} = \mathbb{Z}$.

(iii) The induced isomorphism

$$\theta\colon\ \mathbb{R}/\mathbb{Z} \longrightarrow E(\mathbb{R})^\circ$$

identifies the usual orientation of $\mathbb{R}/\mathbb{Z}$ with the chosen orientation of $E(\mathbb{R})^\circ$.

Furthermore, we have $\mathfrak{b} = \mathcal{O}$ or $\mathfrak{b} = \mathfrak{q}^{-1}$, where $\mathfrak{q}$ is the unique ramified prime of k, and the two cases may be distinguished as follows: If $D = -4$ then $\mathfrak{b}$ is $\mathfrak{q}^{-1}$ or $\mathcal{O}$ according as $E(\mathbb{R})$ has one or two components. If $D \neq -4$ then $\mathfrak{b}$ is $\mathcal{O}$ or $\mathfrak{q}^{-1}$ according as sign (E) is 1 or -1.

PROOF. First we prove uniqueness. If θ and θ' are two such covering maps, then we have $\theta = \theta' \circ \phi$ for some complex analytic homomorphism $\phi\colon\ \mathbb{C} \longrightarrow \mathbb{C}$, necessarily of the form $z \longmapsto cz$. From (i) we get $c \in \mathbb{R}$, and from (ii), $c = \pm 1$; then (iii) implies that $c = 1$, as desired.

Next we prove existence. Let ω be a nonzero regular differential on E defined over $\mathbb{R}$, and let $\mathfrak{b}$ be the period lattice of ω. Then $\mathfrak{b}$ is invariant under complex conjugation. In particular, $\mathfrak{b}$ contains a nonzero real number, so that $\mathfrak{b} \cap \mathbb{R}$ is a lattice in $\mathbb{R}$, and after replacing ω by a nonzero real multiple we may assume that $\mathfrak{b} \cap \mathbb{R} = \mathbb{Z}$. Since $\mathfrak{b}$ is invariant under complex conjugation, the Abel-Jacobi map

$$E(\mathbb{C}) \longrightarrow \mathbb{C}/\mathfrak{b}$$

$$P \longmapsto \int_0^P \omega \mod \mathfrak{L}$$

is a complex analytic isomorphism defined over $\mathbb{R}$, and its inverse determines a complex analytic map θ: $\mathbb{C} \longrightarrow E(\mathbb{C})$ which is defined over $\mathbb{R}$ and has kernel $\mathfrak{L}$. After precomposing θ with $z \longmapsto -z$, we may assume that (iii) is satisfied as well. Now since E has complex multiplication by $\mathcal{O}$, the period lattice $\mathfrak{L}$ is an $\mathcal{O}$-module, and therefore a fractional ideal (since it contains 1).

To see that $\mathfrak{L}$ is either $\mathcal{O}$ or $\mathfrak{d}^{-1}$, observe that a fractional ideal of k which is invariant under complex conjugation is of the form $b\mathcal{O}$ or $b\,\mathfrak{d}^{-1}$, with $b \in \mathbb{Q}$. If $\mathfrak{L} = b\mathcal{O}$ then $b = \pm 1$, because $\mathfrak{L} \cap \mathbb{Q} = \mathbb{Z}$. If $\mathfrak{L} = b\,\mathfrak{d}^{-1}$ then $b = \pm 1$ for the same reason, because

$$b\,\mathfrak{d}^{-1} \cap \mathbb{Q} = b\,(\mathfrak{d}^{-1} \cap \mathbb{Q}) = b\mathbb{Z}.$$

Finally, let us determine when each of the cases $\mathfrak{L} = \mathcal{O}$ and $\mathfrak{L} = \mathfrak{d}^{-1}$ occurs. For $D = -4$ we have $\mathcal{O} = \mathbb{Z}[i]$ and $\mathfrak{d} = (1+i)\mathcal{O}$. If $\mathfrak{L} = \mathcal{O}$ then the inverse image of $E(\mathbb{R})$ under θ has two connected components modulo $\mathfrak{L}$, namely the sets $\mathbb{R}$ and $\mathbb{R} + i/2$, while if $\mathfrak{L} = \mathfrak{d}^{-1}$ one can check that $\mathbb{R}$ is the full inverse image modulo $\mathfrak{L}$. For $D \neq -4$ we observe that we can write $\mathfrak{d} = \kappa\mathcal{O}$ with $\kappa \in \mathbb{R}i$, so that $\mathfrak{d}^{-1} = \kappa^{-1}\mathcal{O}$. Letting g_2 and g_3 denote the usual functions of lattices, we consider the elliptic curves

$$E^+: \quad y^2 = 4x^3 - g_2(\mathcal{O})x - g_3(\mathcal{O})$$

and

$$E^-: \quad y^2 = 4x^3 - g_2(\kappa^{-1}\mathcal{O})x - g_3(\kappa^{-1}\mathcal{O}).$$

The Weierstrass parametrization of these curves, precomposed if necessary with the map $z \longmapsto -z$, satisfies (i), (ii), and (iii), and therefore coincides with the unique parametrization θ. Hence it suffices to show that sign $(E^{\pm}) = \pm 1$. In fact, since $\kappa^2 < 0$, we have sign $(E^-) = -$sign (E^+), and it suffices to show that sign $(E^+) = 1$. For this it is enough to have

$g_2(\mathcal{O}) \geq 0$ and $g_3(\mathcal{O}) > 0$ (with $g_2(\mathcal{O}) = 0$ only if $D = -3$). Now given τ in the upper half-plane, let us write $\mathcal{L}_\tau$ for the lattice $\mathbb{Z} + \tau\mathbb{Z}$ and $g_m(\tau)$ for $g_m(\mathcal{L}_\tau)$ $(m = 2,3)$. For $D = -8$ we wish to show that $g_m(\sqrt{-2}) > 0$ and for D odd we wish to show that $g_m((1+\sqrt{D})/2) \geq 0$ (here $\sqrt{D}$ means $i\sqrt{|D|}$). But in fact $g_m(it) > 0$ for arbitrary $t > 1$ and $g_m(1/2 + it) > 0$ for arbitrary $t > \sqrt{3}/2$, because g_m is nonzero and real-valued over the indicated ranges, and the limiting value for $t \longrightarrow \infty$ is positive.

EXAMPLE. Take E to be $x^3 + y^3 = 1$, so that $D = -3$. The transformation $x \longmapsto (1/2-y)/x$, $y \longmapsto (1/2 + y)/x$ maps E to the curve $12y^2 = 4x^3 - 1$, which is isomorphic over $\mathbb{R}$ to $y^2 = 4x^3 - 1$. Therefore sign $(E) = 1$, and $E(\mathbb{C})$ is isomorphic to $\mathbb{C}/\mathcal{O}$ over $\mathbb{R}$.

3. We turn now to the proof of the theorem stated in the introduction. Let E be an elliptic curve over $\mathbb{Q}$ with complex multiplication by the maximal order $\mathcal{O}$ in an imaginary quadratic field k of class number one. Then E is in particular defined over $\mathbb{R}$, and there is a unique complex analytic homomorphism Θ: $\mathbb{C} \longrightarrow E(\mathbb{C})$ with the properties described in the previous proposition. Let F be the function field of E over $\mathbb{Q}$, and let f and g be elements of $F^\times$ having divisorial support on E_n for some positive integer n. Let $\mathcal{L}$ be the kernel of Θ. Using Θ, we can identify f and g with elliptic functions for $\mathcal{L}$, and then we have

$$\begin{aligned} r(f\bullet g) &= \frac{\int_{\mathbb{C}/\mathcal{L}} \log|f(z)|\, \overline{g'/g(z)}\; d\bar{z} \wedge dz}{\pi i \int_{\mathbb{R}/\mathbb{Z}} dz} \\ &= -\frac{1}{\pi i} \int_{\mathbb{C}/\mathcal{L}} \log|f(z)|\, \overline{g'/g(z)}\; dz \wedge d\bar{z} \\ &= -\sum_{z,w\in\mathbb{C}/\mathcal{L}} (\mathrm{ord}_z f)(\mathrm{ord}_w g)\, \frac{d}{ds} E_1(z-w,s)\Big|_{s=0} \end{aligned}$$

by formula (12). Thus

(13) $\qquad r(f\bullet g) = -M'(0),$

where

(14) $$M(s) = \sum_{z,w\in\mathbb{C}/\mathfrak{b}} (\mathrm{ord}_z f)(\mathrm{ord}_w g)\, E_1(z-w,s).$$

Let us denote the field of algebraic numbers in $\mathbb{C}$ by $\overline{\mathbb{Q}}$. We define an action of $\mathrm{Gal}(\overline{\mathbb{Q}}/\mathbb{Q})$ on $n^{-1}\mathfrak{b}/\mathfrak{b}$ by requiring that

$$\theta(\sigma z) = \sigma\,\theta(z)$$

for $\sigma \in \mathrm{Gal}(\overline{\mathbb{Q}}/\mathbb{Q})$ and $z \in n^{-1}\mathfrak{b}/\mathfrak{b}$. Then the maps $z \longmapsto \mathrm{ord}_z f$ and $w \longmapsto \mathrm{ord}_w g$ are constant on orbits of $\mathrm{Gal}(\overline{\mathbb{Q}}/\mathbb{Q})$ because f and g (viewed as elements of F now) are defined over $\mathbb{Q}$. Writing $[z]$ for the orbit of z under $\mathrm{Gal}(\overline{\mathbb{Q}}/\mathbb{Q})$ and $[z,w]$ for the orbit of (z,w) under the diagonal action of $\mathrm{Gal}(\overline{\mathbb{Q}}/\mathbb{Q})$ on $(n^{-1}\mathfrak{b}/\mathfrak{b}) \times (n^{-1}\mathfrak{b}/\mathfrak{b})$, we express (14) as follows:

(15) $$M(s) = \sum_{[z,w]} (\mathrm{ord}_z f)(\mathrm{ord}_w g) \sum_{(z',w')\in[z,w]} E_1(z'-w',s)$$

$$= \sum_{[z,w]} (\mathrm{ord}_z f)(\mathrm{ord}_w g) \frac{|[z,w]|}{|[z-w]|} \sum_{u\in[z-w]} E_1(u,s).$$

Here $[z,w]$ runs over all orbits of $\mathrm{Gal}(\overline{\mathbb{Q}}/\mathbb{Q})$ on $(n^{-1}\mathfrak{b}/\mathfrak{b}) \times (n^{-1}\mathfrak{b}/\mathfrak{b})$, and the vertical bars denote cardinality. The expression

$$\frac{|[z,w]|}{|[z-w]|}$$

is an integer, for it is the index of the isotropy group of (z,w) in the isotropy group of $z-w$. Given $v \in n^{-1}\mathfrak{b}/\mathfrak{b}$, let us write

(16) $$M_v(s) = \sum_{u\in[v]} E_1(u,s),$$

so that (15) becomes

(17) $$M(s) = \sum_{[z,w]} (\mathrm{ord}_z f)(\mathrm{ord}_w g) \frac{|[z,w]|}{|[z-w]|} M_{z-w}(s).$$

We shall prove that

(18) $$M_v(s) = A_v(s)\, L(s,E),$$

where $A_v(s)$ is holomorphic at $s = 0$ and $A_v(0) \in n^{-1}\mathbb{Z}$. Granting this for the moment, and putting $a_v = A_v(0)$, we have

(19) $$M_v'(0) = a_v L'(0,E)$$

because $L(0,E) = 0$. A comparison of (13), (17), and (19) now yields

$$r(f\otimes g) = a_{f,g}L'(0,E)$$

with

(20) $$a_{f,g} = -\sum_{[z,w]} (\mathrm{ord}_z f)(\mathrm{ord}_w g) \frac{\|[z,w]\|}{\|[z-w]\|} a_{z-w} .$$

The proof of (18) depends on the fact that $L(s,E) = L(s,\chi)$, where χ is the Hecke character of k determined by E. Fix an element $v \in \mathfrak{n}^{-1}\mathfrak{b}/\mathfrak{b}$ and write $\mathfrak{n}$ for the annihilator of v in $\mathcal{O}$. Let $E_{\mathfrak{n}}$ be the subgroup of E annihilated by $\mathfrak{n}$, let $k_{\mathfrak{n}}$ be the extension of k generated by $E_{\mathfrak{n}}$, and let $\mathfrak{f}$ be the conductor of χ. Then k is an abelian extension of k unramified outside $\mathfrak{n}\mathfrak{f}$, and if $\mathfrak{a}$ is an ideal of $\mathcal{O}$ relatively prime to $\mathfrak{n}\mathfrak{f}$, then

(21) $$\theta(u)^{(\mathfrak{a},k_{\mathfrak{n}}/k)} = \theta(\chi(\mathfrak{a})u) \qquad (u \in \mathfrak{n}^{-1}\mathfrak{b}/\mathfrak{b}),$$

where $(\mathfrak{a},k_{\mathfrak{n}}/k)$ is the Artin symbol of $\mathfrak{a}$ (cf. [5], pp. 211-212). The proof of (18) now splits into two cases, according as $\mathfrak{f}$ does or does not divide $\mathfrak{n}$. In both cases we use the fact that

(22) $$\chi(\alpha\mathcal{O}) = \varepsilon(\alpha)\alpha \qquad (\alpha \in \mathcal{O},\ \alpha \text{ prime to } \mathfrak{f}),$$

where ε comes from a homomorphism $(\mathcal{O}/\mathfrak{f})^\times \longrightarrow \mathcal{O}^\times$ which does not factor through $(\mathcal{O}/\mathfrak{d})^\times$ for any proper divisor $\mathfrak{d}$ of $\mathfrak{f}$.

First suppose that $\mathfrak{f}$ does not divide $\mathfrak{n}$. Then there exists $\alpha \in \mathcal{O}$ relatively prime to $\mathfrak{n}\mathfrak{f}$ such that $\alpha \equiv 1 \bmod \mathfrak{n}$ but $\varepsilon(\alpha) \neq 1$. Let us write δ for the root of unity $\varepsilon(\alpha)$. From (22) we have

$$\chi(\alpha\mathcal{O}) \equiv \delta \bmod \mathfrak{n},$$

and then (21) yields the implication

$$u \in [v] \Longrightarrow \delta u \in [v].$$

On the other hand, we have $E_1(\delta u,s) = \delta\, E_1(u,s)$; thus $M_v(s) = \delta\, M_v(s)$. Since δ is different from 1, we conclude that $M_v(s) = 0$, and (18) holds with $A_v(s) = 0$.

Next suppose that $\mathfrak{p}$ does divide $\mathfrak{m}$. Then the map

$$\begin{aligned}(\mathcal{O}/\mathfrak{m})^\times &\longrightarrow (\mathcal{O}/\mathfrak{m})^\times\\ \alpha \bmod \mathfrak{m} &\longmapsto \chi(\alpha\mathcal{O}) \bmod \mathfrak{m}\end{aligned}$$

is well defined, as follows from (22). Writing U for the image of $\mathcal{O}^\times$ in $(\mathcal{O}/\mathfrak{m})^\times$, we obtain an injection

$$\begin{aligned}(\mathcal{O}/\mathfrak{m})^\times/U &\longrightarrow (\mathcal{O}/\mathfrak{m})^\times\\ \text{class of } \alpha &\longmapsto \chi(\alpha\mathcal{O}) \bmod \mathfrak{m},\end{aligned}$$

because $\chi(\delta\mathcal{O}) = \chi(\mathcal{O}) = 1$ for $\delta \in \mathcal{O}^\times$. Let $\langle v\rangle$ denote the orbit of v under $\mathrm{Gal}(\overline{\mathbb{Q}}/k)$. Referring to (21), we see that the map

$$\begin{aligned}(\mathcal{O}/\mathfrak{m})^\times/U &\longrightarrow \langle v\rangle\\ \text{class of } \alpha &\longmapsto \chi(\alpha\mathcal{O})t\end{aligned}$$

is a bijection for any $t \in \langle v\rangle$. In particular, this holds for $t = \delta/\nu \bmod \mathfrak{b}$, where ν is some fixed generator of $\mathfrak{m}\mathfrak{b}^{-1}$ and δ is the unique element of $\mathcal{O}^\times$ for which $\delta/\nu \bmod \mathfrak{b}$ belongs to $\langle v\rangle$. (If $v = \alpha/\nu \bmod \mathfrak{b}$, with $\alpha \in \mathcal{O}$, then $\delta = \varepsilon(\alpha')$, where α' is an inverse of α modulo $\mathfrak{m}$.) Now let $\{\alpha\}$ be a set of representatives in $\mathcal{O}$ for the elements of $(\mathcal{O}/\mathfrak{m})^\times/U$. Then

$$\begin{aligned}\sum_{u\in\langle v\rangle} E_1(u,s) &= \sum_\alpha E_1(\chi(\alpha\mathcal{O})\delta/\nu, s)\\ &= \delta \sum_\alpha E_1(\chi(\alpha\mathcal{O})/\nu, s)\\ &= (\delta/\nu)|\nu|^{2s} \sum_\alpha \sum_{\beta\in\mathfrak{b}} (\chi(\alpha\mathcal{O})+\nu\beta)|\chi(\alpha\mathcal{O})+\nu\beta|^{-2s}.\end{aligned}$$

Writing

$$\chi(\alpha\mathcal{O}) + \nu\beta = \varepsilon(\alpha)(\alpha + \nu\beta\varepsilon(\alpha)^{-1}),$$

we observe that $\nu\beta$ belongs to $\mathfrak{m}$, and is therefore divisible by $\mathfrak{p}$.

Hence

$$\varepsilon(\alpha) = \varepsilon(\alpha + \nu\beta\varepsilon(\alpha)^{-1}),$$

so that

$$\chi(\alpha\mathfrak{b}) + \nu\beta = \chi((\alpha + \nu\beta\varepsilon(\alpha)^{-1})\mathfrak{b})$$

Furthermore, as β runs over $\mathfrak{b}$, the expression $\nu\beta\varepsilon(\alpha)^{-1}$ runs over $\mathfrak{n}$. We conclude that

$$\sum_{u\in\langle v\rangle} E_1(u,s) = (\delta/\nu)|\nu|^{2s} \sum_{\mathfrak{a}} \chi(\mathfrak{a})\,\mathbb{N}\mathfrak{a}^{-s},$$

where $\mathfrak{a}$ runs over integral ideals of k relatively prime to $\mathfrak{n}$. Equivalently,

$$\text{(23)} \qquad \sum_{u\in\langle v\rangle} E_1(u,s) = (\delta/\nu)|\nu|^{2s} \prod_{\mathfrak{p}|\mathfrak{n}} (1-\chi(\mathfrak{p})\mathbb{N}\mathfrak{p}^{-s}) \cdot L(s,\chi),$$

where the product runs over prime ideals dividing $\mathfrak{n}$.

Now $[v] = \langle v\rangle \cup \langle\bar{v}\rangle$, and $\langle v\rangle$ and $\langle\bar{v}\rangle$ are either equal or disjoint. Put

$$B_v(s) = (\delta/\nu)\ |\nu|^{2s} \prod_{\mathfrak{p}|\mathfrak{n}} (1-\chi(\mathfrak{p})\mathbb{N}\mathfrak{p}^{-s}),$$

and

$$A_v(s) = \begin{cases} B_v(s), \text{ if } \langle v\rangle = \langle\bar{v}\rangle, \\ B_v(s) + \overline{B_v(\bar{s})}, \text{ if } \langle v\rangle \cap \langle\bar{v}\rangle = \phi . \end{cases}$$

Referring to (16), we see that (23) becomes

$$M_v(s) = A_v(s)\ L(s,\chi),$$

because $\overline{E_1(u,\bar{s})} = E_1\ (\bar{u},s)$ and $\overline{L(\bar{s},\chi)} = L(s,\chi)$ (the last equality holding because χ is equivariant with respect to complex conjugation). Since

$$B_v(0) = (\delta/\nu) \prod_{\mathfrak{p}|\mathfrak{n}} (1-\chi(\mathfrak{p}))$$

and ν is a generator of $\mathfrak{n}\mathfrak{b}^{-1}$, we see that $nB_v(0)$ belongs to $\mathfrak{g}^{-1}$

(recall that $\mathfrak{q}$ denotes the ramified prime of k). Therefore $A_v(0)$ belongs to $n^{-1}\mathbb{Z}$. This completes the proof of (18).

To summarize, put $a_v = A_v(0)$ and $b_v = B_v(0)$. Then

$$r(f\otimes g) = a_{f,g}\, L'(0,E)$$

with

(24) $$a_{f,g} = -\sum_{[z,w]} (\mathrm{ord}_z f)(\mathrm{ord}_w g) \frac{|[z,w]|}{|[z-w]|}\, a_{z-w}\ ,$$

where

(25) $$a_v = \begin{cases} 0, & \text{if } \mathfrak{f} \text{ does not divide } \mathfrak{n}, \\ b_v, & \text{if } \mathfrak{f} \text{ divides } \mathfrak{n} \text{ and } \langle v\rangle = \langle \bar{v}\rangle, \\ b_v + \overline{b_v}, & \text{if } \mathfrak{f} \text{ divides } \mathfrak{n} \text{ and } \langle v\rangle \cap \langle \bar{v}\rangle = \phi, \end{cases}$$

and

(26) $$b_v = (\delta/\nu) \prod_{\mathfrak{p}|\mathfrak{n}} (1-\chi(\mathfrak{p})).$$

Here $\mathfrak{n}$ is the annihilator of v in $\mathcal{O}$, ν is some fixed generator of $\mathfrak{n}\mathfrak{b}^{-1}$, and $\delta \in \mathcal{O}^\times$ is the unique root of unity such that $\delta/\nu \bmod \mathfrak{b}$ belongs to $\langle v\rangle$. It remains to show that for some choice of f and g we have $a_{f,g} \neq 0$. We shall use the fact that $\mathfrak{b}^{-1}$ divides $\mathfrak{f}$.

Let p be a rational prime which remains prime in k and does not divide $\mathfrak{f}$, and let c be a positive integer which is divisible by $\mathfrak{f}\mathfrak{b}^{-1}$ and divides some power of $\mathfrak{f}$. Put $n = cp$, and consider the divisors

$$\mathrm{div}_c = |[1/c]|\,(0) - \sum_{u\in[1/c]} (u)$$

and

$$\mathrm{div}_p = |[1/p]|\,(0) - \sum_{u\in[1/p]} (u)\ ,$$

where for simplicity we write $1/c$ and $1/p$ instead of $1/c \bmod \mathfrak{b}$ and $1/p \bmod \mathfrak{b}$. Both of these divisors are supported on the torsion subgroup of E and have degree zero, so there exist rational functions f and g on E over $\mathbb{Q}$ such that

$$\text{div }(f) = \text{positive multiple of } \text{div}_c,$$
$$\text{div }(g) = \text{positive multiple of } \text{div}_p.$$

We shall prove that with this choice of f and g we have $a_{f,g} \neq 0$.

To begin with, we claim that the right-hand side of (24) reduces to a sum over just four orbits, namely $[0,0]$, $[1/c,0]$, $[0,1/p]$, and $[1/c,1/p]$. In other words, we claim that for any σ,σ' in Gal $(\overline{\mathbb{Q}}/\mathbb{Q})$ there exists σ'' in Gal $(\overline{\mathbb{Q}}/\mathbb{Q})$ such that

$$\sigma'' \, 1/c = \sigma \, 1/c \quad \text{and} \quad \sigma'' \, 1/p = \sigma' \, 1/p$$

In proving this statement we may assume that σ and σ' belong to Gal $(\overline{\mathbb{Q}}/k)$, because complex conjugation acts trivially on $1/c$ and $1/p$. Therefore it suffices to show that k_c and k_p are linearly disjoint over k, where k_m denotes the extension of k generated by E_m. Now the compositum of k_c and k_p is k_n, and we can check that

$$[k_n:k] = [k_c:k][k_p:k]$$

by viewing both sides as orders of ray class groups: indeed, k_n and k_c are the ray class fields of k modulo n and c respectively, while $[k_p:k] = |(\mathcal{O}/p\mathcal{O})^\times|$. Now since $\mathcal{O}^\times$ maps into $(\mathcal{O}/c\mathcal{O})^\times$ injectively (by (22) and the fact that $\mathfrak{k}$ divides c), the ray class group modulo n is isomorphic to the product of the ray class group modulo c and the group $(\mathcal{O}/p\mathcal{O})^\times$. This establishes the claim. We note additionally that each of the four orbits $[z,w]$ listed above satisfies

$$|[z,w]| = |[z-w]|;$$

this follows from the fact that $1/c$ and $1/p$ have relatively prime orders in $n^{-1}\mathcal{L}/\mathcal{L}$.

Returning to (24), we conclude that $a_{f,g}$ is a nonzero multiple of the expression

(27) $\qquad |[1/c]||[1/p]|\, a_0 - |[1/p]|\, a_{1/c} - |[1/c]|\, a_{-1/p} + a_{1/c\,-1/p}$.

Before evaluating this expression, we mention the following fact, which is a consequence of the equivariance of χ with respect to complex conjugation: If n is a positive integer relatively prime to c, then

$$\varepsilon(n) = \left(\frac{D}{n}\right),$$

where D is the discriminant of k. In particular, $\varepsilon(p) = -1$. We now assert:

(i) $a_0 = a_{-1/p} = 0$

(ii) $a_{1/c} = 1/c$

(iii) $a_{1/c\,-\,1/p} = -\,(1+p)/n$

(iv) $|[1/p]| = p^2 - 1$

Indeed, (i) falls under the first case in (25). Statement (ii) falls under the second case in (25), and we have $\mathfrak{w} = c\,\mathfrak{b}$. Hence we may choose $\nu = c$, and (ii) follows from (26) (with $\delta = 1$). Statement (iii) also falls under the second case in (26), with $\nu = n$. Since $\varepsilon(p) = -1$, we have

$$\frac{1}{c} - \frac{1}{p} = \frac{p-c}{n} = -\,\chi((p-c)\mathcal{O})/n.$$

Therefore we must take $\delta = -1$ in (26); then (iii) follows. Finally, (iv) amounts to the statement that $[k_p:k] = |(\mathcal{O}/p\mathcal{O})^\times|$. Assembling our data, we see that (27) is equal to $-\,(1 + p^3)/n$, which is not zero.

EXAMPLE. We take E to be $x^3 + y^3 = 1$, so that $k = \mathbb{Q}(i\sqrt{3})$, $\mathfrak{b} = \mathcal{O}$, and $\mathfrak{f} = 3\mathcal{O}$. The map $t \longmapsto (t^{1/3}, (1-t)^{1/3})$ gives an identification of $\mathbb{R} \cup \{\infty\}$ with $E(\mathbb{R})$, and we choose the orientation of $E(\mathbb{R})$ which corresponds to the usual orientation of $\mathbb{R} \cup \{\infty\}$. Putting $f = (x - 1)/(x + y)$ and $g = (y - 1)/(x + y)$, we have

$$\operatorname{div}(f) = 3\,(P - 0)$$

and

$$\operatorname{div}(g) = 3\,(Q - 0),$$

where $P = (1,0)$, $Q = (0,1)$, and $0 = (1, -1, 0)$. The points P, Q, and 0 correspond to $2/3$, $1/3$, and 0 respectively in $3^{-1}\mathcal{O}/\mathcal{O}$. Applying our formulas, we obtain, after a calculation,

$$r(f\otimes g) = -9\, L'(0,E).$$

REFERENCES

1. A. Beilinson, Higher regulators and values of L-functions of curves, J. Func. Anal. Appl. 14 (1980), 116-118.
2. S. Bloch, Algebraic K-theory and zeta functions of elliptic curves, Proc. Int. Congress of Mathematicians, Helsinki (1978), 511-515.
3. G. Faltings, Calculus on arithmetic surfaces, Annals of Math. 119 (1984), 387-424.
4. D.S. Kubert and S. Lang, Modular Units, Springer-Verlag, 1981.
5. G. Shimura. Introduction to the Arithmetic Theory of Automorphic Forms, Iwanami Shoten and Princeton University Press, 1971.

Department of Mathematics
Rutgers University
New Brunswick, NJ 08903

Canadian Mathematical Society
Conference Proceedings
Volume 7 (1987)

FOURIER SERIES AND SPECIAL VALUES OF HECKE L-FUNCTIONS

Michael Rosen*

Introduction: Let k be an algebraic number field with ring of integers A and let χ be a non-trivial ray class character for k. Set $L(s,\chi) = \sum \chi(B) \mathbb{N}B^{-s}$, where the sum is over all integral ideals prime to the conductor of χ and $\mathbb{N}B = [A : B]$. $L(s,\chi)$ is absolutely convergent for $\mathrm{Re}(s) > 1$ and extends analytically to a neighborhood of $s = 1$. The value $L(1,\chi)$ is of great arithmetic interest, being directly related to the class number of abelian extensions of k. There are two cases where this value is well understood; $k = \mathbb{Q}$, the rational numbers, and k imaginary quadratic. The main result in these two cases is due respectively to Dirichlet and Kronecker.

An analogous situation exists in function fields. Let k be a function field in one variable over a finite field, ∞ a prime of k, and A the ring of functions regular away from ∞. If χ is a non-trivial ray class character for k and we form $L(s,\chi)$ as above, one can again ask for a satisfactory formula for $L(1,\chi)$. When k is a rational function field such a formula was found by S. Galovich and the author; see [1]. This result was generalized completely by D. Hayes in [3].

The common feature of all three cases is that the global field k has exactly one prime at ∞, k_∞ is locally compact, and A imbeds in k_∞ as a discrete, co-compact subring. It will be shown how the use of abstract harmonic analysis in this situation leads to a uniform proof of all three cases.

This paper is a report on joint work with Benedict Gross [2]. For the most part we will just sketch the proofs and refer to [2] for details. On the other hand we will treat in some detail the more general case of Hecke characters in the hope that the resulting formulas will be helpful in attacking conjectures, due to Beilinson, Lichtenbaum, and others, about special values of L-functions attached to elliptic curves. The reader whose interest is mainly

* This work was partially supported by a grant from the National Science Foundation.

in ray class characters can get the essential ideas by reading Sections 1, 4, and 5.

This paper is an elaboration of my talk at the CMS seminar. My thanks to the organizers of the conference and to the CMS for its support.

SECTION 1: A THEOREM ON LOCALLY COMPACT FIELDS.

Let k_∞ be a locally compact field and A a discrete co-compact subring. An A-lattice B is an A module contained in k_∞ which is discrete and co-compact. Let dx denote a Haar measure on k_∞ normalized by the condition $\int_{k_\infty/A} dx = 1$. For any A-lattice B we define the norm of B, $\mathbb{N}B$, by the formula $\mathbb{N}B = \int_{k_\infty/B} dx$. Then $\mathbb{N}B$ is a positive real number and if $B \subseteq A$, $\mathbb{N}B = [A:B]$. If $a \in k_\infty^*$ then $d(ax) = \|a\|dx$, where $\|a\|$ is the normalized valuation on k_∞. It is easily seen that $\mathbb{N}(Aa) = \|a\|$.

Let $\psi : k_\infty \to T^1$ be a non-trivial additive character and define $\langle a,b \rangle = \psi(ab)$ for $a,b \in k_\infty$. This pairing puts k_∞ into self-duality. For B an A-lattice the Pontryagin dual of k_∞/B is identified with $B^\perp = \{a \in k_\infty \mid \langle a,b \rangle = 1 \text{ for all } b \in B\}$ via this pairing. $B^\perp$ is discrete and co-compact and $(B^\perp)^\perp = B$.

The simple example to keep in mind is $A = \mathbb{Z}$, $k = \mathbb{Q}$, and $k_\infty = \mathbb{R}$. Here dx is Lebesque measure, and $\psi(x)$ can be chosen to be $\psi(x) = e^{2\pi i x}$.

Let $\chi_\infty : k_\infty^* \to \mathbb{C}^*$ be a quasi-character (continuous homomorphism). There is a real number σ such that $|\chi_\infty(x)| = \|x\|^\sigma$ for all $x \in k_\infty^*$. By definition, the real part of χ_∞, $\mathrm{Re}(\chi_\infty)$, is the number σ.

We can now define our principal object of study and state the main theorem.

MAIN DEFINITION: Let B be an A-lattice and χ_∞ a quasi-character. $V_A(\chi_\infty)$ is the complex vector space of complex-valued functions g on $k_\infty - B$ such that

i) g is continuous on $k_\infty - B$

ii) g is periodic with respect to B

iii) As a function on k_∞/B, g is integrable, i.e., in $L^1(k_\infty/B)$.

iv) For $a \in A$, $a \neq 0$, we have

$$g(ax) = \chi_\infty(a)^{-1} \sum_{\gamma \in a^{-1}B/B} g(x+\gamma) .$$

As we will see, functions in $V_A(\chi_\infty)$ arise in the evaluation of certain Hecke L-functions at $s = 1-2\sigma$.

MAIN THEOREM 1.2: *If* $\operatorname{Re}(\chi_\infty) \leq 0$ *then* $V_B(\chi_\infty)$ *has dimension* 1.

If $\operatorname{Re}(\chi_\infty) < 0$ *then all elements of* $V_B(\chi_\infty)$ *extend to continuous functions on all of* k_∞. *If* χ_∞ *is trivial and* $0 \neq g(x) \in V_B(\chi_\infty) \doteq V_B$, *then there is a constant* $c \neq 0$ *such that* $g(x) \sim c \log\|x\|$ *as* $x \to 0$. *(We say* $g(x)$ *has logarithmic singularities at the points of* B).

The proof depends on the use of Fourier analysis on k_∞, and actually shows how to construct non-zero elements of $V_B(\chi_\infty)$ as Fourier series.

Define $\hat{\chi}_\infty \quad \chi_\infty^{-1}\|\ \|$, and

$$(1.3) \qquad g_B(x) = \sum_{0\neq\beta\in B^\perp} \frac{\mathbb{N}B^\perp\,\overline{\langle x,\beta\rangle}}{\hat{\chi}_\infty(\beta)}$$

If $\operatorname{Re}(\chi_\infty) < 0$ it is easy to see the above series converge uniformly and absolutely. It follows that $g_B(x)$ extends to a continuous function on all of k_∞ in this case. If $\operatorname{Re}(\chi_\infty) = 0$ the analysis is more delicate since the convergence is conditional and we must specify an order of summation. We define $g_B(x) = \lim_{n\to\infty} \mathbb{N}B^\perp \sum_{0\neq\beta\in S(n)} \langle x,\beta\rangle\, \hat{\chi}_\infty(\beta)^{-1}$ where $S(n) = \{x \in k_\infty \mid \|x\| \leq n\, \mathbb{N}B^\perp\}$.

In all cases $g_B(x)$ yields a non-zero element of $V_B(\chi_\infty)$.

Note for future reference the following homogeneity property.

$$(1.4) \qquad g_{\lambda B}(\lambda x) = \chi_\infty^{-1}(\lambda) g_B(x) .$$

We sketch the proof that $\dim_{\mathbb{C}} V_B(\chi_\infty) \leq 1$.

LEMMA 1.4: *Let* $g \in V_B(\chi_\infty)$ *and let* $\hat{g}(\beta)$, $\beta \in B^\perp$, *denote the Fourier transform of* g. *If* $a \in A$, $a \neq 0$ *then* $\hat{\chi}_\infty(a)\hat{g}(a\beta) = \hat{g}(\beta)$ *for all* $\beta \in B^\perp$.

The proof goes by using the identities in part iv) of the main definition. One multiplies both sides by $\overline{\langle ax,\beta\rangle}$ and integrates over a fundamental domain for A. Manipulation of these integrals results in the new identities $\hat{\chi}_\infty(a)\hat{g}(a\beta) = \hat{g}(\beta)$.

COROLLARY 1.5: *If* $\chi_\infty \neq \|\ \|$, *then* $\hat{g}(0) = 0$. *Moreover, there is a constant* $c(g)$ *such that for* $\beta \in B^\perp$, $\beta \neq 0$, *one has* $\hat{g}(\beta)\hat{\chi}_\infty(\beta) = c(g)$.

PROOF: If $\chi_\infty \neq \|\ \|$, then $\hat{\chi}_\infty$ is not trivial, and one can find an $a \in A$ such that $\hat{\chi}_\infty(a) \neq 1$. From the Lemma, $\hat{\chi}_\infty(a)\hat{g}(0) = \hat{g}(0)$ implying $\hat{g}(0) = 0$.

Suppose $\beta, \beta' \in B^\perp - \{0\}$. There exist $a, a' \in A$ such that $a\beta = a'\beta'$. Thus

$$\frac{\hat{g}(\beta)}{\hat{\chi}_\infty(a)} = \hat{g}(a\beta) = \hat{g}(a'\beta') = \frac{\hat{g}(\beta)}{\hat{\chi}_\infty(a')} .$$

Multiply both sides of this identity by $\hat{\chi}_\infty(a)\hat{\chi}_\infty(\beta) = \hat{\chi}_\infty(a')\hat{\chi}_\infty(\beta')$ and one finds $\hat{g}(\beta)\hat{\chi}_\infty(\beta) = \hat{g}(\beta')\hat{\chi}_\infty(\beta')$ as desired.

PROPOSITION 1.6: $\dim_{\mathbb{C}} V_B(\chi_\infty) \le 1$.

PROOF: Suppose f and g are non-zero elements of $V_B(\chi_\infty)$, and $c(f)$, $c(g)$ the constants obtained by the above corollary. Then $h(x) = c(g)f(x) - c(f)g(x) \in V_B(\chi_\infty)$ has all its Fourier coefficients equal to zero. It follows that $h(x)$ is zero almost everywhere (see [4]). Since $h(x)$ is continuous it is zero on $k_\infty - B$ and the Proposition follows.

As already mentioned it is more difficult to show the dimension of $V_B(\chi_\infty)$ is equal to 1. As it turns out the above Proposition is sufficient for our proof of the limit formulas of Dirichlet, Kronecker, and Hayes.

SECTION 2: HECKE L-FUNCTIONS AT $s = 1-2\sigma$

In this section k denotes a global field with exactly one prime at infinity which we denote by the symbol ∞ . Let k_∞ denote the completion of k at ∞ . If k is a number field it must either be $\mathbb{Q}$ or an imaginary quadratic field; $k_\infty \approx \mathbb{R}$ if $k = \mathbb{Q}$ and $k_\infty \approx C$ if k is imaginary quadratic. If k is a function field, $k_\infty \approx \mathbb{F}_q((u))$ where q is the number of elements in the residue class field at ∞ . In all cases A will denote the integers of k , i.e., those elements which are integral at all primes away from infinity.

Let χ be a Hecke character on k . Such a character is a continuous homomorphism from the ideles of k to $\mathbb{C}^*$ which takes the value 1 on the principal ideles. There is a product decomposition $\chi = \prod_v \chi_v$ where v runs through all primes of k and χ_v is a continuous homomorphism from k_v^* to $\mathbb{C}^*$. For each non-archimedean v there is a maximal non-negative integer $f(v)$ such that χ_v is 1 on $1+\hat{P}_v^{f(v)}$. If $f(v) = 0$, i.e., χ_v is 1 on U_v , the local units, we say χ_v is unramified and χ is unramified at v . Let S be the set of non-archimedean primes away from ∞ such that $f(v) > 0$. The ideal $F = \prod_{v\in S} P_v^{f(v)}$ is called the conductor of χ .

There is another way to describe χ which we will review briefly. Suppose χ is unramified at v and that π_v is a local uniformizing parameter there. Set $\chi(P_v) = \chi_v(\pi_v)$. This is well defined. Let I_F be the group of fractional ideals of A which are prime to F . If $B \in I_F$, and $B = \prod_{v\notin S\cup\infty} P_v^{b(v)}$ set $\chi(B) = \prod_{v\notin S\cup\infty} \chi(P_v)^{b(v)}$. Then, $B \to \chi(B)$ is a homomorphism from I_F to $\mathbb{C}^*$. For $\beta \in k^*$ let (β) denote the fractional A ideal generated by β . Suppose $\beta \equiv 1(F)$, i.e., $\mathrm{ord}_v(\beta-1) \ge f(v)$ for all $v \in S$. Then

$$1 = \chi(\beta) = \Big(\prod_{v \notin S \cup \infty} \chi_v(\beta)\Big) \Big(\prod_{v \in S} \chi_v(\beta)\Big) \chi_\infty(\beta) .$$

The left hand product is $\chi((\beta))$ by definition, and the middle product is 1 by the definition of the conductor. Thus, $\beta \equiv 1(F)$ implies $\chi((\beta)) = \chi_\infty(\beta)^{-1}$. We can now think of χ either as an idele character vanishing on the principal ideles or as a character on I_F taking the value $\chi_\infty(\beta)^{-1}$ on principal ideals (β) generated by an element $\beta \equiv 1(F)$. Call this latter group of ideals P_F and set $G_F = I_F/P_F$. G_F is called the ray class group modulo F. Its order is denoted by h_F.

Associated to a Hecke character χ of conductor F is a character of $(A/F)^*$ which will play an important role in our considerations. Let $\beta \in k^*$ be prime to F. Define $\rho(\beta) = \chi((\beta))\chi_\infty(\beta)$. If $\beta \equiv 1(F)$ then $\rho(\beta) = 1$. It follows that ρ is a character on the subgroup of k^* consisting of elements prime to F modulo the subgroup of elements congruent to 1 modulo F. This latter group is clearly isomorphic to $(A/F)^*$.

<u>LEMMA 2.1</u>: <u>The character</u> ρ <u>on</u> $(A/F)^*$ <u>is primitive modulo</u> F.

<u>PROOF</u>: For β prime to F it is easily seen that $\rho(\beta) = \prod_{v \in S} \chi_v(\beta)^{-1}$. Using this description and the weak approximation theorem the result follows.

We will need to do some Fourier analysis on the finite group A/F. Using the notation of Section 1 we note $(A/F)^\perp = F^\perp/A^\perp$. This latter group is free on one generator as an A/F module. Let $\gamma \in F^\perp$ reduce to such a generator. We can determine γ in another way. Let $D = (A^\perp)^{-1}$. By choosing an additive character on k_∞ carefully one can obtain $D = \mathbb{Z}$ when $k = Q$, D equal to the different when k is imaginary quadratic, and D equal to a divisor in the canonical class when k is a function field. Now, choose γ such that γFD is integral and prime to F. Then $\bar{\gamma}$ will generate $F^\perp/A^\perp$ over A as required.

<u>DEFINITION</u>: Let ρ be a primitive character on $(A/F)^*$ and $\gamma \in k^*$ be such that γFD is integral and prime to F. The Gauss sum $\tau(\rho,\gamma)$ is given by

$$\tau(\rho,\gamma) = \sum_{a \in (A/F)^*} \rho(a)\overline{\langle \gamma, a \rangle} .$$

<u>LEMMA 2.2</u>:

i) <u>If</u> $c \in k^*$, c <u>prime to</u> F, <u>then</u> $\tau(\rho,c\gamma) = \rho(c)^{-1}\tau(\rho,\gamma)$

ii) $|\tau(\rho,\gamma)| = \sqrt{\mathbb{N} F}$

iii) $\tau(\rho,\gamma)\chi(\gamma FD)\chi_\infty(\gamma)$ <u>is invariant under the substitution</u> $\gamma \to c\gamma$ <u>where</u> $c \in k^*$ <u>is prime to</u> F.

PROOF: i) is straight-forward and ii) is proven in [2]. See [5], Chapter 22, for the case $k_\infty = C$. Part iii) follows from i) and the definition of $\rho(c) = \chi((c))\chi_\infty(c)$.

Extend the definition of ρ to all of k^* by defining $\rho(\beta) = 0$ if β is not prime to F .

LEMMA 2.3: For $\beta \in k^*$ we have

$$\rho(\beta) = \frac{\tau(\rho,\gamma)}{\mathbb{N}F} \sum_{a\in A/F} \overline{\rho(a)} \langle \gamma a,\beta \rangle$$

PROOF: See either [5], Chapter 22, or [2], Lemma 2.8.

LEMMA 2.4: If C is a fractional ideal prime to F , then $|\chi(C)| = \mathbb{N}C^{-\sigma}$ where $\sigma = \mathrm{Re}(\chi_\infty)$.

PROOF: $C^{h_F} = (\lambda)$ where $\lambda \equiv 1(F)$. Thus $\chi(C)^{h_F} = \chi_\infty(\lambda)^{-1}$ and $|\chi(C)|^{h_F} = \|\lambda\|^{-\sigma}$. Also, $\mathbb{N}C^{h_F} = \mathbb{N}(\lambda) = \|\lambda\|$. Thus $|\chi(C)|^{h_F} = \mathbb{N}C^{-\sigma h_F}$ and we're done.

Having dispensed with the preliminaries we can proceed to discuss the Hecke L-function $L(s,\chi) = \sum \chi(B)\mathbb{N}B^{-s}$ where the sum is over all integral B prime to F . From Lemma 2.4 we see that $L(s,\chi)$ converges uniformly and absolutely when $\mathrm{Re}(s) > 1-\sigma+\delta$, where δ is any constant greater than zero. Thus $L(s,\chi)$ is analytic for $\mathrm{Re}(s) > 1-\sigma$. Moreover $L(s,\chi)$ can be analytically continued to all of $\mathbb{C}$ and satisfies a functional equation relating $L(s,\chi)$ and $L(1-2\sigma-s,\bar{\chi})$. We will indicate how this is done in the next section. We will be interested in the special values $L(1-2\sigma,\bar{\chi})$, and $L(0,\chi)$ or $L'(0,\chi)$. If χ_∞ is trivial we will consider $L(1,\bar{\chi})$ and $L'(0,\chi)$ where χ is a ray class character on G_F .

In the remainder of this section we will find a formula for $L(1-2\sigma,\bar{\chi})$ in terms of special values of the functions $g_B(x) \in V_B(\chi_\infty)$ defined by equation (1.3).

Let $\{B_i\}$ be a set of integral ideals prime to F which represent the ideal classes of A . Denote by h the class number of A and by w the order of the unit group of A (this is finite by our assumptions on k) . In the usual way we derive

$$(2.5) \qquad L(s,\bar{\chi}) = \frac{1}{w} \overline{\sum_{i=1}^{h} \chi(B_i)^{-1}\mathbb{N}B_i^{s}} \left(\sum_{\substack{\beta\in B_i \\ \beta\neq 0}} \chi((\beta)) \|\beta\|^{-s} \right)$$

From the definitions one finds that $\overline{\chi((\beta))} = \overline{\rho(\beta)}\chi_\infty(\beta)\|\beta\|^{-2\sigma}$. Thus

$$(2.6)\qquad L(s,\bar{\chi}) = \frac{1}{w}\sum_{i=1}^{h}\overline{\chi(B_i)^{-1}\mathbb{N}B_i^s}\left(\sum_{\substack{\beta\in B_i\\ \beta\neq 0}}\frac{\overline{\rho(\beta)}}{\hat{\chi}_\infty(\beta)\,\|\beta\|^{s+2\sigma-1}}\right)$$

Substitute $s = 1-2\sigma$. When $\sigma = 0$ this step must be justified by a process akin to Abel summation.

$$(2.7)\qquad L(1-2\sigma,\bar{\chi}) = \frac{1}{w}\sum_{i=1}^{h}\overline{(\chi(B_i)\,\mathbb{N}B_i^{2\sigma})^{-1}}\left(\sum_{\substack{\beta\in B_i\\ \beta\neq 0}}\frac{\mathbb{N}B\,\overline{\rho(\beta)}}{\hat{\chi}_\infty(\beta)}\right)$$

By Lemma 2.4, $\overline{\chi(B_i)}\,\mathbb{N}B_i^{2\sigma} = \chi(B_i)^{-1}$. Use Lemma 2.3 to express $\overline{\rho(\beta)}$ as a sum of additive characters. Equation (2.7) can then be rewritten as

$$(2.8)\qquad L(1-2\sigma,\bar{\chi}) = \frac{\overline{\tau(\chi,\gamma)}}{w\ \mathbb{N}F}\sum_{i=1}^{h}\chi(B_i)\sum_{a\in(A/F)^*}\rho(a)\sum_{\substack{\beta\in B_i\\ \beta\neq 0}}\frac{\mathbb{N}B_i\,\langle-\gamma a,\beta\rangle}{\hat{\chi}_\infty(\beta)}$$

Replace a by $-a$ and recall the definition of $g_B(x)$. The inner two summations become

$$(2.9)\qquad \rho(-1)\sum_{a\in(A/F)^*}\rho(a)\,g_{B_i^\perp}(a\gamma)$$

Now, $g_{B_i^\perp}(a\gamma) = \chi_\infty(a)^{-1}g_{a^{-1}B_i^\perp}(\gamma) = \chi_\infty(a)^{-1}g_{(aB_i)^\perp}(\gamma)$. Since $\rho(a) = \chi((a))\chi_\infty(a)$ equation (2.9) becomes

$$(2.10)\qquad \rho(-1)\sum_{a\in(A/F)^*}\chi((a))\,g_{(aB_i)^\perp}(\gamma)\ .$$

As a varies over $(A/F)^*$ and B_i through the class group of A the ideals aB_i varies over a set of integral representation for G_F with w/w_F repititions, where w_F is the number of units in A congruent to 1 modulo F . Thus, with integral ideals C ,

$$(2.11)\qquad L(1-2\sigma,\bar{\chi}) = \frac{\tau(\rho,\gamma)\rho(-1)}{w_F\mathbb{N}F}\sum_{C\in G_F}\chi(C)\,g_{C^\perp}(\gamma)\ .$$

It is reassuring to note that the individual terms $\chi(C)g_{C^\perp}(\gamma)$ are unchanged if C is replaced by $(\beta)C$ where $\beta\in A$ and $\beta\equiv 1(F)$. Equation (2.11) is almost the final form. It remains to eliminate the dependence on γ . We again use the homogeneity property (1.4) to show

(2.12) $g_{C^\perp}(\gamma) = g_{\gamma^{-1}C^\perp}(1) = g_{\gamma^{-1}C^{-1}D^{-1}}(1)$

where, to recall, $D = (A^\perp)^{-1}$.

Write $\gamma^{-1}C^{-1}D^{-1}$ as $(\gamma FD)^{-1}FC^{-1}$. Since γFD is integral and prime to F , $(\gamma FD)C$ runs through a set of integral representations for G_F as C does. We can now state

PROPOSITION 2.13: Let χ be a Hecke character such that $\sigma = \mathrm{Re}(\chi_\infty) \le 0$. For B a fractional ideal prime to F , the conductor of χ , let $g_B(x) \in V(\chi_\infty)$ be the function defined by the Fourier series (1.3). Then

$$L(1-2\sigma,\bar{\chi}) = \frac{\overline{\tau(\rho,\gamma)\rho(-1)}}{w_F \, \mathbb{N}F \, \chi(\gamma FD)\chi_\infty(\gamma)} \sum_{\substack{C\in G_F \\ C \text{ integral}}} \chi(C) g_{FC^{-1}}(1) .$$

Note that the factor in front of the summation is independent of γ because of Lemma 2.2, part iii).

In order for Proposition 2.13 to be really useful it is necessary to identify $g_B(x)$ as a known function whose special values have arithmetic significance. This can be done when χ_∞ is trivial, as we will show later. In other cases, the investigation of $g_B(x)$ remains a challenge.

Instead of pursuing the analysis at $s = 1-2\sigma$ further we shall transfer our attention to $s = 0$, i.e., to the evaluation of $L(0,\chi)$ and $L'(0,\chi)$ if $L(0,\chi) = 0$. This will necessitate introducing some new functions and showing how they can be analytically continued to all of $\mathbb{C}$.

SECTION 3: THETA FUNCTIONS

In this brief section we indicate how to form theta functions on k_∞ analogous to those introduced by Riemann when $k_\infty = \mathbb{R}$. When $k_\infty = \mathbb{C}$ these functions are somewhat novel and lead to a proof of the functional equation for $L(s,\chi)$ a bit different from the original proof of Hecke. The real novelty comes in the function field case where our functions on $k_\infty \approx \mathbb{F}_q((u))$ are an effective analogue of theta functions in this new context. It would be interesting to pursue their study in more detail.

For $f \in L^1(k_\infty)$ define the Fourier transform of f , $\hat{f}$, by the equation

(3.1) $$\hat{f}(y) = \int_{k_\infty} f(x)\overline{\langle y,x\rangle}\,dx$$

As in section 1, let $B \subset k_\infty$ be a discrete, co-compact A module.

DEFINITION: Suppose f is in the Schwartz space of k_∞. Let $t \in k_\infty^*$, and $v,w \in k_\infty$. Define

$$\theta_{B,f}(t;v,w) = \sum_{b\in B} f((v+b)t)\langle w,b\rangle .$$

This series converges absolutely to a continuous function of t.

By use of the Poisson summation formula one can prove the following functional equation for these functions.

PROPOSITION 3.2:

$$\theta_{B,f}(t;v,w) = \frac{\overline{\langle v,w\rangle}}{\mathbb{N}B\|t\|}\,\theta_{B^\perp,\hat{f}}(t^{-1};-w,v) .$$

In the case $A = \mathbb{Z}$, $k = \mathbb{Q}$, $k_\infty = \mathbb{R}$ one can choose $\langle v,w\rangle = e^{2\pi ivw}$. Then $\mathbb{Z}^\perp = \mathbb{Z}$. Choose $f(t) = e^{-\pi t^2}$. Then $\theta(t;v,w) \overset{\text{def}}{=} \theta_{\mathbb{Z},f}(\sqrt{t};v,w) =$

$$\sum_{n\in\mathbb{Z}} e^{-\pi(n+v)^2 t + 2\pi inw}$$

was introduced by Riemann and satisfies $\theta(t;v,w) = e^{-2\pi ivw}t^{-\frac{1}{2}}\theta(t^{-1};-w,v)$. This was used by de la Vallee Poussin in his proof of the functional equation for $L(s,\chi)$ where χ is an even Dirichlet character on $\mathbb{Z}$. In a similar way Proposition 2.2 can be used to prove the functional equation for $L(s,\chi)$ where χ is a Hecke character on any of our fields k_∞. When χ_∞ is trivial, complete details are given in Sections 1 and 3 of [2]. Here we simply state the result.

If χ_∞ is trivial define $L(s,\chi_\infty)$ to be $\pi^{-s/2}\Gamma(s/2)$ if $k_\infty = \mathbb{R}$, $(2\pi)^{1-s}\Gamma(s)$ if $k_\infty = \mathbb{C}$, and $(1-q^{-s})^{-1}$ if $k_\infty = \mathbb{F}_q((u))$. Let $\Lambda(s,\chi) = L(s,\chi_\infty)L(s,\chi)$.

PROPOSITION 3.3: $\Lambda(s,\chi)$ _extends to an analytic function on all of_ $\mathbb{C}$. _If_ $\chi \neq 1$, $\Lambda(s,\chi)$ _is holomorphic, and if_ $\chi = 1$ $\Lambda(s,\chi)$ _is holomorphic except for simple poles at_ $s = 0$ _and_ $s = 1$. $\Lambda(s,\chi)$ _satisfies the functional equation_

$$\Lambda(s,\chi) = W(\chi)(\mathbb{N}DF)^{1/2-s}\Lambda(1-s,\bar{\chi})$$

where $W(\chi)$ _is a constant of absolute value_ 1.

In this case one can specify $W(\chi)$ more precisely; $W(\chi) = \dfrac{\tau(\rho,\gamma)\chi(\gamma FD)}{\sqrt{\mathbb{N}F}}$

We will write out the functional equation in one other case. Suppose n is a positive integer, $k_\infty = \mathbb{C}$ and $\chi_\infty(x) = x^{-n}$. Hecke characters with this component at ∞ occur in the study of elliptic curves with complex multiplication. Note that $\sigma = \mathrm{Re}(\chi_\infty) = -\frac{n}{2}$ so that $1-2\sigma = 1+n$. In this case define $L(s,\chi_\infty) = (2\pi)^{1-s}\Gamma(s)$ and $\Lambda(s,\chi) = L(s,\chi_\infty)L(s,\chi)$. Then

PROPOSITION 3.4: $\Lambda(s,\chi)$ extends to a holomorphic function on all of $\mathbb{C}$ and

$$\Lambda(s,\chi) = W(\chi)(\mathbb{N}DF)^{\frac{n+1}{2}-s}\Lambda(n+1-s,\bar{\chi})$$

where $W(\chi)$ is a constant of absolute value 1.

In both cases under consideration $L(s,\chi_\infty)$ has a pole as $s = 0$. If $\chi \neq 1$ $\Lambda(s,\chi)$ is holomorphic at zero implying $L(0,\chi) = 0$. Thus if $\chi \neq 1$ we are looking for a formula for $L'(0,\chi)$.

SECTION 4: FORMULAS AT $s = 0$

Using Proposition 3.3 and 3.4 we could transfer our results at $s = 1-2\sigma$ to $s = 0$ in the two cases considered. Instead we will work directly at $s = 0$ using an entirely different approach.

Let $\chi_\infty : k_\infty^* \to \mathbb{C}$ be a continuous homomorphism, $B \subset k_\infty$ a discrete, co-compact A module, and $x \in k_\infty - B$. Finally, let $s \in \mathbb{C}$.

DEFINITION: $H_B(s,x;\chi_\infty) = \sum_{b\in B} \frac{\chi_\infty(x+b)^{-1}}{\|x+b\|^s}$ and $H_B^o(s,x,\chi_\infty) = H(s,x,\chi_\infty) - \frac{\chi_\infty(x)^{-1}}{\|x\|^s}$

Both series converge absolutely and uniformly on $\mathrm{Re}\,s \geq 1-\sigma+\delta$ where $\delta > 0$ is a constant. Consequently both are analytic in $\mathrm{Re}\,s > 1-\sigma$. By means of the theta functions introduced in Section 3 we can analytically continue both functions to all of $\mathbb{C}$. For fixed x the resulting functions of s are holomorphic unless $\chi_\infty(x) = \|x\|^{s_o}$ for some $s_o \in \mathbb{C}$ in which case they have a simple pole at $s = 1+s_o$. In particular, if χ_∞ is trivial there is a simple pole at $s = 1$.

These functions arise in the evaluation of $L(s,\chi)$ as follows. If χ is a Hecke character on k with infinite component χ_∞ and conductor $F \neq A$ we see

$$L(s,\chi) = \sum \chi(B)\mathbb{N}B^{-s} =$$

$$\frac{1}{w_F}\sum_{\substack{\alpha\in C_i^{-1}\\ \alpha\equiv 1(F)}}\sum_{i=1}^{h_F} \chi(C_i\alpha)\mathbb{N}(C_i\alpha)^{-s}$$

where $\{C_i\}$ is a set of integral ideals of A representing the cosets of $G_F = I_F/P_F$. Since $\chi((\alpha)) = \chi_\infty(\alpha)^{-1}$ if $\alpha \equiv 1(F)$ we find

$$(4.1) \qquad L(s,\chi) = \frac{1}{w_F}\sum_{i=1}^{h_F}\chi(C_i)\mathbb{N}C_i^{-s}\sum_{\alpha\equiv 1(C_i^{-1}F)}\frac{\chi_\infty(\alpha)^{-1}}{\|\alpha\|^s}$$

Write $\alpha = 1+\beta$ with $\beta \in C_i^{-1}F$ and this result takes the form

$$(4.2) \qquad L(s,\chi) = \frac{1}{w_F}\sum_{i=1}^{h_F}\chi(C_i)\mathbb{N}C_i^{-s}H_{C_i^{-1}F}(s,1;\chi_\infty)$$

By equation (2.5) we find if $F = A$,

$$(4.3) \qquad L(s,\chi) = \frac{i}{w}\sum_{i=1}^{h}\chi(B_i)^{-1}\mathbb{N}B_i^s\, H^o_{B_i}(s,0;\chi_\infty)$$

where the B_i run through a set of integral ideals representing Pic(A) .

By analytic continuation both formulas (4.2) and (4.3) hold for all $s \in \mathbb{C}$ and in particular at $s = 0$.

For simplicity we make the following assumptions; $F \neq A$, and either χ_∞ is trivial or $\sigma = \mathrm{Re}(\chi_\infty) < 0$. We also assume $L(s,\chi_\infty)$, the local L-function at ∞ , has a pole at $s = 0$. This has as a consequence $H(0,x;\chi_\infty) = 0$. These assumptions are satisfied for χ_∞ trivial, and in the case $k_\infty = \mathbb{C}$, $\chi_\infty(x) = x^{-n}$, n a positive integer. From (4.2) we find

PROPOSITION 4.4: $L(0,\chi) = 0$ and

$$L'(0,\chi) = \frac{1}{w_F}\sum_{i=1}^{h_F}\chi(C_i)H'_{C_i^{-1}F}(0,1;\chi_\infty)$$

To make further progress we need a formula for $H'_B(0,x;\chi_\infty) \stackrel{\mathrm{def}}{=} h_B(x;\chi_\infty)$.

PROPOSITION 4.5: $h_B(x;\chi_\infty) \in V_B(\chi_\infty)$. If $\sigma = \mathrm{Re}(\chi_\infty) < 0$ then $h_B(x;\chi_\infty)$ extends to a continuous function on k_∞ . If χ_∞ is trivial, then $h_B(x) \stackrel{\mathrm{def}}{=} h_B(x;\chi_\infty) \sim -\log\|x\|$ as $x \to 0$.

PROOF: If s is fixed and $\mathrm{Re}(s) > 1-\sigma$ the definition of $H_B(s,x;\chi_\infty)$ shows this function is periodic in x with respect to B . By analytic continuation this holds for all s . It follows that $h_B(s;\chi_\infty)$ is periodic with respect to B .

In the course of proving the analytic continuation one shows both $h_B(s,x;\chi_\infty)$ and $H^o_B(s,x;\chi_\infty)$ and all their derivatives with respect to s are continuous in x for $x \in k_\infty - B$. In the case of $H^o_B(s,x;\chi_\infty)$ this remains true for all x in a small neighborhood of zero (this need not hold if $\mathrm{Re}(\chi_\infty) = 0$ and χ_∞ is not trivial). Differentiating

$$H_B(s,x;\chi_\infty) = \chi_\infty(x)^{-1}\|x\|^{-s} + H_B^o(s,x;\chi_\infty)$$

we find

$$h_B(x;\chi_\infty) = -\chi_\infty(x)^{-1}\log\|x\| + \frac{d}{ds}(H_B^o(s,x;\chi_\infty)\Big|_{s=0}$$

If χ_∞ is trivial this shows $h_B(x) \sim -\log\|x\|$ as $x \to 0$. If $\mathrm{Re}(\chi_\infty) < 0$ we see $\chi_\infty(x)^{-1}\log\|x\| \to 0$ as $x \to 0$ and it follows $h_B(x;\chi_\infty)$ can be extended to a continuous function on k_∞. In all cases, $h_B(x;\chi_\infty) \in L^1(k_\infty/B)$.

It remains to prove that for $a \in A$, $a \neq 0$, we have

$$h_B(ax;\chi_\infty) = \chi_\infty(a)^{-1} \sum_{\gamma\in a^{-1}B/B} h_B(x+\gamma;\chi_\infty) .$$

This follows from the easily established identity,

$$H_B(s,ax;\chi_\infty) = \chi_\infty(a)^{-1}\|a\|^{-s}\sum_{\gamma\in a^{-1}B/B} H_B(s,x+\gamma;\chi_\infty)$$

We put the last two propositions together.

PROPOSITION 4.6: <u>With the above assumptions on</u> χ <u>and</u> χ_∞, <u>there are functions</u> $h_B(x;\chi_\infty) \in V_B(\chi_\infty)$ <u>such that</u>

$$L'(0,\chi) = \frac{1}{w_F}\sum_{i=1}^{h_F}\chi(C_i)h_{C_i^{-1}F}(1;\chi_\infty)$$

<u>where the</u> C_i <u>range through a set of integral representatives for</u> $G_F = I_F/P_F$.

Notice the similarity of the formulas in Proposition 2.13 and the above Proposition 4.6. A detailed examination of the functional equation would yield the precise constant relating the functions $g_B(x)$ defined by equation (1.3) and the function $h_B(x;\chi_\infty)$.

As already mentioned we do now know much about either function unless χ_∞ is trivial. In this case we can push our analysis to a satisfactory conclusion which is the task of the next, and final, section.

SECTION 5: THE LIMIT FORMULAS OF DIRICHLET, KRONECKER, AND HAYES

From now on we assume χ_∞ is trivial, which implies that χ can be considered as a primitive character on G_F. We will also assume that F, the conductor of χ, is not trivial. Full details for the material presented here are given in [2], Section 7, where the case $F = A$ is also worked out.

We recall the standard exact sequence (where $A_F^* = \{a \in A^* | a \equiv 1(F)\}$)

(5.1) $\qquad (0) \to A_F^* \to A^* \to (A/F)^* \to G_F \to \mathrm{Pic}(A) \to 1 .$

The character ρ which we have associated to any Hecke character χ is, in this situation, simply the restriction of χ to $(A/F)^*$.

Because the formulas are simpler at $s = 0$ we confine ourself to the evaluation of $L'(0,\chi)$.

Propositions 4.5 and 4.6 specialize to

(5.2) $\qquad L'(0,\chi) = \frac{1}{w_F} \sum_{C \in G_F} \chi(C) h_{C^{-1}F}(1)$. Thus

where $h_B(x) \in V_B$ and $h_B(x) \sim -\log\|x\|$ as $x \to 0$.

The Main Theorem of Section 1 says that V_B has dimension one over $\mathbb{C}$. We will construct functions in V_B and compare them to $h_B(x)$ using the asymptotic behavior as $x \to 0$.

Since $h_B(x)$ has logarithmic singularities at the points of B it is reasonable to suspect $h_B(x)$ has the form $-\log\|G_B(x)\|$ where $G_B(x)$ is a "nice" function having simple zeros at the points of B . Such functions are not hard to construct. Define $e_B(x)$ in each of our three cases as follows

$$e_B(x) = \begin{cases} x \prod\limits_{\substack{b \in B \\ b \neq 0}} \left(1 - \frac{x}{b}\right) e^{x/b} & k_\infty = \mathbb{R} \\ x \prod\limits_{\substack{b \in B \\ b \neq 0}} \left(1 - \frac{x}{b}\right) e^{x/b + x^2/2b^2} & k_\infty = \mathbb{C} \\ x \prod\limits_{\substack{b \in B \\ b \neq 0}} \left(1 - \frac{x}{b}\right) & k_\infty = \mathbb{F}((u)) \end{cases}$$

The first is the well known infinite product expansion for the sine function, and the second is the Weierstrass σ-function for the lattice B . The third is the exponential function attached to the lattice B , which is fundamental in the theory of Drinfeld modules. These functions do not quite solve our problem, but with minor modifications they do.

Suppose $k_\infty = \mathbb{R}$ and $B = \mathbb{Z}t$. Define $\Delta(B) = (2\pi/t)^2$ and $G_B(x) = \Delta(B)^{1/2} e_B(x) = 2 \sin \frac{\pi x}{t}$.

<u>PROPOSITION 5.3</u>: <u>The function</u> $g_B(x) = \log\|G_B(\alpha)\| = \log|1-e^{2\pi i x/t}|$ <u>lies in</u> V_B <u>and is equal to</u> $-h_B(x)$.

The proof that $\log|1-e^{2\pi i x/t}|$ is in V_B is relatively straight forward. This function is asymptotic to $\log\|x\|$ as $x \to 0$. Since $h_B(x) \sim -\log\|x\|$ and $\dim V_B = 1$ we conclude $h_B(x) = -g_B(x)$.

The pattern is the same in the remaining two cases.

Suppose $k_\infty = \mathbb{C}$, A the ring of integers in an imaginary quadratic number field, and $B \subset \mathbb{C}$ a discrete co-compact A module. B is a two dimensional $\mathbb{Z}$ lattice. Let $\Delta(B)$ denote the usual discriminant function. $\zeta_B(x) = \frac{d}{dx}\log e_B(x)$ is the Weierstrass zeta function. For $b \in B$ and $x \notin B$ one has $\zeta_B(x+b) - \zeta_B(x) = \eta_B(b)$ independent of x. Let $\eta_B(x) : \mathbb{C} \to \mathbb{C}$ be the unique $\mathbb{R}$-linear extension of $\eta_B(b)$. The Siegel function $G_B(x)$ is defined up to a twelfth root of unity by the formula

$$G_B(x) = \Delta(B)^{1/12} e^{-x\eta_B(x)/2} e_B(x) \tag{5.4}$$

PROPOSITION 5.5: Let $G_B(x)$ be the Siegel function attached to B and $g_B(x) = \log\|G_B(x)\|$. Then $g_B(x) \in V_B$ and $h_B(x) = -g_B(x)$.

It is somewhat harder in this case to verify that $g_B(x) \in V_B$. Once this is established, however, the proof that $h_B(x) = -g_B(x)$ follows directly from the Main Theorem and the asymptotic behavior as $x \to 0$.

Finally we come to the function field case where $k_\infty = \mathbb{F}_q((u))$. The product $e_B(x) = x\prod\left(1 - \frac{x}{b}\right)$ converges everywhere on k_∞ and extends to a homomorphism $e_B ; \tilde{k}_\infty \to \tilde{k}_\infty$ of the algebraic closure with kernel equal to B. For $a \in A$, $a \neq 0$, the two functions $e_B(ax)$ and $\prod_{\gamma \in a^{-1}B/B} [e_B(x) - e_B(\gamma)]$ have the same zero set, $a^{-1}B$, so they differ by a constant Δ_a. Since $e_B(x) - e_B(\gamma) = e_B(x-\gamma)$ we can write (the minus sign doesn't matter)

$$e_B(ax) = \Delta_a \prod_{\gamma \in a^{-1}B/B} e_B(x+\gamma) \tag{5.6}$$

LEMMA 5.7: If a and a' are non-zero elements of A we have

$$\Delta_a^{\|a'\|-1} = \Delta_a^{\|a\|-1} .$$

PROOF: The above remarks show $e_B(ax)$ is a polynomial in $e_B(x)$ of degree $\|a\|$ with leading coefficient Δ_a. The result follows by comparing the leading coefficient of $e_B(a(a'x))$ with that of $e_B(a'(ax))$.

DEFINITION: For $a \in A$, $\|a\| \neq 1$, define $\Delta(B) = \Delta_a^{(q-1)/\|a\|-1}$. This is independent of the choice of $a \in A$ and yields a well defined element of k_∞^*.

One can check $\Delta(tB) = t^{1-q}\Delta(B)$ so that Δ behaves like a modular form of weight $q-1$. Define

$$G_B(x) = \Delta(B)^{1/(q-1)} e_B(x) .$$

PROPOSITION 5.8: The function $g_B(x) = \log\|G_B(x)\|$ lies in V_B and $h_B(x) = -g_B(x)$.

PROOF: $e_B(x)$ is already periodic modulo B since $e_B(x+b) = e_B(x) + e_B(b) = e_B(x)$. Also $e_B(x)$ is locally constant and thus continuous for $x \notin B$. Clearly $g_B(x) \sim \log\|x\|$ as $x \to 0$. Together this shows $g_B(x)$ is in $L^1(k_\infty/B)$. The requisite identities follow from equation (5.6) and the definition of $\Delta(B)$. Finally, the proof that $h_B(x) = -g_B(x)$ follows as in the previous two cases.

We have defined a function $G_B(x)$ in all three cases. Define the Ramachandra invariant of an integral ideal C prime to F by the formula

$$(5.9) \qquad \Phi_F(C) = C_{FC^{-1}}(1)^{m_F}$$

where $m_F = 2$ if $k_\infty = \mathbb{R}$, $m_F = 12N$ if $k_\infty = C$ where N is an exponent for A/F , and $m_F = q-1$ when $k_\infty = \mathbb{F}_q((u))$. $\Phi_F(C)$ depends only on the class of C in G_F . Combining equation (5.2) with Propositions 5.3, 5.5, and 5.8 yields

THEOREM 5.10: If χ is a primitive ray class character of conductor $F \neq A$, then $L(0,\chi) = 0$ and

$$L'(0,\chi) = \frac{-1}{w_F m_F} \sum_{C \in G_F} \chi(C)\Phi_F(C) .$$

In the cases $k_\infty = \mathbb{R}$, $\mathbb{C}$, and $\mathbb{F}_q((u))$ this theorem encompasses the limit formula of Dirichlet, Kronecker, and Hayes respectively. As mentioned previously in the latter two cases one can treat by the same method the case of unramified characters.

The arithmetic aspect of these formulas can be described as follows. Let K_F be the ray class field of conductor F over k in which ∞ splits completely. Then $G_F \approx Ga(K_F/K)$ and the number of roots of unity in K_F divides m_F . It is known from the theory of complex multiplication that the Ramachandra invariants $\Phi_F(C)$ are all integers in $K_F \subset k_\infty$ which are supported on the primes dividing F (and are units if F has two or more distinct prime factors). If $\sigma_D \in Gal(K_F/k)$ is the Artin symbol it is known that $\Phi_F(C)^{\sigma_D} = \Phi_F(CD)$. If we set $\varepsilon = \Phi_F(A)$ we obtain the final version of the second limit formula:

$$L'(0,\chi) = \frac{-1}{m_F w_F} \sum_{C \in G_F} \chi(C)\log\|\varepsilon^{\sigma_C}\| .$$

BIBLIOGRAPHY

1. Galovich, S. and Rosen, M., The class number of cyclotomic function fields, J. of Number Theory, Vol. 13, No. 3, (1981), pp. 363-375.

2. Gross, B. and Rosen, M., Fourier series and special values of L-functions, preprint, submitted for publication.

3. Hayes, D., Analytic class number formulas in function fields, Invent. Math., 65, (1981), pp. 49-69.

4. Katznelson, Y., An Introduction to Harmonic Analysis, Dover Publications, Inc., New York, (1981).

5. Lang, S., Elliptic Functions, Addison-Wesley Publishing Company, Reading, Massachusetts, (1973).

Canadian Mathematical Society
Conference Proceedings
Volume 7 (1987)

p-ADIC L-FUNCTIONS AND DESCENT ON NON-CM ELLIPTIC CURVES

Karl Rubin*

1. INTRODUCTION.

This paper is the first step in an attempt to apply the techniques of Coates and Wiles ([2], see also [8]) to study elliptic curves _without_ complex multiplication (CM). Stark [10] observed that much of the information used in their proof for CM curves carries over to the general case. I believe that the main theorem of this paper (Theorem 1), while not the final word, shows what sort of result one can expect. In particular these methods do _not_ prove, for any non-CM curve E, that $L(E,1) = 0$ if $E(\mathbb{Q})$ is infinite.

We first state a weakened but simplified version of the theorem. Fix an elliptic curve E defined over $\mathbb{Q}$ which has integral j-invariant. (This latter condition is not essential, but it simplifies the bookkeeping substantially.) Also fix a rational prime $p > 3$ where E has good, ordinary reduction.

Choose a nonzero point x of order p in $E(\overline{\mathbb{Q}})$, and let C be the subgroup generated by x. Let $F = \mathbb{Q}(x)$ and $K = \mathbb{Q}(C)$ be the fields of rationality of x and C, respectively. Unfortunately F is not in general a Galois extension of $\mathbb{Q}$; however F is a Galois extension of K. Writing $\Delta = \mathrm{Gal}(F/K)$ there is a canonical p-adic character κ of Δ defined by

$$\kappa: \Delta \hookrightarrow \mathrm{Aut}(C) \simeq (\mathbb{Z}/p\mathbb{Z})^* \to \mathbb{Z}_p^*$$

i.e. $x^\gamma = \kappa(\gamma)x$ for any γ in $\mathrm{Aut}(\overline{\mathbb{Q}}/K)$.

1980 _Mathematics Subject Classification_ (1985 _Revision_) 11G05,11G40.

* partially supported by NSF grant DMS-8501937

For any $\mathbb{Z}_p[\Delta]$-module A (or even a $\mathbb{Z}[\Delta]$-module A, which we implicitly replace with $A \otimes \mathbb{Z}_p$) and any character $\chi: \Delta \to \mathbb{Z}_p^*$ define $A^\chi = \{a \in A: a^\gamma = \chi(\gamma)a \text{ for all } \gamma \in \Delta\}$.

Let E' denote the quotient of E by C, and ϕ the canonical isogeny from E to E'. Then E' and ϕ are defined over K.

THEOREM 1'. Suppose $y \in E'(F)^\chi$ where $\chi = \kappa^i$ with i odd and $\chi \neq \kappa$ or κ^{-1}, and suppose that $F(\phi^{-1}(y))/F$ is not everywhere unramified. Then $\mathcal{L}_E(\chi) \equiv 0 \pmod p$, where $\mathcal{L}_E(\chi)$ is a value of the Katz [5] one-variable p-adic L-function attached to E. (For a more precise definition and a formula for $\mathcal{L}_E(\chi)$ see §5.)

REMARKS. 1) The restriction to odd eigenspaces is unforturnate but seems necessary. The reason is that the Kubert-Lang units of F constructed in §4, which play a crucial role in the proof, lie only in even eigenspaces. In some special cases (such as when E has CM or when E has a rational point of order 3) one can construct units in odd eigenspaces as well, and extend the theorem to even eigenspaces.

2) If x is such that $F = K$ then Δ is trivial, so $\chi = \kappa = 1$ and the theorem is vacuous. However, for any E there is some choice of x for which Δ has order $p-1$ (see §3).

3) If E has CM, choose x so that K is the field of CM and x is killed by π, a generator of one of the two primes of K above p. Then $E' \simeq E$, ϕ is multiplication by π, $E'(F)/\phi(E(F)) \simeq E(F)/\pi E(F)$, and $\mathcal{L}_E(\chi)$ is "roughly" $L(E,\chi,1)$. In this case the theorem is exactly what the Coates-Wiles theorem gives (in odd eigenspaces) <u>over the field</u> F. To get the full Coates-Wiles theorem one needs an analogous result for the fields $\mathbb{Q}(x_n)$, where x_n has order p^n and $n > 1$. Such a result should also exist in the non-CM case, involving Katz' two-variable p-adic L-function.

4) To go further towards the Birch and Swinnerton-Dyer conjecture (with these methods) one would need:

a) to understand the relation between $\mathcal{L}_E(\chi)$ and $L(E,\chi,1)$, which currently is only known in the CM case; and/or

b) to "find" more units in the field $\mathbb{Q}(E_p)$. Since all the units we know how to construct fall into fields $\mathbb{Q}(x)$, $x \in E_p$, we only get information about $E'(F)/\phi(E(F))$ rather than $E(\mathbb{Q}(E_p))/pE(\mathbb{Q}(E_p))$.

2. THE SELMER GROUP. Keep the notation of §1. For any field L, write G_L for $\mathrm{Gal}(\overline{L}/L)$. If $K \subset L$ the exact sequence

$$0 \to C \to E(\overline{L}) \overset{\phi}{\to} E'(\overline{L}) \to 0$$

gives rise to a cohomology exact sequence

$$(1) \qquad 0 \to E'(L)/\phi(E(L)) \to H^1(G_L,C) \to H^1(G_L, E(\overline{L}))_\phi \to 0$$

where $H^1(G_L,E(\overline{L}))_\phi$ denotes the kernel of the map $H^1(G_L,E(\overline{L})) \overset{\phi}{\to} H^1(G_L,E'(\overline{L}))$.

We can define the Selmer group over F relative to the isogeny ϕ, S_ϕ, as follows. Namely, S_ϕ is the kernel of the map

$$\mathrm{Hom}(G_F,C) \to \bigoplus_v H^1(G_{F_v},E(\overline{F_v}))$$

where the sum runs over all primes v of F. By restricting (1) with $L = F$, S_ϕ fits into a short exact sequence

$$(2) \qquad 0 \to E'(F)/\phi(E(F)) \to S_\phi \to \text{Ш}_\phi \to 0$$

where Ш_ϕ is the kernel of the map induced by ϕ from the Tate-Safarevic group of E over F to that of E'.

We will only be able to study a quotient of this Selmer group. Write $F \otimes \mathbb{Q}_p = \bigoplus_{i=1}^{n} \Phi_i$, and let U_i denote the group of units of Φ_i. By class field theory there is a canonical restriction map

$$g:\ \mathrm{Hom}(G_F, C) \to \mathrm{Hom}\,(\bigoplus_{i=1}^{n} U_i, C).$$

Define $\mathcal{S}_\phi = g(S_\phi)$, so $\mathcal{S}_\phi$ is a quotient of S_ϕ which only contains information on the inertia groups of primes above p. $\mathcal{S}_\phi$ is the largest largest quotient of S_ϕ which we can study locally at p. The main result of this paper is the following theorem.

THEOREM 1. Suppose $\chi = \kappa^i$ with i odd, and $\chi \neq \kappa$ or κ^{-1}. Then $\mathcal{L}_E(\chi)\ \mathcal{S}_\phi^\chi = 0$.

We first compare more closely S_ϕ and $\mathcal{S}_\phi$. In particular Proposition 3 below and (2) show that Theorem 1 implies Theorem 1'.

PROPOSITION 2. E has good reduction everywhere over F.

PROOF. Let v be a prime of F not dividing p. By the criterion of Néron-Ogg-Safarevic [9] it is enough to show that v is unramified in $F(E_{p^\infty})/F$, or equivalently v is unramified in $F(E_{p^\infty})/F(\zeta)$ where ζ is a pth root of unity.

Let $G = \mathrm{Gal}(F(E_{p^\infty})/F(\zeta))$ and I_v be the inertia group of v in G. Since $j(E)$ is integral at v, I_v is finite [9]. We will show that G is torsion-free, so I_v is trivial.

There is an injection $\rho: G \hookrightarrow \mathrm{Aut}(T_p(E)) \simeq GL_2(\mathbb{Z}_p)$ where $T_p(E)$ is the p-adic Tate module of E and the isomorphism with $GL_2(\mathbb{Z}_p)$ depends on a choice of basis for $T_p(E) \simeq \mathbb{Z}_p^2$. Using the point $x \in E(F)$ we can pick a basis so that $\rho(G)$ is contained in $H = \{\left(\begin{smallmatrix} a & b \\ c & d \end{smallmatrix}\right) \in GL_2(\mathbb{Z}_p)\colon a \equiv 1$ and $c \equiv 0 \pmod p\}$. Using the Weil pairing a standard computation shows that for $\sigma \in G$, $\zeta = \zeta^\sigma = \zeta^{\det \rho(\sigma)}$, so in fact $\rho(G) \subset H_1 = \{\left(\begin{smallmatrix} a & b \\ c & d \end{smallmatrix}\right) \in GL_2(\mathbb{Z}_p)\colon$

$a \equiv d \equiv 1$ and $c \equiv 0 \pmod{p}\}$. Since H_1 is a pro-p-group, the only possible torsion is p-power torsion. However, since $p > 3$, $GL_2(\mathbb{Z}_p)$ contains no elements of order p, so G is torsion-free. Thus E has good reduction at all $v \nmid p$. We have also assumed that E has good reduction at p, so the proposition follows.

Now let v be any prime of F not dividing p. By Proposition 2 and [1] Lemma 4.1, $s \in \mathrm{Hom}(G_F, C)$ maps to zero in $H^1(G_{F_v}, E(\bar{F}_v))$ if and only if s is unramified at v. Thus $S_\phi \subset \mathrm{Hom}(X,C)$ where $X = \mathrm{Gal}(M/F)$ and M is the maximal abelian p-extension of F which is unramified outside p. We have the diagram

$$
(3) \qquad \begin{array}{ccccc}
 & & \mathrm{Hom}(X,C) & & \\
 & & \downarrow g_1 & \overset{h}{\searrow} & \\
\bigoplus_{i=1}^{n} \phi^{-1}(E'(\Phi_i))/E(\Phi_i) & \xrightarrow{e} & \bigoplus_{i=1}^{n} \mathrm{Hom}(\Phi_i^*, C) & \longrightarrow & \bigoplus_{i=1}^{n} H^1(G_{\Phi_i}, E(\bar{\Phi}_i)) \\
 & \underset{f = \oplus_i f_i}{\searrow} & \downarrow g_2 & & \\
 & & \bigoplus_{i=1}^{n} \mathrm{Hom}(U_i, C) & &
\end{array}
$$

where g_1, g_2 are induced by restriction maps on the ideles of F, and the middle row (obtained from (1) with $L = \Phi_i$) is exact. Here $\phi^{-1}(E'(\Phi_i))$ is the inverse image of $E'(\Phi_i)$ inside $E(\bar{\Phi}_i)$ and the maps f_i are given by $f_i(y)\colon u_i \to y^{[u_i, \Phi_i]} - y$ where $[\ ,\ ]$ is the local Artin symbol.

From the observation above we see that $S_\phi = \ker(h)$, so $g_1(S_\phi) = \mathrm{image}(g_1) \cap \mathrm{image}(e)$, and (recall $\mathfrak{S}_\phi = g(S_\phi)$ where $g = g_2 \circ g_1$)

$$
(4) \qquad \mathfrak{S}_\phi \subset \mathrm{image}(g) \cap \mathrm{image}(f).
$$

Also we deduce immediately:

PROPOSITION 3. $\mathbb{S}_\phi \simeq S_\phi/(S_\phi \cap \mathrm{Hom}(A,C))$ where $A = \mathrm{Gal}(H/F)$ and H is the maximal <u>unramified</u> abelian p-extension of F.

3. LOCAL METHODS. The idea of the proof is as follows. By class field theory, the image of the map $g = g_2 \circ g_1$ of (3) is the subgroup of $\mathrm{Hom}(\oplus_i U_i, C)$ consisting of those homomorphisms which are trivial on the closure of $\mathcal{O}_F^*$ in $\oplus_i U_i$, where $\mathcal{O}_F$ denotes the ring of integers of F. If $s \in \mathbb{S}_\phi$, then by (4) $s \in \mathrm{image}(g)$ and $s = f(y)$, $y = \oplus y_i \in \oplus \phi^{-1}E'(\Phi_i)$. The observation above shows that for $u \in \mathcal{O}_F^*$, $f(y)(u) = 0$. By writing down specific $u \in \mathcal{O}_F^*$ and using explicit reciprocity laws for the pairing $\langle y_i, u_i\rangle = f_i(y_i)(u_i) = y_i^{[u_i,\Phi_i]} - y_i$ we will obtain relations in $\mathbb{S}_\phi$ that imply Theorem 1.

Let $\tilde{E}$ denote the reduction of E modulo p, and write $y \to \tilde{y}$ for the reduction map from $E(\bar{\mathbb{Q}}_p)$ to $\tilde{E}(\bar{\mathbb{F}}_p)$. For any i write α_i for the image of x in $E(\Phi_i)$.

PROPOSITION 4. If $\tilde{\alpha}_i \neq 0$ then the map $f_i : \phi^{-1}(E'(\Phi_i)) \to \mathrm{Hom}\,(U_i, C)$ is zero.

PROOF. By definition, $f_i(y)(u) = y^{[u,\Phi_i]} - y$. Since $[u,\Phi_i]$ belongs to the inertia group of Φ_i^{ab}/Φ_i, $\widetilde{y^{[u,\Phi_i]}} = \tilde{y}$ or $\widetilde{f_i(y)(u)} = 0$. But $f_i(y)(u)$ is a multiple of α_i and $\tilde{\alpha}_i \neq 0$, so $f_i(y)(u) = 0$ for all u in U_i, y in $\phi^{-1}(E'(\Phi_i))$.

LEMMA 5. Let $\hat{E}$ be the formal group giving the kernel of reduction mod p on E. Then $\hat{E}$ is a Lubin-Tate formal group over $\mathbb{Q}_p$.

PROOF. Since E has ordinary reduction at p, $\hat{E}$ is a height-one formal group over $\mathbb{Q}_p$, and every such formal group is a Lubin-Tate formal group ([4], Chapter 4). In fact, $\hat{E}$ is a Lubin-Tate group for the uniformizing parameter which is a root of the characteristic polynomial of Frobenius on $\tilde{E}$, but we do not need this.

LEMMA 6. There is at most one i such that $\tilde{\alpha}_i = 0$.

PROOF. Since E has ordinary reduction at p, $E(\overline{\mathbb{Q}}_p)$ has exactly $p - 1$ nonzero points of order p which reduce to zero on $\tilde{E}$. Thus there are at most $p - 1$ embeddings of F into $\overline{\mathbb{Q}}_p$ which send x to one of these points, so

$$\sum_{\substack{i \\ \tilde{\alpha}_i = 0}} [\Phi_i : \mathbb{Q}_p] \leq p-1.$$

But if $\tilde{\alpha}_i = 0$ then $\Phi_i = \mathbb{Q}_p(\alpha_i) = \mathbb{Q}_p(v)$ where v is a nonzero p-division point of the formal group $\hat{E}$. By Lemma 5 and the theory of Lubin-Tate formal groups, $[\mathbb{Q}_p(v):\mathbb{Q}_p] = p - 1$. This proves the proposition.

If $\tilde{\alpha}_i \neq 0$ for every i, then by Proposition 4 and (4), $\mathcal{S}_\Phi = 0$ so Theorem 1 is vacuous. From now on suppose that some $\tilde{\alpha}_i = 0$ and write $\alpha = \alpha_i$, $\Phi = \Phi_i$, $U = U_i$, $\mathfrak{m}$ = maximal ideal of Φ, $\overline{\mathfrak{m}}$ = maximal ideal of $\overline{\Phi}$, k = residue field of Φ. As observed above in the proof of Proposition 6, $\Phi = \mathbb{Q}_p(v)$ where v is a p-division point of the formal group $\hat{E}$, so $\Phi/\mathbb{Q}_p$ is a totally ramified cyclic extension of degree $p-1$. Also, in this case the subgroup generated by α is rational over $\mathbb{Q}_p$, so the map $F \to \Phi$ sends K into $\mathbb{Q}_p$, we get an isomorphism $\mathrm{Gal}(\Phi/\mathbb{Q}_p) \simeq \Delta$, and κ maps Δ onto the $(p - 1)^{st}$ roots of unity in $\mathbb{Z}_p^*$.

PROPOSITION 7. $\phi^{-1}(E'(\Phi))/E(\Phi) \simeq [p]^{-1}(\hat{E}(\mathfrak{m}))/(\hat{E}(\mathfrak{m}))$ where $[p]$ denotes multiplication by p on $\hat{E}$.

PROOF. Let $\hat{E}'$ be the formal group of the kernel of reduction modulo $\mathfrak{m}$ on E'. There are natural isomorphisms between $\hat{E}(\overline{\mathfrak{m}})$ (resp. $\hat{E}'(\overline{\mathfrak{m}})$) and the kernel of the reduction map $E(\overline{\Phi}) \to \tilde{E}(\overline{k})$ (resp. $E'(\overline{\Phi}) \to \tilde{E}'(\overline{k})$) [11]. This leads to the following diagram with exact rows:

$$\begin{array}{ccccccccc} 0 & \to & \hat{E}(\overline{\mathfrak{m}}) & \to & E(\overline{\Phi}) & \to & \tilde{E}(\overline{k}) & \to & 0 \\ & & \downarrow\hat{\phi} & & \downarrow\phi & & \downarrow\tilde{\phi} & & \\ 0 & \to & \hat{E}'(\overline{\mathfrak{m}}) & \to & E'(\overline{\Phi}) & \to & \tilde{E}'(\overline{k}) & \to & 0 \\ & & \downarrow\hat{\mu} & & \downarrow\mu & & \downarrow\tilde{\mu} & & \\ 0 & \to & \hat{E}(\overline{\mathfrak{m}}) & \to & E(\overline{\Phi}) & \to & \tilde{E}(\overline{k}) & \to & 0 \end{array} \tag{5}$$

Here $\hat{\phi}$, $\tilde{\phi}$ are the maps induced by ϕ and $\hat{\mu}$, $\tilde{\mu}$ are induced by the dual isogeny μ. Thus $\hat{\mu} \circ \hat{\phi}$, $\tilde{\mu} \circ \tilde{\phi}$ and $\mu \circ \phi$ are all multiplication by p. All vertical maps are surjective because we work over the algebraic closure. Since E has ordinary reduction at p, $\#|\ker \hat{\mu} \circ \hat{\phi}| = p$. Since $\tilde{\alpha} = 0$, $\#|\ker \hat{\phi}| = p$ and $\#|\ker \tilde{\phi}| = 1$ so we conclude that $\#|\ker \hat{\mu}| = 1$.

Write $\hat{\mu}$ as a power series $\hat{\mu}(T) = \sum_{i>1} a_i T^i \in \mathbb{Z}_p[[T]]$. By [4] (18.3.12), $a_1 \neq 0$, and combining the Weierstrass Preparation Theorem with the fact that $\hat{\mu}$ is injective on $\overline{\mathfrak{m}}$ we conclude that $a_1 \in \mathbb{Z}_p^*$. Therefore $\hat{\mu}$ is a formal group isomorphism, so (since $\tilde{\phi}$ is an isomorphism in (5))

$$\begin{aligned} \phi^{-1}(E'(\Phi))/E(\Phi) &\simeq \hat{\phi}^{-1}(\hat{E}'(\mathfrak{m}))/\hat{E}(\mathfrak{m}) \\ &\simeq [p]^{-1}\hat{E}(\mathfrak{m})/\hat{E}(\mathfrak{m}) . \end{aligned}$$

Write $\hat{f}$ for the map $[p]^{-1}\hat{E}(\mathfrak{m})/\hat{E}(\mathfrak{m}) \to \mathrm{Hom}(U,C)$ given by $\hat{f}(y)\colon u \to y^{[u,\phi]} - y$.

COROLLARY 8. Suppose $s \in \mathcal{S}_\phi$. Then $s = \hat{f}(y)$ where $y \in [p]^{-1}\hat{E}(\mathfrak{m})$ and $\hat{f}(y)(u) = 0$ for all $u \in \mathcal{O}_F^*$.

PROOF. By (4), s lies in the images of the maps f and g of (3). Being in the image of g implies (by class field theory) that $s(u) = 0$ if u belongs to the image of $\mathcal{O}_F^*$ in $\bigoplus_i U_i$. The corollary now follows from Propositions 4 and 7 and Lemma 6.

4. KUBERT-LANG UNITS AND LOGARITHMIC DERIVATIVES. Fix an embedding of F into $\mathbb{C}$. Fix a Weierstrass model of $E/_{\mathbb{Q}}$ or equivalently a lattice $L \subset \mathbb{C}$ such that $(\wp, \wp')$ gives an analytic isomorphism from $\mathbb{C}/L$ to $E(\mathbb{C})$, where $\wp(z,L)$ is the Weierstrass $\wp$-function associated to L. Let $\tau \in p^{-1}L$ be a point which is sent to our chosen point x under this map. Also, write $E_1^*(z,L)$ for the weight one Eisenstein series given by the analytic continuation of $\sum_{\omega \in L} (\bar{z} + \bar{\omega})|z + \omega|^{-2s}$ at $s = 1$.

For any integers r, $s \in \mathbb{Z}$ prime to p write

(6) $$\Theta^*(z,L,r) = \prod_{\substack{\omega \in r^{-1}L/L \\ \omega \neq 0}} (\wp(z,L) - \wp(\omega,L))^{-6}$$

and

(7) $$u = u(r,s) = \Theta^*(\tau,L,r)/\Theta^*(s\tau,L,r)\ .$$

It is easily checked that $u \in F$.

PROPOSITION 9. (Kubert-Lang) $u \in \mathcal{O}_F^*$.

PROOF. [6], Theorem 2.2, Chapter 10 and Theorem 2.2, Chapter 2.

This is the unit we will use in applying Corollary 8. When E has CM, u is an elliptic unit.

Let v be the p-division point of $\hat{E}$ corresponding to x. Substituting the formal group expansion [11] of $\wp(z)$ into (6) and (7) yields a power series $h(T) \in \mathbb{Z}_p[[T]]$ such that $h(v) = u$. Define the "logarithmic derivative" of u

$$\delta(u) = (1/\lambda'(T)) \frac{d}{dT} \log h(T)|_{T=v}$$

where λ is the logarithm map from $\hat{E}$ to $\mathbb{G}_a$.

THEOREM 10. (Wiles) Explicit Reciprocity Law. Suppose $y \in [p]^{-1}\hat{E}(\mathfrak{m})$ and $p\lambda(y) \equiv 0 \pmod{\mathfrak{m}^2}$. Then $y^{[u,\Phi]} - y = [\mathrm{Tr}_{\Phi/\mathbb{Q}_p} \delta(u)\lambda(y)]v$

PROOF. [12].

We need to evaluate $\delta(u)$. Choose $\rho, \sigma \in \Delta$ so that $\kappa(\rho) \equiv r$ and $\kappa(\sigma) \equiv s \pmod p$.

PROPOSITION 11. (a) $E_1^*(\tau,L) \in F$;

(b) If $\gamma \in \Delta$ then $E_1^*(\tau,L)^\gamma = E_1^*(c\tau,L)$ where $c \equiv \kappa(\gamma) \pmod p$;

(c) $\delta(u) = r(r-\rho)(1-s\sigma) E_1^*(\tau,L)$, viewing $r(r-\rho)(1-s\sigma)$ $\mathbb{Z}[\Delta]$.

PROOF. We have defined $h(T) = \Theta^*(z,L,r)/\Theta^*(sz,L,r)$ by setting $T = -2\wp(z)/\wp'(z)$ [11]. Inverting this gives $z = \lambda(T)$, so

$$(8) \qquad \begin{aligned} \delta(u) &= \lambda'(T)^{-1} d/dT \log h(T)|_{T=v} \\ &= d/dz \log (\Theta^*(z,L,r)/\Theta^*(sz,L,r))|_{z=\tau}. \end{aligned}$$

From (6) we see that this quantity lies in $\mathbb{Q}(\wp(\tau,L), \wp'(\tau,L)) = F$. Using formulas from §1 of [3] we obtain

$$d/dz \log (\Theta^*(z,L,r)/\Theta^*(sz,L,r)) = \tag{9}$$
$$r^2E_1^*(z,L) - rE_1^*(rz,L) - r^2sE_1^*(sz,L) + rsE_1^*(rsz,L).$$

Applying (8) and (9) with $z = \tau$, $r = s = p+1$ yields

$$\delta(u) = -p^2(p+1)E_1^*(\tau,L)$$

which proves (a). Similarly, if $\gamma \in \Delta$ and $c \equiv \kappa(\gamma) \pmod p$,

$$\delta(u)^\gamma = d/dz \log (\Theta^*(z,L,r)/\Theta^*(sz,L,r))|_{z=c\tau}$$
$$= -p^2(p+1)E_1^*(c\tau,L)$$

by (9), which proves (b). Now for any r,s (c) follows from (8), (9), and (b).

5. p-ADIC L-FUNCTIONS. Let Ω be the completion of $\overline{\mathbb{Q}}_p$, and let $\mathcal{O}$ be the ring of integers of a complete subfield of Ω.

Suppose A is an elliptic curve over $\mathcal{O}$ with ordinary reduction, and ω is a holomorphic differential on A, defined over $\mathcal{O}$. Write $\hat{A}$ for the formal group giving the kernel of reduction on A, and let β be a "trivialization" of A: an isomorphism from $\hat{A}$ to $\mathbb{G}_m$, defined over $\mathcal{O}$ (such an isomorphism always exists if $\mathcal{O}$ is large enough; it is unique up to $\mathrm{End}(\mathbb{G}_m) \simeq \mathbb{Z}_p^*$). To such a triple (A,ω,β) Katz [5] associates an $\mathcal{O}$-valued measure $\mathcal{H} = \mathcal{H}(A,\omega,\beta)$ (the Eisenstein measure) on $\mathbb{Z}_p^*$, and for any character χ on $\mathbb{Z}_p^*$ we define

$$\mathcal{L}_A(\chi) = \mathcal{L}_{A,w,\beta}(\chi) = \int_{\mathbb{Z}_p^*} \chi \, d\,\mathcal{H} \in \mathcal{O}[\chi]$$

(This is a slightly different normalization from the one in [5], §3.7.)

In our case, take A to be E, ω the differential corresponding to our chosen Weierstrass model, and fix some trivialization β. This can be done if $\mathcal{O}$ is the ring of integers of the completion of the maximal unramified extension of $\mathbb{Q}_p$. If χ is any character of Δ we can view it as a character of $\mathbb{Z}_p^*$ via the isomorphism $\kappa: \Delta \to \mathbb{Z}_p^*/(1+p\mathbb{Z}_p)$, and we obtain the values $\mathcal{L}_E(\chi) \in \mathcal{O}$. From [5] we have the formula, for any nontrivial character χ of Δ

$$\mathcal{L}_E(\chi) = -\frac{1}{2} \sum_{\gamma \in \Delta} \hat{\chi}(\gamma) E_1^*(\tau, L)^{\gamma} \tag{10}$$

where $\hat{\chi}(\gamma)$ denotes the Gauss sum $\frac{1}{p} \sum_{\delta \in \Delta} \chi(\delta^{-1}\gamma)\zeta^{\kappa(\gamma)}$ and ζ is the pth root of unity identified with x by β.

6. COMPLETION OF THE PROOF. Let ord_p be the additive valuation on $\overline{\mathbb{Q}}_p$ with $\mathrm{ord}_p(p) = 1$.

LEMMA 12. Suppose $\chi = \kappa^i$, $1 \le i \le p-1$, and $y \in \hat{E}(\mathfrak{m})^{\chi}$. Then

(a) $\mathrm{ord}_p \lambda(y) \ge i/(p-1)$;

(b) If $y \notin [p]\hat{E}(\mathfrak{m})$ and $i \ne 1$ then $\mathrm{ord}_p \lambda(y) = i/(p-1)$;

(c) $\mathrm{ord}_p\left(\sum_{\delta \in \Delta} \chi(\delta^{-1}) \zeta^{\kappa(\delta)} \right) = i/(p-1)$.

PROOF. By the convergence properties of λ, $\lambda(y) \in \mathfrak{m}$ so we can write $\lambda(y) = \sum_{j \ge j_0} a_j v^j$ with $a_i \in \mathbb{Z}_p$, $j_0 \ge 1$ and $a_{j_0} \in \mathbb{Z}_p^*$. (Recall that v is a p-division point of $\hat{E}$, so $\mathrm{ord}_p(v) = 1/(p-1)$.) Then for all $\gamma \in \Delta$,

$$\kappa^{i}(\gamma)\lambda(y) = \lambda(y)^{\gamma} = \sum_{j\geq j_0} a_j (v^{\gamma})^{j} = \sum_{j\geq j_0} a_j([\kappa(\gamma)](v))^{j}$$

$$\equiv \kappa(\gamma)^{j_0} a_{j_0} v^{j_0} \qquad (\text{mod } v^{j_0+1})$$

$$\equiv \kappa(\gamma)^{j_0}\lambda(y) \qquad (\text{mod } v^{j_0+1})$$

Thus $j_0 \equiv i \pmod{p-1}$, which proves (a). If $j_0 > i$ and $i > 1$ then $j_0 \geq p+1$, so the exponential map converges on $\frac{1}{p}\lambda(y)$, and $y \in [p]\hat{E}(\mathfrak{m})$. That proves (b), and (c) is a standard fact about Gauss sums: see for example [7], Theorem 2.1, Chapter 1.

PROOF OF THEOREM 1. Fix a character $\chi = \kappa^{i}$ with i odd, $1 < i < p-2$, and suppose $\mathcal{S}_{\Phi}^{\chi} \neq 0$. By Corollary 8 we can find $y \in [p]^{-1}\hat{E}(\mathfrak{m})^{\chi}$, $y \notin \hat{E}(\mathfrak{m})$ such that $y^{[u,\Phi]} - y = 0$. By Lemma 12, $\text{ord}_p(p\lambda(y)) = i/(p-1)$ so we can apply the Explicit Reciprocity Law (Theorem 10) to obtain

$$y^{[u,\Phi]} - y = [\text{Tr}_{\Phi/\mathbb{Q}_p}(\delta(u)\lambda(y))]v$$

and we conclude

$$\text{Tr}_{\Phi/\mathbb{Q}_p}(\delta(u)\lambda(y)) \in p\mathbb{Z}_p . \tag{11}$$

Using Proposition 11 (c) we can expand this

$$\text{Tr}_{\Phi/\mathbb{Q}_p}(\delta(u)\lambda(y)) = \sum_{\gamma\in\Delta} \delta(u)^{\gamma}\lambda(y)^{\gamma}$$

$$= \lambda(y)\sum_{\gamma} \chi(\gamma) r(r-\rho)(1-s\sigma)\gamma E_1^{*}(\tau,L)$$

$$= r\lambda(y)(r-\chi^{-1}(\rho))(1-s\chi^{-1}(\sigma))\sum_{\gamma}\chi(\gamma)E_1^{*}(\tau,L)^{\gamma} .$$

Since $\chi \neq \kappa$ or κ^{-1} we can finally choose r,s so that $(r-\chi^{-1}(\rho))$ and $(1-s\chi^{-1}(\sigma))$ are units in $\mathbb{Z}_p$. Then (11) becomes

$$\lambda(y) \sum_{\gamma} \chi(\gamma) E_1^*(\tau,L)^{\gamma} \in p\mathbb{Z}_p$$

and multiplying by $\sum_{\delta} \chi(\delta^{-1})\zeta^{\kappa(\delta)}$ yields

$$\lambda(y) \sum_{\gamma,\delta} \chi(\gamma)\chi(\delta^{-1})\zeta^{\kappa(\delta)} E_1^*(\tau,L)^{\gamma} = p\lambda(y) \sum_{\gamma} \hat{\chi}(\gamma) E_1^*(\tau,L)^{\gamma}$$

$$= p\lambda(y) \mathcal{L}_E(\chi) \in p\Big(\sum_{\delta} \chi^{-1}(\delta)\zeta^{\kappa(\delta)}\Big)\mathbb{Z}_p .$$

By Lemma 12, $\mathrm{ord}_p(p\lambda(y)) = \mathrm{ord}_p(\sum_{\delta} \chi^{-1}(\delta)\zeta^{\kappa(\delta)})$, so we deduce $\mathcal{L}_E(\chi) \equiv 0 \pmod{p}$, as desired.

BIBLIOGRAPHY

1. Cassels, J.W.S.: Arithmetic on curves of genus 1 (VIII). J. Reine Angew. Math. 217, 180 - 199 (1965)

2. Coates, J., Wiles, A.: On the conjecture of Birch and Swinnerton-Dyer. Invent. Math. 39, 223 - 251 (1977)

3. Goldstein, C., Schappacher, N.: Séries d'Eisenstein et fonctions L de courbes elliptiques à multiplication complexe. J. Reine Angew. Math. 327, 184 - 218 (1981)

4. Hazewinkel, M.: Formal Groups and Applications. Academic Press, New York (1978)

5. Katz, N.: The Eisenstein measure and p-adic interpolation. Amer. J. Math. 99, 238 - 311 (1977)

6. Kubert, D., Lang, S.: Modular Units. Springer-Verlag, New York (1981)

7. Lang, S.: Cyclotomic Fields. Springer-Verlag, New York (1978)

8. Rubin, K.: Elliptic curves with complex multiplication and the conjecture of Birch and Swinnerton-Dyer. Invent. Math. 64, 455 - 470 (1981)

9. Serre, J-P., Tate, J.: Good reduction of abelian varieties. Ann. of Math. 88, 492 - 517(1968)

10. Stark, H.: The Coates-Wiles theorem revisited. In: Number Theory Related to Fermat's Last Theorem. Prog. Math 26, 349 - 362, Birkhauser, Boston (1982)

11. Tate, J.: Algorithm for determining the type of a singular fiber in an elliptic pencil. In: Modular Functions of One Variable (IV). Lect. Notes Math. 476, Springer-Verlag, New York (1975)

12. Wiles, A.: Higher explicit reciprocity laws. Ann. of Math. 107, 235 - 254 (1978)

DEPARTMENT OF MATHEMATICS
THE OHIO STATE UNIVERSITY
231 W. 18TH AVENUE
COLUMBUS, OHIO 43210

Canadian Mathematical Society
Conference Proceedings
Volume 7 (1987)

MODULAR FORMS AND RELATED OBJECTS

H. M. STARK

0. Introduction. This paper represents the beginnings of an attempt to bridge the chasm separating the classical and modern points of view regarding modular forms and Dirichlet series. This bridge practically constructs itself when we look at the subject from the point of view of the ring of S-integers of a number field. In the first Section we introduce complex, quaternionic, and p-adic upper half planes and the groups acting on them. We then turn to product spaces and discrete groups. In Section Two, we study fundamental domains for discrete groups. With these preliminaries accomplished, in Section Three we consider modular forms defined on these new spaces and show how to rephrase the classical theory in this new setting. In Section Four, we show the advantages of the method by returning to Hecke's proof of the analytic continuation and functional equation of the Dedekind zeta function. The proof here is his but without the complications which seem to so trouble some writers. As a bonus, we get a direct analytic proof of the S-unit theorem as well as a direct evaluation of the first non-vanishing derivative at 0 in terms of the S-regulator. In the last Section, we introduce p-adic L-series from a classical point of view.

All topics considered in this paper are still in their preliminary stages. For this reason, we will stick to only the simplest instances throughout. In many places, the generalizations are obvious and present no extra difficulties at all; in other places, there are still obstacles which I have not been able to surmount.

1. Groups and Upper Half Planes. We will be dealing with groups acting on spaces. Later we will treat products of groups acting on product spaces, but let us now set up the individual building blocks. In each instance we will have a group G of $2n$ by $2n$ matrices (almost always with $n = 1$) acting on a subset $\mathfrak{h}$ of some matrix or division algebra $\mathfrak{A}$. In all cases, $g = \left(\begin{smallmatrix} a & b \\ c & d \end{smallmatrix}\right)$ in G will have a, b, c, d in $\mathfrak{A}$ and we define for z in $\mathfrak{h}$, and sometimes just in $\mathfrak{A}$,

$$\begin{pmatrix} a & b \\ c & d \end{pmatrix} \cdot z = (az + b)(cz + d)^{-1}.$$

We give four illustrations of such groups and spaces although in these lectures, we will concentrate primarily upon the first three cases:

Case 1. $G = G_{\mathbf{R}} = \mathrm{GL}_2(\mathbf{R})^+$ (positive determinant 2 by 2 real matrices), $\mathfrak{h} = \mathfrak{h}_{\mathbf{R}} = \{x + yi \mid x, y \in \mathbf{R}, y > 0\}$ (the classical upper half plane).

Case 2. $g = G_{\mathbf{C}} = \mathrm{GL}_2(\mathbf{C})^+$ (positive real determinant 2 by 2 complex matrices), $\mathfrak{h} = \mathfrak{h}_{\mathbf{C}} = \{x + yk \mid x \in \mathbf{C}, y \in \mathbf{R}^+\}$ (the quaternionic upper half plane consisting of quaternions with no j-component and positive k-component).

Supported in part by NSF

Case 3. $G = G_p = \mathrm{GL}_2(\mathbf{Q}_p)$, (for simplicity we take p *odd*)
$\mathfrak{h} = \mathfrak{h}_p = \{x + y\sqrt{\Delta} \mid x, y \in \mathbf{Q}_p, y \neq 0\}$ where Δ is a unit in $\mathbf{Q}_p$ and is not a square in $\mathbf{Q}_p$.

Case 3a. (A variation on Case 3. Again with p *odd*).
$G = G_p^0$, the subgroup of G_p of matrices whose determinants are units in $\mathbf{Q}_p$.
$\mathfrak{h} = \mathfrak{h}_p^0$, the subset of elements $x + y\sqrt{\Delta}$ in $\mathfrak{h}_p$ such that y has $\mathrm{ord}_p(y)$ even.

Case 4. $G = \mathrm{Sp}_n(\mathbf{R})$, the group of $2n$ by $2n$ symplectic matrices,
$\mathfrak{h} = \mathfrak{h}^{(n)} = \left\{X + Yi \;\middle|\; \begin{array}{c} X \text{ and } Y \text{ are symmetric real } n \text{ by } n \\ \text{matrices with } Y \text{ positive definite} \end{array}\right\}$
(the Siegel upper half plane).

When everything is defined (i.e. $(cz+d)^{-1}$, etc. exist), $(g_1 g_2) \cdot z = g_1 \cdot (g_2 \cdot z)$ for z in $\mathfrak{A}$. In order to uniformly sketch the proof that these groups act transitively upon the indicated spaces, we will make use of an antiautomorphism of $\mathfrak{A}$ to be denoted by $'$. It has the properties that

$$(a+b)' = a' + b', \qquad (ab)' = b'a'.$$

It will also be the case that

$$a' = {}^{\mathrm{t}}a, \quad b' = {}^{\mathrm{t}}b, \quad c' = {}^{\mathrm{t}}c, \quad d' = {}^{\mathrm{t}}d,$$

where ${}^{\mathrm{t}}$ denotes transpose. Of course this is automatic when $n = 1$. Further for $g = \left(\begin{smallmatrix} a & b \\ c & d \end{smallmatrix}\right)$ in any of our groups, the symplectic conditions,

$$a'c \text{ and } b'd \text{ are symmetric, and } a'd - c'b = sI$$

where s is in the center of $\mathfrak{A}$ and I is the n by n identity matrix, hold. (For symplectic matrices, $sI = I$, and then these conditions are an unilluminating alternate definition of the symplectic group.)

For any $g = \left(\begin{smallmatrix} a & b \\ c & d \end{smallmatrix}\right)$ in G and z and w in $\mathfrak{h}$ (or even in $\mathfrak{A}$ so long as $(cz+d)^{-1}$ and $(cw+d)^{-1}$ exist), we have,

$$g \cdot w - (g \cdot z)' = (z'c' + d')^{-1} s(w - z')(cw + d)^{-1}, \tag{1}$$

as an easy calculation using the properties of $'$ listed above shows. For the sake of uniformity, we will write $z = x + y\sqrt{\Delta}$ in the first three cases, where we may take $\Delta = -1$ in Cases 1 and 2, with an unusual square root used in Case 2. For $A = \left(\begin{smallmatrix} a & b \\ c & d \end{smallmatrix}\right)$ in G we also write

$$z_A = x_A + y_A\sqrt{\Delta} = A \cdot z.$$

We first show that G takes $\mathfrak{h}$ into $\mathfrak{h}$ in all four cases. Cases 1 and 3 are very similar in that the algebra $\mathfrak{A}$ is a field which is a quadratic extension of the field containing the entries of elements of G. By (1), with $w = z$ and the antiautomorphism $'$ given by $z' = \bar{z} = x - y\sqrt{\Delta}$, we see that

$$2y_A\sqrt{\Delta} = \frac{(ad - bc)2y\sqrt{\Delta}}{(cz+d)(c\bar{z}+d)}$$

and so,

$$y_A = \frac{(ad - bc)y}{(cz+d)(c\bar{z}+d)}.$$

This shows that G takes $\mathfrak{h}$ into $\mathfrak{h}$ in Cases 1 and 3. For Case 3a, we need to say a word about the normalized p-adic absolute value which we will use. Naturally, we determine the absolute value on $\mathbf{Q}_p$ by taking $|p|_p = p^{-1}$. For our purposes, it is convenient to extend $| \ |_p$ to $\mathbf{Q}_p(\sqrt{\Delta})$ so that $|p|_p$ is still p^{-1}; this is not the usual normalized valuation of $\mathbf{Q}_p(\sqrt{\Delta})$ which is our valuation squared. In particular, we have,

$$|(cz+d)(c\bar{z}+d)|_p = |(cx+d)^2 - (cy)^2\Delta|_p = \max(|cx+d|_p^2, |cy|_p^2) \tag{2}$$

since Δ is not a square in $\mathbf{Q}_p$ AND p is *odd*. Hence in Case 3a, $\operatorname{ord}(y_A) \equiv \operatorname{ord}(y) (\operatorname{mod} 2)$ which shows that here too, G takes $\mathfrak{h}$ into $\mathfrak{h}$.

Case 2 may be less familiar to the reader. Here $\mathfrak{A}$ is made up of the quaternions and is noncommutative. We first take $' = '^{(j)}$ to be given by

$$(r+si+tj+uk)' = r+si-tj+uk.$$

It can either be directly checked that this is an antiautomorphism or it can be observed that this is composition of the well known antiautomorphism denoted by $^-$, which changes the signs of the i, j and k components, and the automorphism over $\mathbf{C}$ which changes the signs of the i and k components only. Thus $(z - z')$ is twice the j-component of the quaternion z times j. Therefore, by (1), with $w = z$, we see that $A \cdot z$ has no j-component when z has no j-component.

Now take $' = '^{(k)}$, the antiautomorphism which changes only the sign of the k component. With $w = z$, we then get

$$\begin{aligned} y_A k &= (ad-bc)(z'c+d)^{-1}(yk)(cz+d)^{-1} \\ &= (ad-bc)(yk)(\overline{cz+d})^{-1}(cz+d)^{-1}. \end{aligned}$$

(Note that for a complex number ς, $\varsigma k = k\bar{\varsigma}$.) In all three cases, we therefore have the fundamental formula

$$y_A = \frac{(ad-bc)y}{(cz+d)(\overline{cz+d})} \tag{3}$$

and in particular, this completes the proof that G takes $\mathfrak{h}$ into $\mathfrak{h}$ in Case 2.

The same considerations work in the symplectic Case 4. First, we take $'$ to be the transpose operation and $w = z$ in $\mathfrak{h}$. By (1), the result is that z_A is also symmetric. Next, we take $'$ to be $^{t-}$, the conjugate transpose, and $w = z$; by (1) we see that

$$y_A = {}^{t}(\overline{cz+d})^{-1} y (cz+d)^{-1}$$

is also positive definite. This formula is the symplectic analogue of (3). The method presented here has been abstracted from the presentation of Siegel [5, Vol II, p. 128] and can be used to show that $(cz+d)^{-1}$ exists in Case 4 as well.

To verify the transitivity of the action of G on $\mathfrak{h}$, observe that

$$\begin{pmatrix} y & x \\ 0 & 1 \end{pmatrix} \cdot \sqrt{\Delta} = x + y\sqrt{\Delta}.$$

This shows the transitivity directly in Cases 1, 2 and 3. In Case 3a, the matrix $\left(\begin{smallmatrix} y & x \\ 0 & 1 \end{smallmatrix}\right)$ isn't in G necessarily but since $\operatorname{ord}_p(y)$ is even, the matrix can be multiplied by a scalar so as to make its determinant a unit in $\mathbf{Q}_p$. The rescaled matrix still takes $\sqrt{\Delta}$ to $x + y\sqrt{\Delta}$ and so G acts transitively on $\mathfrak{h}$ in Case 3a also. The symplectic Case 4 has the same problem with the same sort of solution. If $y = T \cdot {}^{t}T$, then

$$\begin{pmatrix} y & x \\ 0 & 1 \end{pmatrix} = \begin{pmatrix} T & x\,{}^{t}T^{-1} \\ 0 & {}^{t}T^{-1} \end{pmatrix} \begin{pmatrix} {}^{t}T & 0 \\ 0 & {}^{t}T \end{pmatrix}$$

and the rescaled matrix

$$\begin{pmatrix} T & x^{t}T^{-1} \\ 0 & {}^{t}T^{-1} \end{pmatrix}$$

is symplectic and takes iI to $x + iy$. Therefore, G acts transitively on $\mathfrak{h}$ in Case 4 also.

From the point of view of symmetric spaces, we can now identity $\mathfrak{h}$ with G/K where K is the subgroup of G fixing some chosen special point of $\mathfrak{h}$. We will take this point to be $\sqrt{\Delta}$, where as above, this means i in Case 1 and k in Case 2. A choice of a different point in $\mathfrak{h}$ would just replace K by a group conjugate to K in G. Although it is not our intention to explore the symmetric space point of view too far, it will be useful to explicitly determine K. Suppose that $A = \left(\begin{smallmatrix} a & b \\ c & d \end{smallmatrix}\right)$ is in G and that $A \cdot \sqrt{\Delta} = \sqrt{\Delta}$. We therefore have in Cases 1, 2, and 3,

$$a\sqrt{\Delta} + b = \sqrt{\Delta}(c\sqrt{\Delta} + d) = \bar{c}\Delta + \bar{d}\sqrt{\Delta},$$

where the conjugate signs are only necessary in Case 2. We thus see that A takes the form,

$$A = \begin{pmatrix} a & \bar{c}\Delta \\ c & \bar{a} \end{pmatrix},$$

and conversely, any such matrix works so long as $a, c \neq 0, 0$, although in Case 3a, A may not be in G_p^0. It is very useful to have K be *compact*. This requires shrinking the group G thanks to the presence of scalar matrices in K. In Cases 1, 2 and 3a, given a matrix A in G, there is a scalar matrix sI in G with s real in Cases 1 and 2 and s in $\mathbf{Q}_p$ in Case 3a, such that $\det(sA)$ has absolute value 1 where "absolute value" means the appropriate absolute value for the situation. Both A and sA act the same upon every element of $\mathfrak{h}$. In this manner, we get a subgroup G' of elements $\{sA\}$ of G which still acts transitively on $\mathfrak{h}$. In Cases 1, 2 and 3a, we see that G' is $\mathrm{SL}_2(\mathbf{R}), \mathrm{SL}_2(\mathbf{C})$, and G_p^0, respectively. In Case 3, we can rescale A via scalar matrices from $\mathbf{Q}_p$ to have $|\det(A)|_p$ being 1 or p, say. This is not good enough since the resulting set of matrices is not a group. However, if we temporarily allow scalar matrices of the form $\sqrt{pI}$ then we can rescale any A in G to a matrix all of whose elements are in $\mathbf{Q}_p$ or all of whose elements are in $\mathbf{Q}_p\sqrt{p}$ and whose determinant is a unit in $\mathbf{Q}_p$. We let the set of all such rescaled matrices be G'. We denote by K' the stabilizer of $\sqrt{\Delta}$ in G'. The group K' now is compact. Indeed it is just SO_2 in Case 1, SU_2 in Case 2, isomorphic to the group of units of $\mathbf{Q}_p(\sqrt{\Delta})$ in Case 3a and somewhat messier in Case 3. However, in Cases 3 and 3a, *K' is not a maximal compact subgroup of G'.*

The groups G' are given a topology by using the usual absolute value on the fields $\mathbf{R}$, $\mathbf{C}$, and $\mathbf{Q}_p$:

$$\mathrm{dist}((a_{ij}), (b_{ij})) = \max_{i,j}(|a_{ij} - b_{ij}|).$$

This in turn gives rise to a topology on G'/K' via the definition,

$$\mathrm{dist}(g_1K, g_2K) = \min_{\substack{k_1 \in K' \\ k_2 \in K'}} (\mathrm{dist}(g_1k_1, g_2k_2))$$

where the minimum exists since K is compact and is non-zero if the two cosets g_1K and g_2K are distinct. (The choice of the letter "k" here is too convenient to avoid; however it will not appear again in this paper as an element of K.) It is easily checked that the topology given to G'/K' in this manner is the same topology as that already existing on $\mathfrak{h}$ by setting $\mathrm{dist}(z_1, z_2) = |z_1 - z_2|$.

We say that a subgroup Γ of G acts *discontinuously* on $\mathfrak{h}$ if for every pair of points w and z in $\mathfrak{h}$, there is a neighborhood N of w such that only finitely many of the images $A \cdot z$ of z under elements A of Γ lie in N. In other words, for any z in $\mathfrak{h}$, there is no limit point in $\mathfrak{h}$ in the set $\{A \cdot z\}$ where A runs through Γ, in the strong sense that the same point cannot occur infinitely often among the $A \cdot z$. For groups Γ acting discontinuously, we will be able to create a *fundamental domain* for the action of Γ on $\mathfrak{h}$. It is therefore very important to have a criterion for determining if a group Γ acts discontinuously on $\mathfrak{h}$. This is furnished by the concept of a discrete subgroup of G: A subgroup Γ of G is said to be *discrete* if the identity is not a limit of elements of Γ. There is a classical theorem relating these two concepts.

PROPOSITION 1. *If $\mathfrak{h} = G/K$, where K is compact, then a subgroup Γ of G acts discontinuously on $\mathfrak{h}$ if and only if Γ is discrete.*

The proof is not too difficult. In many sources, it is given when $K = \{I\}$ so that $\mathfrak{h} = G$, but the proof is only slightly complicated for general compact K. The hypothesis that K is compact is vital since there are counterexamples otherwise. This is the reason that we made such a fuss with the rescaling procedure above. We are now in position to give several examples.

Example 1. $\mathrm{SL}_2(\mathbf{Z})$ is discrete in $\mathrm{GL}_2(\mathbf{R})$.

Example 2. If K is a complex quadratic field and $\mathfrak{O} = \mathfrak{O}_K$ is its ring of integers, then $\mathrm{SL}_2(\mathfrak{O})$ is discrete in $\mathrm{GL}_2(\mathbf{C})^+$ since any integer of K "near" 0 is 0.

Example 3. If $K \neq \mathbf{Q}$ is a real field (embedded in $\mathbf{R}$), then $\mathrm{SL}_2(\mathfrak{O}_K)$ is *not* discrete. Take any non-zero integer α with $|\alpha| < 1$; we then have a sequence approaching the identity:

$$(4) \qquad \begin{pmatrix} 1+\alpha^{2n} & \alpha^n \\ \alpha^n & 1 \end{pmatrix} \longrightarrow \begin{pmatrix} 1 & 0 \\ 0 & 1 \end{pmatrix} \quad \text{as } n \longrightarrow \infty.$$

For the same reason, if K is complex but not complex quadratic, $\mathrm{SL}_2(\mathfrak{O}_K)$ is not discrete in $\mathrm{GL}_2(\mathbf{C})$.

Example 4. We "repair" Example 3 by introducing $G = G_{\mathbf{R}}^{r_1} \times G_{\mathbf{C}}^{r_2}$ and $\mathfrak{h} = \mathfrak{h}_{\mathbf{R}}^{r_1} \times \mathfrak{h}_{\mathbf{C}}^{r_2}$, where r_1 and $2r_2$ are the number of real and complex conjugate fields of K ordered as always so that for $r_1 < j \leq r_1 + r_2$, $K^{(j+r_2)}$ is the complex conjugate embedding of $K^{(j)}$. We also embed $\Gamma = \mathrm{SL}_2(\mathfrak{O}_K)$ into G by sending A in Γ to $(A^{(1)}, \ldots, A^{(r_1+r_2)})$ in G. Here Γ is discrete in G since if all the conjugates of an algebraic integer are less than 1 in absolute value, that integer is 0. Note that if just one of the $r_1 + r_2$ conjugates were missing, we could find non-zero algebraic integers with all the remaining conjugates having absolute values less than 1 and then we could use a sequence of the form in (4) to see that the resulting embedded version of Γ would not be discrete.

Example 5. The group $\Gamma = \mathrm{GL}_2(\mathbf{Z}[1/p])$ is not discrete in $\mathrm{GL}_2(\mathbf{R})$. The reason is the same as in Example 3. Take $\alpha = 1/p$ and use the sequence in (4).

Example 6. We repair Example 5 in the same manner we repaired Example 3. Namely, we take for our group G the group $G_{\mathbf{R}} \times G_p$, and $\mathfrak{h} = \mathfrak{h}_{\mathbf{R}} \times \mathfrak{h}_p$. We also embed Γ into G in the obvious manner. Now Γ is discrete in G since, if $|\alpha|_\infty < 1$ and $|\alpha|_p < 1$ for some α in $\mathbf{Z}[1/p]$, then $\alpha = 0$.

Example 7. More generally, if S is a finite set of primes of K including the infinite primes, we take G to be the product of all the $G_\mathfrak{p}$ and $\mathfrak{h}$ to be the product of all the $\mathfrak{h}_\mathfrak{p}$ and embed $\Gamma = \mathrm{SL}_2(\mathfrak{O}_K(S))$ in G in the manner indicated above. Here $\mathfrak{O}_K(S)$ denotes the set of S-integers of K. Then Γ is discrete in G. The main technical advantage to this is that by using a carefully chosen S, we can make the class-number of $\mathfrak{O}_K(S)$ equal to one and we can also eliminate problems with the different if we desire.

2. Fundamental Domains. We now turn to the consideration of fundamental domains for the action of our discrete groups. In principle, a *fundamental domain* for the action of Γ on $\mathfrak{h}$ consists of a subset of points $\mathfrak{F}$ of $\mathfrak{h}$ such that every point of $\mathfrak{h}$ is equivalent to exactly one point of $\mathfrak{F}$ under the action of Γ. In practice, one usually desires a nice set of points for $\mathfrak{F}$. For example, $\mathfrak{F}$ should be measurable at the very least. Most fundamental domains that arise in print are connected. There is also the problem of the boundary. In classical examples, describing which points on the boundary are to be included and which are to be excluded is already sometimes awkward. For this reason, we frequently see simply a description of the closure of $\mathfrak{F}$ in $\mathfrak{h}$. These problems multiply when we add p-adic components to our groups and spaces.

There are two standard methods of finding fundamental domains. The first is a method often used in Riemann surface theory which I call the *perpendicular bisector method*. It goes as follows. Select a point w in $\mathfrak{h}$. The closure of the fundamental domain $\mathfrak{F}$ of $\Gamma \setminus \mathfrak{h}$ consists of all points z of $\mathfrak{h}$ which are at least as close to w as to any $A \cdot w$ for A in Γ. Here "closeness" is measured by an invariant distance. It is called the perpendicular bisector method because in $\mathfrak{h}_{\mathbf{R}}$, it consists of drawing the non-Euclidean perpendicular bisector of the non-Euclidean line segment joining w and $A \cdot w$ and then forming the intersection of all the resulting non-Euclidean half planes containing w. Although used for compact fundamental domains in Riemann surface theory, the method works just as well here. For example, let $\Gamma = \mathrm{SL}_2(\mathbf{R})$ and take $w = 2i$. The perpendicular bisector of the line joining w and $\left(\begin{smallmatrix} 0 & -1 \\ 1 & 0 \end{smallmatrix}\right) \cdot w$ is the unit circle and the fundamental domain we obtain is the classic one (see Figure 1).

For our more general spaces, there is not a non-Euclidean geometry involved and this is why we have phrased the method in terms of taking points closer to w than all the $A \cdot w$. However, even when phrased in this manner, it is not so obvious how we intend to measure invariant geodesic distances in our product spaces when there are p-adic components. We circumvent this problem by noting that for $w = u + v\sqrt{\Delta}$ and $z = x + y\sqrt{\Delta}$ in $\mathfrak{h}$ and A in G, we have

$$\left| \frac{(w_A - z_A)^2}{v_A y_A} \right| = \left| \frac{(w - z)^2}{vy} \right|$$

by (1) and (3) in all three Cases. (In Cases 1 and 3, $'$ is the identity while in Case 2, $' = {}'^{(j)}$ which preserves points in $\mathfrak{h}$. In each instance, $|z'c' + d'| = |cz + d|$.) The expression $|(w - z)^2/vy|$ is frequently called a *point pair invariant* and can be used to measure distances.

The second method for finding fundamental domains available to us is what I call the *highest point method*. The method consists of finding a point in a Γ-equivalence class "higher" in $\mathfrak{h}$ (for $\mathfrak{h}_{\mathbf{R}}$, this means greater y) than all the other points in the class. If there are many points of the same greatest height (and for us there will be), then we must also provide a method to decide upon one point among them. This method is *not* suitable to the circle groups of Riemann surface theory where there is no point in an equivalence class of points higher in the fundamental domain than the others. Indeed, for compact (closures of) fundamental domains, there are always higher versions of the entire fundamental domain and this precludes finding a highest point in an equivalence class.

However, the method is beautifully suited to groups containing translations. As a warm up, let us sketch why if Γ is discrete in $\mathrm{SL}_2(\mathbf{R})$ acting on $\mathfrak{h} = \mathfrak{h}_{\mathbf{R}}$, and Γ contains a translation $T = \left(\begin{smallmatrix} 1 & t \\ 0 & 1 \end{smallmatrix}\right)$, then given z in $\mathfrak{h}$ and a bound $\delta > 0$, there exist $(\bmod T)$ only finitely many points z_A equivalent to z under the action of Γ such that $y_A > \delta$. More exactly, this means that there are only finitely many cosets in $\langle T \rangle \setminus \Gamma$ containing matrices A with $y_A > \delta$. Suppose that there are infinitely many A $(\bmod T)$ in Γ mapping z to points z_A with $y_A > \delta$. We may assume that A has been chosen $(\bmod T)$ (with T acting on the left of A) such that z_A is in the strip $0 \leq x < |t|$, say. If the y_A don't tend to ∞, then we get an infinite set of points z_A in a compact subset of $\mathfrak{h}$ and hence there is a limit point. But this cannot happen for discrete groups by Proposition 1 and therefore $y_A \to \infty$. But now look at $w = A^{-1} \cdot (z_A + t)$ which is equivalent to z under the action of Γ. The idea is that z_A and $z_A + t$ are very close in the non-Euclidean geodesic distance sense since y_A is large and therefore z and w are also very close. This in turn makes z and w very close in the Euclidean sense also since z is a fixed point in $\mathfrak{h}$. And indeed as A varies, we would get $w \to z$ which cannot happen. We don't actually need geodesics for this however: If $w = u + iv$ then

$$\left| \frac{(w - z)^2}{vy} \right| = \left| \frac{(z_A + t - z_A)^2}{y_A y_A} \right| = \left(\frac{t}{y_A} \right)^2.$$

Hence v is bounded above since otherwise the left side would be $>> v$, and therefore $|w - z| \to 0$. Q.E.D.

The same method easily shows that if Γ is a discrete subgroup of $\mathrm{SL}_2(\mathbf{C})$ containing two linearly independent translations over $\mathbf{Z}$ (and therefore linearly independent over $\mathbf{R}$ since Γ is discrete), then given z in $\mathfrak{h}_{\mathbf{C}}$ and a bound $\delta > 0$, there exist (mod translations) only finitely many points z_A equivalent to z under the action of Γ such that $y_A > \delta$. We do not have to consider the same problem for a discrete subgroup of G_p acting on $\mathfrak{h}_p$ since no such group can contain a non-zero translation. This is easily seen since if $\left(\begin{smallmatrix} 1 & t \\ 0 & 1 \end{smallmatrix}\right)$ is in Γ then so is

$$\begin{pmatrix} 1 & t \\ 0 & 1 \end{pmatrix}^{p^n} = \begin{pmatrix} 1 & tp^n \\ 0 & 1 \end{pmatrix}$$

which tends to $\left(\begin{smallmatrix} 1 & 0 \\ 0 & 1 \end{smallmatrix}\right)$ as $n \to \infty$, and so Γ is not discrete. Now we will explore the highest point method as it applies to number fields. Let K be a number field and let S be a finite set of primes of K including all the infinite primes. We will let $\mathfrak{O}(S)$ denote the ring of S-integers of K. We take for G the group $G = \prod_{\mathfrak{p} \in S} G_{\mathfrak{p}}$ which acts on $\mathfrak{h} = \prod_{\mathfrak{p} \in S} \mathfrak{h}_{\mathfrak{p}}$. We embed $\mathrm{SL}_2(\mathfrak{O}(S))$ and $\mathrm{GL}_2(\mathfrak{O}(S))^+$ (positive real determinant at all infinite $\mathfrak{p}$) in the usual manner (fixing which embedding we select from each conjugate pair of embeddings once and for all).

Let us define the *absolute norm*, or *S-norm*, appropriate to S-integers. For α in K, we define

$$\mathbf{N}(\alpha) = \mathbf{N}_S(\alpha) = \prod_{1 \leq j \leq r_1} \alpha^{(j)} \cdot \prod_{\substack{\mathfrak{p} \text{ in } S \\ \mathfrak{p} \text{ not real infinite}}} |\alpha|_{\mathfrak{p}}.$$

Clearly, $\mathbf{N}(\alpha)$ is a rational number, possibly negative, $\mathbf{N}(\alpha) = 0$ if and only if $\alpha = 0$, $\mathbf{N}(\alpha)$ is in $\mathbf{Z}$ if α is in $\mathfrak{O}(S)$ and $\mathbf{N}(\alpha) = \pm 1$ if α is an S-unit (i.e., α is in $\mathfrak{O}(S)^\times$). We must carefully distinguish $\mathbf{N}(\alpha)$ from $\mathbf{N}_{K/\mathbf{Q}}(\alpha) = \prod_{j=1}^n \alpha^{(j)}$ ($n = \deg(K)$) which is now just one more instance of a relative norm. We will frequently write, $\mathbf{N}_{K/\mathbf{Q}}(a) = \mathbf{N}_\infty(\alpha)$, and call this the *infinity-norm* of α since it is what results when S consists solely of the infinite primes of K. We may likewise define the absolute norm of S-ideals; it is a positive rational *number*, integral for integral S-ideals and $\mathbf{N}((\alpha)) = |\mathbf{N}(\alpha)|$ as usual.

We extend this idea of the norm to any vector $z = (z_\mathfrak{p})_{\mathfrak{p}\in S}$ indexed by the $\mathfrak{p}$ in S with each $z_\mathfrak{p}$ being complex, quaternionic, or quadratic over $K_\mathfrak{p}$ in the three cases, by defining

$$\mathbf{N}(z) = \prod_{\mathfrak{p} \text{ real}} z_\mathfrak{p} \cdot \prod_{\mathfrak{p} \text{ not real}} |z_\mathfrak{p}|_\mathfrak{p}.$$

Here again, the infinity norm will arise,

$$\mathbf{N}_\infty(z_\infty) = \prod_{\substack{\mathfrak{p}|\infty \\ \mathfrak{p} \text{ real}}} z_\mathfrak{p} \cdot \prod_{\substack{\mathfrak{p}|\infty \\ \mathfrak{p} \text{ not real}}} |z_\mathfrak{p}|_\mathfrak{p},$$

where $z_\infty = (z_\mathfrak{p})_{\mathfrak{p}|\infty}$. We now come to a fundamental formula for $\mathbf{N}(y)$ under the action of an $A = \left(\begin{smallmatrix} a & b \\ c & d \end{smallmatrix}\right)$ in G. We write $z = x + y\sqrt{\Delta}$ and for the $\mathfrak{p}$-component z, $z_\mathfrak{p} = x_\mathfrak{p} + y_\mathfrak{p}\sqrt{\Delta_\mathfrak{p}}$, where $\sqrt{\Delta_\mathfrak{p}}$ is i in Case 1, k in Case 2, and the square root of a non-square unit of $K_\mathfrak{p}$ in Case 3. Also, as usual set $A \cdot z = z_A = x_A + y_A\sqrt{\Delta}$. We find from (3) that

$$\mathbf{N}(y_A) = \frac{\mathbf{N}(ad - bc)\mathbf{N}(y)}{\mathbf{N}((cz + d)(\overline{cz + d}))}.$$

The highest point method consists of trying to maximize $\mathbf{N}(y)$ by transforming under the action of a discrete subgroup Γ of G. We will take $\Gamma = \mathrm{SL}_2(\mathfrak{O}(S))$ and leave $\mathrm{GL}_2((\mathfrak{O}(S)))^+$ to the reader. For each of these groups, if A is in Γ, then $\det(A)$ is a totally positive S-unit and so

$$\mathbf{N}(y_A) = \frac{\mathbf{N}(y)}{\mathbf{N}((cz + d)(\overline{cz + d}))} \qquad (A \text{ in } \Gamma).$$

We will need the stabilizer of ∞,

$$\Gamma_\infty = \left\{ \begin{pmatrix} \varepsilon & b \\ 0 & \varepsilon^{-1} \end{pmatrix} \text{ in } \Gamma \right\}.$$

This subgroup is necessary to us because for A in Γ_∞, $\mathbf{N}(y_A) = \mathbf{N}(y)$. It is therefore natural to split Γ up into cosets $\Gamma_\infty\gamma$ since $\mathbf{N}(y_A)$ is the same for all A in the coset $\Gamma_\infty\gamma$. The highest point method is successful because of the following

THEOREM 1. *Given any point z in $\mathfrak{h}$ and any $\delta > 0$, there are only finitely many cosets $\Gamma_\infty\gamma$ for which $\mathbf{N}(y_A) > \delta$ with A in $\Gamma_\infty\gamma$.*

PROOF. When implementing the highest point method, it is useful to pick one point in each equivalence class of points of $\mathfrak{h}$ $(\mathrm{mod}\,\Gamma_\infty)$ and call it *semi-reduced.* The manner of choosing a semi-reduced point may vary as the need requires. The manner appropriate here is the following: Let U be the group of squares of S-units. For an S-unit ε, ε^2 acts on y by the rule $\varepsilon^2 \cdot y = \varepsilon\bar{\varepsilon}y$. By the S-unit theorem (an analytic proof of which is given in Section 4 below), there is a fundamental domain $\mathfrak{F}_2$ for y $(\mathrm{mod}\, U)$ such that every $|y|_\mathfrak{p}$ is commensurable with every other in the sense that there is a constant $C > 0$ such that

$$C^{-1} < \frac{|y_{\mathfrak{p}'}|_{\mathfrak{p}'}}{|y_\mathfrak{p}|_\mathfrak{p}} < C$$

for all pairs $\mathfrak{p}$ and $\mathfrak{p}'$ in S. In particular, for all $\mathfrak{p}$ in S and y in $\mathfrak{F}_2$,

$$\mathbf{N}(y)^{1/N} < C|y_\mathfrak{p}|_\mathfrak{p},$$

where $N = |S|$, the number of elements in S. In addition there is a fundamental domain $\mathfrak{F}_1$ for x modulo translation by S-integers. Since for a given value of $\mathbf{N}(y)$, the subset of

$\overline{\mathfrak{F}}_2$ with the given value of $\mathbf{N}(y)$ is compact and $\overline{\mathfrak{F}}_1$ is compact and Γ_∞ is discrete, given z in $\mathfrak{h}$, there are only finitely many T in Γ_∞ such that $T \cdot z$ is in $\overline{\mathfrak{F}}_1 \cdot \overline{\mathfrak{F}}_2$. Pick one such T and call the resulting $T \cdot z$ semi-reduced.

Now given z, look at the set of all points $A \cdot z$ for A in Γ such that $A \cdot z$ is semi-reduced and such that $\mathbf{N}(y_A) > \delta$. We see that

$$\delta^{1/N} < C|(y_\mathfrak{p})_A|_\mathfrak{p}$$

and hence

$$|cz_\mathfrak{p} + d|_\mathfrak{p}^2 < C|y_\mathfrak{p}|_\mathfrak{p}\delta^{-1/N}.$$

Since $|cy_\mathfrak{p}|_\mathfrak{p} \leq |cz_\mathfrak{p} + d|_\mathfrak{p}$, we find that each $|c|_\mathfrak{p}$ is bounded and so there are only a finite number of possibilities for c. For each of these, we find in like manner that $|cx_\mathfrak{p} + d|_\mathfrak{p}$ is bounded and so $|d|_\mathfrak{p}$ is bounded as well. Therefore there are only finitely many pairs (c, d) involved in all the A. But by linear algebra, two A's with the same (c, d) are congruent modulo a translation: If $ad - bc = a_0 d - b_0 c$ then for some t

$$\begin{pmatrix} a & b \\ c & d \end{pmatrix} = \begin{pmatrix} 1 & t \\ 0 & 1 \end{pmatrix} \begin{pmatrix} a_0 & b_0 \\ c & d \end{pmatrix}.$$

Since our point z_A is semi-reduced, the pair (c, d) corresponds to only one top row (a, b). Q.E.D.

Incidentally, this same proof applies to $\mathrm{Sp}_n(\mathbf{Z})$ acting on $\mathfrak{h}^{(n)}$. See Siegel [5, Vol. III, pp. 153–158] for details (Siegel uses the word, "semireduced", in a different manner). The closure of the fundamental domain may now be described as the set of all semi-reduced points z such that $|\mathbf{N}(cz + d)| \geq 1$ for all $\left(\begin{smallmatrix} a & b \\ c & d \end{smallmatrix}\right)$ in Γ and the interior of the fundamental domain has $|\mathbf{N}(cz + d)| > 1$ for all $\left(\begin{smallmatrix} a & b \\ c & d \end{smallmatrix}\right)$ in $\Gamma - \Gamma_\infty$ (i.e., $c \neq 0$). In particular, any semi-reduced point $x + y\sqrt{\Delta}$ with $\mathbf{N}(y) > 1$ is in the fundamental domain and hence any reasonable semi-reduction process yields a connected fundamental domain when S has only infinite primes. Specifically, given two points z and z' in the fundamental domain, we have to be able to increase the $y_\mathfrak{p}$ and $y'_\mathfrak{p}$ continuously so that z and z' stay semi-reduced and so that $\mathbf{N}(y)$ and $\mathbf{N}(y')$ increase to some common value greater than 1. The Inequalities $|\mathbf{N}(cz + d)| > 1$ continue to hold for all $\left(\begin{smallmatrix} a & b \\ c & d \end{smallmatrix}\right)$ in Γ with $c \neq 0$. We then have to be able to change y' to y within the chosen fundamental domain for y (mod IU) and x' to x within the chosen fundamental domain for x (mod translations) which requires reasonable fundamental domains for these two actions also.

Thus far, the "highest point method" is hardly a method. Let us investigate to what extent it is a method. If a fundamental domain with boundary matchups has already been provided, there is no trouble in implementing the "method". However, we often wish to use it to find the fundamental domain and when used this way it is more of an art form than a method.

The simplest instance in which to use the highest point method is when the ring $\mathfrak{O}(S)$ is Euclidean with respect to the usual (S-)norm. We will take this to mean that given any $x = (x_\mathfrak{p})_{\mathfrak{p} \in S}$, there is an S-integer t such that $|\mathbf{N}(x - t)| < 1$. (This is easier than saying the same thing for x in K and then worrying about the $<$ sign becoming $\leq$ when we extend to all x.) Given any initial point z, we now proceed to find by a two step process a succession of "higher" points equivalent to z under Γ.

Step 1. Semireduce z using for a fundamental domain for x modulo translations a set of points with $|\mathbf{N}(x)| < 1$.

Step 2. If $|\mathbf{N}(z)| < 1$, replace z by $\left(\begin{smallmatrix} 0 & -1 \\ 1 & 0 \end{smallmatrix}\right) \cdot z$ which raises $\mathbf{N}(y)$.

We repeat this process until it terminates which it must do by Theorem 1. When we come to an end, we will have a point equivalent to the original point z under Γ which is semi-reduced and is on or "above" the unit hyper-mess $|\mathbf{N}(z)| = 1$. For $\mathrm{SL}_2(\mathbf{Z})$, this is the well known translation and inversion process which leads to the usual fundamental domain in the upper half plane (the same domain illustrated in Figure 1 earlier). For Euclidean $\mathfrak{O}(S)$ in general, we have taken any initial point z to an equivalent point $z_A = A \cdot z$ for which $\mathbf{N}(y_A)$ cannot be "too small". In particular, we have found an equivalent z_A "removed from the floor" of $\mathfrak{h}$. This is frequently satisfactory in applications; the fundamental domain itself is often unnecessarily complicated.

However, the question must be asked, have we by translation and inversion arrived at a fundamental domain for $\Gamma \setminus \mathfrak{h}$. In general the answer will be, "no". We illustrate for $K = \mathbf{Q}(\sqrt{-11})$, $S = \{\infty\}$. The first task is to choose a fundamental domain for x modulo translations inside the region $|\mathbf{N}(x)| < 1$. This is possible since $\mathfrak{O} = \mathfrak{O}_K$ is known to be Euclidean. The obvious parallelogram centered at the origin fails to be in the necessary region since two of the corners stick out past the unit circle. So we refer to the standard proof that $\mathfrak{O}$ is Euclidean to find a better fundamental domain for x modulo translations. This fundamental domain, $\mathfrak{F}_1$, is a rectangle centered at the origin of width 1 and height $\sqrt{11}/2$. From the point of view of elliptic curves and complex multiplication, such a fundamental domain for x modulo translations is very disturbing but it is nevertheless legitimate. (It is arrived at by translating by elements of $\mathfrak{O}$ until $|\mathrm{Im}(x)| \leq \sqrt{11}/4$ and then translating by rational integers until $|\mathrm{Re}(x)| \leq 1/2$.) Since $\mathfrak{F}_1$ is entirely contained within the unit circle, the translation and inversion process applied to an initial z will terminate in the region inside the column above $\mathfrak{F}_1$ and above the unit sphere shown in Figure 2.

But this region is *not* the fundamental domain for $\Gamma \setminus \mathfrak{h}$. We discover this easily enough when we try to make the boundary identifications of our purported fundamental domain. The problem lies with the identifications of the sides of $\mathfrak{F}_1$ under translations. Namely, the bottom and top edges of $\mathfrak{F}_1$ split into two equal halves with the right half of the top edge mapping to the left half of the bottom edge and the left half of the top edge mapping to the right half of the bottom edge. From the point of view of the region in Figure 2, this means that a point in the region just above one of the corners (say the lower left corner) is equivalent by a translation to a point inside the unit sphere near the middle of either the front or back face (the back face if we started near the lower left corner). Hence such a point can be translated into the unit sphere and then inverted so as to increase y. Therefore there are sets of equivalent points in the region of Figure 2. The true fundamental domain has four extra spherical caps centered at $(\pm 1 \pm \sqrt{-11})/2$ of radius one (see Figure 3).

This is the fundamental domain which I have been drawing for several years. However, there is another fundamental domain which is indeed of the form which we desire. It is based on still a third version of the fundamental domain for $\mathbf{C}$ modulo translations. It is given in Figure 4. In fact the domain, $\mathfrak{F}_1$, shown there is nothing more than the set of points closer to the origin than to all the translations of the origin. In other words, it is the fundamental domain for $\mathbf{C}$ modulo translations given by the perpendicular bisector method. The fundamental domain for $\Gamma \setminus \mathfrak{h}$ consists of those points outside the unit sphere with x coordinate inside $\mathfrak{F}_1$. I would be surprised for a general Euclidean number field if this can always be done, but I have not attempted to find a counterexample.

Now we give an example of a fundamental domain for a group with a p-adic component. We take $\mathbf{Q}$ to be the number field, and $S = \{\infty, p\}$ where p is an odd prime. We take $G = G_{\mathbf{R}} \times G_p$ which acts on $\mathfrak{h} = \mathfrak{h}_{\mathbf{R}} \times \mathfrak{h}_p$ and for our discrete group, we take $\Gamma = \mathrm{GL}_2(\mathbf{Z}[p^{-1}])^+$. Alternately we will take $\Gamma^0 = \mathrm{SL}_2(\mathbf{Z}[p^{-1}])$ acting on $\mathfrak{h}^0 = \mathfrak{h}_{\mathbf{R}} \times \mathfrak{h}_p^0$. In this context it is convenient to call a point $z = (z_\infty, z_p)$ semireduced if x_p is in $\mathbf{Z}_p$, $|y_p|_p = 1$, and $|x_\infty|_\infty \le 1/2$. Every point of $\mathfrak{h}$ (respectively $\mathfrak{h}^0$) is equivalent to a semireduced point modulo Γ_∞ (respectively, Γ_∞^0) since we can first bring $|y_p|$ to the correct value then translate so that x_p is in $\mathbf{Z}_p$ and then translate by an element of $\mathbf{Z}$ so that $|x_\infty|_\infty \le 1/2$.

We now apply the highest point method to this concept of semireduced. When the process terminates, we must have arrived at a point $z = (z_\infty, z_p)$ which is in $\mathfrak{F}_\infty \times \mathfrak{F}_p$ where $\mathfrak{F}_\infty$ is the usual fundamental domain in $\mathfrak{h}_{\mathbf{R}}$ consisting of points z_∞ with $|z_\infty|_\infty \ge 1$ and $|x_\infty|_\infty \le 1/2$ and $\mathfrak{F}_p$ consists of those points z_p with $|y_p|_p = 1$ and $|x_p|_p \le 1$. It will be handy to say that $\mathfrak{F}_p$ consists of those points of $\mathfrak{h}_p$ *near* to $\sqrt{\Delta}$.

THEOREM 2. *$\mathfrak{F}_\infty \times \mathfrak{F}_p$ is the closure of a fundamental domain for the action of $\mathrm{GL}_2(\mathbf{Z}[p^{-1}])^+$ acting on $\mathfrak{h}_{\mathbf{R}} \times \mathfrak{h}_p$ and for $\mathrm{SL}_2(\mathbf{Z}[p^{-1}])$ acting on $\mathfrak{h}_{\mathbf{R}} \times \mathfrak{h}_p^0$.*

PROOF. We need only prove the latter since for any A in $\mathrm{GL}_2(\mathbf{Z}[p^{-1}])^+$ $\det(A)$ is a totally positive unit in $\mathbf{Z}[p^{-1}]$ and so is a power of p. If the power is even, we can rescale A to be in $\mathrm{SL}_2(\mathbf{Z}[p^{-1}])$ and if the power is odd, A can't take a point near $\sqrt{\Delta}$ to a point near $\sqrt{\Delta}$. So we may deal with $\Gamma^0 = \mathrm{SL}_2(\mathbf{Z}[p^{-1}])$ acting on $\mathfrak{h}^0$. The question which we must answer is, "Does there exist an $A \ne \pm I$ in Γ^0 and a point $z = (z_\infty, z_p)$ in $\mathfrak{F}_\infty \times \mathfrak{F}_p$ such that $A \cdot z$ is also in $\mathfrak{F}_\infty \times \mathfrak{F}_p$?" This question is completely answered by the following Lemma from which the Theorem clearly follows. (Actually, the answer is, "yes", due to boundary identifications in $\mathfrak{F}_\infty$.)

LEMMA 1. *If A is in $\mathrm{SL}_2(\mathbf{Q}_p)$, z is in $\mathfrak{h}_p$, and both z and $A \cdot z$ are near to $\sqrt{\Delta}$ in $\mathfrak{h}_p$, then A is in $\mathrm{SL}_2(\mathbf{Z}_p)$.*

PROOF. Let $A = \left(\begin{smallmatrix} a & b \\ c & d \end{smallmatrix}\right)$ and $z_A = x_A + y_A\sqrt{\Delta} = A \cdot z$ and suppose that we have

$$|y_A| = |y| = 1, \quad |x_A| \le 1, \quad |x| \le 1 \quad \text{where } |\ | = |\ |_p.$$

First we have from $|y_A| = |y|$ that $\max(|cy|, |cx + d|) = 1$ from which we see that c and d are in $\mathbf{Z}_p$ and at least one of them has absolute value 1. In particular $|cz + d| = |cy\sqrt{\Delta} + d| = 1$. Hence,

$$|A \cdot z - A \cdot (y\sqrt{\Delta})| = \frac{|z - y\sqrt{\Delta}|}{|cz + d| \cdot |cy\sqrt{\Delta} + d|} = |x| \le 1.$$

It follows that $x' + y'\sqrt{\Delta} = A \cdot (y\sqrt{\Delta})$ is also near $\sqrt{\Delta}$ (that y' has absolute value 1 follows from the fact that $|cy\sqrt{\Delta} + d| = 1$); we need only the x' part of this statement, namely $|x'| \le 1$. We have

$$x' + y'\sqrt{\Delta} = \frac{ay\sqrt{\Delta} + b}{cy\sqrt{\Delta} + d} \cdot \frac{-cy\sqrt{\Delta} + d}{-cy\sqrt{\Delta} + d}$$

and so

$$|bd - acy^2\Delta| = \frac{|bd - acy^2\Delta|}{|cy\sqrt{\Delta} + d|^2} = |x'| \le 1.$$

We wish to conclude from this that both a and b must be integral.

Since at least one of c and d have absolute value 1, we can find a_0 and b_0 in $\mathbf{Z}_p$ so that $\det\left(\begin{smallmatrix} a_0 & b_0 \\ c & d \end{smallmatrix}\right) = \det\left(\begin{smallmatrix} a & b \\ c & d \end{smallmatrix}\right) = 1$. By linear algebra, there is a t in $\mathbf{Q}_p$ such that

$$\begin{pmatrix} a & b \\ c & d \end{pmatrix} = \begin{pmatrix} 1 & t \\ 0 & 1 \end{pmatrix} \begin{pmatrix} a_0 & b_0 \\ c & d \end{pmatrix}.$$

Hence

$$bd - acy^2\Delta = (b_0 d - a_0 cy^2\Delta) + t(d^2 - c^2y^2\Delta).$$

But the coefficient of t has absolute value 1 and $b_0 d - a_0 cy^2\Delta$ is in $\mathbf{Z}_p$ and therefore t is in $\mathbf{Z}_p$ too. Thus a and b are in $\mathbf{Z}_p$. Q.E.D.

3. Modular Forms. We wish to define modular forms with p-adic components. Such modular forms will have a Fourier expansion at ∞ and before we can even begin to discuss the forms, we must describe the exponential factor in the Fourier expansion. Just as we previously had to define the absolute or S-norm, so we will now have to define the absolute or S-trace. Let S be a finite set of primes of $\mathbf{Q}$, including the prime at ∞ and let K be a number field. By an abuse of terminology, we will say that a prime $\mathfrak{p}$ of K is in S if $\mathfrak{p}$ divides a prime in S. This is reasonable since the S-integers of K are the integral closure of the S-integers of $\mathbf{Q}$. Let $x = (x_\mathfrak{p})_{\mathfrak{p}\in S}$. For p in S, it will also be convenient to let $x_p = (x_\mathfrak{p})_{\mathfrak{p}|p}$. We may then think of x as $(x_p)_{p\in S}$. For p in S, the number

$$t_p = \sum_{\mathfrak{p}|p} \operatorname{tr}_{K_\mathfrak{p}/\mathbf{Q}_p}(x_\mathfrak{p})$$

is in $\mathbf{Q}_p$. We therefore cannot combine the various t_p exactly. However, we can combine them (mod 1). For p finite, we let $\operatorname{tr}_p(x_p)$ in $\mathbf{Z}[p^{-1}]$ denote the fractional part of t_p. Thus $\operatorname{tr}_p(x_p) - t_p$ is in $\mathbf{Z}_p$ and $\operatorname{tr}_p(x_p)$ is well defined (mod $\mathbf{Z}$). For p infinite, we let $\operatorname{tr}_\infty(x_\infty) = t_\infty$; $\operatorname{tr}_\infty(x_\infty)$ is real but need not be in $\mathbf{Q}$.

Definition. The *absolute trace*, or *S-trace* of x in $\prod_{\mathfrak{p}\in S} K_\mathfrak{p}$ is given by

$$\operatorname{tr}(x) = \operatorname{tr}_S(x) = \operatorname{tr}_\infty(x_\infty) - \sum_{\substack{p\in S\\ p\neq\infty}} \operatorname{tr}_p(x_p).$$

This is well defined in $\mathbf{R}/\mathbf{Z}$.

Note the minus sign. Roughly it serves the same purpose as having the p-adic norm of p be p^{-1}. We now have an analogue of the product formula.

LEMMA 2 (The trace formula, preliminary version). *Let ν be an S-integer of K. Then*

$$\operatorname{tr}(\nu) \equiv 0 \pmod 1.$$

The proof is easy and is omitted. For $K = \mathbf{Q}$, the trace formula is the analogue of the partial fraction decomposition of rational functions. The Fourier expansions of modular forms will contain terms of the form $\exp(2\pi i \operatorname{tr}(\nu x))$ where ν is an S-integer of K. This is well defined in spite of the ambiguity in the absolute trace and is also translation invariant by S-integers thanks to the trace formula.

For this preliminary discussion, let us take $S = \{\infty, p\}$ where p is an odd prime and $\mathfrak{O} = \mathfrak{O}(S) = \mathbf{Z}[p^{-1}]$, the ring of S-integers of $\mathbf{Q}$. Let $\mathfrak{h} = \mathfrak{h}_\mathbf{R} \times \mathfrak{h}_p$ and $\Gamma = \mathrm{GL}_2(\mathfrak{O})^+$. In this case, for an S-integer n, and x in $\mathbf{R} \times \mathbf{Q}_p$, by an abuse of notation, we may think of $\exp(2\pi i \operatorname{tr}(nx))$ as just $\exp(2\pi i(nx_\infty - nx_p))$. The abuse comes because in this last expression, we can only evaluate the exponential after replacing nx_p by a number in $\mathbf{Z}[p^{-1}]$ congruent to nx_p (mod $\mathbf{Z}_p$).

It is reasonable to require that a modular form $F(z)$ on $\mathfrak{h}$ of weight k and group Γ should satisfy the condition that for $A = \left(\begin{smallmatrix} a & b\\ c & d\end{smallmatrix}\right)$ in Γ and $z = (z_\infty, z_p)$ in $\mathfrak{h}$, we have

$$F(A\cdot z) = \mathbf{N}(cz+d)^k F(z).$$

We would naturally hope that the theory relating modular forms to Dirichlet series would carry over. In particular, we would hope that from the point of view of S-integers, the classical theory can be reinterpreted in the new language. Suppose that $f(z_\infty)$ is a modular form of weight k (and level 1) on $\mathfrak{h}_\mathbf{R}$ which is an eigenform of the Hecke operator T_p. To make life easy, we will suppose that f is a cusp form,

$$f(z_\infty) = \sum_{n=1}^{\infty} a_n e^{2\pi i n z_\infty}$$

and that $a_1 = 1$. The Dirichlet series $\sum a_n n^{-s}$ corresponding to f may be rewritten as

$$\sum a_n n^{-s} = (1 - a_p p^{-s} + p^{k-1-2s})^{-1} \sum_{(n,p)=1} a_n n^{-s}.$$

The object will be to think of the factor $(1 - a_p p^{-s} + p^{k-1-2s})^{-1}$ as another gamma factor corresponding to p and to write the series $\sum_{(n,p)=1} a_n n^{-s}$ in a manner suitable to the ring $\mathfrak{O}$. We then wish to find a modular form on $\mathfrak{h}$ corresponding to the series.

Handling the series is easy if we think of number fields. The Dirichlet series should be a sum over *ideals* of $\mathfrak{O}$. If n is in $\mathfrak{O}$, $n = \pm p^j m$ where m in $\mathbf{Z}$ is relatively prime to p, we set $a_{(n)} = a_m$. Since $\mathbf{N}((n))^{-s} = |\mathbf{N}(m)|^{-s} = m^{-s}$, we have

$$\sum_{\substack{(m,p)=1\\ m\in\mathbf{Z}}} a_m m^{-s} = \sum_{\substack{(n)\\ n\in\mathfrak{O}}} a_{(n)}\mathbf{N}((n))^{-s},$$

and we have written our series in a more ideal manner. The gamma factor is also easy. We set

$$\Gamma_p(s) = (1 - a_p p^{-s} + p^{k-1-2s})^{-1}$$

and we wish to write this as an integral analogous to

$$\Gamma_\infty(s) = (2\pi)^{-s}\Gamma(s) = \int_0^\infty v^s e^{-2\pi v}\frac{dv}{v}.$$

To do this we define for y in $\mathbf{Q}_p^\times$,

$$e_p(y) = \begin{cases} a_{p^j} & \text{if } |y|_p = p^{-j}, j \geq 0, \\ 0 & \text{otherwise.} \end{cases}$$

Then

$$\Gamma_p(s) = (1 - a_p p^{-s} + p^{k-1-2s})^{-1} = \sum_{j=0}^{\infty} a_{p^j} p^{-js} = \sum_{j=-\infty}^{\infty} e_p(p^j)|p^j|_p^s$$
$$= \int_{\mathbf{Q}_p^\times} |y|_p^s e_p(y)\frac{dy}{|y|_p}.$$

Here $dy/|y|_p$ is a *Haar measure* on the multiplicative group $\mathbf{Q}_p^\times$, normalized so that $\int_{|y|=1} dy/|y|_p = 1$. In this paper, we will be dealing with integrals of functions of $|y|$ and for these integrals, we may give a simple definition of the Haar integral which, absurd though it appears, will be eminently practical later on.

For a real number $t > 0$, define $\{t\}_p$ to be the largest integral power of p less than or equal to t. In other words,

$$\{e^v\}_p = \exp\left(\log(p)\left[\frac{v}{\log(p)}\right]\right).$$

where $[\ ]$ is the greatest integer function. If $g = g(|y|)$ is a function of $|y|_p$ on $\mathbf{Q}_p^\times$, we *define*

$$\int_{\mathbf{Q}_p^\times} g(|y|)\,\frac{dy}{|y|_p} = \frac{1}{\log(p)}\int_{-\infty}^{\infty} g(\{e^v\}_p)\,dv.$$

This definition has the two fundamental properties required of a Haar integral on an abelian group:

$$\int_{\mathbf{Q}_p^\times} g(|ay|)\,\frac{dy}{|y|_p} = \int_{\mathbf{Q}_p^\times} g(|y|)\,\frac{dy}{|y|_p} \quad \text{for all } a \text{ in } \mathbf{Q}_p$$

and

$$\int_{\mathbf{Q}_p^\times} g(|y^{-1}|)\,\frac{dy}{|y|_p} = \int_{\mathbf{Q}_p^\times} g(|y|)\,\frac{dy}{|y|_p}.$$

Suppose n in $\mathfrak{O}$ is totally positive (which just means that $n > 0$ in this simple case). We have

$$\begin{aligned}\Gamma_\infty(s)\Gamma_p(s)\mathbf{N}(n)^{-s} &= [n^{-s}\Gamma_\infty(s)][|n|_p^{-s}\Gamma_p(s)]\\ &= \int_{\mathbf{R}^+\times\mathbf{Q}_p^\times} \mathbf{N}(y)^s e(ny)\,\frac{dy}{|y|}\end{aligned}$$

where

$$\frac{dy}{|y|} = \frac{dy_\infty}{y_\infty}\,\frac{dy_p}{|y_p|_p}$$

and

$$e(ny) = \begin{cases} e^{-2\pi n y_\infty} e_p(ny_p) & \text{if } |ny_p| \le 1,\\ 0 & \text{otherwise.}\end{cases}$$

Hence, if we define

$$\xi(s) = \Gamma_\infty(s)\Gamma_p(s)\sum_{(n)} a_{(n)}\mathbf{N}((n))^{-s},$$

then we get (for $\mathrm{Re}(s)$ large),

$$\xi(s) = \int_{\mathbf{R}^+\times\mathbf{Q}_p^\times} \mathbf{N}(y)^s \sum_{\substack{(n)\\ n>0}} a_{(n)}e(ny)\,\frac{dy}{|y|}.$$

The group, U, of totally positive S-units of $\mathfrak{O}$ is generated by p. Since $\mathbf{N}(y)$ is invariant under $y \mapsto py$, we change the integral in the manner pioneered by Hecke, getting

$$\xi(s) = \int_{\mathbf{R}^+\times\mathbf{Q}_p^\times/U} \mathbf{N}(y)^s \sum_{n>0} a_{(n)}e(ny)\,\frac{dy}{|y|}.$$

This leads us to suspect that our desired modular form (with $x = 0$!!) is given by

$$F(iy_\infty, y_p\sqrt{\Delta}) = \sum_{n>0} a_{(n)}e(ny).$$

As always, we have to guess how to insert the x-variables into the expression for $F(y\sqrt{\Delta})$. Hecke [3] faced this problem with theta functions attached to complex fields when he proved the quadratic reciprocity law. Although he did not state this to be the case, the resulting functions are modular forms [6]. Hecke's experience dictates that corresponding to an ny term in $F(y\sqrt{\Delta})$, there should be an exponential term of the form $\exp(2\pi i\,\mathrm{tr}(nx))$, which in this case is $\exp(2\pi i[nx_\infty - nx_p])$, and we want

$$F(z) = \sum_{n>0} a_{(n)}e(ny)\exp(2\pi i\,\mathrm{tr}(nx)). \tag{5}$$

Our immediate problem is to relate this back to the original $f(z_\infty)$. Suppose that $|y_p|_p = p^{-j}$ and let us replace n in the sum (5) by n/p^j. Since $a_{(n/p^j)} = a_{(n)}$, the $a_{(n)}$ term remains unchanged, however. Now for an individual n-term, let $n = p^h m$ where $|m|_p = 1$. Then

$$\begin{aligned} a_{(n)} e\left(\frac{ny}{p^j}\right) \exp\left(2\pi i \operatorname{tr}\left(\frac{nx}{p^j}\right)\right) &= \begin{cases} a_m \exp\left(\frac{2\pi n y_\infty}{p^j}\right) e_p\left(\frac{n y_p}{p^j}\right) \exp(2\pi i n(x_\infty - x_p)/p^j) & \text{if } h \geq 0, \\ 0 & \text{otherwise} \end{cases} \\ &= \begin{cases} a_m \exp\left(\frac{2\pi n y_\infty}{p^j}\right) a_{p^h} \exp(2\pi i n(x_\infty - x_p)/p^j) & \text{if } h \geq 0, \\ 0 & \text{otherwise.} \end{cases} \\ &= \begin{cases} a_n \exp\left(\frac{2\pi n y_\infty}{p^j}\right) \exp(2\pi i n(x_\infty - x_p)/p^j) & \text{if } n \text{ is in } \mathbf{Z}^+, \\ 0 & \text{otherwise.} \end{cases} \end{aligned}$$

In other words,

$$F(z) = f\left(\frac{z_\infty - x_p}{p^j}\right) \quad \text{where } |y_p|_p = p^{-j}$$

(and x_p in the numerator is taken in $\mathbf{Z}[p^{-1}] (\operatorname{mod} p^j)$ and can thus be subtracted from z_∞).

Given the manner in which $F(z)$ appears in this expression, it seems reasonable that the invariance under $\mathrm{SL}_2(\mathfrak{O})$ has nothing to do with whether or not f is an eigenform of T_p. This is indeed the case as the next theorem will show. However, if $f(z_\infty)$ is not an eigenform of T_p, the Fourier expansion of the resulting $F(z)$ will not be as simple as (5). Rather, the resulting $F(z)$ would be analogous to adding two wave forms of different eigenvalues. The resulting sum transforms correctly, but mixes two spaces that deserve to be kept separate. In this manner, as more primes are added to S, the space of modular forms on $\mathfrak{h}_{\mathbf{R}}$ splits up into subspaces, each ultimately of dimension 1 if S is carefully chosen (the "multiplicity one" theorem).

THEOREM 3. *Let $f(z_\infty)$ be a modular form of weight k for* $\mathrm{SL}_2(\mathbf{Z})$ *and define the function*

$$F(z) = f\left(\frac{z_\infty - x_p}{p^j}\right) \quad \text{where } |y_p|_p = p^{-j} \tag{6}$$

on $\mathfrak{h} = \mathfrak{h}_{\mathbf{R}} \times \mathfrak{h}_p$. Then for $A = \left(\begin{smallmatrix} a & b \\ c & d \end{smallmatrix}\right)$ in $\Gamma = \mathrm{GL}_2(\mathfrak{O})^+$ $(\mathfrak{O} = \mathbf{Z}[p^{-1}])$, we have

$$F(A \cdot z) = \mathbf{N}(cz + d)^k F(z).$$

PROOF. Note that if z_p is near to $\sqrt{\Delta}$ ($|y_p|_p = 1$, $|x_p|_p \leq 1$) then

$$F(z) = f(z_\infty).$$

Therefore for any A in Γ and any w in $\mathfrak{h}$ with w_p near to $\sqrt{\Delta}$ and $z = A \cdot w$, we wish to have

$$F(z) = \mathbf{N}(cw + d)^k f(w_\infty). \tag{7}$$

We now show that we can take (7) as the definition of $F(z)$. For this, we have to prove that (7) is consistent and agrees with (6).

Suppose that $z = A \cdot w = B \cdot w'$ where w_p and w'_p ar near $\sqrt{\Delta}$, and A and B are in Γ. We must show that

$$\mathbf{N}(c_A w + d_A)^k f(w_\infty) = \mathbf{N}(c_B w' + d_B)^k f(w'_\infty). \tag{8}$$

Since $w' = B^{-1}A \cdot w$, $\det(B^{-1}A)$ is an even power of p; we can thus rescale B so that $B^{-1}A$ has unit determinant in $\mathbf{Q}_p$. Since $\mathbf{N}(p) = 1$, this rescaling does not affect (8). Now however, $B^{-1}A$ takes w_p to w'_p and both are points near to $\sqrt{\Delta}$. It follows from Lemma 1 that $B^{-1}A = C$ is in $\mathrm{SL}_2(\mathbf{Z})$. We further have $|c_C w_p + d_C|_p = 1$ and hence $(c_C w_\infty + d_C)^k = \mathbf{N}(c_C w + d_C)^k$. Therefore

$$\begin{aligned}\mathbf{N}(c_B w' + d_B)^k f(w'_\infty) &= \mathbf{N}(c_B w' + d_B)^k f(C \cdot w_\infty)\\ &= \mathbf{N}(c_B C \cdot w + d_B)^k \mathbf{N}(c_C w + d_C)^k f(w_\infty)\\ &= \mathbf{N}(c_{BC} w + d_{BC})^k f(w_\infty)\\ &= \mathbf{N}(c_A w + d_A)^k f(w_\infty).\end{aligned}$$

Thus our extended definition (7) is consistent.

Further, if $y_p = vp^j$ where $|v|_p = 1$ and $|x_p - u|_p \le p^{-j}$ where u is in $\mathbf{Z}[p^{-1}]$, then

$$z_p = x_p + y_p\sqrt{\Delta} = \begin{pmatrix} p^j & u \\ 0 & 1\end{pmatrix} \cdot w_p = A \cdot w_p$$

where

$$w_p = p^{-j}(z_p - u) = \left(\frac{x_p - u}{p^j}\right) + v\sqrt{\Delta}$$

is near $\sqrt{\Delta}$. Therefore with $w = (z-u)/p^j$ and $A = \left(\begin{smallmatrix} p^j & u \\ 0 & 1\end{smallmatrix}\right)$, we see that (7) agrees with (6).

But now the proof of the Theorem is easy. If $B \cdot w = z$ where B is in Γ and w_p is near $\sqrt{\Delta}$ then $AB \cdot w = A \cdot z$ and so we have

$$\begin{aligned}F(A \cdot z) = F(AB \cdot w) &= \mathbf{N}(c_{AB} + d_{AB})^k F(w)\\ &= \mathbf{N}(c_A B \cdot w + d_A)^k \mathbf{N}(c_B w + d_B)^k F(w)\\ &= \mathbf{N}(c_A z + d_A)^k F(z). \quad \text{Q.E.D.}\end{aligned}$$

The theory of Hecke operators for primes other than p presents no difficulties. We will not pursue this topic here. The analytic continuation and functional equation of $\xi(s)$ also follow from $F(-1/z) = \mathbf{N}(z)^k F(z)$ in the usual manner. Instead of completing the proof of the functional equation for $\xi(s)$, we turn now to the functional equation for the Dedekind zeta function.

4. Theta functions and the Dedekind zeta function revisited. We turn now to the functional equation for the Dedekind zeta function from the point of view of S-integers. Since we will deal with ideals and S-ideals at the same time, we need to settle upon a notation that will allow this. Here S is a finite set of primes of $\mathbf{Q}$, including ∞. However, since the x-variable will not be present in the theta functions while deriving the functional equation, we may safely allow 2 to be in S. Suppose now that K is a finite extension of $\mathbf{Q}$. We continue to say that a prime ideal $\mathfrak{p}$ of K (possibly infinite) is in S if $\mathfrak{p}$ divides a prime p in S. The ring of S-integers of K will be denoted by $\mathfrak{O}_K(S)$. We will continue to let $\mathfrak{O}_K$ denote the ring of integers of K which in the current notation corresponds to $S = \{\infty\}$. The relation between the fractional ideals of K and the fractional S-ideals of K is simple. Corresponding to any fractional ideal $\mathfrak{a}$ of K there is a fractional S-ideal $\mathfrak{a}_S$ of K which is the $\mathfrak{O}_K(S)$-module generated by the elements of $\mathfrak{a}$. The mapping $\mathfrak{a} \mapsto \mathfrak{a}_S$ is a homomorphism of the group of fractional ideals of K onto the group of fractional S-ideals of K. The kernel consists precisely of the subgroup of fractional ideals of K generated by the finite primes of K in S. In particular, corresponding to any

fractional S-ideal $\mathfrak{a}_S$ of K is a unique fractional ideal $\mathfrak{a}$ of K having no primes of S in its factorization. The appropriate norms of these two ideals are also equal,

$$\mathbf{N}(\mathfrak{a}_S) = \mathbf{N}_\infty(\mathfrak{a}).$$

Thus we may think of any prime ideal not in S as an S-ideal also, but we must be very careful in describing primes of S which we should not think of as S-ideals.

Let ∂_K and D_K denote the different and discriminant of K. We let $\partial_K(S)$ denote the S-different of K and $D_K(S)$ denote the S-discriminant of K. The different splits into the product of pieces,

$$\partial_K = \prod_{\mathfrak{p}\nmid\infty} \partial_\mathfrak{p},$$

where each $\partial_\mathfrak{p}$ is a power of $\mathfrak{p}$, as does the discriminant,

$$|D_K| = \prod_{p\neq\infty} D_p$$

where

$$D_p = \prod_{\mathfrak{p}|p} \mathbf{N}_\infty(\partial_\mathfrak{p}).$$

The relation between ∂_K and $\partial_K(S)$ is

$$\partial_K(S) = \left(\prod_{\mathfrak{p}\notin S} \partial_\mathfrak{p}\right)_S$$

is the S-ideal generated by the elements of ∂_K. In particular, ∂_K maps to $\partial_K(S)$ under the homomorphism above. Since $\mathbf{Z}(S)$ has infinitely many units, to avoid the problem of $D_K(S)$ being determined only modulo squares of units, we use $\mathbf{N}(\partial_K(S))$ as the definition of $D_K(S)$. Therefore,

$$D_K(S) = \prod_{p\notin S} D_p.$$

The S-different plays the same role for S-integers as the classical different does for integers. In particular, we have the following refined version of the trace formula,

LEMMA 3 (The trace formula). *If ν is in $\partial_K(S)^{-1}$ then*

$$\mathrm{tr}(\nu) \equiv 0 (\mathrm{mod}\ 1).$$

In particular, $\mathrm{tr}_{K/\mathbf{Q}}(\nu)$ is in $\mathbf{Z}(S)$.

Again, the proof is easy and is omitted. With these preliminaries out of the way, we introduce the basic symplectic theta function of several variables.

DEFINITION. For z in $\mathfrak{h}^{(n)}$, u and v n-dimensional complex column vectors, set

$$\Theta\left(Z, \begin{pmatrix} u \\ v \end{pmatrix}\right) = \Theta(Z, u, v) = \sum_m \exp[\pi i^t(m+v)Z(m+v) - 2\pi i^t mu - \pi i^t vu]$$

where m runs through all integral valued n-dimensional column vectors. We call $\Theta(Z, u, v)$ a *theta function* or sometimes a *symplectic theta function.*

This series converges absolutely and indeed with exponential rapidity although we will not need such an estimate in this paper. The mere fact of the absolute convergence will have some very useful consequences, however. We will soon have other theta functions more closely allied with number fields. They will arise as specializations of the symplectic theta function. The messier first notation for $\Theta(Z, u, v)$ is useful when dealing with the transformation formula under the action of the theta subgroup of the symplectic group. In this paper, the easier to write and typeset second notation will suffice.

PROPOSITION 2 (The inversion formula for the symplectic theta function). *For all Z in $\mathfrak{h}^{(n)}$ and complex column vectors u and v in $\mathbf{C}^n$,*

$$\Theta(-Z^{-1}, -v, u) = [\det(-iZ)]^{1/2}\Theta(Z, u, v)$$

where the square root is given by analytic continuation from the principal value when $Z = iY$.

Now we introduce the theta functions attached to the number field K. We number the conjugate fields $K^{(j)}$ of K in the usual manner as in Example 4 of Section 1. We will first take $S = \{\infty\}$, and define the theta function attached to K corresponding to S. This will be the standard theta function which has appeared since Hecke's time and to match the standard usage, we will leave S out of the notation. Let y be in $(\mathbf{R}^+)^{r_1+r_2}$, and u and v be in $\mathbf{R}^{r_1}\mathbf{C}^{r_2}$. We denote the components of y, u and v by y_j, $u^{(j)}$ and $v^{(j)}$ respectively. It is convenient to let j run from 1 to n and so for $r_1 + 1 \leq j \leq r_1 + r_2$, we let $y_{j+r_2} = y_j$, $u^{(j+r_2)} = \overline{u^{(j)}}$ and $v^{(j+r_2)} = \overline{v^{(j)}}$. Ultimately, the numbers $u^{(j)}$ and $v^{(j)}$ will be specialized to be the n conjugates of numbers μ and ν in K and this is the reason for the superscript notation.

DEFINITION. (With $S = \{\infty\}$.) We define for a fractional ideal $\mathfrak{a}$ of K the *theta function*

$$\theta_K\left(y, \begin{pmatrix} u \\ v \end{pmatrix}, \mathfrak{a}\right) = \theta_K(y, u, v, \mathfrak{a}) = \sum_{\alpha \in \mathfrak{a}} \exp\{-\pi\, \mathrm{tr}[y|\alpha + v|^2] - 2\pi i\, \mathrm{tr}[u\alpha + uv/2]\}$$

where $\mathrm{tr}[\cdots] = \mathrm{tr}_\infty[\cdots]$. In the special case $u = v = 0$, we write

$$\theta_K(y, \mathfrak{a}) = \theta_K(y, 0, 0, \mathfrak{a}).$$

Written out in full, the exponential term in the theta function is,

$$-\sum_{j=1}^{n}[\pi y_j|\alpha^{(j)} + v^{(j)}|^2 + 2\pi i u^{(j)}\alpha^{(j)} + \pi i u^{(j)}v^{(j)})].$$

It is to be noted that we have defined our theta function as $\theta_K(y)$ rather than $\theta_K(y\sqrt{\Delta})$ as might have been expected. One can indeed define a theta function of z in $\mathfrak{h}_{\mathbf{R}}^{r_1} \times \mathfrak{h}_{\mathbf{C}}^{r_2}$ and the resulting function is a modular form [6]. However as is well known, the x variable is not necessary for the functional equation of the Dedekind zeta function and thus the use of $\sqrt{\Delta}$ in the notation is unnecessary.

PROPOSITION 3 (The inversion formula with $S = \{\infty\}$). *We have $\theta_K(y, u, v, \mathfrak{a}) = \mathbf{N}(\mathfrak{a})^{-1}|D_K|^{-1/2}\mathbf{N}(y)^{-1/2}\theta_K(y^{-1}, -v, u, \mathfrak{a}^{-1}\mathfrak{d}_k^{-1})$ where by y^{-1} we mean $(\{y_j^{-1}\})$ and of course, $\mathbf{N}(y)^{-1/2} = \prod_{j=1}^{n} y_j^{-1/2}$.*

The function θ_K is essentially that introduced by Hecke. As Hecke discovered, it is a specialization of the symplectic theta function and the proof of the inversion formula is a specialization of the inversion formula above for the symplectic theta function. Of particular note is the fact that the series defining $\theta_K(y, u, v, \mathfrak{a})$ converges absolutely because it is a specialization of the symplectic series.

On the other hand, the following refinement for S-ideals has never before appeared in quite this manner. Now we allow S to be any finite set of primes of $\mathbf{Q}$ including ∞. If $\mathfrak{p}$ is a prime ideal of K in S, we introduce variables $y_\mathfrak{p}, u_\mathfrak{p}, v_\mathfrak{p}$ in $K_\mathfrak{p}$. By y_p, u_p, v_p we mean vector variables $(y_\mathfrak{p})_{\mathfrak{p}|p}$, $(u_\mathfrak{p})_{\mathfrak{p}|p}$, $(v_\mathfrak{p})_{\mathfrak{p}|p}$. These vector variables are analogous to the vectors $(y_1, \cdots, y_{r_1+r_2})$, $(u_1, \cdots, u_{r_1+r_2})$, $(v_1, \cdots, v_{r_1+r_2})$ introduced earlier which correspond to the infinite primes of K and which we now denote by y_∞, u_∞, v_∞. By y, we now mean the collection made up of all the $y_\mathfrak{p}$ for $\mathfrak{p}$ in S, and we attach similar meanings to u and v.

The variables $y_\mathfrak{p}$ for finite $\mathfrak{p}$ will be restricted in $K_\mathfrak{p}^\times$ in a manner similar to the restriction on $y_\mathfrak{p}$ for infinite $\mathfrak{p}$. The latter are restricted to be in $\mathbf{R}^+$, which for the purpose here may be thought of as being the squares of numbers in $\mathbf{R}^\times$. For finite $\mathfrak{p}$, we let $K_\mathfrak{p}^0$ denote the subset of $K_\mathfrak{p}^\times$ given by the condition,

$$\operatorname{ord}_\mathfrak{p}(y_\mathfrak{p}) \equiv \operatorname{ord}_\mathfrak{p}(\partial_\mathfrak{p}) \pmod{2};$$

the variable $y_\mathfrak{p}$ will be restricted to $K_\mathfrak{p}^0$. For uniformity of notation, we will let $K_\mathfrak{p}^0$ be $\mathbf{R}^+$ for infinite $\mathfrak{p}$. At last we can define our generalized theta function.

DEFINITION. For variables y, u, v in $\prod K_\mathfrak{p}$ with y restricted so that each component $y_\mathfrak{p}$ is in $K_\mathfrak{p}^0$, and for a fractional S-ideal $\mathfrak{a}$, let

$$\begin{aligned}\theta_K\left(y, \begin{pmatrix} u \\ v \end{pmatrix}, \mathfrak{a}, S\right) &= \theta_K(y, u, v, \mathfrak{a}, S) \\ &= \sum_{\alpha \in \mathfrak{a}} e_S(-y(\alpha + v)^2) \exp(-2\pi i \operatorname{tr}(u\alpha + uv/2))\end{aligned}$$

where

$$e_S(-y\beta^2) = \begin{cases} \exp[-\pi \operatorname{tr}_\infty(y|\beta|^2)] & \text{if } \operatorname{ord}_\mathfrak{p}(y\beta^2\partial_K) \geq 0 \text{ for all } \mathfrak{p}\nmid\infty \text{ in } S, \\ 0 & \text{otherwise.} \end{cases}$$

In the special case $u = v = 0$, we write

$$\theta_K(y, \mathfrak{a}_S, S) = \theta_K(y, 0, 0, \mathfrak{a}_S, S).$$

We now come to the relation between θ_K and $\theta_K(S)$. In this relationship, ideals will appear as both S-ideals and regular ideals. Of necessity, we will therefore put the "S" into the notation for S-ideals. If $\mathfrak{a}_S$ is a fractional S-ideal of K, we let $\mathfrak{a}$ denote the unique fractional ideal of K corresponding to $\mathfrak{a}_S$ under the homomorphism above which has no primes of S in its prime factorization. This notational convention will hold for the duration of the proof of the following Lemma and Theorem. For convenience, we first take the case $u = v = 0$ since this is the only case we will actually use in this paper.

LEMMA 4. *Let* $\operatorname{ord}_\mathfrak{p}(y_\mathfrak{p}\partial_\mathfrak{p}) = 2b(\mathfrak{p})$ *and set*

$$\mathfrak{b} = \prod_{\substack{\mathfrak{p}\in S \\ \mathfrak{p}\nmid\infty}} \mathfrak{p}^{-b(\mathfrak{p})}.$$

Then

$$\theta_K(y, \mathfrak{a}_S, S) = \theta_K(y_\infty, \mathfrak{a}\mathfrak{b}).$$

Further, the series defining $\theta_K(y, \mathfrak{a}_S, S)$ *converges absolutely.*

PROOF. For α in $\mathfrak{a}_S$, the condition that $e_S(-y\alpha^2) \neq 0$ implies that α is in $\mathfrak{a}\mathfrak{b}$. For α in $\mathfrak{a}\mathfrak{b}$, the series for $\theta_K(y, \mathfrak{a}_S, S)$ is term for term equal to the series for $\theta_K(y_\infty, \mathfrak{a}\mathfrak{b})$ and both parts of the Lemma follow immediately. Q.E.D.

THEOREM 4 (The inversion formula). *For y in $\prod K_\mathfrak{p}^0$ and a fractional S-ideal $\mathfrak{a}_S$, we have*

$$\theta_K(y, \mathfrak{a}_S, S) = \mathbf{N}(\mathfrak{a}_S)^{-1}|D_K(S)|^{-1/2}\mathbf{N}(y)^{-1/2}\theta_K(y^{-1}, \mathfrak{a}_S^{-1}\partial_K^{-1}, S).$$

PROOF. By Lemma 4 and the inversion formula of Proposition 3,

$$\theta_K(y, \mathfrak{a}_S, S) = \mathbf{N}_\infty(\mathfrak{a}\mathfrak{b})^{-1}|D_K|^{-1/2}\mathbf{N}_\infty(y_\infty)^{-1/2}\theta_K(y_\infty^{-1}, \mathfrak{a}^{-1}\mathfrak{b}^{-1}\partial_K^{-1}).$$

Let $\operatorname{ord}_\mathfrak{p}(\partial_\mathfrak{p}) = d(\mathfrak{p})$, $\operatorname{ord}_\mathfrak{p}(y_\mathfrak{p}^{-1}\partial_\mathfrak{p}) = 2c(\mathfrak{p})$, $\mathfrak{c} = \prod_{\mathfrak{p}\in S, \mathfrak{p}\nmid\infty} \mathfrak{p}^{-c(\mathfrak{p})}$ and split ∂_K into two pieces, $\partial_K = \partial_1\partial_2$ where

$$\partial_1 = \prod_{\substack{\mathfrak{p}\in S\\ \mathfrak{p}\nmid\infty}} \partial_\mathfrak{p}, \qquad \partial_2 = \prod_{\mathfrak{p}\notin S} \partial_\mathfrak{p}.$$

Thus $\partial_K(S) = (\partial_2)_S$ and $\mathbf{N}(\partial_K(S)) = \mathbf{N}_\infty(\partial_2)$. We also have $c(\mathfrak{p}) = -b(\mathfrak{p}) + d(\mathfrak{p})$ and $\mathfrak{c} = \mathfrak{b}^{-1}\partial_1^{-1}$. Therefore by Lemma 4,

$$\theta_K(y^{-1}, \mathfrak{a}_S^{-1}\partial_K(S)^{-1}) = \theta_K(y_\infty^{-1}, \mathfrak{a}^{-1}\partial_2^{-1}\mathfrak{b}^{-1}\partial_1^{-1}) = \theta_K(y_\infty, \mathfrak{a}^{-1}\mathfrak{b}^{-1}\partial_K^{-1}).$$

Further, for $\mathfrak{p}\nmid\infty$ in S, $-\operatorname{ord}_\mathfrak{p}(y_\mathfrak{p}) = \operatorname{ord}_\mathfrak{p}(\mathfrak{b}^2\partial_1)$ and hence,

$$\begin{aligned}\mathbf{N}_\infty(y_\infty)\mathbf{N}_\infty(\mathfrak{b})^2|D_K| &= \mathbf{N}_\infty(y_\infty)\mathbf{N}_\infty(\mathfrak{b}^2\partial_1\partial_2)\\ &= \mathbf{N}_\infty(y_\infty)\mathbf{N}_\infty(\partial_2)\prod_{\substack{\mathfrak{p}\in S\\ \mathfrak{p}\nmid\infty}} |y_\mathfrak{p}|_\mathfrak{p}\\ &= D_K(S)\mathbf{N}(y)\end{aligned}$$

and the Theorem follows. Q.E.D.

The following special case of Theorem 4 shows the potential simplifying benefits of this approach. It is precisely this special case we will use in the proof of the functional equation of the Dedekind zeta function.

THEOREM 5. *Let S be such that every prime p dividing D_K is in S. For $\mathfrak{O} = \mathfrak{O}_K(S)$ and y in $\prod K_\mathfrak{p}^0$, we have*

$$\theta_K(y, \mathfrak{O}, S) = \mathbf{N}(y)^{-1/2}\theta_K(y^{-1}, \mathfrak{O}, S).$$

PROOF. We need only note that by hypothesis, $\partial_D(S) = (1)$ and $D_K(S) = 1$. Q.E.D.

We now give a simple proof of a Theorem of Kronecker.

LEMMA 5. *Suppose that we are given real numbers $\delta_\mathfrak{p} > 0$ for each $\mathfrak{p}$ in S. Then there are only finitely many S-integers α such that $|\alpha|_\mathfrak{p} < \delta_\mathfrak{p}$ for all $\mathfrak{p}$ in S.*

PROOF. Choose y in $\prod K_\mathfrak{p}^0$ such that $|y_\mathfrak{p}\partial_\mathfrak{p}|_\mathfrak{p}\delta_\mathfrak{p}^2 < 1$ for all finite $\mathfrak{p}$ in S. The Lemma is now a direct consequence of the absolute convergence of the series for $\theta_K(y, \mathfrak{O}_K(S), S)$. Q.E.D.

THEOREM 6 (Kronecker). *Given K and S, there is a number $\delta > 0$ such that if α is a non-zero S-integer satisfying $|\alpha|_\mathfrak{p} < 1 + \delta$ for all $\mathfrak{p}$ in S, the α is a root of unity.*

PROOF. By Lemma 5, there is a number $\delta > 0$ such that $|\alpha|_\mathfrak{p} < 1 + \delta$ for all $\mathfrak{p}$ in S implies that $|\alpha|_\mathfrak{p} \leq 1$ for all $\mathfrak{p}$ in S. Since α is non-zero, every $|\alpha|_\mathfrak{p}$ must equal 1. The set of all such α forms a group, which is finite by Lemma 5. Hence every such α is a root of unity. Q.E.D.

It should be remarked that any $\delta < 1$ which satisfies this Theorem for one S satisfies it for all S. For us, the importance of Kronecker's Theorem is the following

COROLLARY 1. *Let S' be the set S with one prime $\mathfrak{p}$ of K deleted and let $r(s)$ be the number of $\mathfrak{p}$ in S'. We let an S-unit ε act on v in $\mathbf{R}^{r(S)}$ by*

$$\varepsilon \cdot v = (v_{\mathfrak{p}} + \log(|\varepsilon_{\mathfrak{p}}|_{\mathfrak{p}}))_{\mathfrak{p} \in S'}.$$

This action is a discrete action of $U(K,S)/\mu(K)$ on $\mathbf{R}^{r(S)}$ where $U(K,S)$ is the group of S-units of K and $\mu(K)$ is the group of roots of unity of K.

PROOF. If all $|\varepsilon|_{\mathfrak{p}}$ for $\mathfrak{p}$ in S' are near 1 then the sole remaining $|\varepsilon|_{\mathfrak{p}}$ for $\mathfrak{p}$ in S is near 1 as well. Thus the Corollary follows from Kronecker's Theorem. Q.E.D.

The significance of Corollary 1 is that there is now a fundamental domain for the action of $U(K,S)$ on $\mathbf{R}^{r(S)}$. The key step in proving the unit theorem is to prove that this fundamental domain has finite volume. This we will accomplish below in the process of proving the functional equation of the Dedekind zeta function.

Although we do not need the case where $u, v \neq 0, 0$ in this paper, it would arise if we were to deal with L-functions. We therefore state the analogues of Lemma 4 and Theorem 4.

LEMMA 6. *Let $\operatorname{ord}_{\mathfrak{p}}(y_{\mathfrak{p}}\partial_K) = 2b(\mathfrak{p})$ and set $\mathfrak{b} = \prod_{\mathfrak{p}\in S, \mathfrak{p}\nmid\infty} \mathfrak{p}^{-b(\mathfrak{p})}$. Further, let $\operatorname{ord}_{\mathfrak{p}}(\partial_{\mathfrak{p}}) = d(\mathfrak{p})$, $\operatorname{ord}_{\mathfrak{p}}(\mathfrak{a}) = a(\mathfrak{p})$. Suppose that μ and ν in K satisfy the approximation conditions,*

1. *For finite $\mathfrak{p}$ in S,*

$$\operatorname{ord}_{\mathfrak{p}}(u-\mu) \geq b(\mathfrak{p}) - d(\mathfrak{p}),$$
$$\operatorname{ord}_{\mathfrak{p}}(u-\nu) \geq -b(\mathfrak{p}),$$
$$\operatorname{ord}_{\mathfrak{p}}\left(\frac{(u-\mu)(v-\nu)}{2}\right) \geq -d(\mathfrak{p}).$$

2. *For prime $\mathfrak{p}$ not in S,*

$$\operatorname{ord}_{\mathfrak{p}}(\mu) \geq \max(a(\mathfrak{p}), 0),$$
$$\operatorname{ord}_{\mathfrak{p}}(\nu) \geq \max(-a(\mathfrak{p}) - d(\mathfrak{p}), 0),$$
$$\operatorname{ord}_{\mathfrak{p}}\left(\frac{\mu\nu}{2}\right) \geq -d(\mathfrak{p}).$$

(Note that $a(\mathfrak{p})$ and $d(\mathfrak{p})$ are zero almost always and so μ and ν exist.) Then

$$\theta_K(y,u,v,\mathfrak{a}_S,S) = \exp\left[2\pi i \operatorname{tr}\left(\frac{u\nu - v\mu}{2}\right)\right]\theta_K(y_\infty, u_\infty - \mu, v_\infty - \nu, \mathfrak{a}\mathfrak{b}).$$

The proof consists of replacing α by $\alpha - \nu$ in the definition of θ_K, which is legal since ν is an S-integer and then making careful use of the trace formula in Lemma 3. We leave it to the reader.

THEOREM 7 (The inversion formula). *For y in $\prod K_{\mathfrak{p}}^0$ and a fractional S-ideal $\mathfrak{a}_S$, we have*

$$\theta_K(y,u,v,\mathfrak{a}_S,S) = \mathbf{N}(\mathfrak{a}_S)^{-1}|D_K(S)|^{-1/2}\mathbf{N}(y)^{-1/2}\theta_K(y^{-1}, -v, u, \mathfrak{a}_S^{-1}\partial_K(S)^{-1}, S).$$

The proof is the same as that of Theorem 4 except now we must also keep track of the symmetries in the condition on μ and ν in Lemma 6.

Now we come to the functional equation. Let

$$\zeta_K(s,S) = \sum_{\mathfrak{a}} \mathbf{N}(\mathfrak{a})^{-s}$$

where the sum is taken over all integral S-ideals of K (in other words, over all ideals of the ring $\mathfrak{O}_K(S)$—other than the zero ideal). Thus $\zeta_K(s,S)$ is *the zeta function of the ring* $\mathfrak{O}_K(S)$. When $S = \{\infty\}$, we will drop the "S" in the notation and continue to speak of $\zeta_K(s)$. The Euler product for $\zeta_K(s,S)$ is same as the Euler product for $\zeta_K(s)$ with the Euler $\mathfrak{p}$-factors corresponding to the finite primes in S deleted,

$$\zeta_K(s,S) = \zeta_K(s) \prod_{\substack{\mathfrak{p}\in S \\ \mathfrak{p}\nmid\infty}} (1 - \mathbf{N}_\infty(\mathfrak{p})^{-s})^{-1}. \tag{9}$$

The functional equation for $\zeta_K(s)$ is usually given in terms of the function

$$\xi_K(s) = \left(\frac{|D_K|}{2^{2r_2}\pi^n}\right)^{s/2} \Gamma\left(\frac{s}{2}\right)^{r_1} \Gamma(s)^{r_2} \zeta_K(s)$$

and is just

$$\xi_K(s) = \xi_K(1-s).$$

We will recast the definition of $\xi_K(s)$ in terms of $\zeta_K(s,S)$. Let

$$\Gamma_\mathfrak{p}(s) = \begin{cases} \pi^{-s/2}\Gamma(s/2) & \text{if } \mathfrak{p} \text{ is a real infinite prime} \\ (2\pi)^{-s}\Gamma(s) & \text{if } \mathfrak{p} \text{ is a complex infinite prime} \\ |\partial_K|_\mathfrak{p}^{s/2}(1-\mathbf{N}_\infty(\mathfrak{p})^{-s})^{-1} & \text{if } \mathfrak{p}\nmid\infty. \end{cases}$$

and set

$$G(s,S) = \prod_{\mathfrak{p}\in S} \Gamma_\mathfrak{p}(s).$$

By (9), we may now rewrite the definition of $\xi_K(s)$ as,

$$\xi_K(s) = |D_K(S)|^{s/2} G(s,S)\zeta_K(s,S).$$

Using our more general theta function, the proof of the functional equation for $\xi_K(s)$ is exactly the same as that given by Hecke. However, a judicious choice of S eliminates all the "complications" in the proof that so many current authors decry.

Namely, let us now choose S so that $\mathfrak{O} = \mathfrak{O}_K(S)$ has class-number one and so that all primes of $\mathbf{Q}$ which ramify in K are in S. Thus $\partial_K(S) = (1)$ and $D_K(S) = \mathbf{N}(\partial_K(S)) = 1$. With this choice of S, we will have no need of any auxiliary ideals in the proof. We may also write $\zeta_K(s,S)$ simply as

$$\zeta_K(s,S) = \sum_{\substack{(\alpha) \\ \alpha\in\mathfrak{O}^\times}} |\mathbf{N}(\alpha)|^{-s}.$$

Further notation that we will need is $r = r(S)$, $W(K)$ is the order of $\mu(K)$ (i.e. the number of roots of unity in K), and $U^2 = U(K,S)^2$ is the group of squares of S-units of K.

We wish to write $\xi_K(s)$ as a multidimensional integral of the theta function $\theta_K(y,\mathfrak{O},S)$ defined above. Because of parity conditions on $\mathrm{ord}_\mathfrak{p}(y)$ in the theta function, we will only be integrating on the subset $K_\mathfrak{p}^0$ of $K_\mathfrak{p}^\times$ defined above.

We want to define a Haar integral restricted to $K_\mathfrak{p}^0$ in a manner analogous to that of Section 2. Let $\{e^v\}_\mathfrak{p}$ be the highest even or odd power of $\mathbf{N}_\infty(\mathfrak{p})$ less than or equal to e^v. The choice of "even" or "odd" is that of the parity of $\mathrm{ord}_\mathfrak{p}(\partial_\mathfrak{p})$. For a function $f(|y|_\mathfrak{p})$ of $|y|_\mathfrak{p}$ on $K_\mathfrak{p}^0$, we *define* the integral

$$\int_{K_\mathfrak{p}^0} f(|y|_\mathfrak{p})\frac{dy}{|y|_\mathfrak{p}} = (2\log \mathbf{N}_\infty(\mathfrak{p}))^{-1}\int_\mathbf{R} f(\{e^v\}_\mathfrak{p})dv.$$

The factor of $(2\log \mathbf{N}(\mathfrak{p}))^{-1}$ is to take care of the required normalization.

As in the previous section, we have the fundamental properties,

$$\int_{K_\mathfrak{p}^0} f(|y|_\mathfrak{p})\frac{dy}{|y|_\mathfrak{p}} = \int_{K_\mathfrak{p}^0} f(|a^2y|_\mathfrak{p})\frac{dy}{|y|_\mathfrak{p}}$$

for any a in $K_\mathfrak{p}^\times$ and,

$$\int_{K_\mathfrak{p}^0} f(|y|_\mathfrak{p})\frac{dy}{|y|_\mathfrak{p}} = \int_{K_\mathfrak{p}^0} f(|y^{-1}|_\mathfrak{p})\frac{dy}{|y|_\mathfrak{p}}.$$

Let $\mathrm{ord}_\mathfrak{p}(\partial_K) = b$; then we have for $\mathrm{Re}(s) > 0$,

$$\begin{aligned} q^{bs/2}(1-q^{-s})^{-1} &= q^{bs/2} + q^{(b-2)s/2} + q^{(b-4)s/2} + \cdots \\ &= \int_{K_\mathfrak{p}^0} |t|_\mathfrak{p}^{s/2} e_\mathfrak{p}(t)\frac{dt}{|t|_\mathfrak{p}} \end{aligned}$$

where $q = \mathbf{N}_\infty(\mathfrak{p})$, $e_\mathfrak{p}(t)$ is the characteristic function,

$$e_\mathfrak{p}(t) = \begin{cases} 1 & \text{if } \mathrm{ord}_\mathfrak{p}(\partial t) \geq 0 \text{ (and is even)} \\ 0 & \text{otherwise.} \end{cases}$$

We replace t by $\alpha^2 y_\mathfrak{p}$, and get

$$\begin{aligned} \Gamma_\mathfrak{p}(s)|\alpha|_\mathfrak{p}^{-s} &= q^{bs/2}(1-q^{-s})^{-1}|\alpha|_\mathfrak{p}^{-s} \\ &= \int_{K_\mathfrak{p}^0} |y_\mathfrak{p}|_\mathfrak{p}^{s/2} e_\mathfrak{p}(y_\mathfrak{p}\alpha^2)\frac{dy_\mathfrak{p}}{|y_\mathfrak{p}|_\mathfrak{p}}. \end{aligned}$$

We also wish to make an exponential change of variable in the integrals for $\Gamma_\mathfrak{p}(s)$ with $\mathfrak{p}$ infinite. For such $\mathfrak{p}$, we let $|y_\mathfrak{p}|_\mathfrak{p} = \exp(v_\mathfrak{p})$. Thus for $\mathfrak{p}$ real,

$$\Gamma_\mathfrak{p}(s)|\alpha|_\mathfrak{p}^{-s} = \int_\mathbf{R} |y_\mathfrak{p}|_\mathfrak{p}^{s/2}\exp(-\pi y_\mathfrak{p}\alpha^2)dv_\mathfrak{p}$$

and for $\mathfrak{p}$ complex,

$$\Gamma_\mathfrak{p}(s)|\alpha|_\mathfrak{p}^{-s} = 2^{-1}\int_\mathbf{R} |y_\mathfrak{p}|_\mathfrak{p}^{s/2}\exp(-2\pi y_\mathfrak{p}\alpha\bar{\alpha})dv_\mathfrak{p}.$$

Now let us use these integrals. Let

$$L = 2^{r_2}\prod_{\substack{\mathfrak{p}\in S\\ \mathfrak{p}\nmid\infty}} (2\log \mathbf{N}_\infty(\mathfrak{p})).$$

We have for $\mathrm{Re}(s) > 0$,

$$\begin{aligned} G(s)|\mathbf{N}(\alpha)|^{-s} &= \prod_{\mathfrak{p}\in S}\Gamma_\mathfrak{p}(s)(|\alpha^2|_\mathfrak{p})^{-s/2} \\ &= L^{-1}\int_{\mathbf{R}^{r+1}} \mathbf{N}(y)^{s/2} e_S(-y\alpha^2)\,dv, \end{aligned}$$

where $e_S(-y\alpha^2)$ is given in the definition of $\theta_K(y, \mathfrak{O}, S)$ above.

We have one more variable change to make. If the $\mathfrak{p}$ in S are numbered as $\mathfrak{p}_1, \ldots, \mathfrak{p}_{r+1}$, then it is natural to number the $v_\mathfrak{p}$ and $y_\mathfrak{p}$ as $v_1, \ldots, v_{r+1}$ and $y_1, \ldots, y_{r+1}$ instead of the more cumbersome $v_{\mathfrak{p}_1}$, etc. It is equally reasonable to write $|y_i|_i$ instead of $|y_{\mathfrak{p}_i}|_{\mathfrak{p}_i}$. We will assume that the numbering has been chosen so that $\mathfrak{p}_{r+1}$ is an infinite prime. As a result, $|y_{r+1}|_{r+1}$ varies continuously in $\mathbf{R}^+$. We therefore make a variable change from v_{r+1} to a variable u given by the condition that

$$u = \mathbf{N}(y).$$

In terms of v_{r+1}, we have $dv_{r+1} = du/u$. Hence the last integral for $G(s)$ above may be rewritten as,

$$G(s)|\mathbf{N}(\alpha)|^{-s} = L^{-1}\int_0^\infty u^{s/2}\int_{\mathbf{R}^r} e_S(-y\alpha^2)dv\frac{du}{u}.$$

For $\mathrm{Re}(s) > 1$, the series for $\zeta_K(s,S)$ converge absolutely thanks to the Euler product which allows a comparison with $\zeta(s)^{\deg(K)}$. As a result, for $\mathrm{Re}(s) > 1$, we have,

$$\begin{aligned}
G(s)\zeta_K(s,S) &= \sum_{\substack{(\alpha)\\ \alpha\in\mathfrak{O}_K(S)}} G(s)|\mathbf{N}(\alpha)|^{-s} \\
&= \sum_{\substack{(\alpha)\\ \alpha\in\mathfrak{O}_K(S)}} L^{-1}\int_0^\infty u^{s/2}\int_{\mathbf{R}^r} e_S(-y\alpha^2)dv\frac{dv}{u} \\
&= L^{-1}\int_0^\infty u^{s/2}\int_{\mathbf{R}^r}\sum_{\substack{(\alpha)\\ \alpha\in\mathfrak{O}_K(S)}} e_S(-y\alpha^2)dv\frac{du}{u} \\
&= \frac{1}{LW(K)}\int_0^\infty u^{s/2}\int_{\mathbf{R}^r/U^2}\sum_{\substack{\alpha\in\mathfrak{O}_K(S)\\ \alpha\neq 0}} e_S(-y\alpha^2)dv\frac{du}{u}
\end{aligned}$$

$$(10)\qquad = \frac{1}{LW(K)}\int_0^\infty u^{s/2}\int_{\mathbf{R}^r/U^2}[\theta_K(y,0,0,\mathfrak{O},S)-1]dv\frac{du}{u},$$

the various interchanges being justified by the absolute convergence of the original expression.

We will break the integral on u up into two pieces, $\int_0^\infty = \int_0^1 + \int_1^\infty$.

The second of these integrals, $\int_1^\infty u^{s/2}[\cdots]du/u$ is an entire function of s since it converges absolutely and uniformly on compact sets in the complex s-plane by comparison with the integral at a large real value of s. (Note that it is crucial in this that u is in a range greater than 1.) We now change the integral from 0 to 1 in the manner indicated by Riemann and Hecke. For $\mathrm{Re}(s) > 1$, where we have the absolute convergence of the starting expression, we have

$$\begin{aligned}
&\int_0^1 u^{s/2}\int_{\mathbf{R}^r/U^2}[\theta_K(y,\mathfrak{O},S)-1]dv\frac{du}{u} \\
&= \int_0^1 u^{s/2}\int_{\mathbf{R}^r/U^2}[u^{-1/2}(\mathbf{N}(y)^{1/2}\theta_K(y,\mathfrak{O},S)-1)+(u^{-1/2}-1)]dv\frac{du}{u} \\
&= \int_0^1 u^{(s-1)/2}\int_{\mathbf{R}^r/U^2}[\theta_K(y^{-1},\mathfrak{O},S)-1]dv\frac{du}{u} \\
&\quad + \int_0^1 [u^{(s-1)/2}-u^{s/2}]\int_{\mathbf{R}^r/U^2} 1\,dv\frac{du}{u}
\end{aligned}$$

by Theorem 5. This split of our absolutely convergent integral into two pieces is justified since the first of the two pieces involving the theta function is absolutely convergent and hence the second piece must be absolutely convergent as well. We therefore find that for $\mathrm{Re}(s) > 1$,

$$\int_0^1 u^{s/2} \int_{\mathbf{R}^r/U^2} [\theta_K(y, \mathfrak{O}, S) - 1] dv \frac{du}{u}$$
$$= \int_1^\infty u^{(1-s)/2} \int_{\mathbf{R}^r/U^2} [\theta_K(y, \mathfrak{O}, S) - 1] dv \frac{du}{u} + \left(\frac{2}{s-1} - \frac{2}{s}\right) \int_{\mathbf{R}^r/U^2} 1\, dv.$$

It follows that $\int_{\mathbf{R}^r/U^2} 1\, dv$ is finite. This is the content of the unit theorem. For the moment, we define the *S-regulator* $R_K(S)$ by the expression,

$$2^r R_K(S) = \int_{\mathbf{R}^r/U^2} 1\, dv.$$

The final expression for $\xi_K(s)$ is,

$$\xi_K(s) = \frac{1}{LW(K)} \int_1^\infty [u^{s/2} + u^{(1-s)/2}] \int_{\mathbf{R}^r/U^2} [\theta_K(y, \mathfrak{O}, S) - 1] dv \frac{du}{u}$$
$$+ \frac{2^{r+1} R_K(S)}{LW(K)} \left(\frac{1}{s(s-1)}\right),$$

and this expression provides the desired analytic continuation to the complex s-plane as well as the functional equation,

$$\xi_K(1-s) = \xi_K(s).$$

Although this proof has followed Hecke's, it is in some sense a hybrid of the proofs of Hecke [4] and Tate [9]. For instance besides the obvious similarities, the integration of characteristic functions on $\mathfrak{p}$-adic spaces in order to get the Euler $\mathfrak{p}$-factor of $\zeta_K(s)$ appears in Tate's version.

The residue at $s = 0$ is

$$\operatorname{res}_{s=0}(\xi_K(s)) = -\frac{2^{r+1} R_K(S)}{LW(K)}.$$

On the other hand, the functions $\Gamma_\mathfrak{p}(s)$ have first order poles at $s = 0$ of residues, 2 ($\mathfrak{p}$ real infinite), 1 ($\mathfrak{p}$ complex infinite), $[\log(\mathbf{N}_\infty(\mathfrak{p}))]^{-1}$ ($\mathfrak{p}$ finite) respectively. Hence, $\zeta_K(s, S)$ has a pole at $s = 0$ of order $r(S)$ and

$$\left.\frac{\zeta_K(s, S)}{s^{r(S)}}\right|_{s=0} = -\frac{R_K(S)}{W(K)}.$$

This is the first direct derivation of this formula of which I am aware and is made possible by our definition of the Haar integral. Earlier proofs proceed inductively from the classical case of the behavior of $\zeta_K(s)$ at 0.

We conclude this Section by sketching the completion of the proof of the S-unit theorem. Since $R_K(S)$ is finite, there must be $r = r(S)$ S-units $\varepsilon_1, \ldots, \varepsilon_r$ such that the r vectors $(\log|\varepsilon_i|_j)_{1 \le j \le r}$ are linearly independent over $\mathbf{R}$. Let U' denote the subgroup of $U = U(K, S)$ generated by the ε_i and the roots of unity in K. Thus $U'/\mu(K)$ acts discretely on $\mathbf{R}^r$ and has a fundamental domain of finite volume. Therefore $U'/\mu(K)$ is of finite index in $U/\mu(K)$, the index being the quotient of the volumes of the fundamental domains. Hence there is an integer $N \ge r$ and S-units $\varepsilon_{r+1}, \ldots, \varepsilon_N$ such that the N vectors $(\log|\varepsilon_i|_j)_{1 \le j \le r}$ generate the $\mathbf{Z}$-module $\{(\log|\varepsilon|_j)_{1 \le j \le r} \mid \varepsilon \in U\}$ which is isomorphic to $U/\mu(K)$ by Kronecker's Theorem. But a finitely generated $\mathbf{Z}$-module in

$\mathbf{R}^r$ has at most r generators. Since we have found that at least r are needed, the S-unit theorem follows.

Hecke noted that an analytic proof of the unit theorem is possible but gave no indication of what he had in mind. Twelve years ago, I also indicated that an analytic proof of the unit theorem could be derived in the process of proving the functional equation [7]. The proof which I had in mind was never published and was much messier than the proof presented here. It proceeded by showing directly that the integral analogous to (10) is divergent if there are fewer than r independent units. That proof in turn made use of some of the same elements of the geometry of numbers that are usually used in the proof of the unit theorem. The current proof which works for all S is much more satisfying.

5. P-adic L-functions. We have dealt with the zeta function on the ring of S-integers of a number field. For simplicity let us take the field to be $\mathbf{Q}$, and $S = \{\infty, p\}$ where p is a positive rational prime, possibly 2. So, for complex s, we have,

$$\zeta(s, S) = \sum_{\substack{n=1 \\ (n,p)=1}}^{\infty} n^{-s}$$

where we have once again chosen the obvious representatives for each ideal of $\mathfrak{O} = \mathbf{Z}[p^{-1}]$. Suppose that f is a positive rational integer and that χ is a Dirichlet character (mod f). We will assume that f is relatively prime to p. Then χ gives rise to a Dirichlet character (mod f) on $\mathfrak{O}$ and we have the corresponding L-function,

$$L(s, \chi, S) = \sum_{\substack{n=1 \\ (n,p)=1}}^{\infty} \chi(n) n^{-s}$$

which is just

$$L(s, \chi, S) = L(s, \chi)(1 - \chi(p)^{-s}). \tag{11}$$

We may also break up this L-function into partial zeta functions,

$$L(s, \chi, S) = \sum_{a=1}^{f} \chi(a) f^{-s} \zeta\left(s, \frac{a}{f}, S\right)$$

where,

$$\zeta\left(s, \frac{a}{f}, S\right) = f^{s} \sum_{\substack{n \equiv a (\mathrm{mod}\, f) \\ (n,p)=1}} n^{-s} = \sum_{\substack{n=0 \\ (fn+a,p)=1}}^{\infty} \left(n + \frac{a}{f}\right)^{-s}.$$

Thanks to (11), this may be also expressed as,

$$\zeta\left(s, \frac{a}{f}, S\right) = \zeta\left(s, \frac{a}{f}\right) - p^{-s} \zeta\left(s, \frac{b}{f}\right), \quad bp \equiv a (\mathrm{mod}\, f), \quad 0 < b < f.$$

The Hurwitz zeta function $\zeta(s, x)$ is defined by the series,

$$\zeta(s, x) = \sum_{n=0}^{\infty} (n + x)^{-s} \tag{12}$$

for positive x and $\mathrm{Re}(s) > 1$. However, there is an analytic continuation in both s and x to the complex s-plane with $\mathrm{Re}(x) > 0$ with a first order pole at $s = 1$ with residue

1. (Actually, we can analytically continue to a multivalued function of all x with the obvious singularities at $x = 0$ and the negative integers.) For $\mathrm{Re}(s) > 1$, the function satisfies,

$$\int_0^1 \varsigma(s, x+1)dx = \int_1^\infty x^{-s}dx = \frac{1}{s-1}.$$

By analytic continuation, this holds for all $s \neq 1$ and therefore for $\mathrm{Re}(s) < 1$, we have,

$$\int_0^1 \varsigma(s,x)dx = \int_0^1 [x^{-s} + \varsigma(s, x+1)]dx = \frac{-1}{s-1} + \frac{1}{s-1} = 0. \tag{13}$$

We also have the relation,

$$\frac{d}{dx}\varsigma(s,x) = -s\varsigma(s+1, x) \tag{14}$$

valid initially for $\mathrm{Re}(s) > 1$ where we can differentiate (12) term by term, but then valid for all s by analytic continuation. Thus for example,

$$\frac{d}{dx}\varsigma(0,x) = -\operatorname*{res}_{s=1} \varsigma(s,x) = -1$$

and so $\varsigma(0,x) = -x + C$ for some C. By (13), $C = 1/2$ and $\varsigma(0,x) = -B_1(x)$. Now in general for $k > 0$, we have

$$\frac{d}{dx}\varsigma(-k,x) = k\varsigma(-(k-1),x) \text{ and } \int_0^1 \varsigma(-k,x)dx = 0.$$

These recursions with the initial conditions $\varsigma(0,x) = B_1(x)$ are the same recursions satisfied by the Bernoulli polynomials and determine $\varsigma(-k,x)$ completely. The result is,

$$\varsigma(-k,x) = -\frac{B_{k+1}(x)}{k+1}. \tag{15}$$

This discussion has been taken from [8]. There are also similarities to [1].

Using this, we get

$$\varsigma\left(-k, \frac{a}{f}, S\right) = -\frac{B_{k+1}\left(\frac{a}{f}\right)}{k+1} + p^k \frac{B_{k+1}\left(\frac{b}{f}\right)}{k+1} \tag{16}$$

where $bp \equiv a \pmod f$, $0 < b < f$. From (15) and (16) in turn, we get the evaluation of $L(s,\chi)$ and $L(s,\chi,S)$ at 0 and negative integers in terms of Bernoulli polynomials. Another classical formula comes from the fact that for $k \geq 1$,

$$\varsigma(-k,0) = 0^k + \varsigma(-k,1);$$

namely for $k \geq 1$,

$$B_{k+1}(1) = B_{k+1}(0). \tag{17}$$

The expression,

$$(1 - f^{-s})\varsigma(s) = f^{-s}\sum_{a=1}^{f-1} \varsigma\left(s, \frac{a}{f}\right),$$

gives us still another result. Evaluating at $s = -k$ for $k \geq 0$ and using (17) for $k \geq 1$ (for $k = 0$, the left side is 0) gives for $k \geq 0$,

$$(1 - f^k)B_{k+1}(0) = f^k \sum_{a=1}^{f-1} B_{k+1}\left(\frac{a}{f}\right).$$

This generalizes a well known formula relating $B_{k+1}(1/2)$ and $B_{k+1}(0)$ which is the case with $f = 2$ and $a = 1$. This in turn yields for $k \geq 0$,

$$(18) \qquad (1 - f^{k+1})B_{k+1}(0) = f^k \sum_{a=1}^{f-1} \left(B_{k+1}\left(\frac{a}{f}\right) - B_{k+1}(0) \right),$$

a formula that will be of use shortly. A similarly derived, but somewhat more complicated formula results if the sum on the right of (18) is taken only over a with $(a, f) = 1$.

We have two primes in S, but hitherto have dealt only with complex s. We now remedy this multiyear oversight by letting s be in $\mathbf{Z}_p$. Our goal is to give meaning to the p-adic Dirichlet series,

$$(19) \qquad f^{-s}\zeta_p\left(s, \frac{a}{f}, S\right) = \sum_{\substack{n>0 \\ n\equiv a \ (\mathrm{mod}\, f) \\ (n,p)=1}} n^{-s}.$$

For the benefit of the reader and the author, we begin by recalling what n^{-s} means for n relatively prime to p in $\mathbf{Z}$, or even $\mathbf{Z}_p$, and s in $\mathbf{Z}_p$. There are in fact $p-1$ reasonable definitions. Let $0 \leq J < p-1$; J will be fixed throughout the following discussion. If t runs through a sequence of positive integers congruent to $J \pmod{p-1}$ and tending to $-s$ p-adically, then we define n^{-s} to be the p-adic limit of the sequence of powers (n^t). This makes sense because $n^{(p-1)p^a} \equiv 1 \pmod{p^{a+1}}$ and so the p-adic limit exists. We take note of the basic continuity property that if $|s_1 - s_2|_p \leq p^{-r}$ then $|n^{s_1} - n^{s_2}|_p \leq p^{-r-1}$. We repeat that this definition depends upon the choice of J. In particular, for a rational integer j, n^j is the expected usual value for all n if and only if $j \equiv J \pmod{p-1}$. Worse still, except when $J = 0$, $n^{s_1} \cdot n^{s_2}$ does not always equal $n^{s_1+s_2}$. It is easy to see that for different J's, the powers n^{-s} differ by $(p-1)^{\text{st}}$ roots of unity. We will take $J = 0$ in the sequel; this will allow us to investigate the values of our p-adic series at integers divisible by $p-1$. For other integers, it would be convenient to use another value of J or, as is done elsewhere, to introduce the Teichmuller character which has the same effect as changing J.

So *from this point on, we take* $J = 0$. For a given s, every term in (19) has absolute value 1 and thus for the series to converge, we will have to group the terms. The basis for the grouping chosen is given by the following

LEMMA 7. *Suppose that s is in $\mathbf{Z}_p$, and that $p^a \| s$. Let $v = 1$ for p odd and $v = 2$ for $p = 2$. Then for $r \geq a + v$,*

$$\sum_{\substack{n \ (\mathrm{mod}\, p^r) \\ (n,p)=1}} n^{-s} \equiv \begin{cases} 0 \pmod{p^{r-a-v}} & \text{if } s \neq 0 \\ 0 \pmod{p^{r-1}} & \text{if } s = 0. \end{cases}$$

PROOF. The case where $s = 0$ is trivial since we are just counting the number of elements in a reduced residue system $\pmod{p^r}$. So, let $s \neq 0$. For p odd, let u be a primitive root $\pmod{p^2}$; for $p = 2$, let $u = 5$. We see by induction on a that there is an integer c relatively prime to p such that for all $a \geq 0$,

$$u^{p^a(p-1)} \equiv 1 + cp^{a+v} \pmod{p^{a+v+1}},$$

and in particular this holds for the chosen a. If $t = (p-1)bp^a$ is in $\mathbf{Z}^+$ where $(b, p) = 1$, then

$$u^t \equiv 1 + bcp^{a+v} \pmod{p^{a+v+1}}.$$

But clearly,

$$(u^t - 1) \cdot \sum_{\substack{n \,(\mathrm{mod}\, p^r) \\ (n,p)=1}} n^t \equiv \sum_{\substack{n \,(\mathrm{mod}\, p^r) \\ (n,p)=1}} n^t - \sum_{\substack{n \,(\mathrm{mod}\, p^r) \\ (n,p)=1}} n^t \equiv 0 \pmod{p^r}.$$

The result of the Lemma now follows for $-s = t$ and hence for general $s \neq 0$. Q.E.D.

We should remark that in number fields, where there aren't always primitive roots, we would simply have to pick some u other than a root of unity and use whatever v is necessary.

DEFINITION. For x in $\mathbf{Z}_p$, we define the p-adic Hurwitz zeta function $\varsigma_p(s, x, S)$ by

$$\varsigma_p(s, x, S) = \lim_N \sum_{\substack{0 \leq n < N \\ n+x \in \mathbf{Z}_p^\times}} (n+x)^{-s} \tag{20}$$

where the limit is taken over a sequence of positive integers N tending to $-x$ p-adically. As a notational convenience, we will write $N \underset{p}{\rightarrow} -x$ to express that we are taking a p-adic limit.

The obvious choice of $N \underset{p}{\rightarrow} 0$ would lead to a p-adic limit which would be identically zero in s thanks to Lemma 7.

THEOREM 8. *The function $\varsigma_p(s, x, S)$ is defined and continuous for all s and x in $\mathbf{Z}_p$. In fact for $|s_1 - s_2|_p \leq p^{-r}$ and all x,*

$$|\varsigma_p(s_1, x, S) - \varsigma_p(s_2, x, S)|_p \leq p^{-r-1},$$

while for $|x_1 - x_2|_p \leq p^{-r}$ and all s,

$$|\varsigma_p(s, x_1, S) - \varsigma_p(s, x_2, S)|_p \leq p^{-r}.$$

PROOF. The whole result follows from Lemma 7 and the definition of $(n+x)^{-s}$. Q.E.D.

Now that we have this nice p-adic function, we are entitled to ask whether we can determine its values at 0 and negative integers. In fact we can do so relatively easily but because of our convention on J, we will have to take integers divisible by $p-1$ in this paper.

THEOREM 9. *Given x in $\mathbf{Z}_p$, let u in $\mathbf{Z}$ be chosen so that $0 \leq u < p$ and $x + u$ is divisible by p in $\mathbf{Z}_p$. For k in $\mathbf{Z}$, positive or 0, and divisible by $(p-1)$, we have,*

$$\varsigma_p(-k, x, S) = -\frac{B_{k+1}(x) - B_{k+1}(0)}{k+1} + p^k \frac{B_{k+1}\left(\frac{x+u}{p}\right) - B_{k+1}(0)}{k+1}.$$

PROOF. We first take the case where x is in $\mathbf{Z}^+$. For such an x, we now explicitly sum the series,

$$\sum_{\substack{0 \leq n < N \\ (n+x,p)=1}} (n+x)^k$$

where N is a positive integer chosen so that $N + x = Mp$ is divisible by p. Ignoring the relative primality condition, we have

$$\sum_{0 \leq n < N} (n+x)^k = \varsigma(-k, x) - \varsigma(-k, x+N)$$

$$= -\frac{B_{k+1}(x) - B_{k+1}(x+N)}{k+1}.$$

Now we have to remove the terms divisible by p. If $n+x=mp$ and n is in the range, $0 \le n < N$, then m is in the range, $\frac{x+u}{p} \le m < M$ and hence,

$$\sum_{\substack{0 \le n < N \\ p \mid (n+x)}} (n+x)^k = p^k \sum_{\frac{x+u}{p} \le m < M} m^k$$

$$= p^k \left[\zeta\left(-k, \frac{x+u}{p}\right) - \zeta(-k, M) \right]$$

$$= p^k \left[\frac{B_{k+1}\left(\frac{x+u}{p}\right) - B_{k+1}(M)}{k+1} \right].$$

Therefore,

$$\sum_{\substack{0 \le n < N \\ (n+x,p)=1}} (n+x)^k = -\frac{B_{k+1}(x) - B_{k+1}(x+N)}{k+1} + p^k \frac{B_{k+1}\left(\frac{x+u}{p}\right) - B_{k+1}(M)}{k+1}.$$

Clearly $M \underset{p}{\longrightarrow} 0$ as $N \underset{p}{\longrightarrow} -x$, and so we get the result of the Theorem for $x \in \mathbf{Z}^+$. Finally, as x approaches a limit in $\mathbf{Z}_p$, the value of u is eventually fixed and the Theorem follows from the continuity of both sides. Q.E.D.

THEOREM 10. *Suppose that f is in $\mathbf{Z}^+$, $(f,p)=1$, and that a is in $\mathbf{Z}$, $0 \le a < f$. Choose b in $\mathbf{Z}$, $0 \le b < f$ so that $bp \equiv a \pmod f$. Then,*

$$\zeta_p\left(-k, \frac{a}{f}, S\right) = -\frac{B_{k+1}\left(\frac{a}{f}\right) - B_{k+1}(0)}{k+1} + p^k \frac{B_{k+1}\left(\frac{b}{f}\right) - B_{k+1}(0)}{k+1}. \tag{21}$$

PROOF. We choose $u < p$ such that $a/f + u$ is divisible by p in $\mathbf{Z}p$. Thus there is an integer b such that

$$\frac{a}{f} + u = \frac{bp}{f} \tag{22}$$

and for this value of b, the formula (21) holds by Theorem 9. It remains to show that this b is the same as in the statement of the Theorem. First,

$$0 \le \frac{a}{f} + u < 1 + (p-1) = p$$

and therefore, $0 \le b < f$ as is required. Second, we see from (22) that $a \equiv bp \pmod f$. Q.E.D.

When χ is a non-trivial character $\pmod f$, the function $\sum_{a=1}^{f-1} \chi(a) f^{-s} \zeta_p(s, a/f, S)$ interpolates the values usually used to define $L_p(s,\chi,S)$. The needed congruences regarding Bernoulli polynomials are therefore a consequence of the fact that $L_p(s,\chi,S)$ exists rather than the other way around. When χ is trivial, we see that we have not quite managed to interpolate the values of $\zeta(-k,S)$ thanks to the $-B_{k+1}(0)/k+1$ terms in (21). However, this is easily accomplished using (18). Indeed, we see from (18) and (21) that

$$-f^{-s}(1-p^{-s})(1-f \cdot f^{-s})^{-1} \sum_{a=1}^{f-1} \zeta_p\left(s, \frac{a}{f}, S\right)$$

at $s = -k$ takes on the value $(1-p^k)B_{k+1}(0)$. The latter therefore possesses a continuous interpolation to $\mathbf{Z}_p$ except possibly for a pole at $s = 1$ (which because of our convention on J is a point we have agreed not to discuss). Note also that $f \cdot f^{-s}$ and f^{1-s} are not the same thing!

Let us now investigate the value of $L'_p(0, \chi, S)$. For this purpose, we wish to differentiate term by term. It appears that even in p-adic analysis this requires justification. Thus we will briefly derive the formula for

$$\lim_{\substack{s \to 0 \\ p}} \frac{n^s - 1}{s},$$

paying particular attention to the rate of convergence. Recall that if $n = 1 + cp$ (c in $\mathbf{Z}_p$), the p-adic log of n is given by the usual series,

$$\log_p(n) = \sum_{j=1}^{\infty} \frac{(-1)^{j-1}}{j} c^j p^j.$$

This satisfies the usual rules,

$$\begin{aligned} \log_p(n_1 n_2) &= \log_p(n_1) + \log_p(n_2), \\ \log_p(n^{-1}) &= -\log_p(n), \end{aligned} \tag{23}$$

since these expressions involve formal series rearrangements which work over $\mathbf{C}$. If n is in $\mathbf{Z}_p$, then $n^{p-1} \equiv 1 \pmod p$ and so we define

$$\log_p(n) = \frac{1}{p-1} \log_p(n^{p-1});$$

this agrees with the earlier value if $n \equiv 1 \pmod p$ thanks to (23). Frequently the log function is extended to $\mathbf{Q}_p$ by defining $\log_p(p) = 0$.

For $(n, p) = 1$, $t = (p-1)bp^a$ in $\mathbf{Z}^+$ and $n^{p-1} = 1 + cp$, we have

$$\begin{aligned} n^t = (1+cp)^{bp^a} &= 1 + \sum_{j=1}^{\infty} (bp^a)(bp^a - 1) \cdots (bp^a - j + 1) \frac{(cp)^j}{j!} \\ &\equiv 1 + \sum_{j=1}^{\infty} (bp^a) \left[(-1)^{j-1}(j-1)!\right] \frac{(cp)^j}{j!} \pmod{p^{2a+1}} \end{aligned}$$

since $p^j/j!$ always has a factor of p in the numerator, even when p is 2. Therefore

$$\frac{n^t - 1}{t} \equiv \frac{1}{p-1} \sum_{j=1}^{\infty} \frac{(-1)^{j-1}}{j} (cp)^j \equiv \frac{1}{p-1} [\log_p(n^{p-1})] \equiv \log_p(n) \pmod{p^{a+1}}.$$

From this we see that $dn^{-s}/ds|_{s=0} = -\log_p(n)$ and for $|s|_p \le p^{-a}$ ($s \neq 0$),

$$\left| \frac{n^{-s} - 1}{s} + \log_p(n) \right|_p \le p^{-a-1}. \tag{24}$$

DEFINITION. For x in $\mathbf{Z}_p$, we define the p-adic gamma function Γ_p by

$$\Gamma_p(x)^{-1} = \lim_{\substack{N \to -x \\ p \\ N \in \mathbf{Z}^+}} (-1)^{pN} \cdot \prod_{\substack{n < N \\ |n+x|_p = 1}} (n + x). \tag{25}$$

We may think of this as giving a p-adic analogue of the infinite product for $\Gamma(x)^{-1}$. We will shortly justify this further by deriving an equivalent formula which we will think of as the factorial version of $\Gamma_p(x)$.

LEMMA 8. *The product defining $\Gamma_p(x)^{-1}$ converges on $\mathbf{Z}_p$.*

PROOF. Let $M > N$ and $M \equiv N \pmod{p^r}$. Since the product of all group elements of an abelian group is the unique element of order 2, if such an element exists, and is the identity otherwise, we see that

$$(-1)^{p(M-N)} \prod_{\substack{N \le n < M \\ |n+x|_p = 1}} (n+x) \equiv 1 \pmod{p^r},$$

(when $p = 2$, this requires $r \ge 3$) and the Lemma follows. Q.E.D.

Now we come to the p-adic analogue of the derivative at $s = 0$ of the Hurwitz zeta function in terms of $\log(\Gamma(x)^{-1})$.

THEOREM 11. *For x in $\mathbf{Z}_p$, the derivative of $\zeta_p(s, x, S)$ at $s = 0$ exists and is*

$$\left.\frac{d}{ds}\zeta_p(s, x, S)\right|_{s=0} = \log_p(\Gamma_p(x)).$$

PROOF. Lemma 7 and the fact that the estimate in (24) is uniform in n allows us to conclude that

$$\left.\frac{d}{ds}\zeta_p(s, x, S)\right|_{s=0} = -\lim_{\substack{N \xrightarrow[p]{} -x \\ N \in \mathbf{Z}^+}} \log_p \left[\prod_{\substack{n<N \\ |n+x|_p=1}} (n+x) \right]$$

including the existence of the right hand side, if we did not already know it exists. But in this case, since $\log_p(\pm 1) = 0$, the right hand side is $-\log_p(\Gamma_p(x)^{-1}) = \log_p(\Gamma_p(x))$. Q.E.D.

We now turn to another version of $\zeta_p(s, x, S)$ which is in fact slightly easier to treat than the definition we have already given; however, the definition given provides the closest analogy with the classical complex valued Hurwitz zeta function.

Alternate Definition. For s and x in $\mathbf{Z}_p$, let

$$\zeta_p(s, x, S) = -\lim_{M \xrightarrow[p]{} x} \sum_{\substack{n<M \\ (n,p)=1}} n^{-s}$$

where the limit is taken over a sequence of positive integers M.

The limit exists for all s and x in $\mathbf{Z}_p$ and is continuous in both variables. The agreement with the earlier definition is a consequence of

LEMMA 9. *Suppose that $|N + x|_p = p^{-r}$ where $0 < N < p^r$ and set $M = p^r - N$ (so that $|M - x|_p \le p^{-r}$). Then the set of numbers $\{n + x\}_{0 \le n < p^r} \pmod{p^r}$ splits into the two sets $\{n + x\}_{0 \le n < N}$ and $\{n\}_{0 \le n < M}$.*

PROOF. For $N \le n < p^r$, set $n = N + m$ so that $0 \le m < p^r - N = M$. We have $n + x = m + (N + x)$ and by hypothesis $p^r | (N + x)$ in $\mathbf{Z}_p$. Hence $n + x \equiv m \pmod{p^r}$. Q.E.D.

The numbers relatively prime to p in the first set are of course congruent to the numbers relatively prime to p in the second two sets. Now thanks to Lemmas 7 and 9, the two definitions of $\zeta_p(s, x, S)$ agree on $\mathbf{Z}_p \times \mathbf{Z}_p$. We now recalculate the derivative of

$\varsigma_p(s,x,S)$ at $s=0$:

$$\varsigma_{p'}(0,x,S) = \lim_{\substack{M\to x\\ p}} \log_p \left[\prod_{\substack{n<M\\(n,p)=1}} n\right] = \log_p \left[\lim_{\substack{M\to x\\ p}} (-1)^{Mp} \prod_{\substack{n<M\\(n,p)=1}} n\right].$$

This suggests another formula for $\Gamma_p(x)$ which is the analogue of the factorial version of $\Gamma(x)$ for integeral x and is the function of Morita which appears in Ferraro and Greenberg [2] in their proof of the L-function variant of Theorem 11.

THEOREM 12. *For x in $\mathbf{Z}_p$,*

$$\Gamma_p(x) = \lim_{M\to x} (-1)^{p^M} \prod_{\substack{n<M\\(n,p)=1}} n.$$

PROOF. The Theorem is a direct consequence of Lemma 9 applied to the definition of $\Gamma_p(x)$ in (25) together with the already used property on the product of the elements of an abelian group. Q.E.D.

Generalizations of this Section to number fields are possible and are in progress.

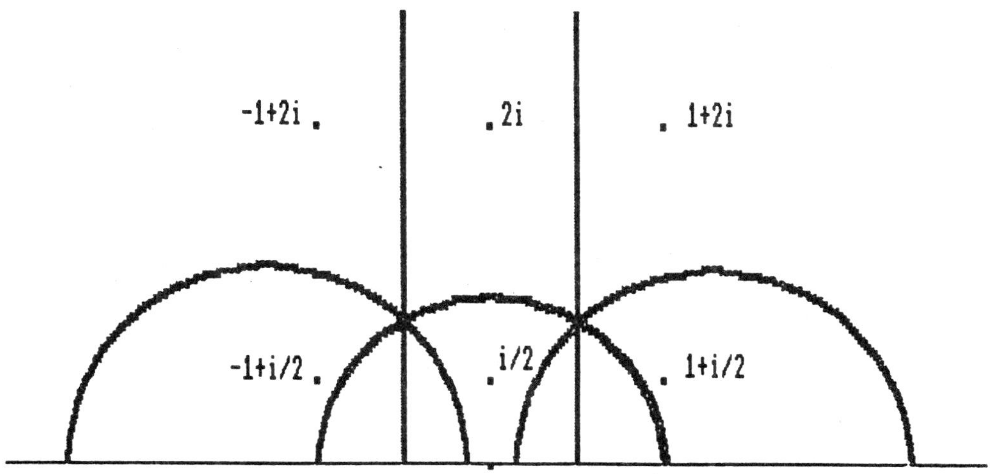

FIGURE 1. The fundamental domain for $SL_2(\mathbf{Z})\backslash\mathfrak{h}_{\mathbf{R}}$ by the perpendicular bisector method. Shown are the perpendicular bisectors of the segments joining $2i$ with the equivalent points $\frac{i}{2}$, $\pm1+2i$, $\pm1+\frac{i}{2}$ (the segments themselves aren't shown)

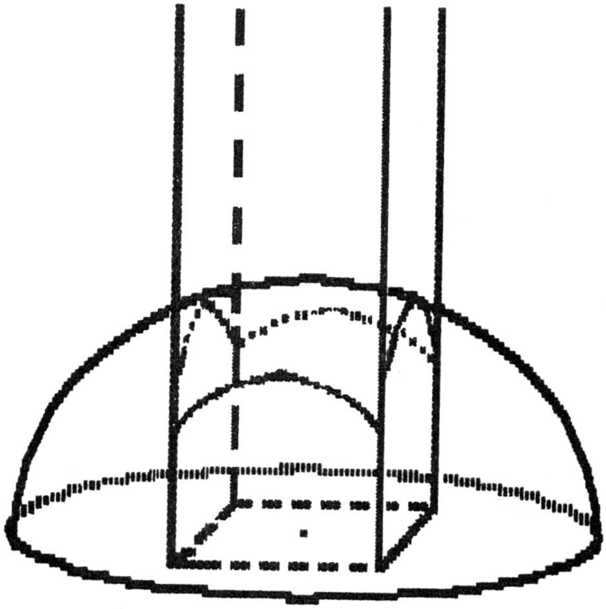

FIGURE 2. The region $x + yk$ with x in the rectangle $|\mathrm{Re}(x)| \le \frac{1}{2}$, $|\mathrm{Im}(x)| \le \sqrt{11/4}$ and $x + yk$ outside the unit sphere.

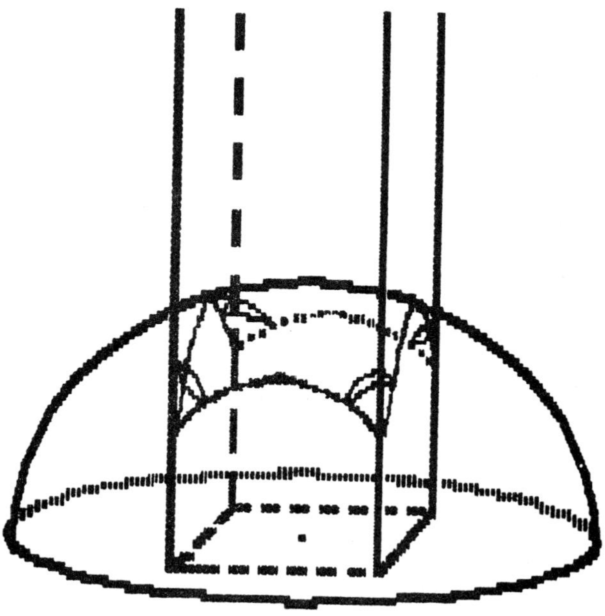

FIGURE 3. The fundamental domain for $\mathrm{SL}_2(\mathfrak{O}_K)\backslash\mathfrak{h}_{\mathbf{C}}$ $[K = \mathbf{Q}(\sqrt{-11})]$ is the region of Figure 2 with four hemispherical caps (radius 1 centered at $(\pm 1 + \sqrt{-11})/2$) cut off the corners.

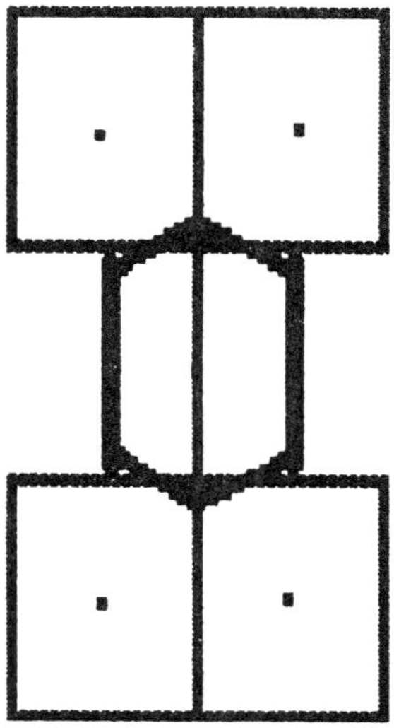

FIGURE 4. An alternate fundamental domain for C modulo translations by $\mathfrak{O}_K$ where $K = \mathbf{Q}(\sqrt{-11})$ [The one given by the perpendicular bisector method.] The set of points $x + yk$ outside the unit sphere with x in the fundamental domain is also a fundamental domain for $\mathrm{SL}_2(\mathfrak{O}_k)\backslash\mathfrak{h}_{\mathbf{C}}$.

REFERENCES

1. Bruce C. Berndt and Ronald J. Evens, Chapter 7 of Ramanujan's second notebook, to appear.
2. Bruce Ferrero and Ralph Greenberg, *On the behavior of p-adic L-functions at s = 0*, Inv. Math. 50 (1978), 91–102.
3. Erich Hecke, *Vorlesungen Über die Theorie der Algebraische Zahlen*, Chelsea, New York, 1948.
4. Erich Hecke, *Über die Zetafunktion beliebiger algebraischer Zahlkorper*, Mathematische Werke, Vandenhoeck and Ruprecht, Gottingen, 1959, 159–171.
5. C. L. Siegel, *Topics in Complex Function Theory*, 3 Volumes, Wiley, New York, 1969, 1971, 1973.
6. H. M. Stark, *Examples of modular forms in number fields*, to be published.
7. H. M. Stark, *The analytic theory of algebraic numbers*, BAMS 81 (1975), 961–972
8. H. M. Stark, *Values of zeta and L-functions*, Abh. Braunschweigischen Wiss. Gesell. 33 (1982), 71–83.
9. J. T. Tate, *Fourier analysis in number fields and Hecke's zeta-functions*, Algebraic Number Theory, Thompson, Washington, D.C. 1967, 305–347.

DEPARTMENT OF MATHEMATICS, MASSACHUSETTS INSTITUTE OF TECHNOLOGY, CAMBRIDGE, MA 02139

DEPARTMENT OF MATHEMATICS, UNIVERSITY OF CALIFORNIA, SAN DIEGO, LA JOLLA, CA 92093

Canadian Mathematical Society
Conference Proceedings
Volume 7 (1987)

CANONICAL LIFTINGS OF FORMAL MODULES

Leon H. Tatevossian[1]

ABSTRACT. Let $K \subset L$ be local fields and M the maximal unramified extension of L. Let $A \subset \mathcal{O} \subset W$ be the associated rings of integers and ω the residue field of W. In a forthcoming paper, Gross uses Lubin-Tate formal groups over $\mathcal{O}$ to identify specific liftings of formal A-modules defined over ω. We investigate the endomorphism rings of the corresponding infinitesimal liftings.

1. In this paper all formal groups will be one-dimensional and commutative. If F is a formal group over a ring B, we have the natural homomorphism $D : \mathrm{End}_B(F) \to B$. (Recall that if B is an integral domain of characteristic zero, then D is an injection, [4], Chapter 1, §3, Theorem 2.)

Let R be a ring and let B be an R-algebra with structure map $i : R \to B$.

DEFINITION 1. A <u>formal R-module over B</u> is a pair (F,f) where

(a) F is a formal group over B.

(b) f is a homomorphism from R to $\mathrm{End}_B(F)$ such that $D \circ f = i$. (The action of R on $\mathrm{Lie}(F)$ is via i.) We let $[r]_F(x)$ denote $f(r)$ for $r \in R$.

DEFINITION 2. An <u>R-module endomorphism of (F,f) over B</u> is an endomorphism $\phi \in \mathrm{End}_B(F)$ such that for any $r \in R$ we have $\phi([r]_F(x)) = [r]_F(\phi(x))$. We denote the (possibly non-commutative) R-algebra of all R-module endomorphisms over B by $\mathrm{End}_{B,R}(F,f)$, or simply $\mathrm{End}_{B,R}(F)$.

Note that $\mathrm{End}_{B,R}(F)$ is the commutant of the image of R in $\mathrm{End}_B(F)$ (via f).

1980 <u>Mathematics Subject Classification</u> (1985 <u>Revision</u>). 11S31, 14L05.

[1]The author thanks the Canadian Mathematical Society for its support during the 1985 CMS Number Theory Conference.

2. Let A be the ring of integers in a complete non-archimedean field K (with finite residue field) and let π be a prime element of A. Let $A/\pi A = k$ and $\#(k) = q = p^n$. Let E be an algebraic extension of k and let (F,f) be a formal A-module over E (via the structure map $A \to k \hookrightarrow E$).

DEFINITION 3. The <u>height</u> of (F,f) (denoted $ht(F,f)$) is defined to be $\begin{cases} \infty & \text{if } [\pi]_F(x) = 0 \\ h & \text{if } [\pi]_F(x) \text{ has first non-zero term in degree } q^h \end{cases}$.

Let C be an algebraic closure of k.

THEOREM 1. (Drinfel'd [3], Lazard [8]) <u>The height is a complete isomorphism invariant for formal A-modules over C</u>.

THEOREM 2. (Dieudonné [1], Lubin [10]) <u>Suppose $ht(F,f) = h < \infty$. Then $End_{C,A}(F)$ is isomorphic to the ring of integers of the central division algebra over K of rank h^2 and invariant $1/h$</u>.

If (F,f) is defined over k and has finite height h, Lubin [9] has shown that $End_{k,A}(F)$ (the "rational" endomorphism algebra) is a free A-module of rank h equal to $A[\tau]$, where τ is the Frobenius A-endomorphism of F, $\tau(x) = x^q$.

Suppose A' is an extension of A and that (J,j) is a formal A-module over A'. The injectivity of the homomorphism D shows that $End_{A',A}(J)$ is commutative. See Lubin [10] for a discussion of the <u>absolute</u> endomorphism ring of such a module.

3. Let L be an algebraic extension of K of degree d. Let $\mathcal{O}$ be the ring of integers in L and let y be a prime element of $\mathcal{O}$. Let M be the maximal unramified extension of L and let W be the ring of integers in M. Then y is a prime element of W and $W/yW = \omega$ is algebraically closed.

Let (G,g) (we write simply G) denote the unique formal A-module of height d over ω. $End_{\omega,A}(G)$ is isomorphic to $\mathcal{D}$, the maximal order in the central division algebra over K of invariant $1/h$. We will identify a lift of G to W.

We choose and embedding $\mathcal{O} \overset{a}{\hookrightarrow} End_{\omega,A}(G) \simeq \mathcal{D}$ for which $D \circ a$ is the reduction map $\mathcal{O} \to \mathcal{O}/y\mathcal{O} \hookrightarrow \omega$. This is analogous to choosing a "normalized" action of the CM field on holomorphic one forms in elliptic curve theory.

Gross [5] and Drinfel'd [3] observe that this construction realizes G as a formal $\mathcal{O}$-module of height 1 over ω.

The Lubin-Tate theory of formal complex multiplication [11] produces a formal $\mathcal{O}$-module G_{LT} (defined over $\mathcal{O}$) such that $G_{LT} \times_{\mathcal{O}} \mathcal{O}/y\mathcal{O}$ has height 1. (In fact, $[y]_{G_{LT}}(x) = yx + x^{q^f}$ where $q^f = \#(\mathcal{O}/y\mathcal{O})$.) Theorem 1 implies that $G_{LT} \times_{\mathcal{O}} \omega$ is isomorphic to G (as formal $\mathcal{O}$-modules). Moreover, the Lubin-Tate theory of formal moduli [12] shows that (up to $\star$-isomorphism) $G_{LT} \times_{\mathcal{O}} W$ is the unique lift of G to W (as a formal $\mathcal{O}$-module). In the category of formal A-modules, however, the space of lifts of G to W is (h-1)-dimensional [12].

DEFINITION 4. (Gross [5]) $G_{LT} \times_{\mathcal{O}} W$ (regarded as a formal A-module with reduction of height h) is the canonical lift of G corresponding to the pair $(\mathcal{O}, a)$.

We will briefly discuss the origin of this terminology at the end of the paper.

We let $G_{LT/W}$ denote $G_{LT} \times_{\mathcal{O}} W$, regarded as a formal A-module. Following Gross, we let R_{n-1} denote $\mathrm{End}_{W/y^nW,A}(G_{LT/W} \times_W W/y^nW)$ for $n \geq 1$. We remark that $R_0 = \mathrm{End}_{\omega,A}(G) \simeq \mathcal{D}$ and that $\mathrm{End}_{W,A}(G_{LT/W}) = \mathcal{O}$. We note that R_0 is a free $\mathcal{O}$-module of rank h (see [10], Lemma 2.1.1).

A result of Lubin [10] shows that the reduction maps on the endomorphism algebras are injections and we get a diagram

$$\mathcal{O} \hookrightarrow \dots \hookrightarrow R_n \hookrightarrow R_{n-1} \hookrightarrow \dots \hookrightarrow R_1 \hookrightarrow R_0$$

THEOREM 3. Suppose L is the unramified extension of K of degree d. Then $R_n = \mathcal{O} + y^n R_0$.

PROOF. We will sketch the proof of the theorem. The details will appear in [15]. Gross has proven this result (without the unramified restriction) in the case d = 2.

a. The main ingredient is the formal cohomology theory of Lubin-Tate. We refer the reader to the original papers [3], [12] for the definitions.

The formal cohomology group $H^2(G_{LT/W}, \mathcal{y}^nW/\mathcal{y}^{n+1}W)$ measures the infinitesimal deformations of $G_{LT/W} \times_W W/\mathcal{y}^nW$ to $W/\mathcal{y}^{n+1}W$. The space of deformations can be thought of as a "homogeneous space" for the action of H^2. (See [15] for some explicit calculations describing this action.)

A 2-dimensional cocycle of the module $G_{LT/W}$ with coefficients in $\mathcal{y}^nW/\mathcal{y}^{n+1}W$ is a set $\{ \Delta(x,y) \in \mathcal{y}^nW/\mathcal{y}^{n+1}W[[x,y]], \ \delta_a(x) \in \mathcal{y}^nW/\mathcal{y}^{n+1}W[[x]]$ for $a \in A \}$ satisfying the cocycle conditions. A coboundary of the series $\Psi(x) \in \mathcal{y}^nW/\mathcal{y}^{n+1}W[[x]]$ is a cocycle $\{ \Delta, \delta_a \}$ where

$$\Delta(x,y) = \Psi(G_{LT/W}(x,y)) - \Psi(x) - \Psi(y) \quad \text{and}$$

$$\delta_a(x) = \Psi([a]_{G_{LT/W}}(x)) - a\Psi(x) \quad \text{for} \quad a \in A .$$

LEMMA 1. <u>If</u> $\{ \Delta, \delta_a \}$ <u>is a coboundary then</u> $\delta_\pi(x)$ <u>is a power series in</u> x^{q^d}.

PROOF. Recall that $[\mathcal{y}]_{G_{LT/W}}(x) = \mathcal{y}x + x^{q^f}$ where $f = \#(\mathcal{O}/\mathcal{y}\mathcal{O})$. Since $L \supset K$ is unramified we can assume that $\mathcal{y} = \pi$. Of course we have $f = d$. The Lemma is then a direct application of the coboundary condition. □

b. Lubin and Tate show that $H^2(G_{LT/W}, \mathcal{y}^nW/\mathcal{y}^{n+1}W)$ is a $W/\mathcal{y}W$ vector space of dimension $h-1$ and explicitly give an injection

$$\alpha_n : R_{n-1}/R_n \hookrightarrow H^2(G_{LT/W}, \mathcal{y}^nW/\mathcal{y}^{n+1}W) \qquad \text{for} \quad n \geq 1 .$$

This shows that R_{n-1}/R_n is an $\mathcal{O}/\mathcal{y}\mathcal{O}$ vector space of dimension $\leq h-1$. (Note that we can now conclude that R_n is an order in R_0.)

Recall the assumption $\mathcal{y} = \pi$. In this case G_{LT} is defined over A and thus $G_{LT} \times_W \omega$ admits the Frobenius endomorphism $\mathrm{Frob} : x \to x^q$. We calculate the map α_1 on $(\mathrm{Frob}), (\mathrm{Frob})^2, \ldots, (\mathrm{Frob})^{h-1}$ and apply the Lemma to conclude that R_0/R_1 is $(h-1)$-dimensional over $\mathcal{O}/\mathcal{y}\mathcal{O}$.

c. Gross [5] shows that the injections α_n give a commutative diagram

$$\begin{array}{ccccc} \alpha_n & : & R_{n-1}/R_n & \to & H^2(G_{LT/W}, y^nW/y^{n+1}W) \\ & & [y] \downarrow & & y \downarrow \approx \\ \alpha_{n+1} & : & R_n/R_{n+1} & \to & H^2(G_{LT/W}, y^{n+1}W/y^{n+2}W) \end{array}$$

The vertical map on the left is composition with the endomorphism $[y]_{G_{LT}}$ and the vertical map on the right is the isomorphism given by multiplication by y "on the coefficients." The Theorem follows from our remarks in part b concerning R_0/R_1. □

4. Let R be an Artinian local ring of characteristic zero and let k be its residue field. (We assume that k is perfect and that $\text{char}(k) = p > 0$.) Let Φ be an abelian variety over k and let Φ^* be its p-divisible group. Serre and Tate have shown that lifting Φ to an abelian scheme over R is equivalent to lifting Φ^* [13].

If Φ is an ordinary abelian variety, there is a canonical lifting of Φ^* to R. (This follows from the general theory of group schemes of p-power order over a field of characteristic p.) Thus if Φ has the maximum number of points of order p, there is a <u>canonical lifting</u> of Φ to an abelian scheme over R. (For details, see also [7], [14]..)

If Φ is supersingular we can identify Φ^* with the formal group of Φ. Thus the infinitesimal deformation theory of supersingular elliptic curves is described in the Lubin-Tate article [12].

5. Let j be the elliptic modular function. A <u>singular modulus</u> is the value of j at a number that generates an imaginary quadratic extension of the rational numbers. These numbers are algebraic integers and possess many important arithmetic properties. In [6], Gross and Zagier prove remarkable formulas for the factorization of the absolute

norm of the difference of two singular moduli. (This work is continued by Dorman [2] .) The determination of these formulas led Gross and Zagier to consider the infinitesimal liftings of CM elliptic curves that preserve the module structure. The results parallel the Serre-Tate theory discussed in Section 4 and this similarity is the origin of the terminology adopted by Gross in [5] and used in Section 3 of this article.

This problem was a suggestion of Professor Benedict Gross and the author would like to thank Professor Gross and Professors Jonathan Lubin and Karl Zimmermann for many helpful discussions. He extends a special thanks to his advisor, Professor Michael Rosen, for his insight, encouragement, and friendship.

BIBLIOGRAPHY

1. Dieudonné, J., "Groupes de Lie et Hyperalgèbres de Lie sur un Corps de Caractéristique $p > 0$: VII," Math. Ann., 134 (1957).

2. Dorman, D., "Special Values of the Elliptic Modular Function and Factorization Formulas," to appear.

3. Drinfel'd, V., "Elliptic Modules," Math. USSR Sbornik, 23 (1974).

4. Fröhlich, A., Formal Groups (Lecture Notes in Mathematics 74), Springer-Verlag, New York, 1968.

5. Gross, B., "On Canonical and Quasi-canonical Liftings," to appear.

6. Gross, B. and Zagier, D., "On Singular Moduli," J. f. Reine u. Ang. Math. (Crelle), 355 (1985).

7. Katz, N., "Serre-Tate Local Moduli," in Surfaces Algébriques (Lecture Notes in Mathematics 868), Springer-Verlag, New York, 1981.

8. Lazard, M., "Sur les Groupes de Lie Formels a un Paramètre," Bull. Soc. Math. France, 83 (1955).

9. Lubin, J., "Formal A-Modules defined over A," Istituto Nazionale di Alta Matematica, Volume III (1970).

10. Lubin, J., "One-parameter Formal Lie Groups over P-adic Integer Rings," Annals of Math., 80 (1964).

11. Lubin, J. and Tate, J., "Formal Complex Multiplication in Local Fields," Annals of Math., 81 (1965).

12. Lubin, J. and Tate, J., "Formal Moduli for One-parameter Formal Lie Groups," Bull. Soc. Math. France, 94 (1966).

13. Lubin, J., Serre, J.-P., and Tate, J., "Elliptic Curves and Formal Groups," Woods Hole Summer Institute (mimeographed notes), 1964.

14. Messing, W., Appendix to The Crystals Associated to Barsotti-Tate Groups: with Applications to Abelian Schemes (Lecture Notes in Mathematics 264), Springer-Verlag, New York, 1972.

15. Tatevossian, L. "Liftings of Formal Modules," Thesis, Brown University, 1986.

MATHEMATICS DEPARTMENT
BROWN UNIVERSITY
PROVIDENCE, RHODE ISLAND 02912

Current Address:
Mathematics Department
Union College
Schenectady, New York 12308

Canadian Mathematical Society
Conference Proceedings
Volume 7 (1987)

INTEGRAL MATRICES IN ALGEBRAIC NUMBER THEORY

Olga Taussky

For orders generated by an algebraic integer of degree n in an algebraic number field of degree n, the theorem of Latimer and MacDuffee gives a $1-1$ correspondence between ideal classes and $n \times n$ matrix classes (see O. Taussky, On a theorem of Latimer and MacDuffee, *Canadian Journal of Mathematics*, **1** (1949)). For a quadratic field, the maximal order is of this type always.

However, the theorem itself does not give all that ideal classes provide. In particular, while suitable pairs of binary quadratic forms allow composition and corresponding ideal classes in quadratic fields allow multiplication, the product of corresponding matrix classes seems only to have been considered by this author. Earlier results were the fact that the principal ideal class corresponds to the matrix class generated by the companion matrix and the fact that the class of the transposed matrix corresponds to the inverse ideal class (see *Illinois J. Math.*, **1** (1957)). Recent work by the author (*Linear and Multilinear Algebra*, **10** (1981)) solves this problem for $n \times n$ matrices. In fact, for $n = 2$, an explicit formula is obtained. The main idea is to replace a pair of primitive quadratic forms by a pair A, B of 2×2 integral matrices by the correspondence

$$(*) \qquad X = \begin{pmatrix} x_{11} & x_{12} \\ x_{21} & x_{22} \end{pmatrix} \quad \begin{matrix} \rightarrow \\ \leftarrow \end{matrix} \quad x_{21}x^2 + (x_{22} - x_{11})xy - x_{12}y^2.$$

Here the elements in the diagonal of X have to be chosen in such a way that A and B have the same characteristic polynomial, and hence the same discriminant as the two quadratic forms. (Assume the characteristic polynomial to define the maximal order). The form

$$f(x, y) = a_{21}x^2 + (a_{22} - a_{11})xy - a_{12}y^2$$

is the quadratic form which corresponds to the ideal class attached to the matrix class of A.

A square of this form can be expressed in the following way

$$(**) \qquad a(x, y) = (a_1a_3 - a_2^2)x^2 + (a_1\tilde{a}_3 - 2a_2\tilde{a}_2 + a_3\tilde{a}_1)xy + (\tilde{a}_1\tilde{a}_3 - \tilde{a}_2^2)y^2.$$

Here $a_1, a_2, a_3; \tilde{a}_1, \tilde{a}_2, \tilde{a}_3$ are two independent solutions of the linear diophantine equation

$$a_{21}x_1 + (a_{22} - a_{11})x_2 - a_{12}x_3 = 0.$$

A completely different proof was obtained by van der Waerden, (see Ternary Quadratic Forms and Norms, edited by the author, Dekker 1982).

Via the correspondence (*) between matrices and quadratic forms, the form (**) leads to the following matrix which is then an element in the squared matrix class

$$\begin{pmatrix} \frac{1}{2}(trA + a_1\tilde{a}_3 - 2a_2\tilde{a}_2 + a_3\tilde{a}_1) & \tilde{a}_1\tilde{a}_3 - \tilde{a}_2^2 \\ -(a_1a_3 - a_2^2) & \frac{1}{2}(trA - a_1\tilde{a}_3 + 2a_2\tilde{a}_2 - a_3\tilde{a}_1) \end{pmatrix}.$$

The method of composition developed by the author can be displayed by the classical problem of composition of the sum of two squares. This was already solved by Gauss, via the field of complex numbers. Instead of this, the matrix

$$\begin{pmatrix} 0 & 1 \\ -1 & 0 \end{pmatrix}$$

can be employed. It is the companion matrix of the polynomial $x^2 + 1$. This simplifies the procedure greatly. The most general transformation of the given matrix into itself is a linear polynomial

$$\begin{pmatrix} \Lambda & M \\ -M & \Lambda \end{pmatrix}$$

in

$$\begin{pmatrix} 0 & 1 \\ -1 & 0 \end{pmatrix}.$$

The same holds for the two factors, one of which is an adjugate.

$$\begin{pmatrix} \Lambda & M \\ -M & \Lambda \end{pmatrix} = \begin{pmatrix} \lambda_1 & \mu_1 \\ -\mu_1 & \lambda_1 \end{pmatrix} \begin{pmatrix} \lambda_2 & -\mu_2 \\ \mu_2 & \lambda_2 \end{pmatrix}$$

leading to

$$\Lambda^2 + M^2 = (\lambda_1\lambda_2 + \mu_1\mu_2)^2 + (\mu_1\lambda_2 - \lambda_1\mu_2)^2.$$

M. Rosen raised the question about the role of the different in connection with the matrix representation. In the case where the maximal order is given as $\mathbf{Z}[\alpha]$, the different is a principal ideal. If the maximal order is not obtainable as $\mathbf{Z}[\alpha]$, then the matrix classes have to replaced by $n \times n$ integral representation classes of the maximal order under unimodular similarity. This was treated in 'A result concerning classes of matrices,' *Journal of Number Theory,* **6** (1974) 64-71. (See also, E. Bender, Classes of matrices over an integral domain, *Illinois J. Math.,* **11** (1967) 697-702.) The representation can be obtained from the ideal matrices via the equation

$$A \begin{pmatrix} \omega_1 \\ \vdots \\ \omega_n \end{pmatrix} = \begin{pmatrix} \alpha_1 \\ \vdots \\ \alpha_n \end{pmatrix}$$

where A is an integral $n \times n$ matrix, the ω_i form a basis for the order, the α_i form a basis for the ideal. The inverse of the different is attached to one of the representations.

To obtain the representation for the different itself, the transposed representation has to be taken. The inverse of the different has a basis given by the dual with respect to the basis of the order.

MATHEMATICS DEPARTMENT
California Institute of Technology
Pasedena, California 91125

ABCDEFGHIJ — 8987